# Communications
# in Computer and Information Science

2166

## Rationale

The CCIS series is devoted to the publication of proceedings of computer science conferences. Its aim is to efficiently disseminate original research results in informatics in printed and electronic form. While the focus is on publication of peer-reviewed full papers presenting mature work, inclusion of reviewed short papers reporting on work in progress is welcome, too. Besides globally relevant meetings with internationally representative program committees guaranteeing a strict peer-reviewing and paper selection process, conferences run by societies or of high regional or national relevance are also considered for publication.

## Topics

The topical scope of CCIS spans the entire spectrum of informatics ranging from foundational topics in the theory of computing to information and communications science and technology and a broad variety of interdisciplinary application fields.

## Information for Volume Editors and Authors

Publication in CCIS is free of charge. No royalties are paid, however, we offer registered conference participants temporary free access to the online version of the conference proceedings on SpringerLink (http://link.springer.com) by means of an http referrer from the conference website and/or a number of complimentary printed copies, as specified in the official acceptance email of the event.

CCIS proceedings can be published in time for distribution at conferences or as post-proceedings, and delivered in the form of printed books and/or electronically as USBs and/or e-content licenses for accessing proceedings at SpringerLink. Furthermore, CCIS proceedings are included in the CCIS electronic book series hosted in the SpringerLink digital library at http://link.springer.com/bookseries/7899. Conferences publishing in CCIS are allowed to use Online Conference Service (OCS) for managing the whole proceedings lifecycle (from submission and reviewing to preparing for publication) free of charge.

## Publication process

The language of publication is exclusively English. Authors publishing in CCIS have to sign the Springer CCIS copyright transfer form, however, they are free to use their material published in CCIS for substantially changed, more elaborate subsequent publications elsewhere. For the preparation of the camera-ready papers/files, authors have to strictly adhere to the Springer CCIS Authors' Instructions and are strongly encouraged to use the CCIS LaTeX style files or templates.

## Abstracting/Indexing

CCIS is abstracted/indexed in DBLP, Google Scholar, EI-Compendex, Mathematical Reviews, SCImago, Scopus. CCIS volumes are also submitted for the inclusion in ISI Proceedings.

## How to start

To start the evaluation of your proposal for inclusion in the CCIS series, please send an e-mail to ccis@springer.com.

Ngoc-Thanh Nguyen · Bogdan Franczyk ·
André Ludwig · Manuel Nunez · Jan Treur ·
Gottfried Vossen · Adrianna Kozierkiewicz
Editors

# Advances in Computational Collective Intelligence

16th International Conference, ICCCI 2024
Leipzig, Germany, September 9–11, 2024
Proceedings, Part II

Springer

*Editors*
Ngoc-Thanh Nguyen ⓘ
Wrocław University of Science
and Technology
Wrocław, Poland

André Ludwig ⓘ
University of Leipzig
Leipzig, Germany

Jan Treur ⓘ
Vrije Universiteit Amsterdam
Amsterdam, The Netherlands

Adrianna Kozierkiewicz ⓘ
Wrocław University of Science
and Technology
Wrocław, Poland

Bogdan Franczyk ⓘ
University of Leipzig
Leipzig, Germany

Manuel Nunez ⓘ
Universidad Complutense de Madrid
Madrid, Spain

Gottfried Vossen ⓘ
University of Münster
Münster, Germany

ISSN 1865-0929 ISSN 1865-0937 (electronic)
Communications in Computer and Information Science
ISBN 978-3-031-70258-7 ISBN 978-3-031-70259-4 (eBook)
https://doi.org/10.1007/978-3-031-70259-4

This Springer imprint is published by the registered company Springer Nature Switzerland AG
The registered company address is: Gewerbestrasse 11, 6330 Cham, Switzerland

If disposing of this product, please recycle the paper.

# Preface

This volume contains the first part of the proceedings of the 16th International Conference on Computational Collective Intelligence (ICCCI 2024), held in Leipzig, Germany from 9–11 September 2024. The conference was organized in a hybrid mode which allowed for both on-site and online paper presentations. The conference was hosted by Leipzig University, Germany and jointly organized by Wrocław University of Science and Technology, Poland in cooperation with IEEE SMC Technical Committee on Computational Collective Intelligence, European Research Center for Information Systems (ERCIS), and International University-VNU-HCM (Vietnam).

Following the successes of the 1st ICCCI (2009), held in Wrocław, Poland, the 2nd ICCCI (2010) in Kaohsiung - Taiwan, the 3rd ICCCI (2011) in Gdynia - Poland, the 4th ICCCI (2012) in Ho Chi Minh City - Vietnam, the 5th ICCCI (2013) in Craiova - Romania, the 6th ICCCI (2014) in Seoul - South Korea, the 7th ICCCI (2015) in Madrid - Spain, the 8th ICCCI (2016) in Halkidiki - Greece, the 9th ICCCI (2017) in Nicosia - Cyprus, the 10th ICCCI (2018) in Bristol - UK, the 11th ICCCI (2019) in Hendaye - France, the 12th ICCCI (2020) in Da Nang - Vietnam, the 13th ICCCI (2021) in Rhodes - Greece, the 14th ICCCI (2022) in Hammamet - Tunisia, and the 15th ICCCI (2023) in Budapest - Hungary, this conference continued to provide an internationally respected forum for scientific research in computer-based methods of collective intelligence and their applications.

Computational collective intelligence (CCI) is most often understood as a subfield of artificial intelligence (AI) dealing with soft computing methods that facilitate group decisions or processing knowledge among autonomous units acting in distributed environments. Methodological, theoretical, and practical aspects of CCI are considered as the form of intelligence that emerges from the collaboration and competition of many individuals (artificial and/or natural). The application of multiple computational intelligence technologies such as fuzzy systems, evolutionary computation, neural systems, consensus theory, etc., can support human and other collective intelligence, and create new forms of CCI in natural and/or artificial systems. Three subfields of the application of computational intelligence technologies to support various forms of collective intelligence are of special interest but are not exclusive: the Semantic Web (as an advanced tool for increasing collective intelligence), social network analysis (as a field targeted at the emergence of new forms of CCI), and multi-agent systems (as a computational and modeling paradigm especially tailored to capture the nature of CCI emergence in populations of autonomous individuals).

The ICCCI 2024 conference featured a number of keynote talks and oral presentations, closely aligned to the theme of the conference. The conference attracted a substantial number of researchers and practitioners from all over the world, who submitted their papers for the main track and nine special sessions.

The Main Track, covering the methodology and applications of CCI, included: collective decision-making, data fusion, deep learning techniques, natural language processing, data mining and machine learning, social networks and intelligent systems, optimization, computer vision, knowledge engineering and application, as well as Internet of Things: technologies and applications. The Special Sessions, covering some specific topics of particular interest, included: cooperative strategies for decision making and optimization, security and reliability of information, networks and social media, anomalies detection, machine learning, deep learning, digital image processing, artificial intelligence, speech communication, IOT applications, natural language processing, and innovative applications in data science.

We received 234 papers submitted by authors coming from 45 countries around the world. Each paper was reviewed by at least three members of the international Program Committee (PC) and by the Meta-reviewing Committee. The final decision regarding acceptance for all submissions to the Main Track and Special Sessions was made by the ICCCI Program Chairs to select the best papers and ensure the highest quality of the chosen papers. Our review model ensured that papers from both the Main Track and Special Sessions were held to an equally high standard.

Finally, we selected 59 papers for oral presentation and publication in two volumes of the Lecture Notes in Artificial Intelligence series and 67 papers for oral presentation and publication in two volumes of the Communications in Computer and Information Science series.

We would like to express our thanks to the keynote speakers: Jarosław Jankowski from West Pomeranian University of Technology in Szczecin (Poland), Klaus Solberg Söilen from Halmstad University (Sweden), Sören Auer from Leibniz University Hannover (Germany), and Krzysztof Czarnecki from University of Waterloo (Canada)

Many people contributed to the success of the conference. First, we would like to recognize the work of the PC co-chairs and Special Sessions organizers for taking good care of the organization of the reviewing process, an essential stage in ensuring the high quality of the accepted papers. The Special Session chairs deserve a special mention for the evaluation of the proposals and the organization and coordination of the work of nine Special Sessions. In addition, we would like to thank the PC members for performing their reviewing work with diligence. We thank the Local Organizing Committee chairs, Publicity chairs, Web chair, and Technical Support chairs for their fantastic work before and during the conference. Finally, we cordially thank all the authors, presenters, and delegates for their valuable contribution to this successful event. The conference would not have been possible without their support.

Our special thanks are also due to Springer for publishing the proceedings and to all the other sponsors for their kind support.

It is our pleasure to announce that the ICCCI conference series continues to have a close cooperation with the Springer journal Transactions on Computational Collective Intelligence, and the IEEE SMC Technical Committee on Transactions on Computational Collective Intelligence.

Finally, we hope that ICCCI 2024 contributed significantly to the academic excellence of the field and will lead to the even greater success of ICCCI events in the future.

September 2024

Ngoc Thanh Nguyen
Bogdan Franczyk
André Ludwig
Manuel Núñez
Jan Treur
Gottfried Vossen
Adrianna Kozierkiewicz

# Organization

## Organizing Committee

### Honorary Chairs

Eva Inés Obergfell     Leipzig University, Germany
Arkadiusz Wójs     Wrocław University of Science and Technology, Poland
Piotr Jędrzejowicz     Gdynia Maritime University, Poland

### General Chairs

Ngoc Thanh Nguyen     Wrocław University of Science and Technology, Poland
Bogdan Franczyk     Leipzig University, Germany

### Program Chairs

André Ludwig     Leipzig University, Germany
Manuel Núñez     Universidad Complutense de Madrid, Spain
Jan Treur     Vrije Universiteit Amsterdam, The Netherlands
Gottfried Vossen     University of Münster, Germany

### Steering Committee

Ngoc Thanh Nguyen (Chair)     Wrocław University of Science and Technology, Poland
Piotr Jędrzejowicz     Gdynia Maritime University, Poland
Shyi-Ming Chen     National Taiwan University of Science and Technology, Taiwan
Kiem Hoang     University of Information Technology, VNU-HCM, Vietnam
Dosam Hwang     Yeungnam University, South Korea
Lakhmi C. Jain     University of South Australia, Australia
Geun-Sik Jo     Inha University, South Korea
Janusz Kacprzyk     Polish Academy of Sciences, Poland
Ryszard Kowalczyk     University of South Australia, Australia

Yannis Manolopoulos        Open University of Cyprus, Cyprus
Toyoaki Nishida        Kyoto University, Japan
Manuel Núñez        Universidad Complutense de Madrid, Spain
Klaus Solberg Söilen        Halmstad University, Sweden
Khoa Tien Tran        International University-VNUHCM, Vietnam

## Organizing Chairs

Philippe Krajsic        Leipzig University, Germany
Marcin Pietranik        Wrocław University of Science and Technology,
                        Poland

## Special Session Chairs

Adrianna Kozierkiewicz        Wrocław University of Science and Technology,
                              Poland
Paweł Sitek        Kielce University of Technology, Poland
Patrick Zschech        Leipzig University, Germany

## Doctoral Track Chairs

Marek Krótkiewicz        Wrocław University of Science and Technology,
                         Poland
Rainer Unland        University of Duisburg-Essen, Germany

## Publicity Chairs

Andreas Barton        Leipzig University, Germany
Marcin Jodłowiec        Wrocław University of Science and Technology,
                        Poland
Rafal Palak        Wrocław University of Science and Technology,
                   Poland

## Webmaster

Marek Kopel        Wrocław University of Science and Technology,
                   Poland

## Local Organizing Committee

| | |
|---|---|
| Martin Schieck | Leipzig University, Germany |
| Martin Max Röhling | Leipzig University, Germany |
| Christian Alverman | Leipzig University, Germany |
| Patient Zihisire Muke | Wrocław University of Science and Technology, Poland |
| Thanh-Ngo Nguyen | Wrocław University of Science and Technology, Poland |
| Jose Fabio Ribeiro Bezerra | Wrocław University of Science and Technology, Poland |

## Keynote Speakers

| | |
|---|---|
| Jaroslaw Jankowski | West Pomeranian University of Technology in Szczecin, Poland |
| Klaus Solberg Söilen | Halmstad University, Sweden |
| Krzysztof Czarnecki | University of Waterloo, Canada |
| Sören Auer | University of Hannover, Germany |

## Special Session Organizers

### ADMDL2024: Special Session on Anomalies Detection using Machine and Deep Learning

| | |
|---|---|
| Yousra Chabchoub | Institut Supérieur d'Electronique de Paris, France |
| Maurras Togbe | Institut Supérieur d'Electronique de Paris, France |

### AISC2024: Special Session on AI and Speech Communication

| | |
|---|---|
| Ualsher Tukeyev | Al-Farabi Kazakh National University, Kazakhstan |
| Orken Mamyrbayev | Institute of Information and Computational Technologies, Kazakhstan |

**CCINLP2024: Special Session on Computational Collective Intelligence and Natural Language Processing**

Ismail Biskri                          University of Québec a Trois-Rivieres, Canada
Nadia Ghazzali                      University of Québec a Trois-Rivieres, Canada

**CSDMO2024: Special Session on Cooperative Strategies for Decision Making and Optimization**

Piotr Jędrzejowicz              Gdynia Maritime University, Poland
Dariusz Barbucha               Gdynia Maritime University, Poland
Ireneusz Czarnowski          Gdynia Maritime University, Poland

**DICV2024: Special Session on Recent Advances of Deep Learning and Internet of Things in Computer Vision-Related Applications**

Wadii Boulila                       Prince Sultan University, KSA
Maha Driss                           Prince Sultan University, KSA
Anis Koubaa                        Prince Sultan University, KSA
Jawad Ahmad                      Edinburgh Napier University, UK
Faisal Saeed                         Birmingham City University, UK

**DIPMAI2024: Special Session on Digital Image Processing for Medical and Automotive Industry**

Debora Gil                           Universitat Autònoma de Barcelona, Spain
Mihail Gaianu                      West University of Timisoara, Romania

**IADS2024: Special Session on Innovative Applications in Data Science**

Małgorzata Przybyła-Kasperek     University of Silesia in Katowice, Poland
Agnieszka Wosiak               Lodz University of Technology, Poland
Agnieszka Duraj                  Lodz University of Technology, Poland
Rafał Skinderowicz             University of Silesia in Katowice, Poland
Wiesław Paja                       University of Rzeszów, Poland

**MLRWD2024: Special Session on Machine Learning in Real-World Data**

Jan Kozak                            University of Economics in Katowice, Poland
Mikhail Moshkov                King Abdullah University of Science and
                                              Technology, KSA
Artur Kozłowski                  Łukasiewicz Research Network, Poland

Przemysław Juszczuk                Polish Academy of Sciences, Poland
Barbara Probierz                   University of Economics in Katowice, Poland

*SIRENE2024: Special Session on Security and Reliability of Information, Networks and Social Media*

Rafal Kozik                        Bydgoszcz University of Science and Technology,
                                     Poland
Adrianna Kozierkiewicz             Wroclaw University of Science and Technology,
                                     Poland
Marcin Pietranik                   Wroclaw University of Science and Technology,
                                     Poland
Marek Pawlicki                     Bydgoszcz University of Science and Technology,
                                     Poland
Wojciech Mazurczyk                 Warsaw University of Science and Technology,
                                     Poland
Michal Choraś                      Bydgoszcz University of Science and Technology,
                                     Poland

## Senior Program Committee

Plamen Angelov                     Lancaster University, UK
Costin Badica                      University of Craiova, Romania
Nick Bassiliades                   Aristotle University of Thessaloniki, Greece
Maria Bielikova                    Slovak University of Technology in Bratislava,
                                     Slovakia
Abdelhamid Bouchachia              Bournemouth University, UK
David Camacho                      Universidad Autónoma de Madrid, Spain
Richard Chbeir                     University of Pau and Pays de l'Adour, France
Shyi-Ming Chen                     National Taiwan University of Science and
                                     Technology, Taiwan
Paul Davidsson                     Malmö University, Sweden
Mohamed Gaber                      Birmingham City University, UK
Daniela Godoy                      ISISTAN Research Institute, Argentina
Manuel Grana                       University of the Basque Country, Spain
William Grosky                     University of Michigan, USA
Francisco Herrera                  University of Granada, Spain
Tzung-Pei Hong                     National University of Kaohsiung, Taiwan
Dosam Hwang                        Yeungnam University, South Korea
Lazaros Iliadis                    Democritus University of Thrace, Greece
Mirjana Ivanovic                   University of Novi Sad, Serbia

| Piotr Jędrzejowicz | Gdynia Maritime University, Poland |
| Geun-Sik Jo | Inha University, South Korea |
| Kang-Hyun Jo | University of Ulsan, South Korea |
| Janusz Kacprzyk | Polish Academy of Sciences, Poland |
| Ryszard Kowalczyk | Swinburne University of Technology, Australia |
| Ondrej Krejcar | University of Hradec Kralove, Czech Republic |
| Hoai An Le Thi | University of Lorraine, France |
| Edwin Lughofer | Johannes Kepler University Linz, Austria |
| Yannis Manolopoulos | Aristotle University of Thessaloniki, Greece |
| Grzegorz J. Nalepa | AGH University of Science and Technology, Poland |
| Toyoaki Nishida | Kyoto University, Japan |
| Manuel Núñez | Universidad Complutense de Madrid, Spain |
| George A. Papadopoulos | University of Cyprus, Cyprus |
| Radu-Emil Precup | Politehnica University of Timişoara, Romania |
| Leszek Rutkowski | Częstochowa University of Technology, Poland |
| Tomasz M. Rutkowski | University of Tokyo, Japan |
| Ali Selamat | Universiti Teknologi Malaysia, Malaysia |
| Edward Szczerbicki | University of Newcastle, Australia |
| Ryszard Tadeusiewicz | AGH University of Science and Technology, Poland |
| Muhammad Atif Tahir | National University of Computer and Emerging Sciences, Pakistan |
| Jan Treur | Vrije Universiteit Amsterdam, The Netherlands |
| Serestina Viriri | University of KwaZulu-Natal, South Africa |
| Bay Vo | Ho Chi Minh City University of Technology, Vietnam |
| Gottfried Vossen | University of Munster, Germany |
| Lipo Wang | Nanyang Technological University, Singapore |
| Michał Woźniak | Wrocław University of Science and Technology, Poland |

## Program Committee

| Muhammad Abulaish | South Asian University, India |
| Sharat Akhoury | University of Cape Town, South Africa |
| Stuart Allen | Cardiff University, UK |
| Ana Almeida | GECAD-ISEP-IPP, Portugal |
| Bashar Al-Shboul | University of Jordan, Jordan |
| Adel Alti | University of Setif, Algeria |
| Taha Arbaoui | University of Technology of Troyes, France |

| | |
|---|---|
| Mehmet Emin Aydin | University of the West of England, Bristol, UK |
| Thierry Badard | Laval University, Canada |
| Amelia Badica | University of Craiova, Romania |
| Hassan Badir | École Nationale des Sciences Appliquées de Tanger, Morocco |
| Paulo Batista | Universidade de Évora, Portugal |
| Khalid Benali | University of Lorraine, France |
| Morad Benyoucef | University of Ottawa, Canada |
| Szymon Bobek | Jagiellonian University, Poland |
| Leon Bobrowski | Bialystok University of Technology, Poland |
| Grzegorz Bocewicz | Koszalin University of Technology, Poland |
| Urszula Boryczka | University of Silesia, Poland |
| Mariusz Boryczka | University of Silesia, Poland |
| János Botzheim | Eötvös Loránd University, Hungary |
| Peter Brida | University of Žilina, Slovakia |
| Ivana Bridova | University of Žilina, Slovakia |
| Krisztian Buza | Budapest University of Technology and Economics, Hungary |
| Aleksander Byrski | AGH University of Science and Technology, Poland |
| Alberto Cano | Virginia Commonwealth University, USA |
| Roberto Casadei | Università di Bologna, Italy |
| Amine Chohra | Paris-East Créteil University (UPEC), France |
| Kazimierz Choros | Wrocław University of Science and Technology, Poland |
| Robert Cierniak | Częstochowa University of Technology, Poland |
| Mihaela Colhon | University of Craiova, Romania |
| Antonio Corral | University of Almería, Spain |
| Jose Alfredo Ferreira Costa | Universidade Federal do Rio Grande do Norte, Brazil |
| Rafal Cupek | Silesian University of Technology, Poland |
| Ireneusz Czarnowski | Gdynia Maritime University, Poland |
| Camelia Delcea | Bucharest University of Economic Studies, Romania |
| Shridhar Devamane | Global Academy of Technology, India |
| Muthusamy Dharmalingam | Bharathiar University, India |
| Tien V. Do | Budapest University of Technology and Economics, Hungary |
| Márk Domonkos | Eötvös Loránd University, Hungary |
| Nadia Essoussi | University of Tunis, Tunisia |
| Rim Faiz | University of Carthage, Tunisia |
| Marcin Fojcik | Western Norway University of Applied Sciences, Norway |

| | |
|---|---|
| Anna Formica | IASI-CNR, Italy |
| Bogdan Franczyk | University of Leipzig, Germany |
| Dariusz Frejlichowski | West Pomeranian University of Technology in Szczecin, Poland |
| Naoki Fukuta | Shizuoka University, Japan |
| Mauro Gaspari | University of Bologna, Italy |
| K. M. George | Oklahoma State University, USA |
| Janusz Getta | University of Wollongong, Australia |
| Daniela Gifu | Romanian Academy - Iaşi Branch, Romania |
| Arkadiusz Gola | Lublin University of Technology, Poland |
| Foteini Grivokostopoulou | University of Patras, Greece |
| László Gulyás | Eötvös Loránd University, Hungary |
| Petr Hajek | University of Pardubice, Czech Republic |
| Kenji Hatano | Doshisha University, Japan |
| Marcin Hernes | Wrocław University of Economics, Poland |
| Huu Hanh Hoang | Hue University, Vietnam |
| Jeongky Hong | Yeungnam University, South Korea |
| Frédéric Hubert | Laval University, Canada |
| Zbigniew Huzar | Wrocław University of Science and Technology, Poland |
| Fethi Jarray | University of Gabes, Tunisia |
| Joanna Jedrzejowicz | University of Gdansk, Poland |
| Gordan Jezic | University of Zagreb, Croatia |
| Ireneusz Jóźwiak | Wrocław University of Science and Technology, Poland |
| Przemysław Juszczuk | University of Economics in Katowice, Poland |
| Arkadiusz Kawa | Poznań School of Logistics, Poland |
| Attila Kiss | Eötvös Loránd University, Hungary |
| Marek Kopel | Wrocław University of Science and Technology, Poland |
| Petia Koprinkova-Hristova | Bulgarian Academy of Sciences, Bulgaria |
| Ivan Koychev | University of Sofia "St. Kliment Ohridski", Bulgaria |
| Jan Kozak | University of Economics in Katowice, Poland |
| Dalia Kriksciuniene | Vilnius University, Lithuania |
| Stelios Krinidis | Centre for Research and Technology Hellas (CERTH), Greece |
| Dariusz Krol | Wrocław University of Science and Technology, Poland |
| Marek Krotkiewicz | Wrocław University of Science and Technology, Poland |
| Jan Kubicek | VSB - Technical University of Ostrava, Czech Republic |

| | |
|---|---|
| Elzbieta Kukla | Wrocław University of Science and Technology, Poland |
| Julita Kulbacka | Wrocław Medical University, Poland |
| Marek Kulbacki | Polish-Japanese Academy of Information Technology, Poland |
| Piotr Kulczycki | Polish Academy of Science, Systems Research Institute, Poland |
| Kazuhiro Kuwabara | Ritsumeikan University, Japan |
| Florin Leon | "Gheorghe Asachi" Technical University of Iaşi, Romania |
| Doina Logofatu | Frankfurt University of Applied Sciences, Germany |
| Aphilak Lonklang | Eötvös Loránd University, Hungary |
| Juraj Machaj | University of Žilina, Slovakia |
| George Magoulas | Birkbeck, University of London, UK |
| Bernadetta Maleszka | Wrocław University of Science and Technology, Poland |
| Marcin Maleszka | Wrocław University of Science and Technology, Poland |
| Adam Meissner | Poznań University of Technology, Poland |
| Manuel Méndez | Universidad Complutense de Madrid, Spain |
| Jacek Mercik | WSB University in Wrocław, Poland |
| Radosław Michalski | Wrocław University of Science and Technology, Poland |
| Peter Mikulecky | University of Hradec Kralove, Czech Republic |
| Miroslava Mikusova | University of Žilina, Slovakia |
| Jean-Luc Minel | Université Paris Ouest Nanterre La Défense, France |
| Javier Montero | Universidad Complutense de Madrid, Spain |
| Dariusz Mrozek | Silesian University of Technology, Poland |
| Manuel Munier | University of Pau and Pays de l'Adour, France |
| Phivos Mylonas | Ionian University, Greece |
| Anand Nayyar | Duy Tan University, Vietnam |
| Filippo Neri | University of Napoli Federico II, Italy |
| Linh Anh Nguyen | University of Warsaw, Poland |
| Sinh Van Nguyen | International University – Vietnam National University, HCMC, Vietnam |
| Loan Thuy Thi Nguyen | International University Ho Chi Minh City, Vietnam |
| Adam Niewiadomski | Lodz University of Technology, Poland |
| Adel Noureddine | University of Pau and Pays de l'Adour, France |
| Alberto Núñez | Universidad Complutense de Madrid, Spain |
| Mieczyslaw Owoc | Wrocław University of Economics, Poland |

| | |
|---|---|
| Marcin Paprzycki | Systems Research Institute, Polish Academy of Sciences, Poland |
| Marek Penhaker | VSB -Technical University of Ostrava, Czech Republic |
| Isidoros Perikos | University of Patras, Greece |
| Elias Pimenidis | University of the West of England, Bristol, UK |
| Nikolaos Polatidis | University of Brighton, UK |
| Piotr Porwik | University of Silesia, Poland |
| Paulo Quaresma | Universidade de Évora, Portugal |
| David Ramsey | Wrocław University of Science and Technology, Poland |
| Mohammad Rashedur Rahman | North South University, Bangladesh |
| Ewa Ratajczak-Ropel | Gdynia Maritime University, Poland |
| Virgilijus Sakalauskas | Vilnius University, Lithuania |
| Ilias Sakellariou | University of Macedonia, Greece |
| Khouloud Salameh | University of Pau and Pays de l'Adour, France |
| Imad Saleh | Université Paris 8, France |
| Sana Sellami | Aix-Marseille University, France |
| Yeong-Seok Seo | Yeungnam University, South Korea |
| Andrzej Sieminski | Wrocław University of Science and Technology, Poland |
| Dragan Simic | University of Novi Sad, Serbia |
| Stanimir Stoyanov | University of Plovdiv "Paisii Hilendarski", Bulgaria |
| Grażyna Suchacka | University of Opole, Poland |
| Piotr Sulikowski | West Pomeranian University of Technology, Poland |
| Libuse Svobodova | University of Hradec Kralove, Czech Republic |
| Martin Tabakov | Wrocław University of Science and Technology, Poland |
| Yasufumi Takama | Tokyo Metropolitan University, Japan |
| Joe Tekli | Lebanese American University, Lebanon |
| Trong Hieu Tran | VNU-University of Engineering and Technology, Vietnam |
| Maria Trocan | Institut Superieur d'Electronique de Paris, France |
| Krzysztof Trojanowski | Cardinal Stefan Wyszyński University in Warsaw, Poland |
| Ualsher Tukeyev | al-Farabi Kazakh National University, Kazakhstan |
| Olgierd Unold | Wrocław University of Science and Technology, Poland |
| Serestina Viriri | University of KwaZulu-Natal, South Africa |
| Adam Wojciechowski | Lodz University of Technology, Poland |

| Krystian Wojtkiewicz | Wrocław University of Science and Technology, Poland |
| Sadok Ben Yahia | University of Tallinn, Estonia |
| Drago Zagar | University of Osijek, Croatia |
| Constantin-Bala Zamfirescu | "Lucian Blaga" University of Sibiu, Romania |
| Katerina Zdravkova | University Ss Cyril and Methodius, Macedonia |
| Haoxi Zhang | Chengdu University of Information Technology, China |
| Adam Ziebinski | Silesian University of Technology, Poland |

# Contents – Part II

**Computational Intelligence for Digital Content Understanding**

**Knowledge Engineering and Application for Industry 4.0**

**Collective Intelligence in Healthcare**

# Contents – Part I

## Natural Language Processing

## Data Mining and Machine Learning

## Social Networks and Intelligent Systems

# Cybersecurity, Blockchain Technology, and Internet of Things

# Enhanced Intrusion Detection Based Hybrid Meta-heuristic Feature Selection

Ali Hussein Ali[1] , Boudour Ammar[2(✉)] , Maha Charfeddine[2] ,
and Bassem Ben Hamed[3]

[1] REGIM-Lab: REsearch Groups in Intelligent Machines, National School of
Electronics and Telecommunications of Sfax, University of Sfax, Sfax , Tunisia
[2] REGIM-Lab: REsearch Groups in Intelligent Machines, National Engineering
School of Sfax (ENIS), University of Sfax, Sfax , Tunisia
{boudour.ammar,maha.charfeddine}@enis.usf.tn
[3] Laboratory of Signals, systeMs, aRtificial Intelligence and neTworkS (SM@RTS),
National School of Electronics and Telecommunications of Sfax, University of Sfax,
Sfax , Tunisia
bassem.benhamed@enetcom.usf.tn

**Abstract.** As technology advances and becomes more complex, the
frequency and complexity of cyber-attacks also increase. Unscrupulous
hackers and cybercriminals are developing innovative methods to pen-
etrate computer systems and unlawfully steal crucial data. To address
these threats, organizations must implement effective Intrusion Detection Systems (IDSs) to detect and respond to assaults swiftly. IDSs have
recently attracted significant attention due to their effectiveness and precision in detecting abnormal patterns in network traffic using machine
learning techniques. This work aims to create a model that can identify intrusions by employing various machine learning algorithms on the
chosen features obtained from the modelling process. Evolutionary algorithms and local search techniques are merged in hybrid mode to build
new models to select the most suitable features and prepare them for
the proposed investigation. The SMOTE method is used for addressing
the issue of imbalanced datasets. The utilization of imbalance techniques
is highly effective. It may effectively rectify the uneven distribution of
classes, hence enhancing the accuracy of a machine learning model in
identifying the underrepresented class. A study is conducted to investigate various machine learning techniques and assess the effectiveness
of the proposed methodology. The evaluation of research on the widely
recognized Intrusion Detection benchmark CSE-CIC-IDS2018 and KDD
CUP 99 datasets demonstrates significant promise.

**Keywords:** -Intrusion Detection Systems · Machine Learning
Techniques · Synthetic Minority Oversampling Techniques ·
Metaheuristics Methods

N.-T. Nguyen et al. (Eds.): ICCCI 2024, CCIS 2166, pp. 3–15, 2024.
https://doi.org/10.1007/978-3-031-70259-4_1

## 1   Introduction

Network intrusion detection plays a crucial role in contemporary cybersecurity. Conventional intrusion detection systems depend on detection approaches based on signatures, which have limitations in identifying novel and emerging threats. Machine Learning (ML) algorithms provide an intriguing alternative by facilitating the identification of previously unidentified dangers by examining network traffic data [1]. By utilising ML models trained on labelled network traffic datasets, it becomes feasible to discern patterns that signify malevolent behaviour and employ them to promptly detect and counter network infiltration. With sophisticated ML algorithms, an up-to-date Intrusion Detection System (IDS) should be able to automate the identification of complicated and persistent threats. These algorithms are engineered to track changing threats over long periods, enabling the identification of highly congruent patterns significantly quicker than what humans are capable of [2].

Scalability in today's complicated systems enables large enterprises to have extensive coverage without needing to deploy additional resources for evaluating alerts. This can save significant money that would otherwise be spent on ongoing maintenance. Applying ML and Deep Learning (DL) techniques to manage network infiltration is increasingly prevalent. It is gaining popularity as a security measure and symbolizes a potentially effective method for improving computer security [3].

Contemporary research employs ML and DL techniques in real-time distributed computing to manage network intrusions. Using distributed computing to handle network intrusion is essential in modern cybersecurity environments, especially in large and complex networks where traditional, centralized solutions may have challenges in scaling up. By assigning tasks and leveraging the collective capabilities of multiple nodes, enterprises can enhance their ability to detect and mitigate security issues effectively. Distributed computing in networks refers to employing interconnected computers or nodes to detect, mitigate, and resolve security vulnerabilities and breaches. In network security, distributed computing is employed to enhance the capabilities of IDS and intrusion prevention systems (IPS) [4].

This paper utilizes various machine learning (ML) algorithms, including Decision Tree Classifier (DT), Random Forest (RF), K-Nearest Neighbor (KNN), and Multilayer Perceptron (MLP). These algorithms are applied to two widely recognized datasets in the field of cyber security: KDD CUP99 and CSE-CIC-IIDS2018. The study also introduces a new method for choosing features to build an accurate model. An exhaustive examination is performed on all specified algorithms to select the algorithm that produces the most precise results.

The remainder of this paper is structured as follows. Section 2 provides an overview of existing methods on intrusion detection for classification. Next, the proposed methodology is detailed in Sect. 3. The experimental settings, as well as the obtained results and their analysis, are presented in Sect. 4. At last, the 5th section concludes the paper and highlights the future work.

## 2   Related Works

IDSs have become a significant research focus in cybersecurity and academia. A multitude of academic papers on this topic have been published recently. In this section, we will provide an overview of some critical studies that specifically address imbalanced datasets related to IDS.

The study [5] employed the KDDCUP'99 and CIC-MalMem-2022 datasets to assess and develop an IDS. The authors propose improving detection rates and introducing a novel hybrid model that integrates ML with DL. This study yields outstanding results when examined on two datasets, KDDCUP'99 and CIC-MalMem-2022. KDDCUP'99 achieves an accuracy of 99.99%, while CIC-MalMem-2022 achieves a 100% accuracy. Additionally, no overfitting is detected.

The study of [6] examines the categorization analysis of intrusion detection by employing several supervised learning techniques, such as Support Vector Machine (SVM), Naive Bayes (NB), Logistic Regression (LR), k-nearest neighbors (KNN), Random forest (RF), and Decision Tree (DT), on the NSL-KDD dataset. The findings demonstrate the exceptional efficacy of a certain approach in terms of both precision and computational speed.

The authors of this study [7] constructed an IDS model utilizing the CSE-CIC-IDS 2018 dataset, which comprises the latest prevalent network threats. They utilize deep-learning methodologies and construct a convolutional neural network (CNN) framework for CSE-CIC-IDS 2018 after assessing its efficacy compared to a recurrent neural network (RNN) model. The empirical findings indicate that the efficacy of their CNN model surpasses that of the RNN model when employed in the CSE-CIC-IDS 2018 dataset.

The objective of this research [8] is to detect and categorize cyber-attacks by employing diverse DL techniques and improving the data characteristics with a metaheuristic method. The authors provide an optimization technique for artificial root foraging based on a Restricted Boltzmann Machine inspired by nature. They utilise a dataset accessible to the public from the Oak Ridge National Laboratory Notebook at Mississippi State University. Conventional supervised machine learning techniques like ANN, CNN, and SVM are contrasted with the suggested algorithm to showcase its efficacy. Simulations demonstrate that the suggested algorithm yields superior outcomes, attaining an accuracy of 97.8% for binary classification, 95.6% for three-class classification, and 94.3% for multiclass classification, thus surpassing its comparable algorithms in accuracy, precision, recall, and f1 score.

This study [9] employs ML approaches to detect intrusions. The research suggests a combined feature selection method using the pearson correlation coefficient and the random forest model. The ML model utilizes the DT, AdaBoost, and KNN algorithms to train and test the TON IoT dataset. Regarding deep learning (DL), both MLP and LSTM are trained and tested. Assessment is conducted based on accuracy, precision, and recall criteria. The results indicate that the DT and MLP offer the highest level of accuracy while minimizing FP and FN rates.

## 3   Proposed Methodology

Several studies on the advancement of Intrusion Detection Systems (IDS) have
been conducted over time, with developers mostly concentrating on improving
detection precision, regarded as the most critical measure. Figure 1 depicts the
sequential stages of the proposed approach.

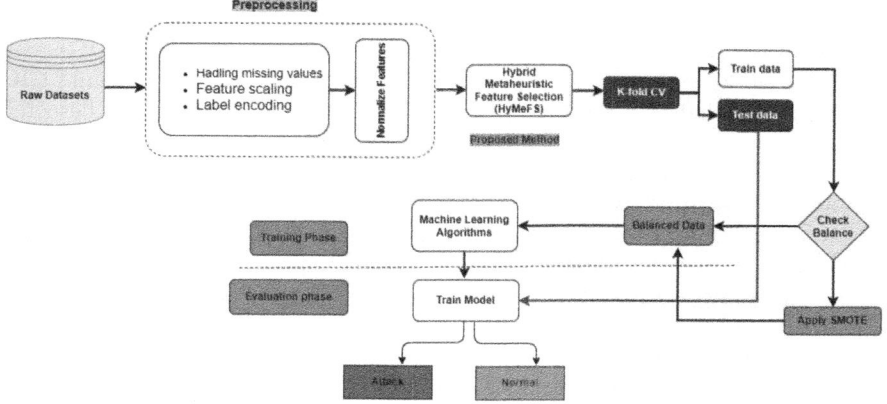

**Fig. 1.** Flow chart illustrating the sequence of steps in the proposed method.

### 3.1   Datasets

Owing to many variables, datasets offer valuable information for identifying malicious activities or suspicious behaviour on networks and systems. This work uses
two datasets: KDDCUP'99 and CSE-CIC-IDS2018.

– KDDCUP'99 Dataset: The dataset employed in the Knowledge Discovery
  and Data Mining (KDD) Cup 1999 is a standardized dataset provided by the
  US Department of Defence Advanced Research Projects Agency (DARPA).
  The information was generated to construct a predictive model distinguishing
  between detrimental connections, called intrusions or attacks, and authentic
  connections [11]. It consists of an organized collection of verified data covering various virtual invasions in a military network setting. Since it has been
  available for a long time, this dataset has seen heavy use in evaluating network intrusion detection systems. With 41 attributes in the input and 23
  subcategories in the output, we can classify the assaults into five types and
  give each one of those types two names [12].

– CSE-CIC-IDS2018: The Communications Security Establishment (CSE) and the Canadian Institute for Cybersecurity (CIC) collaborated on a project to generate the data set offered for the CSE-CIC-IDS-2018. Originating as a research evaluation tool for intrusion detection systems, it has since become a standard dataset. The extensive variety of scenarios available for analysis in this dataset is the consequence of its careful curation and development to mimic actual cyber threats and attacks. Academics and practitioners can use it to evaluate and enhance intrusion detection systems because it can simulate complex network settings [13]. The dataset comprises seven distinct attack scenarios: brute force, Heartbleed, Botnet, DoS, DDoS, Web assaults, and network infiltration from within. The dataset contains each computer's network traffic and system logs and 80 features [14].

## 3.2  Data Preparation Stage

In this work, we will delve into the importance of preprocessing and how it impacts the performance of ML models.

– Handling Missing values: A missing value in the dataset might be attributed to either data corruption or improper data recording. We address missing values in our dataset by eliminating rows that contain nan (null value), -inf, and duplicate entries.
– Feature Scaling using Standardization: We have implemented a standardization technique to rescale the numerical values of the features in the dataset. Applying a process of Standardization to the values of the features enhances the precision of the model [15].
– Label Encoding: It transforms categorical data into numerical values to construct a model using a machine-learning approach.

## 3.3  Features Selection

Feature selection is a set of methods for choosing the most relevant input features to predict the target variable. The dimensionality of data refers to the number of input features in a dataset. However, having more dimensions in space can result in a dataset representation that is more sparse and representative of samples in that space. This stimulates feature selection to eliminate irrelevant and duplicate input information, decreasing predictive performance [16].

This study presents a feature selection approach based on the Hybrid Metaheuristic (HyMeFS). Our strategy aims to enhance classification performance and expedite searching for significant feature subsets. The filter approach adjusts the Genetic Algorithm (GA) solutions population by incorporating or removing features depending on multivariate feature information. Metaheuristic algorithms are optimization techniques designed to find the best response to optimization issues. These algorithms are derivative-free methods with simplicity, versatility, and the capacity to circumvent local optima [17].

The HyMeFS method frequently restricts the feasibility of Metaheuristic algorithms when utilized for extensive real-world applications where processing time is crucial. The Genetic Algorithm (GA) randomly initializes the population, where each chromosome represents a potential subset of traits. Subsequently, a local search or meme is implemented on the elite chromosomes. This improvement process is designed to achieve a local optimum or enhance the solution. GA is used to create the next population. This procedure persists until the specified termination criteria are satisfied. The local search method extensively examines comprehensive solutions to improve their quality. As a result, there is an enhancement in the potential solution under consideration. The goal of the local search is to continually provide superior candidate solutions at each phase, thus maximizing exploitation.

The main steps of the proposed method are presented in the algorithm.

---

**Algorithm 1.** The procedure of Proposed Method (HyMeFS).

---

1: **Begin**
2: **Initialize:** Initialize randomly from feature subsets;
3: **While** (stop if the condition is not satisfied)
4: Examine the level of acceptability of each feature subset represented in the population;
5: Find E best feature subset in the population and put them into elite pop;
6: **For** (each subset in elite pop )
7: Conduct a local search and replace it with a new subset of features;
8: **End For**
9: Assess the effectiveness of newly generated solutions produced by local search;
10: Choose the optimal solution depending on the fitness function;
11: Execute evolutionary operators, specifically selection, crossover, and mutation;
12: **EndWhile**
13: **End**

---

### 3.4   Class Imbalance Processing

In an actual network, the number of samples for certain atypical traffic may be quite limited, which will significantly impact the efficacy of our model. Specifically, it is challenging for samples from minority classes to identify the accurate boundary between classes, hence complicating the classification process of determining the region and limit of each class. To address this problem, the Synthetic Minority Over-sampling Technique (SMOTE) method was applied in this study. Below is a brief explanation of this method [18]. SMOTE is a technique for over-sampling that creates synthetic examples of the minority class. The process involves selecting the class with fewer instances and augmenting the dataset by introducing additional examples until the number of instances in both classes is balanced. Nevertheless, it does this task not merely by replicating the existing data. Instead, it generates novel synthetic data comprising plausible values that are proximate to the "feature space" of the minority class through the process of data augmentation [19].

### 3.5  Machine Learning Algorithms

Several ML techniques, detailed below, were employed to accomplish multilabel classification on the datasets utilized in this study.

1. Random Forest (RF): RF is a meta-approximation that uses averaging to improve accuracy. To avoid the problem of over-fitting, different sub-trials of the dataset are fitted with different decision tree classifiers. It integrates multiple uncut DTs obtained from various training data bootstrap samples; additionally, it samples each attribute subset independently from the real feature space [21]. The model's prediction is based on the class most trees predict, estimated by the class and each tree individually. It takes the features from the training data set and uses them to build multiple decision trees; then, it uses a vote to fuse them into one classification model [22].

2. Decision Tree (DT):
   DT is most commonly employed for classification tasks. In addition, DT is a popular classification model in the Data Mining field [23]. The three primary nodes within DT are the root, branches, and leaf nodes. The root node represents partitioning the dataset into two or more homogenous sets; characteristics or attributes are combined at the branches, and processing is stopped at the output nodes, called leaf nodes. Decision trees are versatile and useful in many contexts due to their accuracy and ease of analysis with different kinds of data [24].

3. K-Nearest Neighbour (KNN):
   KNN is a data classification method that uses the estimated distance between an unlabeled data instance and a set of similar marked examples to place the data instance in the set [25]. Since KNN uses the Euclidean distance equation to determine connections, it is easy to implement even without prior knowledge about the data distribution. In addition, it can make accurate recommendations to users by classifying things simply using distance or similarity [9].

4. The Multilayer Perceptron (MLP):
   MLP is a popular artificial neural network (ANN) design that uses stacked layers of neurons and the connections between them. It utilizes an activation function to produce a signal transmitted to the subsequent neuron after measuring the weighted sum of its inputs [26]. One or more hidden layers separate the input and output layers. Neurons are organized in layers, with connections being directed from lower to higher levels. Additionally, neurons within the same layer are not connected. In the input layer, the quantity of neutrons is proportional to the pattern query's measurement count. In the output layer, the amount of neurons is proportional to the class count [27].

## 4    Experiment Results

The present study investigates the machine learning algorithms' efficiency in detecting intrusions. The available datasets (KDD CUP99 and CSE-CIC-IDS2018) are used for training and tests. The implemented algorithms have default parameters. To evaluate and observe the improvement and accuracy of the proposed method. The proposed systems' performance evaluation is conducted through various metrics, including accuracy, precision, recall, and f1-Score [10]. These metrics are calculated according to Equations (1-4).

$$Accuracy = \frac{TP + TN}{TP + TN + FP + FN} \tag{1}$$

$$Precision = \frac{TP}{TP + FP} \tag{2}$$

$$Recall = \frac{TP}{TP + FN} \tag{3}$$

$$F1\_Score = 2 * \frac{Precision * Recall}{Precision + Recall} \tag{4}$$

Where:

- True Positive (TP): The classifier recognizes data objects as attacks.
- False Negative (FN): Incorrectly identified as Normal.
- False Positives (FP): Instances in the data that were wrongly identified as Attacks.
- True Negative (TN): Standard instances were designated.

Table 1 presents the results of three metrics, Precision, Recall and F1_Score, after applying four ML algorithms named DT, RF, KNN, and MLP. The DT algorithm shows high results with the KDD-cup99 and CSE-CIC-IDS2018 datasets by all metrics, while the remaining algorithms (RF, KNN, and MLP) give lower results.

**Table 1.** Evaluate results of the proposed method.

| Algorithm | Dataset Name | Precision | Recall | F1_Score |
|---|---|---|---|---|
| DT | KDD CUP99 | 99.98 | 99.96 | 99.98 |
| | CSE-CIC-IDS 2018 | 99.94 | 99.92 | 99.94 |
| RF | KDD CUP99 | 98.21 | 98.21 | 98.21 |
| | CSE-CIC-IDS 2018 | 99.87 | 99.86 | 99.89 |
| KNN | KDD CUP99 | 98.62 | 98.58 | 98.58 |
| | CSE-CIC-IDS 2018 | 99.93 | 99.96 | 99.95 |
| MLP | KDD CUP99 | 98.09 | 96.64 | 96.47 |
| | CSE-CIC-IDS 2018 | 99.83 | 99.81 | 99.85 |

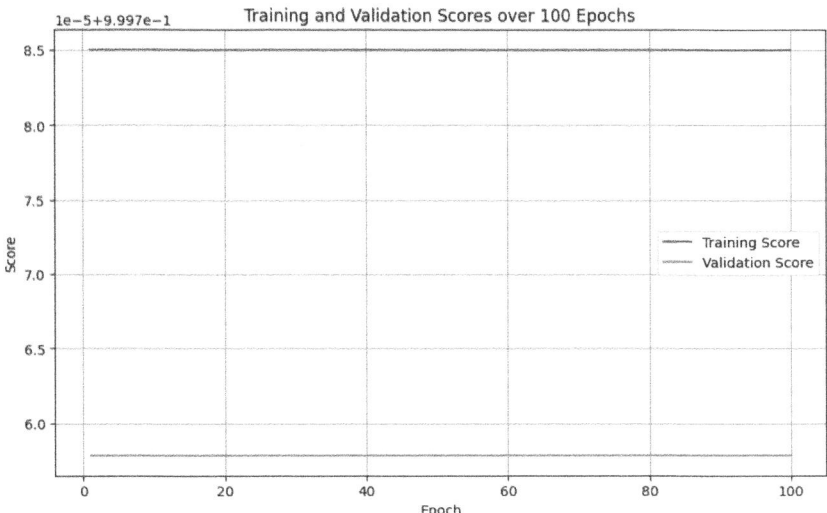

**Fig. 2.** Accuracy curves of RF training and validation scores over 100 epochs for cse-cic-ids2018 dataset

**Table 2.** Comparison with similar works regarding accuracy and f1_score.

| Ref. | Algorithm | Dataset Name | Results | |
|---|---|---|---|---|
| | | | Accuracy | F1_Score |
| [28] | RF-GWO-RF | KDD CUP99 | 99.14 | 99.14 |
| | RF-GWO-DT | | 99.61 | 99.2 |
| [29] | Bagging | CSE-CIC-IDS 2018 | 98.8 | 99.2 |
| | MLP | | 94.5 | 96.7 |
| [30] | RF | CSE-CIC-IDS 2018 | 98.91 | 97.24 |
| | DT | | 97.53 | 95.72 |
| | KNN | | 95.72 | 95.73 |
| | MLP | | 97.23 | 96.87 |
| [31] | MLP-PSO | CSE-CIC-IDS 2018 | 95.32 | 97.6 |
| | MLP-BP | | 98.41 | 99.2 |
| Our Method | DT | KDD CUP99 | 99.8 | 99.78 |
| | | CSE-CIC-IDS 2018 | 99.96 | 99.94 |
| | RF | KDD CUP99 | 99.98 | 99.96 |
| | | CSE-CIC-IDS 2018 | 99.87 | 99.87 |
| | KNN | KDD CUP99 | 99.94 | 99.93 |
| | | CSE-CIC-IDS 2018 | 99.98 | 99.96 |
| | MLP | KDD CUP99 | 99.64 | 99.31 |
| | | CSE-CIC-IDS 2018 | 99.85 | 99.83 |

Table 2 presents an accuracy and F1 Score comparison between our proposed approach and a set of prior efforts. The results demonstrate that the proposed approach achieved superior outcomes in all the specified scenarios.

Figure 2 illustrates the progression of accuracy across 100 epochs for the cse-cic-ids2018 dataset. It indicates that our model does not display overfitting, and the validation data is closely similar to the patterns observed in the training data for accuracy in RF technique for classification. Figures 3 and 4 display the most important results from applying the selected algorithms with the proposed method for feature selection on the two datasets: KDD-CUP99 and CSE-CIC-IDS2018.

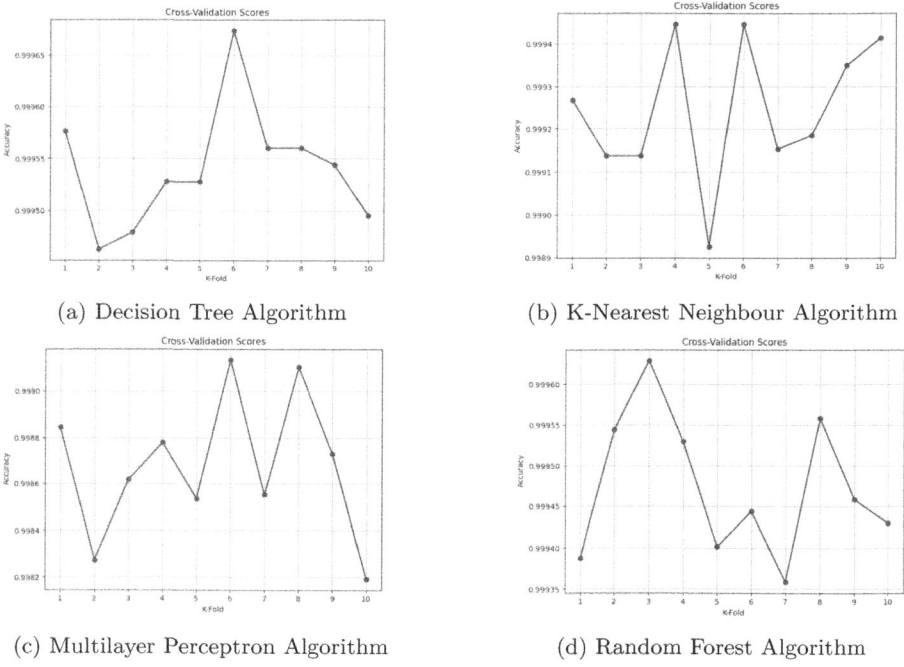

(a) Decision Tree Algorithm

(b) K-Nearest Neighbour Algorithm

(c) Multilayer Perceptron Algorithm

(d) Random Forest Algorithm

**Fig. 3.** Accuracy with cross-validation of proposed Algorithms with CSE-CIC-IDS2018 Dataset.

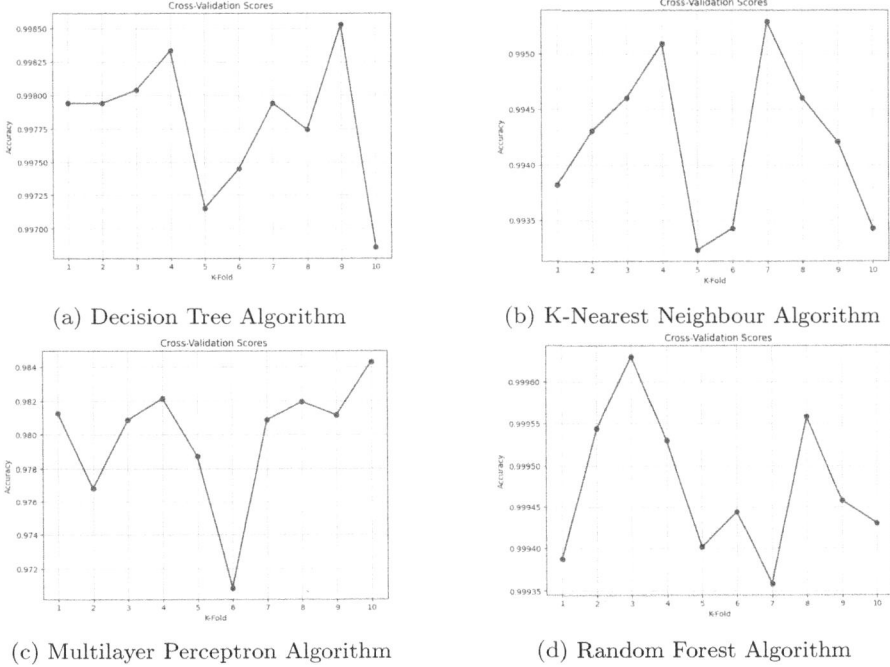

(a) Decision Tree Algorithm

(b) K-Nearest Neighbour Algorithm

(c) Multilayer Perceptron Algorithm

(d) Random Forest Algorithm

**Fig. 4.** Accuracy with cross-validation of proposed Algorithms with KDD-CUP99 Dataset.

## 5 Conclusion

Four different ML algorithms were developed in this work, Decision Tree (DT), Random Forest (RF), K-Nearest Neighbours (KNN) and Multilayer Perceptron (MLP), using established datasets KDD CUP99 and CSE-CIC-IDS2018. To address the imbalance ratio, the SMOTE approach was used to increase the data volume of the minority groups. Next, a hybrid metaheuristic HyMeFS algorithm chooses the optimal subset of features. The experimental results showed that the generated models achieved superior performance in metrics, such as accuracy, precision, recall, and F1 Score, compared to earlier studies. ML-based methods are more efficient and demand fewer computer resources and labeled data. While ML can accurately identify known attacks, it has limits when recognizing unknown attacks. We can in that case utilize deep learning methods in future studies to exceed this limitation. In the future, we intend to use different contemporary datasets and various preprocessing techniques, feature selection methods, and hybrid imbalance strategies.

# References

1. Alkasassbeh, M., Al-Haj Baddar, S.: Intrusion detection systems: a state-of-the-art taxonomy and survey. Arab. J. Sci. Eng. **48**(8), 10021–10064 (2023)
2. Rizvi, S., Scanlon, M., McGibney, J., Sheppard, J.: Deep learning based network intrusion detection system for resource-constrained environments. In: International Conference on Digital Forensics and Cyber Crime, pp. 355-367. Cham, Springer Nature Switzerland (2022). https://doi.org/10.1007/978-3-031-36574-4_21
3. Sarhan, M., Layeghy, S., Moustafa, N., Gallagher, M., Portmann, M.: Feature extraction for machine learning-based intrusion detection in IoT networks. Digital Commun. Netw. (2022)
4. Samunnisa, K., G. Sunil Vijaya Kumar, and K. Madhavi. "Intrusion detection system in distributed cloud computing: Hybrid clustering and classification methods." Measurement: Sensors 25 (2023)
5. Talukder, M.A., et al.: A dependable hybrid machine learning model for network intrusion detection. J. Inf. Secur. Appl. **72**, 103405 (2023)
6. Rastogi, S., Shrotriya, A., Singh, M.K., Potukuchi, R.V.: An analysis of intrusion detection classification using supervised machine learning algorithms on NSL-KDD dataset. J. Comput. Res. Innovation **7**(1), 124–137 (2022)
7. Kim, J., Shin, Y., Choi, E.: An intrusion detection model based on a convolutional neural network. J. Multimedia Inf. Syst. **6**(4), 165–172 (2019)
8. Diaba, S.Y, Shafie-Khah, M., Elmusrati, M.: Cyber security in power systems using meta-heuristic and deep learning algorithms. IEEE Access **11**, 18660–18672 (2023)
9. Hidayat, I., Ali, M.Z., Arshad, A.: Machine learning-based intrusion detection system: an experimental comparison. J. Comput. Cogn. Eng. **2**(2), 88–97 (2023)
10. Al-Omari, M., Rawashdeh, M., Qutaishat, F., Alshira'H, M., Ababneh, N.: An intelligent tree-based intrusion detection model for cyber security. J. Netw. Syst. Manage. **29**, 1–18 (2021)
11. Kumar, S., Gupta, S. and Arora, S.: A comparative simulation of normalization methods for machine learning-based intrusion detection systems using KDD Cup'99 dataset. J. Intell. Fuzzy Syst. **42**(3), 1749–1766 (2022)
12. Serinelli, B.M., Collen, A. and Nijdam, N.A.: On the analysis of open source datasets: validating IDS implementation for well-known and zero-day attack detection. Procedia Comput. Sci. **191**, 192–199 (2021)
13. Farhan, B.I. and Jasim, A.D.: Performance analysis of intrusion detection for deep learning model based on CSE-CIC-IDS2018 dataset. Indonesian J. Electr. Eng. Comput. Sci. **26**(2), 1165–1172 (2022)
14. Karatas, G., Demir, O. and Sahingoz, O.K.: Increasing the performance of machine learning-based IDSs on an imbalanced and up-to-date dataset. IEEE access **8** , 32150–32162 (2020)
15. Fki, Z., Ammar, B., Fourati, R., Fendri, H., Hussain, A., Ben Ayed, M.: A novel IoT-based deep neural network for COVID-19 detection using a soft-attention mechanism. Multimedia Tools Appl. (2023)
16. Hussein, A., Charfeddine, M.,Ammar, B., Ben Hamed, B.: Intrusion detection schemes based on synthetic minority oversampling technique and machine learning models. In: Conference 27th IEEE International Symposium on Real-Time Distributed Computing, pp. 1–8. IEEE (2024)
17. Fki, Z., Ammar, B., Ayed, M.B.: Towards automated optimization of residual convolutional neural networks for electrocardiogram classification. Cognitive Computation (2023)

18. Zhao, T., Zheng, Y., Wu, Z.: Feature selection-based machine learning modelling for distributed model predictive control of nonlinear processes. Comput. Chem. Eng. **169**, 108074 (2023)

19. Mirjalili, S., Mirjalili, S.M., Lewis, A.: Grey wolf optimizer. Adv. Eng. Soft. **69**, 46-61 (2014)

20. Kaur, S., Kumar, Y., Koul, A., Kumar Kamboj, S.: A systematic review on meta-heuristic optimization techniques for feature selections in disease diagnosis: open issues and challenges. Arch. Comput. Methods Eng. **30**(3), 1863-1895 (2023)

21. Singh, A., Jang-Jaccard, J.: Autoencoder-based unsupervised intrusion detection using multi-scale convolutional recurrent networks. arXiv preprint arXiv:2204.03779 (2022)

22. Wang, C., Sun, Y., Wang, W., Liu, H., Wang, B.: Hybrid intrusion detection system based on a combination of random forest and autoencoder. Symmetry **15**(3), 568 (2023)

23. Maseer, Z.K., Yusof, R., Bahaman, N., Mostafa, S.A., Foozy, C.F.M.: Benchmarking of machine learning for anomaly based intrusion detection systems in the cicids2017 dataset. IEEE Access **9**, 22351-22370 (2021)

24. Azam, Z., Islam, M.M., Huda, M.N.: Comparative analysis of intrusion detection systems and machine learning based model analysis through decision tree. IEEE Access (2023)

25. Mohy-eddine, M., Guezzaz, A., Benkirane, S., Azrour, M.: An efficient network intrusion detection model for IOT security using k-NN classifier and feature selection. Multimedia Tools Appl. 1–19, (2023)

26. Al-Safaar, D., Al-Yaseen, W.L.: Hybrid AE-MLP: hybrid deep learning model based on autoencoder and multilayer perceptron model for the intrusion detection system. Int. J. Intell. Eng. Syst. **16**(2) (2023)

27. Dao, T.-N., Van Le, D., Tran, X.N.: Optimal network intrusion detection assignment in multi-level IOT systems. Comput. Netw. **232**, 109846 (2023)

28. Shanbhag, A., Vincent, S., Gowda, S.B., Kumar, O.P., Francis, S.A.J.: Leveraging metaheuristics for feature selection with machine learning classification for malicious packet detection in computer networks. IEEE Access (2024)

29. Najafi Mohsenabad, H., Tut, M. A.: Optimizing cybersecurity attack detection in computer networks: a comparative analysis of bio-inspired optimization algorithms using the CSE-CIC-IDS2018 2018 dataset. Appl. Sci. **14**(3), 1044 (2024)

30. Abbas, Q., Hina, S., Sajjad, H., Zaidi, K.S., Akbar, R.: Optimization of the predictive performance of intrusion detection system using hybrid ensemble model for secure systems. PeerJ Comput. Sci. **9**, e1552 (2023)

31. Alzughaibi, S., El Khediri, S.: A cloud intrusion detection systems based on DNN using backpropagation and PSO on the CSE-CIC-IDS2018 dataset. Appl. Sci. **13**(4), 2276 (2023)

# Data Distribution-Based Change Detection Framework in SWaT Security Monitoring

Máté Hekfusz[ID], Vrushali Mahajan[ID], Adolf Kamuzora[(✉)][ID],
and Imre Lendák[ID]

Department of Data Science and Engineering, Institute of Industry - Academia
Innovation, ELTE Eötvös Loránd University, Pázmány Péter sétány 1/C,
1117 Budapest, Hungary
{ej97dd,y2hse8,adolfnfsp,lendak}@inf.elte.hu

**Abstract.** Networks of sensors and Internet of Things have led to an
ever-increasing amount of data that is now more commonly available in
streaming settings. Often it is assumed that the process generating these
streams is stationary, however, in real world scenarios, systems are evolving and dynamic in nature. This results in degradation of trained model
predictions and decision making process. Thus, methods and approaches
to be able to detect when there is a change or drift in the environment
are necessary. In this research project, we proposed a change detection
framework and analyzed various data distribution-based change detection algorithms for real-time analysis and change detection in Secure
Water Treatment. Results show that the framework is promising, able to
effectively monitor the cyber-physical system, and out of the algorithms
we experimented with, the Kolmogorov-Smirnov WINdowing algorithm
performed better in terms of changes detected and detection delay, albeit
with a high false alarm rate.

**Keywords:** Open-source technologies · change detection ·
cyber-physical systems · secure water treatment

## 1 Introduction

Nowadays, huge amount of data and streams of data are generated from all kinds
of devices, in different formats, from independent or connected applications. It
can be foreseen that Internet of Things and various network sensors will raise
the scale of data to an unprecedented level. In real world, systems generating
these streams of data are dynamic in nature, resulting in degradation of model
predictions and decision-making process. Approaches to detect changes in the
environment are necessary to ameliorate these challenges [4].

---

Supported by Eötvös Loránd Tudományegyetem.

N.-T. Nguyen et al. (Eds.): ICCCI 2024, CCIS 2166, pp. 16–28, 2024.
https://doi.org/10.1007/978-3-031-70259-4_2

In this research project, we have built a complete stream processing pipeline and analyzed data distribution-based change detection methods on the Secure Water Treatment (SWaT) cyber-physical system. Within the field of water treatment systems, quick and accurate identification of changes, anomalies, and evolving patterns is important for operational security and resilience. Our work aspires to contribute a framework that facilitates insights into the efficacy of change detection algorithms.

Data distribution-based drift detection consists usage of a distance function to quantify dissimilarity between the distribution of historical data and new data. If the difference is statistically significant enough, drift is detected and the learning model is updated. While they are less commonly used than error rate-based change detection methods like Drift Detection Method and ADaptive WINdowing, data distribution-based methods can operate in an unsupervised or semi-supervised manner on unlabelled data, making them well-suited in this case [11].

The linchpin of our research project is Docker, a containerization platform which provides a base environment for our array of open-source technologies, one that anyone can replicate. Within Docker, we have setup and configured diverse services such as Apache Kafka, Apache Spark, Telegraf, InfluxDB, Grafana, and Chronograf to work in tandem, each tool fitting into our stream processing pipeline.

InfluxDB, a time-series database, is another part of our framework. The SWaT stream is persisted in InfluxDB, making it available for comprehensive exploration and insightful visualization that single-pass systems cannot give us. Beyond persisting the data stream, InfluxDB also provides a structured foundation for batch training.

This research work details our proposed system architecture, the technologies we have used, and the algorithms we have implemented. We also show the results of our experiments with several different change detection methods on the SWaT dataset. We aim to highlight the necessity of stream processing frameworks for today's complex, data-heavy systems and improve understanding of data distribution-based change detection on the specific field of water treatment infrastructure.

The remainder of this research work is organized as follows: Sect. 2 describes related work and high level concepts, Sect. 3 presents a description of the dataset and approach for the framework, with results of the conducted experiments provided in Sect. 4. Finally Sect. 5 concludes the paper and introduce future research direction.

## 2   State-of-the-Art

Data distribution-based change detection methods aim to identify shifts or alterations in the statistical properties of the data distribution [6]. In the context of Secure Water Treatment (SWaT) dataset, which represents the operational data of a water treatment facility, detecting changes in the data distribution is crucial for identifying abnormal behavior, potential attacks, or shifts in the underlying system. Algorithms described were implemented using the open-source libraries Menelaus[1] and River[2]

## 2.1    K-Dimensional-Quad-Tree

The K-Dimensional-Quad-tree (KDQ-tree) algorithm is a robust statistical method designed for detecting changes or drifts in high-dimensional streaming data. Unlike traditional methods, KDQ-tree utilizes a combination of bootstrapping techniques and statistical tests to continuously monitor the evolving distribution of the data. In the context of Secure Water Treatment dataset, the algorithm dynamically adapts to variations in the operational characteristics of the water treatment system. As streaming data is consumed from a streaming system, the KDQ-tree algorithm analyzes the features, and when a significant change or drift is identified, it triggers a detection event. This adaptive approach allows the algorithm to differentiate between normal operational fluctuations and anomalous events, contributing to the system's resilience in identifying potential attacks or deviations from expected behavior. The KDQ-tree's ability to operate on streaming data in real-time makes it particularly well-suited for applications where prompt detection of changes is crucial for maintaining the security and integrity of the underlying system [3].

## 2.2    CUmulative SUM

The Cumulative Sum (CUSUM) algorithm is a statistical technique employed for identifying shifts or alterations in the mean of a chronologically ordered set of observations. CUSUM involves the calculation of a running sum representing the cumulative deviations of individual data points from an anticipated mean. When this cumulative sum surpasses a predefined threshold, it signals the presence of a change point, indicative of a significant shift in the underlying process [1] [13]. The algorithm initiates by computing the mean of the initial observations, serving as the baseline or reference value. With each new observation in the time series, the algorithm computes the deviation of the observed value from the expected mean. The cumulative sum is continuously updated based on these calculated deviations. The cumulative sum is then compared against a predetermined threshold. Surpassing this threshold indicates a substantial departure from the expected mean. Once the cumulative sum exceeds the threshold, the algorithm identifies a change point. Threshold is a parameter which dictates the significance level for considering a change point. When a change point is detected, the algorithm outputs pertinent details such as timestamp, value, and the nature of the change (true positive or false positive based on the attack label). The flexibility to adjust parameters, including the threshold, caters to the specific sensitivity and specificity requirements of the monitoring system without compromising on originality and integrity in reporting [5].

---

[1] Available at https://github.com/mitre/menelaus.
[2] Available at https://github.com/online-ml/river.

## 2.3   Crámer-von Mises

This algorithm uses the Crámer-von Mises (CvM) test and magnitude-based criteria. The algorithm monitors a time-ordered sequence of observations with the aim to detect deviations or anomalies. The algorithm used is a change point detection algorithm based on two main components: the Crámer-von Mises (CvM) test and a magnitude-based criteria. The CvM test is a statistical test used to assess the goodness of fit of a dataset to a theoretical distribution. The algorithm dynamically evaluates the goodness of fit of the observed data to a normal distribution using the CvM test. It then combines this statistical assessment with a magnitude-based criterion to identify significant changes in the values of sensors in water treatment process. The incorporation of attack labels ensures the accuracy of the detection system by distinguishing between genuine attacks and other process variations. The performance metrics such as detection delay and false alarms provide a quantitative evaluation of the system's effectiveness in detecting relevant events [2].

## 2.4   Principal Component Analysis-Change Detection

Principal Component Analysis-Change Detection (PCA-CD) algorithm is used for real-time detection of changes or anomalies in a streaming dataset. PCA-CD leverages the principles of principal component analysis, a dimensionality reduction technique, to identify significant variations in the data. In this implementation, the algorithm is initialized with parameters such as the window size, divergence metric (utilizing Kullback-Leibler divergence), and a delta value. The PCA-CD algorithm is continually updated with incoming records. This method evaluates whether the detected change corresponds to a normal operation or an attack, updating detection-related metrics such as actual detections, detected attacks, detection delays, and false alarms. This implementation offers a real-time and dynamic approach to monitoring the Secure Water Treatment dataset, providing insights into the effectiveness of the PCA-CD algorithm for change detection in a streaming environment [12].

## 2.5   Kolmogorov-Smirnov WINdowing

Kolmogorov-Smirnov WINdowing (KS-WIN) algorithm is used for change detection in a streaming dataset. KS-WIN is a drift detection method designed to identify shifts or anomalies in data distribution over time. The KS-WIN algorithm functions by maintaining a sliding window of data observations during an initial training phase. In this phase, the algorithm builds a reference window to establish the statistical characteristics of the normal data distribution. Subsequently, as new data points arrive, KS-WIN continuously updates its internal statistics and compares them to the reference window. When a significant deviation is detected, signaling a potential change or drift in the data distribution, the algorithm triggers a warning or drift state. KS-WIN is a univariate detection method; in our case, it monitors the 'LIT 301' feature which we found to give the best results for detecting changes and attacks on the entire dataset. During the initial records, the algorithm learns the baseline behavior. After the training, it dynamically assesses incoming data points and raises warnings or identifies drifts when observed values deviate significantly from the established normal distribution [8].

### 2.6   One-Class Support Vector Machine

The One-Class Support Vector Machine (one-class SVM) is a machine learning algorithm used for anomaly detection, particularly in scenarios where the majority of the data belongs to one class, and anomalies are the exceptions. The algorithm learns the characteristics of the normal class during the training phase, constructing a boundary that encapsulates normal instances. During the testing or detection phase, data points falling outside this boundary are considered anomalies [8]. The key parameter 'nu' controls the proportion of training errors and serves as an upper bound on the fraction of margin errors and a lower bound of the fraction of support vectors. One-Class SVM is enhanced with a Quantile Filter, introducing a threshold for anomaly detection. As streaming data is consumed from a streaming system, the One-Class SVM continually updates its model, dynamically adapting to changes in the data distribution. This method evaluates the accuracy of detection by comparing them with the known attack labels [9].

## 3   Proposed Framework and Dataset

### 3.1   Secure Water Treatment Dataset

For this research project we used the A4 & A5 versions of the Secure Water Treatment (SWaT) dataset provided by iTrust [7]. The SWaT testbed can be viewed as a smaller version of a real industrial water treatment plant in Singapore. It is a cyber-physical system (CPS), where each part of the system is controlled by Programmable Logic Controllers (PLC). The sensors on each machine can change how the system operates by telling the computer to take certain actions [14].

The A4 & A5 dataset version, captures the operation of the water treatment plant testbed during a single run on July 20, 2019. The testbed was run for about 4 h, recording sensor data at a frequency of one observation per second. In total, the dataset comprises 14,996 records, 78 columns, consisting of values and states of 28 sensors at every part of the water treatment process. The temporal structure of the dataset unfolds as follows: a predominant period of normal operation which spans approximately three hours, setting the baseline for routine system behavior. Subsequently, the dynamics shift in the last hour, during which the system experiences a series of six attacks, at various stages and points. These attacks are strategically arranged to target various parts of the water treatment system, injecting anomalies and disturbances into the otherwise stable operational environment.

The deliberate introduction of attacks in the final hour serves to challenge the robustness and resilience of the water treatment processes, simulating real-world scenarios where unexpected events or malicious activities may compromise the integrity of the system. This intentional variation in operational conditions provides a unique and valuable testing ground for the development and evaluation of anomaly detection and change detection algorithms. Researchers and data analysts can leverage this SWaT dataset to explore, analyze, and develop methodologies that enhance the security and reliability of water treatment systems [10].

### 3.2 Proposed Framework

Our proposed system architecture as shown in Fig. 1 consists of a complete streaming pipeline within Docker environment, creating a unified framework for thorough data processing and analysis. Detailed implementation of the framework and the experiments are available in our GitHub[3] repository.

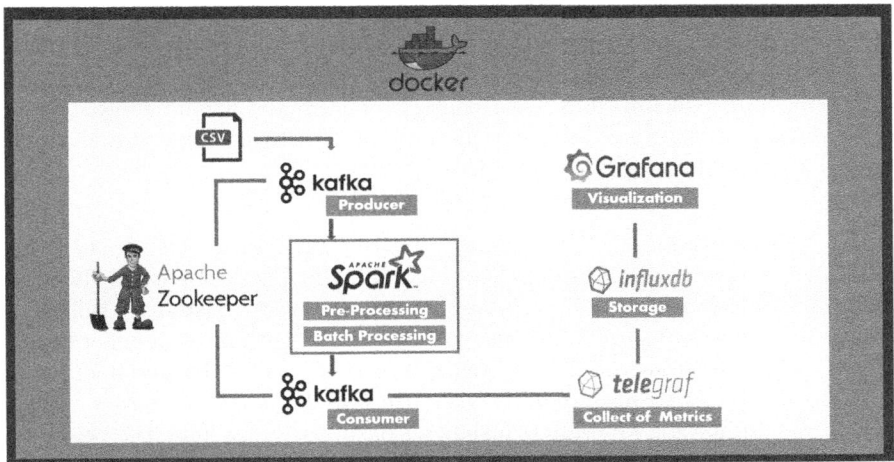

**Fig. 1.** System architecture framework.

**Phase 1: Data Ingestion and Preprocessing.** The process commenced with Kafka, a platform for data ingestion. Here, we convert our dataset into a streaming format and the relevant columns (the ones that are attacked at some point of the run, plus the timestamp) are ingested into Kafka, leveraging its resilience and scalability. To monitor the flow of our data, we deployed CMAK, a Kafka cluster manager. The dataset, loaded from a sheet, underwent preprocessing in

---

[3] Available at https://github.com/mateotis/swat-change-detection.

Spark Streaming. Each record, uniquely distinguished by its timestamp, was formatted and labelled according to whether it belongs to normal operation or an attack. Aggregate statistics (mean, standard deviation, min/max, etc.) were calculated on data received over 5 and 10 min windows. This enriched data then flows back into Kafka under a different topic, setting the stage for subsequent analysis.

**Phase 2: Change Detection.** We employed data distribution-based change detection and anomaly detection algorithms to experiment with the preprocessed data to detect changes or drifts and anomalies. We implemented these algorithms: KDQ-tree, PCA-CD, 1-class-SVM, CUSUM, Kolmogorov-Smirnov WINdowing, C  ́amer-von Mises test. We ran each of them on the entirety of the SWaT data stream, recording true and false detections, as well as detection delays for every algorithm. We then plot graphs to visualize the performance of each algorithm, shown in our Results section.

**Phase 3: Data Storage and Batch Training.** At this stage, Telegraf took the preprocessed Kafka records and delivered to InfluxDB for time-series data persistence. Each record's timestamp in InfluxDB aligns with its original timestamp. In batch training, we implemented a Decision Tree model taken from the popular open-source scikit-learn package. The model is trained on the initial three hours of normal data and the first three attacks, then tested on the remainder of the data with the last three attacks.

**Phase 4: Visualization.** We implemented the TIG-stack: picking up Kafka records with Telegraf and storing them in InfluxDB. The last part of the stack is Grafana, which we used to visualize the data now persisted in InfluxDB. Grafana is connected to InfluxDB through an authentication token generated by the latter. We also integrated Chronograf into this part of the pipeline, giving us a greater ability to meaningfully visualize our data. Using queries written in Flux and InfluxQL, we crafted insightful dashboards to show the change of our sensor values over time, highlighting correlations between them, and take a visual look at the attacks themselves.

## Batch Processing and Training

We implemented batch training for a machine learning classifier using historical data stored in InfluxDB. In batch training, the machine learning model learns from a fixed dataset, allowing for an efficient and thorough exploration of historical patterns. The data was retrieved from InfluxDB using a time-range query, focusing on relevant features and the target variable (attack or normal record). This historical dataset was then divided into two sets: a training set used for model training and a testing set used for model evaluation. The Decision Tree classifier, chosen for its simplicity and interpretability, was initialized, trained on the training set, and subsequently evaluated on the testing set. The accuracy score was computed, providing a quantitative measure of the model's performance in correctly classifying instances as normal or indicative of an attack. This approach reflects the batch learning paradigm, where the model was trained offline on a static dataset, making it well-suited for scenarios where real-time updating is not a requirement. The use of InfluxDB as the data source underscores its role in facilitating efficient historical data retrieval for machine learning tasks. In our case, the training set was the initial three hours of normal data and the first three attacks, while the test set was the remainder of the data which included the last three attacks. The result was an accuracy of approximately 71%. While this is better than the Random Forest and SVM models we also tested, it also shows the difficulty of detecting the various-length (some only lasting a few minutes) attacks which all target different points of the system.

**Stream Processing**
We implemented sliding window aggregation on streaming data for real-time analysis in Spark. The key steps involve defining a window configuration with a size of 10 min and a sliding interval of 5 min. The streaming Data Frame was then grouped by both the specified time window and the attack label, allowing computation of various aggregations, including count, mean, minimum, maximum, standard deviation, rate, and 10-minute average for each numerical column. The use of a watermark on the timestamp ensured that late data was appropriately handled. The results of these aggregations provided a continuous and evolving summary of key statistical measures within sliding time windows. This approach enabled monitoring of dynamic trends and patterns in water treatment process over time.

## 4    Results and Discussion

### 4.1    Comparative Analysis of Change Detection Algorithms

The change detection algorithms were experimented over the entire SWaT data stream and recorded their performance. One-class-SVM performed poorly, likely because of a bad fit to the dataset (which we were unable to compensate for with parameter tuning) and thus, excluded from the graphs. The detection performance of the remaining five is shown in Fig. 2 and Table 1 shows the corresponding parameters used during experimentation. KDQ-Tree detected a single change event while generating six false alarms. PCA-CD was more effective and detected four attacks but reported eleven false alarms. Surprisingly, the univariate KS-WIN performed the best, detecting five of the six attacks, though it also produced 56 false alarms. CUSUM detected two change events but reported 54 false alarms. Lastly, the CvM algorithm detected nine change events with nine false alarms. Notably, the algorithms exhibited varying performance in detecting specific attacks, with detection delays ranging from 19.0 to 585.0 time units, with their averages (total delays divided by number of attacks detected) shown in Fig. 3.

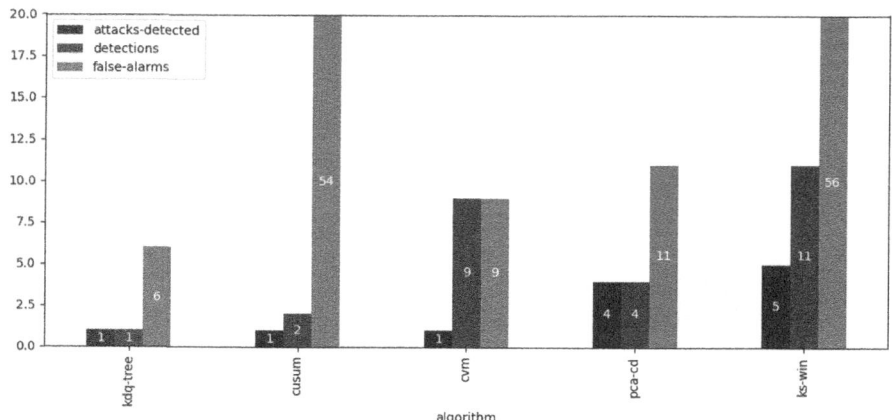

**Fig. 2.** Attack and change detection with false alarms.

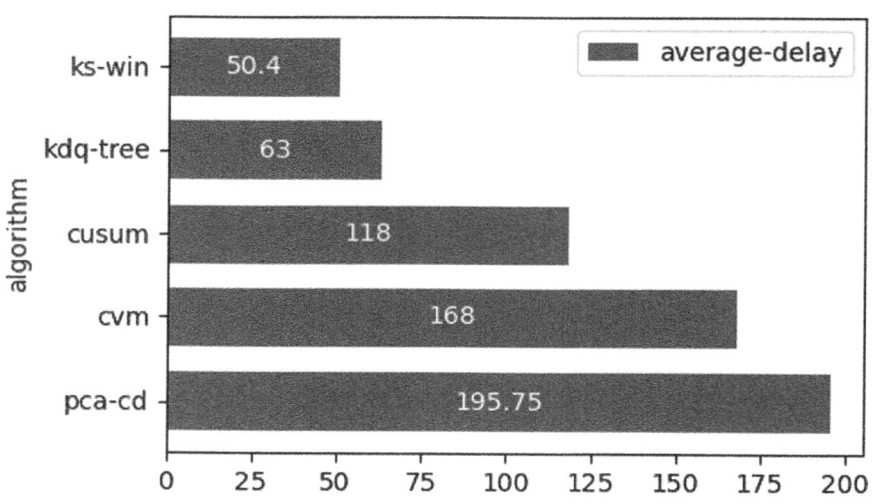

**Fig. 3.** Average detection delays.

**Table 1.** Parameters for the change detection algorithms.

| Sn. | Algorithm | Parameter values |
|---|---|---|
| 1. | KDQ-tree | $window = 1000$, $\alpha = 0.05$, $bootstrap = 500$, $unbound = 100$ |
| 2. | CUSUM | $\lambda = 1$, $\delta = 0.005$ |
| 3. | CvM | $window = 10$, $p = 0.05$, $\lambda = 10$ |
| 4. | PCA-CD | $window = 50$, $divergence =$'kl', $\delta = 0.1$ |
| 5. | KS-WIN | $\alpha = 0.0001$, $window = 200$, $statsize = 100$ |

## 4.2   Dashboards

The Chronograf dashboard shown in Fig. 4 shows the visualization of the LIT 301 and MV 501 columns of the SWaT dataset. The plotted Timed Moving Average of the MV 501 provides a dynamic representation of the average value, giving more weight to recent data points. This emphasizes the trends and changes in MV 501 over time, offering a responsive insight into the column's evolving patterns. Mean of both column values were plotted using the gauge tool as shown on the dashboard. Sampling periods of MV 501 column is also shown.

Figure 5 is a Grafana dashboard showing several graphs including a pie chart illustrating the ratio of normal and attacked records in the data. The 10 s aggre-

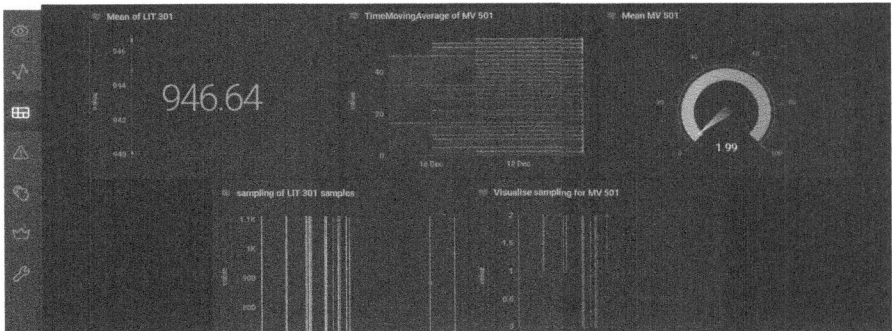

**Fig. 4.** Chronograf dashboard of Timed Moving Average.

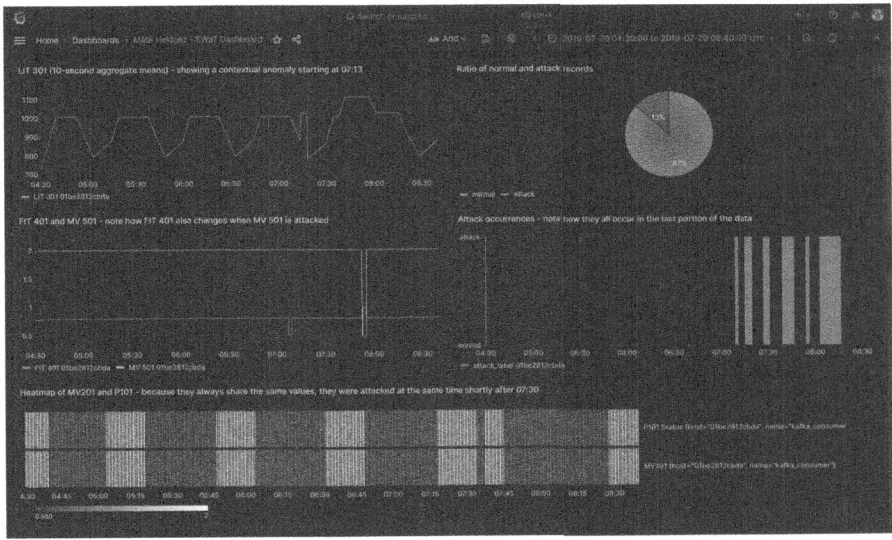

**Fig. 5.** Grafana dashboard of normal to attack ratio in the dataset.

gate mean of LIT 301 was plotted showing a contextual anomaly (in other words, a change that is classified as such by its surrounding context, as the values themselves aren't extreme) starting at 7:30. This dashboard also shows when exactly each attack occurs. The line plot of the MV 501 and FIT 401 shows that the latter changes even when the former is attacked, highlighting the correlation between them. Finally, a heatmap at the bottom of the dashboard shows the columns MV 201 and P101. They were attacked at the same time, and the heatmap shows why: at every point in time, their values are the same, so an efficient attack had to target both, or else it would be easily detected.

## 5   Conclusion and Future Work

In this research work we established a robust and scalable framework for real-time analysis and change detection in the Secure Water Treatment (SWaT) dataset. Leveraging Docker, Kafka, Spark, and the TIG-stack open-source technologies, we have crafted an integrated environment that seamlessly processes streaming data, implements multiple change detection algorithms, and provides insightful visualizations. The Kolmogorov-Smirnov WINdowing algorithm was able to detect changes early relative to KDQ-tree, with others being worse-off. The comparative analysis of algorithms sheds light on their performance across various criteria, providing valuable insights into their strengths and weaknesses. Up on moving forward, the project laid the groundwork for further research and advancements in the domain of secure water treatment, emphasizing the importance of continuous monitoring and adaptive analysis in safeguarding critical infrastructure.

Potential future directions for our work involve expanding the repertoire of change detection algorithms, incorporating both traditional and machine learning-based approaches for a more comprehensive evaluation and comparing the resulting framework with other approaches and frameworks. Enhancing the system's adaptability to evolving attack scenarios and continuous monitoring of algorithms performance are critical considerations for future iterations. Moreover, exploring and integrating other open-source technologies, such as Apache Flink or Apache Beam, could provide alternative avenues for stream processing and change detection. Additionally, advancements in cloud-native solutions and edge computing present opportunities to optimize scalability and resource efficiency in the architecture. These future endeavors aim to fortify the project's technological landscape and ensure its relevance in the evolving domain of secure water treatment systems.

# References

1. Callegari, C., Giordano, S., Pagano, M., Pepe, T.: WAVE-CUSUM: Improving cusum performance in network anomaly detection by means of wavelet analysis. Comput. Secur. **31**(5), 727–735 (2012). ISSN 0167-4048. https://doi.org/10.1016/j.cose.2012.05.001, URL https://www.sciencedirect.com/science/article/pii/S0167404812000788

2. Zhou, C., van Nooijen, R., Kolechkina, A., Hrachowitz, M.: Comparative analysis of nonparametric change-point detectors commonly used in hydrology. Hydrol. Sci. J. **64**(14), 1690–1710 (2019). https://doi.org/10.1080/02626667.2019.1669792, URL https://doi.org/10.1080/02626667.2019.1669792

3. Dasu, T., Krishnan, S., Venkatasubramanian, S., Yi, K.: An information-theoretic approach to detecting changes in multi-dimensional data streams. In: Proc. Symposium on the Interface of Statistics, Computing Science, and Applications (Interface) (2006)

4. Ditzler, G., Roveri, M., Alippi, C., Polikar, R.: Learning in nonstationary environments: a survey. IEEE Comput. Intell. Mag. **10**(4), 12–25 (2015). https://doi.org/10.1109/MCI.2015.2471196

5. Flynn, T., Yoo, S.: Change detection with the kernel cumulative sum algorithm. In: 2019 IEEE 58th Conference on Decision and Control (CDC), pp. 6092–6099 (2019). https://doi.org/10.1109/CDC40024.2019.9029854

6. Gama, J., Žliobaitė, I., Bifet, A., Pechenizkiy, M., Bouchachia, A.: A survey on concept drift adaptation. ACM Comput. Surv. **46**(4) (2014), ISSN 0360-0300. https://doi.org/10.1145/2523813

7. Goh, J., Adepu, S., Junejo, K.N., Mathur, A.: A dataset to support research in the design of secure water treatment systems. In: Havarneanu, G., Setola, R., Nassopoulos, H., Wolthusen, S. (eds.) Critical Information Infrastructures Security, pp. 88–99. Springer International Publishing, Cham (2017), ISBN 978-3-319-71368-7

8. Hu, H., Kantardzic, M., Sethi, T.S.: No free lunch theorem for concept drift detection in streaming data classification: a review. WIREs Data Min. Knowl. Discovery **10**(2), e1327 (2020)

9. Krawczyk, B., Woźniak, M.: One-class classifiers with incremental learning and forgetting for data streams with concept drift. Soft. Comput. **19**(12), 3387–3400 (2015)

10. Lamshöft, K., Neubert, T., Krätzer, C., Vielhauer, C., Dittmann, J.: Information hiding in cyber physical systems: challenges for embedding, retrieval and detection using sensor data of the swat dataset. In: Proceedings of the 2021 ACM Workshop on Information Hiding and Multimedia Security, pp. 113-124. IH&MMSec '21, Association for Computing Machinery, New York, NY, USA (2021), ISBN 9781450382953. https://doi.org/10.1145/3437880.3460413

11. Lu, J., Liu, A., Dong, F., Gu, F., Gama, J., Zhang, G.: Learning under concept drift: a review. IEEE Trans. Knowl. Data Eng. **31**(12), 2346–2363 (2019). https://doi.org/10.1109/TKDE.2018.2876857

12. Qahtan, A.A., Alharbi, B., Wang, S., Zhang, X.: A PCA-based change detection framework for multidimensional data streams: change detection in multidimensional data streams. In: Proceedings of the 21th ACM SIGKDD International Conference on Knowledge Discovery and Data Mining, pp. 935-944, KDD '15, Association for Computing Machinery, New York, NY, USA (2015), ISBN 9781450336642. https://doi.org/10.1145/2783258.2783359

13. Vaswani, N.: The modified CUSUM algorithm for slow and drastic change detection in general HMMs with unknown change parameters. In: Proceedings. (ICASSP '05). IEEE International Conference on Acoustics, Speech, and Signal Processing, 2005, vol. 4, pp. iv/701–iv/704 (2005). https://doi.org/10.1109/ICASSP.2005.1416105
14. Yoong, C.H., Heng, J.: Framework for continuous system security protection in swat. In: Proceedings of the 2019 3rd International Symposium on Computer Science and Intelligent Control, ISCSIC 2019, Association for Computing Machinery, New York, NY, USA (2020), ISBN 9781450376617, https://doi.org/10.1145/3386164.3387297

# Fuzzy Rule-Based Anomaly Explanation in Micro-electromechanical Systems

Hunor István Lukács[1,2]([envelope]) [ORCID], Tamás Fischl[2] [ORCID], and János Botzheim[1] [ORCID]

[1] Department of Artificial Intelligence, Faculty of Informatics, ELTE Eötvös Loránd University, Pázmány P. Sétány 1/A, 1117 Budapest, Hungary
{lukacs.hunor,botzheim}@inf.elte.hu
[2] Robert Bosch Kft, Gyömrői út 104,Budapest 1103, Hungary
tamas.fischl@hu.bosch.com

**Abstract.** This paper illustrates the application of a fuzzy rule-based system to enhance explainability, thereby reducing the time required to identify anomalous data behavior. The proposed algorithm, integrated with an autoencoder model, is employed in the field of anomaly detection, which consists of searching for small amounts of anomalies within extensive datasets.

The data analyzed is sourced from the production of a MEMS (Micro-Electromechanical System) based inertial sensor employed in the automotive industry. Therefore, the paper addresses real-world challenges encountered in industry, with the suggested method's results validated by domain experts.

The article places a strong emphasis on explainability through the utilization of a fuzzy rule-based system, which greatly facilitates and shortens the time of decision and intervention in the industrial environment.

The fuzzy system is trained by the bacterial memetic algorithm, which combines the bacterial evolutionary algorithm with the Levenberg-Marquardt local search technique, thus providing an efficient optimization for the model. Leveraging the white-box behavior of the fuzzy system, the trained model is then utilized to generate comprehensive linguistic interpretations, which can be readily understood by MEMS experts.

**Keywords:** Fuzzy rule-based system · Anomaly detection · MEMS inertial sensor

## 1 Introduction

With the spread of digitalization in industry, more and more data is becoming available in manufacturing, product development and services. Analyzing and evaluating this data is also a big task, and there are numerous methods for solving it. Today, the use of artificial intelligence to analyze industrial data is becoming important in the field of understanding the correlation in the data.

N.-T. Nguyen et al. (Eds.): ICCCI 2024, CCIS 2166, pp. 29–42, 2024.
https://doi.org/10.1007/978-3-031-70259-4_3

This article analyses data from the production of a MEMS (Micro-Electromechanical Systems) based inertial sensor used in the automotive industry. Within that, it deals with the topic of anomaly detection, which consists of searching for small amounts (ppm) of anomalies in the data. The article places emphasis on explainability, demonstrating how anomalies can be described using fuzzy rules. By defining hyper-cuboids and employing linguistic interpretation, the anomalies are associated with comprehensible descriptions.

The nature of the effects in the data set (and in data sets in general) can also be diverse. They can be the result of random measurement error (not a real error), or the result of systematic behavior either from the product or the environment, or a combination of both. Overall, however, the identification of these deviations, like an early activation, is of paramount importance for the manufacture of the product. If it is considered that an initial deviation that is within the measurement limits, due to long time drift, may lead to an out-of-specification deviation later in the product life cycle, then the detection and identification of these deviations should be part of the quality assurance process.

This article heavily draws from the research outlined in the "Bacterial Evolutionary Algorithm Based Autoencoder Architecture Search for Anomaly Detection" [7] article. As implied by its title, the referenced study aimed to develop an optimal autoencoder architecture specifically tailored for this anomaly detection task. While in the current scenario, the anomaly detection problem is addressed using an autoencoder model with an optimal architecture derived from the aforementioned article, it remains challenging to offer a comprehensive explanation regarding the specific nature of the detected anomalous data and how they deviate.

The article [1] introduces a method utilizing Shapley Additive Explanations to interpret anomalies identified by an autoencoder, with an emphasis on the connection between features exhibiting high reconstruction errors. In contrast, the article [5] describes a technique that employs Quantile Regression Forests to model the dependencies between features. The method outlined in this article offers a white-box model that characterizes anomalous data by directly extracting fuzzy rules from the trained model.

Thus, this article aims at the bacterial memetic optimization of the fuzzy rule-based system on the chosen MEMS based inertial sensor production data set, and by exploiting the white-box behavior of the trained fuzzy model, explaining why a data point is labeled as anomaly or inlier.

The objective of this experiment is to develop an anomaly detector with high sensitivity to both known and unknown anomalies. Furthermore, to explore methods for transparently representing and explaining the origins of these anomalies to MEMS experts.

## 2    MEMS-Based Inertial Sensor

Micro-electromechanical systems (MEMS) [8] are silicon based electro-mechanical systems, approximately 1 to 100 microns ($10^{-6}$) integrated devices.

The MEMS systems combine electrical and mechanical components into small structures in the micrometer scale. The MEMS system consist of the following major components: sensing mechanism, mechanical elements, and the application-specific integrated circuit (ASIC). The ASIC drives the mechanical element electrically and evaluate their signals. The MEMS-based inertial sensor consists of an acceleration chip, a gyroscope, and an ASIC. The accelerometers (acceleration chip) measure linear acceleration in one or more axis, while the gyroscopes measure the angular motion. MEMS-based inertial sensors are used in automotive industry, and as consumer sensor, for example in mobile phones. In order to have accurate sensors, a lot of signal must be tested under different conditions. MEMS-based gyroscopes must meet countless requirements like vibration, temperature, humidity, etc., some of which are provided by design, while others must be adjusted through calibration and the key parameters of the product must be provided to the customer. These parameters are e.g. the offset, sensitivity, noise parameters and their temperature dependence.

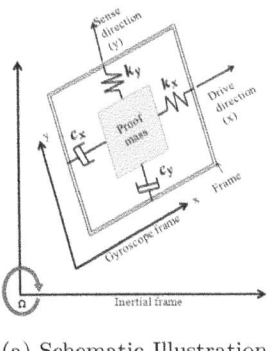

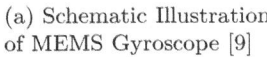

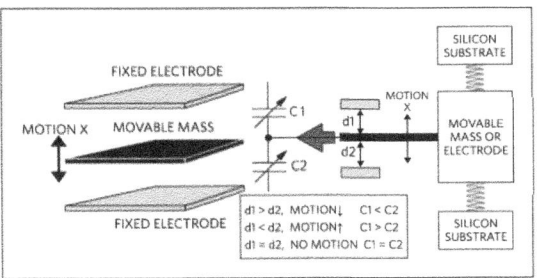

(a) Schematic Illustration of MEMS Gyroscope [9]

(b) Acceleration with a single moving mass [3]

**Fig. 1.** Illustrations of MEMS Gyroscope and Accelerometer

A MEMS gyroscope sensor [9] is used in measuring the angular rate of an object by means of Coriolis acceleration. In order to sense angular velocity, the majority of these gyroscopes use vibrating mechanical elements. The schematic illustration of such MEMS vibratory gyroscope is presented in Fig. 1a. The gyroscope structure movement is modeled while it rotates, thus the motion is represented based on a stationary frame (gyroscope frame) and a non-stationary frame (inertial frame). At the core, it has a vibrating (proof) mass, suspended by flexible beams. The device is vibrated at natural frequency, which is known as drive mode. When the gyroscope experiences an angular rotation, a Coriolis force is induced, which causes an energy transfer between the drive mode and sense mode. Thus, the sense mode is the appearing new vibration when the structure begins to rotate and due to the Coriolis force acting on the proof mass, changing the direction of its natural vibration. The movement of the proof mass in the sense direction, caused by the Coriolis force, is proportional to the applied angular rotation. By the use of integrated comb electrodes, based on differential capacitor techniques, this angular rotation can be measured.

The MEMS accelerometer sensors [3] measure the displacement of a mass with an interface circuit. Then the measurement is converted into a digital electrical signal for digital processing. The basic operation of an accelerometer is based on the Newton's second law of motion, which says that the acceleration of a body is directly proportional to, and in the same direction as, the net force acting on the body, and inversely proportional to its mass. In accelerometers, a commonly used sensing approach is capacitor sensing, in which acceleration is based on the change of the capacitor of a moving mass. Figure 1b shows an accelerometer with capacitor arranged as a differential pair. It consists of a movable mass placed between two, fixed reference silicon substrates or electrodes. The movement of the mass (Motion x) is relative to the fixed electrodes (d1, d2), therefore causes change in capacitors (C1, C2). The displacement of our mass and its direction can be derived by calculating the difference between the two capacitors.

## 3    Anomaly Detection by Fuzzy Systems

### 3.1    Fuzzy Rule-Based System (FRBS)

The model [2] consists of a rule base: $R = \{R_i\}$, where $R_i$ are the fuzzy rules. Each rule has the following form:

$R_i$ : if $(x_1$ is $A_{i1})$ and $(x_2$ is $A_{i2})$ and ... and $(x_n$ is $A_{in})$ then $(y$ is $B_i)$

The $A_{ij}$ antecedent membership function belongs to the $i$th rule and $j$th input variable and has 4 parameters (breakpoints): $a_{ij}$, $b_{ij}$, $c_{ij}$, $d_{ij}$. The $B_i$ consequent membership function belongs to the $i$th rule, and it has 4 parameters as well: $a_i$, $b_i$, $c_i$, $d_i$. The relative importance of the $j$th variable in the $i$th rule is as follows:

$$\mu_{ij}(x_j) = \frac{x_j - a_{ij}}{b_{ij} - a_{ij}} N_{i,j,1}(x_j) + N_{i,j,2}(x_j) + \frac{d_{ij} - x_j}{d_{ij} - c_{ij}} N_{i,j,3}(x_j)$$

where $a_{ij} <= b_{ij} <= c_{ij} <= d_{ij}$ must hold, and

$$N_{i,j,1}(x_j) = \begin{cases} 1, & \text{if } x_j \in [a_{ij}, b_{ij}] \\ 0, & \text{otherwise} \end{cases} \qquad N_{i,j,2}(x_j) = \begin{cases} 1, & \text{if } x_j \in (b_{ij}, c_{ij}) \\ 0, & \text{otherwise} \end{cases}$$

$$N_{i,j,3}(x_j) = \begin{cases} 1, & \text{if } x_j \in [c_{ij}, d_{ij}] \\ 0, & \text{otherwise} \end{cases}$$

The activation degree of the $i$th rule (the $t$-norm is the minimum):

$$w_i = \min_{j=1}^{n} \mu_{ij}(x_j). \tag{1}$$

Here $n$ is the number of input dimensions, $w_i$ is the importance of the $i$th rule if the input vector is $\mathbf{x}$, and $\mu_{ij}(x_j)$ is the importance of the $j$th element of the input vector in the $i$th rule.

The prediction of the fuzzy rule-based system, which uses Mamdani inference and center of sums (COS) as a defuzzification method, can be written in the following explicit form (as presented in Article [2]):

$$y_c(\mathbf{x}) = \frac{1}{3} \frac{\sum_{i=1}^{R} 3w_i(d_i^2 - a_i^2)(1-w_i) + 3w_i^2(c_id_i - a_ib_i) + w_i^3(c_i - d_i + a_i - b_i)(c_i - d_i - a_i + b_i)}{\sum_{i=1}^{R} 2w_i(d_i - a_i) + w_i^2(c_i + a_i - d_i - b_i)}. \quad (2)$$

This method is applicable to classification tasks as well. In such scenarios, the output $y_c(\mathbf{x})$ indicates the extent to which the input $\mathbf{x}$ belongs to a particular class. In the specific application discussed in this article, the output $y_c(\mathbf{x})$ denotes the degree to which an observation $\mathbf{x}$ is considered an inlier. When anomalies are detected, the output is 0; otherwise, its value exceeds 0.

## 3.2   Bacterial Memetic Algorithm (BMA)

The Bacterial Memetic Algorithm [4] is a population-based stochastic optimization algorithm, that combines global and local search algorithms.

As a global search technique, the BMA algorithm uses the bacterial mutation and the gene transfer operators. The bacterial mutation aims to optimize the chromosome, while the gene transfer allows a part of the encoded data to be copied among the bacteria. As a local search technique, the Levenberg-Marquardt method [6] is applied with a given probability for each individual. In each generation, the bacterial mutation, the gene transfer and the Levenberg-Marquardt methods are called, and the BMA algorithm stops when the maximum number of generations, denoted by the hyperparameter $N_{gen}$, is reached.

## 3.3   BMA Applied on FRBS

One fuzzy rule-based system corresponds to one bacterial individual in the population. The encoded data are the four breakpoints of the trapezoidal membership functions for all antecedents and consequents. An individual is evaluated according to its ability to approximate the ground truth values. There are $N_{ind}$ individuals in the population.

## 3.4   Bacterial Mutation

Figure 2 depicts the bacterial mutation operator, which is applied for each individual in the population. It creates $N_{clones}+1$ number of clones for each individual. One group of genes is randomly selected and changed in $N_{clones}$ clones, with one remaining unchanged. The best clone from $N_{clones}+1$ clones is selected to transfer the mutated genes into the other $N_{clones}$ clones. The same process executes again with the yet-unselected group of genes, and repeats until all the genes in the bacterial chromosome are selected exactly once. The best clone replaces the original individual in the population, discarding the other clones.

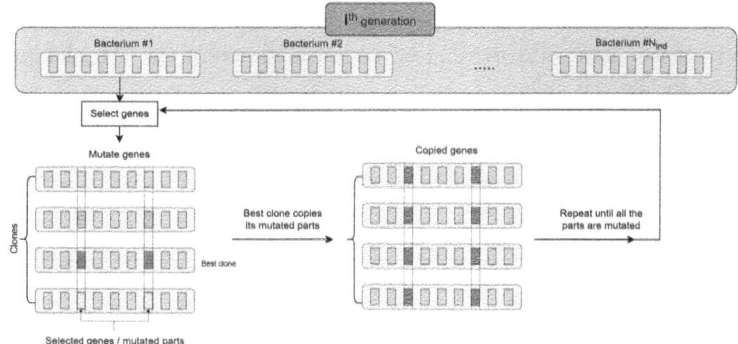

**Fig. 2.** Bacterial Mutation

## 3.5   Gene Transfer

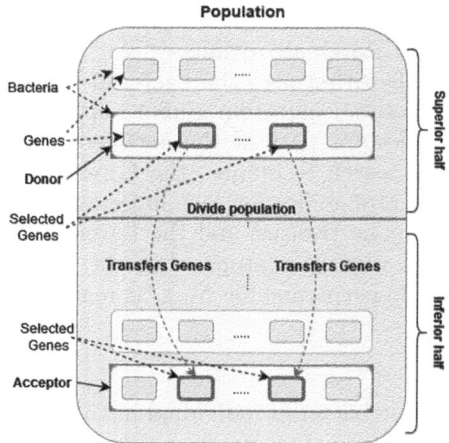

The gene transfer operator, presented in the Fig. 3, works on the population level. The bacterial population is split into two halves according to error definition, where the individuals with a low loss value will belong to the Superior Half, while the worst individuals to the Inferior Half. A randomly chosen individual from the Superior Half of the population transfers genes to another randomly chosen bacterium from the Inferior Half.

After transferring the genes, the population is sorted again, and the procedure is repeated $N_{inf}$ times.

**Fig. 3.** Gene Transfer

## 3.6   Levenberg-Marquardt Method

The Levenberg-Marquardt method [4] is applied for each bacterium by a given probability, indicated by the $LM_{prob}$ parameter. This algorithm uses the update rule below in order to optimize the model, encoded by a bacterium.

The update-vector ($s_k$):

$$s_k = -\left(\mathbf{J}_k^T \mathbf{J}_k + \gamma_k \mathbf{I}\right)^{-1} \mathbf{J}_k^T e_k$$

The Jacobian matrix ($\mathbf{J}_k$):

$$\mathbf{J}_k = \left[\frac{\partial y(\boldsymbol{x}^{(p)}, \boldsymbol{b}_k)}{\partial \boldsymbol{b}_k^T}\right]$$

where $\boldsymbol{b}_k$ stands for the parameters of the fuzzy model, the encoded FRBS, and $k$ is the step index.

Bravery factor ($\gamma$):

$$\gamma_{k+1} = \begin{cases} 4\gamma_k, & \text{if } r_k < 0.25 \\ \gamma_k/2, & \text{if } r_k > 0.75 \\ \gamma_k, & \text{otherwise} \end{cases}$$

Then, the new bacterium:

Trust region ($r_k$):

$$r_k = \frac{||\boldsymbol{b}_k||^2 - ||\boldsymbol{b}_k + \boldsymbol{s}_k||^2}{||\boldsymbol{b}_k||^2 - ||\mathbf{J}_k\boldsymbol{s}_k + \boldsymbol{e}_k||^2}$$

$$\boldsymbol{b}_{k+1} = \begin{cases} \boldsymbol{b}_k + \boldsymbol{s}_k, & \text{if } ||\boldsymbol{b}_k + \boldsymbol{s}_k||^2 < ||\boldsymbol{b}_k||^2 \\ \boldsymbol{b}_k, & \text{otherwise,} \end{cases}$$

where $\boldsymbol{b}_k + \boldsymbol{s}_k$ is the parameter vector of the updated fuzzy model.

# 4   Anomaly Detection by Fuzzy Systems in MEMS Based Sensor Production

The article "Bacterial Evolutionary Algorithm Based Autoencoder Architecture Search for Anomaly Detection" [7] introduces a technique for determining an optimal autoencoder architecture tailored for anomaly detection tasks optimized by an evolutionary algorithm. This article will primarily focus on providing transparent explainability, which will aid in describing the identified anomalies effectively.

## 4.1   Dataset

The dataset is sourced from the production of MEMS-based inertial sensors, initially comprising 400 thousand data points and 64 features. These features encompass measurement data collected across the entire production process, starting from individual components and culminating in the final product inspection.

Using a dataset and leveraging expert knowledge, certain data points are initially labeled by domain experts as anomalies, serving as the ground truth anomalies.

For illustrative purposes, a smaller dataset is used, sampled from the original normalized dataset. The reduced dataset consists of 1 thousand sampled data points and 6 features, containing 3 ground truth anomalies.

The selected features are as follows: $Feature_1$, ..., $Feature_6$. Due to the sensitive nature of the dataset, only these numerical feature names are utilized for security reasons, instead of the original feature names. The features were selected such that the ground truth anomalies can be discovered (Fig. 6), furthermore the narrowed dataset also contains other real anomalies in manageable quantity that are hidden in the multidimensional correlations.

First, the dataset is labeled by an autoencoder in an unsupervised manner, splitting the data into inliers and outliers. Defining an optimal autoencoder architecture for this task is presented more in details in the article "Bacterial Evolutionary Algorithm Based Autoencoder Architecture Search for Anomaly

Detection" [7]. The threshold $\Sigma$ value is chosen such that the three ground truth anomalies lie in the outlier zone. The labeled dataset contains 994 inliers and 6 outliers according to the autoencoder.

Figure 4 illustrates the dataset projected onto 2D planes defined by specific features. Inliers are represented by green points, while anomalies are denoted by red, purple, and cyan points. Note, that due to the sampling process, the data does not cover the entire [0, 1] interval.

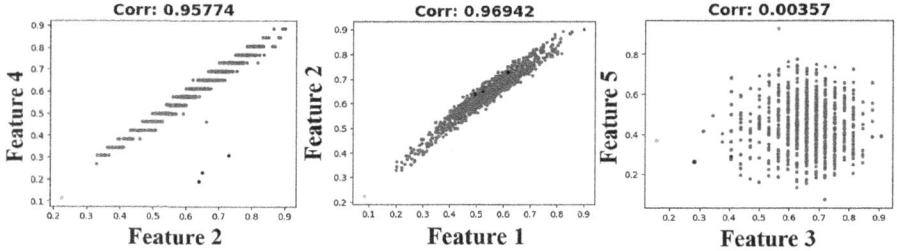

**Fig. 4.** Feature correlations and projections

## 4.2 Rule Initialization by Exploiting Domain Knowledge

The fuzzy rules can be associated with linguistic interpretation, by construction if-then expression. By leveraging this capability, the domain expert defined the rules for identifying ground truth anomalies as follows:

*"IF Feature_2 is high AND Feature_4 is low THEN anomaly"*

Please note that adjectives such as 'low' and 'high' are inherently fuzzy values. The ranges for these adjectives can be chosen arbitrarily and independently for each feature dimension. In this example, for $Feature_2$, the 'low' value corresponds to the interval $[0.2, 0.6]$, and the 'high' value to $[0.6, 0.9]$. Conversely, for $Feature_4$, 'low' refers to the interval $[0.1, 0.4]$, while 'high' pertains to $[0.4, 0.9]$.

The rules of the fuzzy systems presented in this article focuses on finding and describing the inliers, as the consequent part suggests, but it can be also used for describing anomalies. Rephrasing the fuzzy expression above from describing anomaly to inlier, the initialized rules for the fuzzy system are as follows:

*Rule_1:*
"**IF** *Feature_1* is *anything*
**AND** *Feature_2* is *anything*
**AND** *Feature_3* is *anything*       OR
**AND** *Feature_4* is **high**
**AND** *Feature_5* is *anything*
**AND** *Feature_6* is *anything*
**THEN** *inliner"*

*Rule_2:*
"**IF** *Feature_1* is *anything*
**AND** *Feature_2* is **low**
**AND** *Feature_3* is *anything*
**AND** *Feature_4* is *anything*
**AND** *Feature_5* is *anything*
**AND** *Feature_6* is *anything*
**THEN** *inliner"*

The inclusion of the adjective 'anything' in the rules above indicates that at the initialization, the respective feature does not play a role in describing anomalies. Note that these rules align with those described by the domain expert. If $Feature_2$ is classified as 'high' while $Feature_4$ is classified as 'low', both rules would be deemed false, indicating an anomaly.

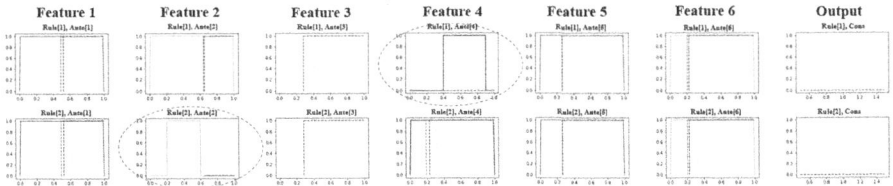

**Fig. 5.** Initializing the rules of FRBS based on domain expert knowledge.

According to the linguistic rules, the initialized fuzzy rules are depicted in Fig. 5. Using the presented fuzzy system, in case of inliers the FRBS model has an output value greater than 0, while in case of anomalies the output will be 0. In Fig. 5 the fuzzy membership functions in the Consequent are centered around 1, which means that if there is even a small activation in each Antecedent, the output of the model will be 1. The output will be 0, if at least one Antecedent with 0 activation degree exists in each Rule. It's worth noting that the 'anything' attribute in the fuzzy membership function level is represented as trapezoids with full activation across the whole $[0, 1]$ interval. The circled membership functions pertain to the rules outlined by the domain expert.

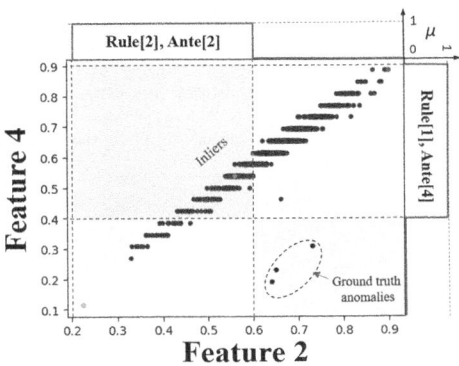

The dashed cyan, magenta, and purple lines represent the three ground truth anomalies as observations and their corresponding activation degree in the initialized fuzzy system. Given that the activation in the circled membership functions is 0, the ground truth anomalies will indeed be identified as anomalies by the fuzzy system.

In Fig. 6, the data projection onto the plane spanned by $Feature_4$ and $Feature_2$ is depicted. It illustrates how the initialized fuzzy rules distinguish between the inliers and the ground truth anomalies.

**Fig. 6.** The initial rules identify the ground truth anomalies within the projected data.

## 4.3   Training and Results

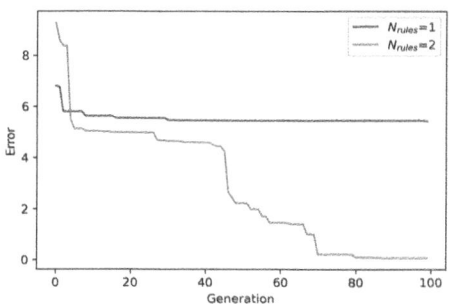

**Fig. 7.** Learning curve. Finding the optimal rule number

In this experiment[1], the used dataset contains measurements of MEMS inertial sensors, and it consists of 1 thousand data points and 6 features.

Figure 7 presents the training curve of the fuzzy model, how the error decreases by generations of the BMA algorithm. It also reveals that if the number of rules is set to 1, then the derived model is not complex enough, thus it tends to underfit, while using 2 rules the model's complexity is enough to separate the dataset according to the labels provided by the autoencoder.

The labeled dataset contains 994 inliers and 6 outliers according to the autoencoder. The task in this case is to use a fuzzy model on this classification task and see how the anomalies can be described according to the fuzzy rules. The parameters of BMA used in this experiment are as follows: $N_{gen} = 100, N_{ind} = 10, N_{clone} = 3, N_{inf} = 10, LM_{prob} = 0.1$.

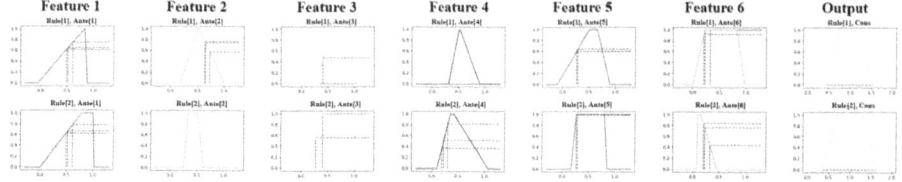

(a) Ground truth anomalies as observations in the trained Fuzzy system

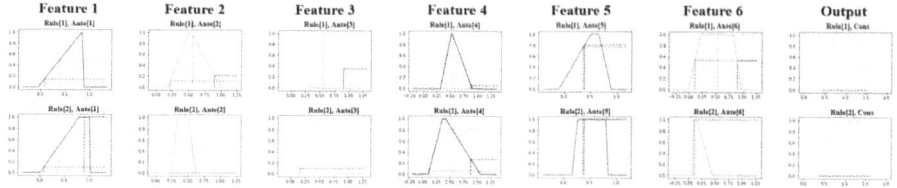

(b) Real anomalies as observations in the trained Fuzzy system

**Fig. 8.** Comparison of ground truth and real anomalies

Figure 8a depicts the trained fuzzy rule-based system, and illustrates how the decisions are made in the case of the three ground truth anomalies. The system

---

[1] System configuration: Intel(R) Xeon(R) Gold 6248R CPU @ 3.00GHz, 3 x NVIDIA Tesla T4 (16 GB) GPU, 256 GB RAM.

consists of 2 rules, each has 6 antecedents and 1 consequent, while with dashed lines the observations are represented. The dashed red lines show the ground truth anomalies as observations in the FRBS. The image in Fig. 8b illustrates that for the first rule (Rule[1]) at the fourth antecedent (Ante[4]), and for the second rule (Rule[2]) at the second antecedent (Ante[2]), the relative importance is 0 for all three ground truth anomalies observed. The fourth antecedent corresponds to $Feature_4$, while the second antecedent corresponds to $Feature_2$. The image depicted in Fig. 8b illustrates the relative importance on the trained fuzzy system for three newly discovered real anomalies, distinguished by cyan, magenta, and purple dashed lines.

**Table 1.** The rules of the trained FRBS

| Rule Id | $Feature1$ | | | | $Feature2$ | | | | $Feature3$ | | | |
|---|---|---|---|---|---|---|---|---|---|---|---|---|
| 1 | [−0.045 | 0.841 | 0.844 | 0.882] | [0.186 | 0.5 | 0.526 | 1.002] | [0.309 | 0.514 | 0.659 | 1.042] |
| 2 | [0.0005 | 0.79 | 1.002 | 1.024] | [0.2562 | 0.337 | 0.501 | 0.62] | [0.129 | 0.404 | 0.432 | 0.447] |

| $Feature4$ | | | | $Feature5$ | | | | $Feature6$ | | | | Output |
|---|---|---|---|---|---|---|---|---|---|---|---|---|
| [0.317 | 0.514 | 0.528 | 0.905] | [−0.098 | 0.51 | 0.661 | 0.899] | [0.001 | 0.219 | 0.831 | 0.983] | $Inlier$ |
| [0.098 | 0.356 | 0.42 | 1.064] | [0.165 | 0.264 | 0.8 | 0.853] | [0.084 | 0.089 | 0.154 | 0.454] | $Inlier$ |

**Table 2.** Values of anomalies in each dimension

| Anomaly | $Feature1$ | $Feature2$ | $Feature3$ | $Feature4$ | $Feature5$ | $Feature6$ |
|---|---|---|---|---|---|---|
| $1^{st}$ ground truth anomaly | 0.614 | 0.729 | 0.406 | 0.307 | 0.289 | 0.325 |
| $2^{nd}$ ground truth anomaly | 0.523 | 0.65 | 0.281 | 0.23 | 0.266 | 0.227 |
| $3^{rd}$ ground truth anomaly | 0.494 | 0.64 | 0.281 | 0.192 | 0.262 | 0.202 |
| cyan, real anomaly | 0.401 | 0.56 | 0.562 | 0.538 | 0.927 | 0.544 |
| purple, real anomaly | 0.899 | 0.899 | 0.906 | 0.884 | 0.393 | 0.902 |
| magenta, real anomaly | 0.082 | 0.224 | 0.156 | 0.115 | 0.37 | 0.116 |

Table 1 presents the antecedent breakpoints of the trained fuzzy system, while Table 2 shows the values of the discovered anomalous data points across each feature dimension.

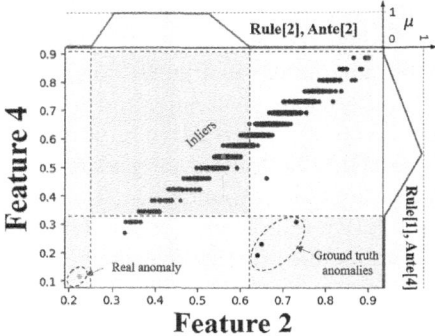

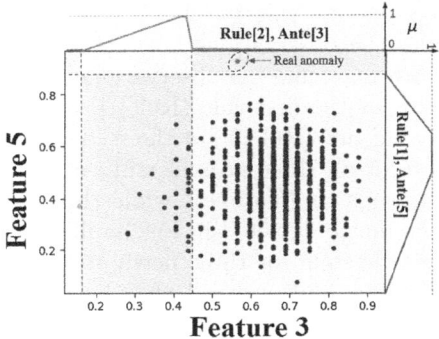

**Fig. 9.** Fuzzy rules revealing real and ground truth anomalies

**Fig. 10.** Fuzzy rules revealing a new real anomaly

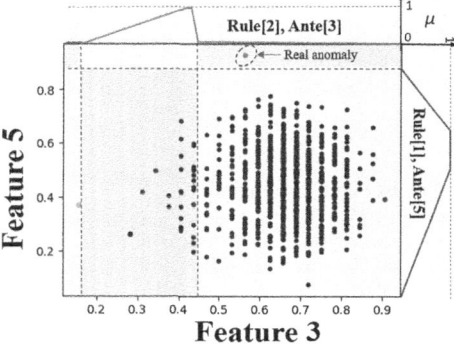

**Fig. 11.** Fuzzy rules revealing a new real anomaly

Figure 9 shows the data projected onto the two-dimensional plane defined by $Feature_2$ and $Feature_4$. The upper and right axes display the respective membership functions and their role in decision-making. If a data point is covered by at least one of the membership functions, it is deemed an inlier. Conversely, if it is not covered by any of the membership functions, it is identified as an anomaly. The two antecedent membership functions distinctly identify the three ground truth anomalies depicted in red, as well as another real anomaly depicted in magenta. By setting up intervals in each dimension, a possible linguistic interpretation can be associated to these fuzzy rules. A possible option for establishing intervals for each feature: $Feature_2$: *low*: $[0, 0.25]$, *medium*: $[0.25, 0.62]$, *high*: $[0.62, 1]$; $Feature_4$: *low*: $[0, 0.31]$, *high*: $[0.31, 1]$. Then the corresponding rules has the following linguistic interpretation:

"*IF    Feature_4    is    low    AND    Feature_2    is    high    THEN    anomaly*"

"*IF    Feature_4    is    low    AND    Feature_2    is    low    THEN    anomaly*"

Note, that the first rule is very similar to the rule that the domain expert described at the model initialization. Additionally, it's worth noting that the determination of the number of regions and their corresponding intervals can be arbitrarily defined.

Figure 10 illustrates the detection of another real anomaly, shown in cyan. This anomaly has zero activation in Rule[1]-Ante[5], associated with $Feature_5$, as well as in Rule[2]-Ante[3], associated with $Feature_3$. A possible option for establishing intervals for each feature: $Feature_5$: *low*: $[0, 0.9]$, *high*: $[0.9, 1]$; $Feature_3$: *low*: $[0, 0.13]$, *medium*: $[0.13, 0.44]$ *high*: $[0.44, 1]$. The corresponding linguistic interpretation:

"*IF   Feature_5   is   high   AND   Feature_3   is   low   THEN   anomaly*"

"*IF   Feature_5   is   high   AND   Feature_3   is   high   THEN   anomaly*"

In Fig. 11, the detection of another real anomaly is demonstrated, depicted in purple. This anomaly has zero activation in Rule[1]-Ante[1], corresponding to $Feature_1$, and in Rule[2]-Ante[2], corresponding to $Feature_2$. An option for defining intervals for each feature: $Feature_1$: *low*: $[0, 0.88]$, *high*: $[0.88, 1]$. The corresponding linguistic interpretation:

"*IF   Feature_1   is   high   AND   Feature_2   is   high   THEN   anomaly*"

These linguistically interpreted fuzzy rules enable expressing and describing anomalies to domain experts in a way that is understandable for humans, easy to grasp and follow.

It also demonstrates that for a given anomalous data point, adjusting the value of a specific dimension can potentially transform it into an inlier.

Please note that in the presented use-case within this article, the anomalous data contained in the dataset, when projected onto certain 2-dimensional plains, were linearly separable. As a result, the fuzzy rules were strict in this context, meaning that an output greater than 0 from the fuzzy system denotes an inlier, while an output precisely equal to 0 indicates an anomaly. However, if these data points are not linearly separable, the output of the fuzzy system can still provide certainty regarding the associated class.

# 5   Conclusion

The article demonstrates the application of a fuzzy rule-based model for anomaly detection tasks. Leveraging the model's transparent decision-making process, it provides insight into how decisions are reached. An added advantage of the proposed method is the optimization technique utilized: the bacterial memetic algorithm. This approach integrates the bacterial evolutionary algorithm with the Levenberg-Marquardt local search technique, enhancing the learning efficiency of the fuzzy system.

Regarding explainability, the article illustrates how the rules generated by the trained model in the fuzzy rule-based system inherently depict the relevant feature dimensions due to its transparent nature. The fuzzy rules can be interpreted as linguistic "if-then" statements, thereby enabling a linguistic interpretation. Furthermore, the article exploits the fact that fuzzy systems are effectively trainable expert systems, enabling the incorporation of human experts' decision-making and reasoning processes through if-then constructs.

As a trainable expert system, the trained fuzzy system makes it possible to determine the anomalous regions within the dataset. However, if new data is included containing as-yet-undiscovered types of anomalies (e.g., in a different region), then new training might be needed. For future work, we intend to combine autoencoder-based anomaly detection and fuzzy-based explainability into a single architecture.

# References

1. Antwarg, L., et al.: Explaining anomalies detected by autoencoders using Shapley Additive Explanations. Expert Syst. Appl. **186**, 115736 (2021). https://doi.org/10.1016/j.eswa.2021.115736
2. Botzheim, J., et al.: Fuzzy rule extraction by bacterial memetic algorithms. Int. J. Intell. Syst. **24**(3), 312-339 (2009). https://doi.org/10.1002/int.20338
3. Dadafshar, M.: Accelerometer and Gyroscopes Sensors: Operation, Sensing, and Applications (2019)
4. Horváth, C.M., et al.: Bacterial memetic algorithm trained fuzzy system-based model of single weld bead geometry. IEEE Access **8**, 164864–164881 (2020). https://doi.org/10.1109/ACCESS.2020.3021950
5. Li, Z., Leeuwen, M.: Explainable Contextual Anomaly Detection using Quantile Regression Forests (2023). https://doi.org/10.48550/arXiv.2302.11239
6. Lourakis, M.: A Brief Description of the Levenberg-Marquardt Algorithm Implemened by levmar. In: A Brief Description of the Levenberg- Marquardt Algorithm Implemented by Levmar 4 (Jan 2005)
7. Lukács, H.I., Fischl, T., Botzheim, J.: Bacterial evolutionary algorithm based autoencoder architecture search for anomaly detection. In: Advances in Computational Collective Intelligence, pp. 560-572. Springer Nature Switzerland (2023). isbn: 978-3-031-41774-0. https://doi.org/10.1007/978-3-031-41774-0_44
8. Maenaka, K.: MEMS inertial sensors and their applications. In: 2008 5th International Conference on Networked Sensing Systems, pp. 71–73 (2008). https://doi.org/10.1109/INSS.2008.4610859.
9. Patel, C., McCluskey, P.: Modeling and simulation of the MEMS vibratory gyroscope. In: 13th InterSociety Conference on Thermal and Thermomechanical Phenomena in Electronic Systems, pp. 928-933 (2012). isbn: 978-1-4244- 9533-7. https://doi.org/10.1109/ITHERM.2012.6231524.

# Analysis of Network Intrusion Detection and Potential Botnets Identification Using Selected Machine Learning Techniques

Patryk Zabawa and Michal Kedziora[✉][iD]

Wroclaw University of Science and Technology, Wroclaw, Poland
michal.kedziora@pwr.edu.pl

**Abstract.** The paper presented here is centered around the analysis of network attack detection using machine learning. It starts by examining the development and categorization of network attacks, and then provides an overview of conventional detection methods. Additionally, it conducts a thorough analysis of the ninth scenario in the CTU-13 dataset and outlines the steps involved in preparing for the experiment using this dataset and conducting feature engineering. The results section primarily focuses on the Random Forest and Decision Tree algorithms, as well as the classification of botnets using a Multilayer Perceptron (MLP), comparing the machine learning approaches with traditional methods of network attack detection.

**Keywords:** Machine Learning · AI · Network Attack Detection

## 1 Introduction

Due to the swift progress in technology and the ever-expanding network infrastructure, cyber attacks have not only become more frequent but also more sophisticated and unpredictable. Virtual spaces hold a significant amount of valuable data, which is both invaluable and vulnerable to potential threats. The loss or compromise of this data can have severe repercussions. While traditional defense mechanisms like IDS or IPS, which rely on known attack signatures, offer reliable protection, they may fall short when it comes to "zero-day" attacks, where the attack's unique signature is unknown [3,8,9]. One of the notable instances of a zero-day attack was the Equifax data breach that occurred between May and July 2017, affecting the private data of 150 million individuals. Exploiting a vulnerability in the Apache Struts MVC Framework, hackers gained access to Equifax servers, extracting sensitive information like social security numbers and driver's licenses. The breach lasted for 76 d before being detected, exposing weaknesses in Equifax's security infrastructure and delays in communicating the breach [4]. This attack underscored the critical need for robust cybersecurity measures and transparency in incident management, prompting shifts in cybersecurity protocols and raising societal awareness on this issue [6]. In the face

N.-T. Nguyen et al. (Eds.): ICCCI 2024, CCIS 2166, pp. 43–53, 2024.
https://doi.org/10.1007/978-3-031-70259-4_4

of these novel challenges, the focus shifts towards exploring the possibilities of machine learning in the field of cybersecurity [1,10]. By harnessing the advancing abilities of computers to process large volumes of data instantaneously, machine learning emerges as a pivotal component in safeguarding networks in the future. Its capacity to adjust and constantly acquire knowledge from emerging attack patterns, such as zero-day attacks, positions it as a potentially powerful defensive tool in today's landscape teeming with sophisticated cyber threats [15].

The primary objective of this paper is to analyze the efficiency of different techniques for detecting network attacks, with a particular emphasis on the use of advanced machine learning methods. The research aims to validate the effectiveness of machine learning models that are specifically developed to identify, classify, and mitigate various types of network attacks.

Moreover, a major aim of this study is to contrast contemporary techniques with conventional methods for detecting attacks. The examination will focus on discerning the advantages and drawbacks of each approach, taking into account factors like efficacy, operational velocity, and adaptability to the changing nature of cyber risks. The outcomes of the research seek to heighten awareness about network protection and the effective integration of machine learning into defense mechanisms. Additionally, these findings may offer backing for subsequent efforts aimed at strengthening network security systems.

## 2     Related Works

The use of machine learning techniques is becoming more common in the field of cybersecurity as a response to the increasing number of threats [16] [5]. This field has a wide range of scientific literature, including research studies and review articles, which showcase the various approaches and trends in this area [2]. This chapter aims to provide a comprehensive analysis of this field by conducting an extensive literature review, which includes two review papers and two empirical studies. A review article by Huseyin Ahmetoglu and Resul Das [1] discusses approaches to machine learning methods used for attack detection. The authors analyzed the detection, classification, clustering, and analysis of anomalies in network traffic. The focus was on cybersecurity, machine learning methods, and the datasets used in individual studies. The authors examined the feature selection or dimensionality reduction methods applied to the datasets used in these studies. The types of classifications conducted in these studies were described in detail, comparing the methods used with others, the performance metrics utilized, and the results presented in tables. Additionally, datasets concerning network attacks available as open sources were analyzed, proposing a basic taxonomy of cyber attacks. A review article authored by Ozlem Yavanoglu and Murat Aydos [16] concentrates on datasets used in artificial intelligence and machine learning techniques, which are key tools for analyzing network traffic and detecting irregularities. In the Machine Learning domain, it describes the definition, objectives, and types of algorithms such as supervised, unsupervised, and Reinforcement Learning, along with popular techniques like Linear

Regression, SVM, or Random Forest. It also emphasizes diverse AI applications such as knowledge representation, speech recognition, robotics, expert systems, and cybersecurity. The article provides detailed information about various popular datasets like KDD Cup 1999, ECML-PKDD 2007, or CTU-13, used for analyzing network anomalies, detecting DDoS attacks or botnets, serving as significant sources of knowledge for researchers involved in data analysis and cybersecurity. The project by Antoine Delplace, Sheryl Hermoso, and Kristofer Anandita [5] focused on building models to detect botnets in real network traffic represented by Netflow datasets (CTU-13 dataset) [13,14]. Significant features were extracted and subjected to selection, yet all proved significant and were included in the training. Various algorithms were tested, with the Random Forest being the most effective, achieving accuracy above 95% in most scenarios. Subsequently, efforts were concentrated on improving effectiveness in more challenging cases, using bootstrap methods to increase data volume and achieving over 55% accuracy in certain scenarios. Although improving results for some scenarios proved more difficult, it is possible they require more advanced techniques, such as deep neural networks, to better reflect botnet behaviors. In the work of Souhail Meftah, Tajjeeddine Rachidi, and Nasser Assem [12], an analysis of the intrusion detection process in networks using the UNSW-NB15 dataset was conducted. Employing a two-step method, the first stage focused on selecting optimal dataset features through techniques like Recursive Feature Elimination and Random Forests, then on binary classification to differentiate between invasive and normal traffic using various data mining techniques including Logistic Regression, Gradient Boost Machine, and Support Vector Machine. It demonstrated that the Support Vector Machine classifier achieved the highest accuracy (82.11%). The next step was using the results of this classifier for a series of multiclass classifiers, where the Decision Trees method (C5.0) obtained the highest accuracy (74%) and F1 score (86%). The two-step hybrid classification increased the accuracy of results by 12%, reaching 86.04% multiclass accuracy

## 3   Methedology

CTU-13 is a compilation of data on botnet traffic collected at the CTU University in the Czech Republic in 2011 [7]. The primary objective of this data set was to capture and analyze a significant volume of actual botnet traffic, which was intertwined with regular network traffic and background communication. The CTU-13 dataset comprises 13 captures, referred to as scenarios, that represent different instances of botnet activity. In each scenario, specific malicious software was executed, using various network protocols and performing different actions. What sets the CTU-13 dataset apart is that each scenario was manually analyzed and labeled.

The main topic of the research will be aligned with the ninth scenario, which involved the use of malicious software called Neris. The data capture for this scenario lasted a total of 5.18 h. During this time, the botnet used HTTP6-based C&C channels to send spam and participate in click fraud activities. The NetFlow file for this scenario has a size of 273MB and includes a total of 2,753,884 bidirectional communications. These communications are described using 15 features, similar to the other scenarios.

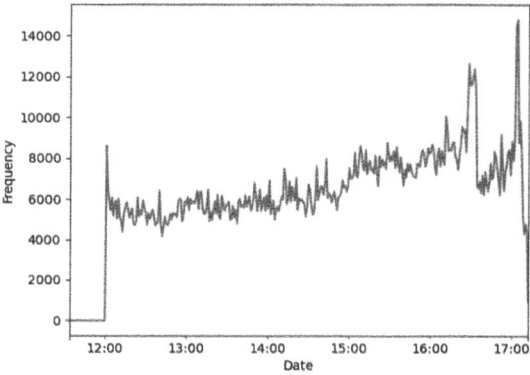

**Fig. 1.** Frequency of Flow

It appears that the start times of the NetFlow streams were fairly evenly distributed during the experiment, as can be observed in Fig. 1. However, a significant increase in NetFlow stream starts was observed between 16:00 and 17:00.

**Feature Engineering.** To compare and optimize the model results, we conducted the Feature Engineering process following the methodology proposed by Quan Luu on GitHub [11]. The data were divided into a training set and a test set using the train_test_split function from the scikit-learn library [16], with the test data size being 30% of the entire dataset.

Standardizing data using MinMaxScaler from the scikit-learn library is a process that alters the range of numerical feature values to fit within a specified range. This operation aims to ensure that all features in the training dataset have a mean of zero and a standard deviation of one. This process is exclusively applied to selected numerical columns of the training set, x_train, according to the defined set of num_col.

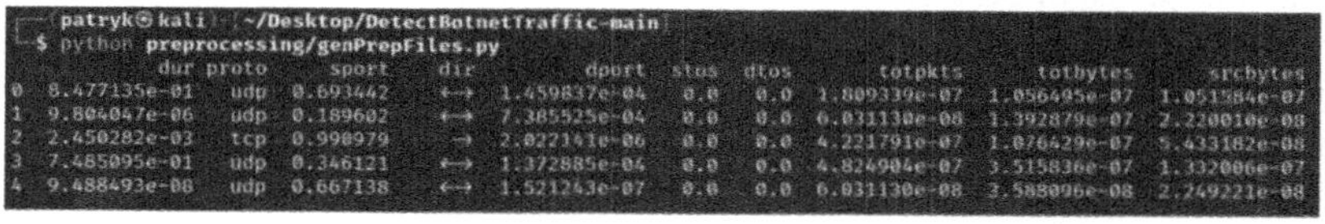

```
┌──(patryk㉿kali)-[~/Desktop/DetectBotnetTraffic-main]
└─$ python preprocessing/genPrepFiles.py
          dur proto    sport    dir       dport   stos  dtos      totpkts       totbytes      srcbytes
0  8.477135e-01   udp  0.693442  ←→  1.459837e-04  0.0  0.0  1.809339e-07  1.056495e-07  1.051984e-07
1  9.804047e-06   udp  0.189602  ←→  7.385525e-04  0.0  0.0  6.031130e-08  1.392879e-07  2.220010e-08
2  2.450282e-03   tcp  0.990979   →  2.022141e-06  0.0  0.0  4.221791e-07  1.076429e-07  5.433182e-08
3  7.485095e-01   udp  0.346121  ←→  1.372885e-04  0.0  0.0  4.824904e-07  3.515836e-07  1.332006e-07
4  9.488493e-08   udp  0.667138  ←→  1.521243e-07  0.0  0.0  6.031130e-08  3.588096e-08  2.249221e-08
```

**Fig. 2.** Presentation of dataset after standardization

Standardization is an important process in machine learning that involves bringing the features of a dataset into a consistent value range. This helps to eliminate any potential issues that may arise from differences in the scales of the features. By standardizing the data, machine learning algorithms such as Logistic Regression, Decision Trees, or Neural Networks are able to perform better, as they are more effective when working with data that has similar value spreads. This leads to the creation of more stable and reliable models. One of the main reasons for standardization is to avoid the dominance of features with larger values over those with smaller values during the learning process. If standardization is not applied, the models may primarily focus on the features with larger values, while neglecting the smaller ones. This can result in incorrect or misleading conclusions.

Converting categorical columns into numerical ones is essential because machine learning models cannot handle textual data and instead require numerical data. One technique used for this purpose is One-Hot Encoding, which converts categorical values into numerical form. In One-Hot Encoding, a new column is created for each unique value in a categorical column. For each observation, a value of 1 is assigned to the corresponding categorical value and 0 to the remaining new columns, allowing the representation of data in numerical form. To perform One-Hot Encoding, OneHotEncoder objects are created for each categorical column, such as 'stos', 'dtos', 'dir', and 'proto'. These encoders are then created using the fit() method for each categorical column. The data transformation using One-Hot Encoding is done by applying the transformation iteratively for each categorical column. The data is converted into a numerical matrix using the encoding objects created earlier and then combined with the original dataset by adding the newly created numerical columns. This process results in a training dataset (x_train) with additional numerical columns after the One-Hot Encoding transformation. In addition, important parameters and objects used during this process are saved in order to apply the same transformation process to the test set. An example portion of the data set after standardization can be observed in Fig. 2.

The preparation of the data processing pipeline for test data is based on previously prepared parameters and objects used during the training process. The test data undergo the same transformation processes as the training data to be used to test the trained predictive model. The clean_data function initiates the data preparation process in the same way as it was performed on the training data, by removing columns, handling missing values, converting hexadecimal data to decimal numbers, standardizing and encoding categorical columns. Subsequently, the test data are processed using previously saved objects. The standardization function standardizes the data by mean and scales the variance using the scaler object, which was previously trained on the training data. Meanwhile, the ohencoding function performs one-hot encoding of categorical columns, utilizing previously trained encoding objects (ohe_dir, ohe_proto, ohe_stos, ohe_dtos) on the training data. By following these procedures, we generate a collection of test data that is identical in format to the training data after undergoing processing. This implies that the values in the test data are scaled and encoded in the same manner as in the training data, guaranteeing consistency in the range of values and data structure across the test and training sets. This standardized set of test data is prepared for immediate utilization in evaluating the effectiveness and generalizability of the trained predictive model, facilitating an accurate assessment of its performance on new, unfamiliar data.

### 3.1   Results and Analysis of Experiments

To compare the performance of different algorithms, it is necessary to select metrics. A common practice is to evaluate the number of false positives (instances of normal traffic wrongly classified as botnets) and false negatives (instances of botnets labeled as regular traffic). Additionally, the accuracy of the multi-classification model in identifying botnets after detecting abnormal traffic is crucial. It is important to consider that, in the context of identifying malicious software, having a low recall is more problematic than having a low precision. Low recall means that while we are good at accurately identifying botnet communications (precision), many actual botnet communications go undetected (recall). Achieving high recall may seem simple, such as labeling everything as botnet, but this approach is not practical. Therefore, our approach aims to strike a balance by maximizing the f1 score while ensuring that the recall does not drop too low.

**Algorithms.** In the process of creating a Decision Tree classifier , class weights (0:0.05, 1:0.95) were used to address the issue of unbalanced classes within the data set. The hyperparameter_tuning function was applied, utilizing GridSearchCV and 3-fold cross-validation to optimize the hyperparameters max_depth and min_samples_split. Subsequently, after selecting the best Decision Tree estimator, focus shifted to fine-tuning the hyperparameter min_samples_leaf, testing various values (9, 11, 13) to find the optimal configuration.

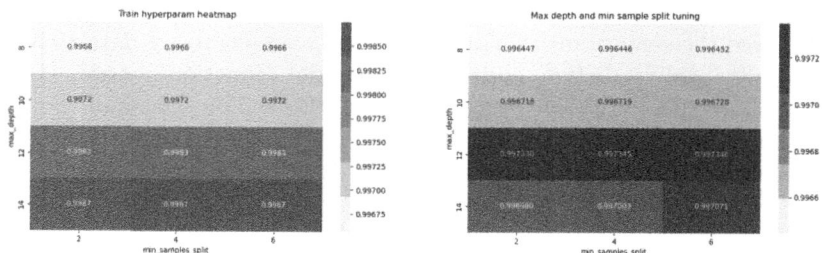

**Fig. 3.** Max depth and min sample split tuning

The model's effectiveness primarily depends on the 'max_depth' parameter, while the other two parameters have a lesser impact on the results. As can be seen in Fig. 3. There is a noticeably small difference between the results on the training and test sets, indicating a lack of overfitting of the model. The first step in optimizing a Random Forest classifier is tuning the 'n_estimators' parameter, which defines the number of trees in the decision forest. Using the cross_validation function, cross-validation is performed for values of 10, 15, and 20 to find the optimal number of trees in the model as can be observed on Fig. 4.

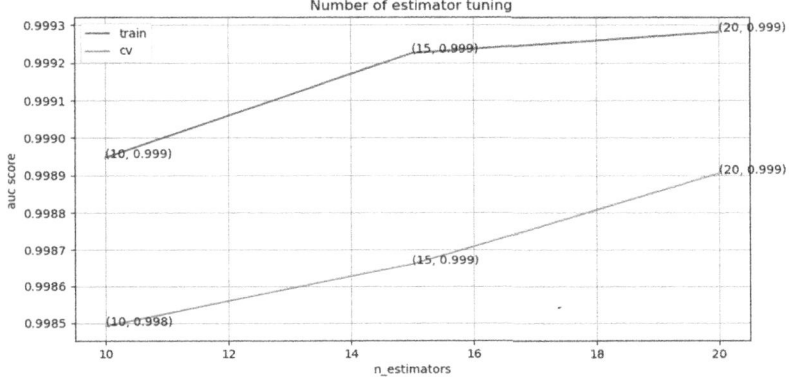

**Fig. 4.** Number of estimator tuning

The next step in tuning involves adjusting the 'criterion' parameter, which determines the function used to measure the quality of a split when constructing a Decision Tree in Random Forest (Fig. 5). This process involves testing two criterion values: 'gini' and 'entropy'. The Random Forest classifier is trained and tested for each of these values as part of cross-validation.

| Dataset | Model | AUC | f1-score | False Alarm Rate |
|---------|-------|-----|----------|------------------|
| Train | Random Forest | 0.95743 | 0.92851 | 0.04315 |
| Test | Random Forest | 0.94159 | 0.90163 | 0.06027 |

**Fig. 5.** Score on training and test set of Random Forest

The overall effectiveness of the model heavily relies on two main parameters, namely 'n_estimators' and 'max_depth', while the impact of the other two parameters is relatively minor. It has been observed that the results obtained from the training and test sets are quite similar. However, there is some discrepancy when compared to previous models, indicating a potential issue of overfitting the model to the training data. Nevertheless, this difference is relatively small, suggesting that the problem of overfitting is not significant. The performance of this model is not satisfactory compared to the Decision Tree model.

Both the number of false positives and false negatives have increased, indicating that this model is less effective in identifying both positive and negative cases compared to the previously used Decision Tree. The results obtained from the Decision Tree and Random Forest models exhibit similarities to the findings of Quan Luu [11]. By comparing both sets of results, it can be observed that for both Decision Trees and Random Forests, We achieved better AUC, F1-score, and a lower False Alarm Rate. This confirms the robustness and stability of both models - both decision trees and random forests appear to be effective in making predictions on this dataset.

**Botnet Classification.** The additional purpose of this study was to classify different types of botnets using a multilayer neural network called Multiple Layer Perceptron (MLP). The main goal was to categorize and identify botnets based on the available data, which consisted only of samples from botnets. The analysis focused on identifying specific botnet types such as Murlo, Virut, Rbot, and Neris. Although other botnets like NSIS, SoGou, and DonBot were recognized as potential threats, they were not included in the subset of data analyzed. The study utilized varying numbers of samples for each of the selected botnets in the analyzed dataset. - Murlo: 6147 samples - Virut: 185182 samples - Rbot: 40481 samples - Neris: 33549 samples The applied MLP model comprised 6 hidden layers with varying numbers of neurons in each layer: 128, 1024, 512, 128, 128, 10. The ReLU activation function was applied in the hidden layers, while the output layer utilized the Softmax function to determine the probability distribution for all botnet classes (Fig. 6).

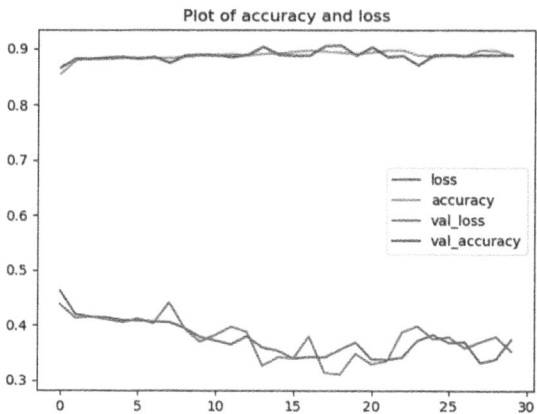

**Fig. 6.** Plot of accuracy and loss of MLP model

The outcomes suggest that the model performs well in categorizing various botnets, both in the training and test datasets. The model exhibited remarkable accuracy, precision, and recall rates for each botnet class, demonstrating its capability to accurately identify and classify network threats.

# 4 Conclusion and Future Work

The objective of this paper was to assess the effectiveness of machine learning techniques in detecting network attacks and the identification of potential botnets. The study aimed to explore how machine learning models can be utilized to identify and counter prevalent threats in computer networks. Additionally, the study compared traditional methods with machine learning-based techniques to determine their respective advantages and limitations. Initially, a comprehensive overview of network attacks and their prevention methods was provided, covering both traditional and machine learning approaches. As part of the analysis, one of the CTU-13 dataset scenarios was thoroughly examined, and a Feature Engineering process was performed on the entire dataset. The Feature Engineering process involved cleaning and preparing the data, removing unnecessary features, standardization, and encoding of categorical features.

The prepared dataset was ultimately tested using algorithms like Decision Tree and Random Forest. The study also involved botnet classification, which compared traditional methods with machine learning methods and explored the potential of hybrid approaches. These components and actions collectively contributed to the study's main objective. The work primarily focused on two algorithms, namely Decision Tree and Random Forest, which were utilized for pattern recognition in network traffic and the identification of potential botnets. The performance of both models was evaluated using various metrics, including AUC, f1 score, and False Alarm Rate (FAR). The Decision Tree model demonstrated strong results on both the training and testing data, exhibiting high accuracy, precision, and recall. This suggests that it is effective in detecting both normal traffic and botnets. The effectiveness of the Decision Tree model was significantly influenced by parameters, particularly the 'max_depth' parameter.

However, even after optimization attempts, the Random Forest model exhibited slightly inferior performance in comparison to the Decision Tree model. Our analysis revealed an increased number of errors in both false positives and false negatives, indicating a reduced capability in accurately identifying positive and negative cases in contrast to the Decision Tree model. Subsequently, our attention shifted towards utilizing a multilayer perceptron (MLP) for botnet classification. Encouragingly, this model demonstrated promising outcomes in distinguishing various botnet types, attaining high levels of accuracy, precision, and recall for both the training and testing datasets. The assortment of performance metrics for distinct botnet classes showcases the model's proficiency in precisely detecting and categorizing network threats.

Examining the disparities between conventional techniques and machine learning-based approaches reveals a clear illustration of their benefits and limitations. Traditional methods such as IDS are proficient at recognizing familiar attack patterns, thereby reducing the likelihood of missing incidents and generating false alarms. Conversely, machine learning models, which learn from diverse data, may exhibit heightened sensitivity in detecting novel threats. Nevertheless, their more generalized approach can result in reduced specificity and an increased number of false alarms. In the realm of cybersecurity, the ability to adapt to variable conditions becomes a crucial aspect. Hybrid models, which combine the rapid identification of known attacks with the capability to detect new threats, represent an innovative approach to safeguarding networks. Their capacity to dynamically adjust and switch between methods, depending on the attack type or available training data, allows for the effective utilization of the advantages of both approaches. Such models not only mitigate the risk of overlooking new attack patterns but also respond effectively to familiar threats, making them a promising avenue for advancement in the field of cybersecurity, particularly in the face of constantly evolving threats.

Enhancing the effectiveness of network attack detection can be achieved by optimizing machine learning models. This aspect holds great importance and offers ample opportunities for further investigation. One potential avenue for improvement involves conducting a comprehensive analysis of the hyperparameters associated with algorithms like Random Forest or Decision Tree. By experimenting with various parameter combinations and optimizing the learning process, the models' ability to detect network attacks can be significantly improved.

In addition, improving the quality of predictions may rely on feature engineering and selecting significant attributes from network data. Enhancing the extraction of relevant information from the analyzed data or removing unnecessary features can enhance the efficiency of models, leading to better attack detection. Exploring advanced machine learning techniques is also worth considering. The implementation of neural networks, particularly convolutional (CNN) or recurrent (RNN) networks, could provide a fresh perspective on detecting network attacks. Additionally, employing ensemble learning techniques such as bagging or boosting may effectively enhance the precision of models. Evaluating the temporal and performance aspects of attack detection is crucial as well. Investigating the potential for real-time attack detection and optimizing models for computational efficiency is particularly important in resource-constrained environments. Exploring different datasets and analyzing their impact on model effectiveness is another avenue to explore. Understanding how models perform with diverse network data and testing their resilience against manipulations or adversarial attacks can significantly strengthen and protect models against various threats.

# References

1. Ahmetoglu, H., Das, R.: A comprehensive review on detection of cyber-attacks: data sets, methods, challenges, and future research directions. Internet Things **20**, 100615 (2022)
2. Al Lail, M., Garcia, A., Olivo, S.: Machine learning for network intrusion detection - a comparative study. Future Internet **15**(7), 243 (2023)
3. Aljabri, M., et al.: Intelligent techniques for detecting network attacks: review and research directions. Sensors **21**(21), 7070 (2021)

4. Bond, M., Human, K., Kwon, N.: Analysis and implications for equifax data breach (2022)
5. Delplace, A., Hermoso, S., Anandita, K.: Cyber attack detection thanks to machine learning algorithms. arXiv preprint arXiv:2001.06309 (2020)
6. Erickson, S.L., Stone, M., Serdar, G., Pfeffer, B.: When crisis victims are not customers: SCCT and the equifax data breach. J. Manag. Issues **35**(2) (2023)
7. Garcia, S., Uhlir, V.: The CTU-13 dataset. a labeled dataset with botnet, normal and background traffic. S. Lab, Ed., ed (2014)
8. Kedziora, M., Gawin, P., Szczepanik, M., Jozwiak, I.: Malware detection using machine learning algorithms and reverse engineering of android java code. Int. J. Netw. Secur. Its Appl. (IJNSA) **11** (2019)
9. Kedziora, M., Gawin, P., Szczepanik, M., Jozwiak, I., et al.: Android malware detection using machine learning and reverse engineering. Comput. Sci. Inf. Technol. (CS&IT) 95–107 (2018)
10. Krolik, L., Kedziora, M., Mizera-Pietraszko, J., Jozwiak, I.: Detecting attacks on computer networks using artificial intelligence algorithms. In: Proceedings of the 14th International Conference on Management of Digital EcoSystems, pp. 110–114 (2022)
11. Luu, Q.: Detect botnet traffic (2023). https:// github.com/lmquan1609/ detect-botnettraffic
12. Meftah, S., Rachidi, T., Assem, N.: Network based intrusion detection using the UNSW-NB15 dataset. Int. J. Comput. Digital Syst. **8**(5), 478–487 (2019)
13. Sarhan, M., Layeghy, S., Moustafa, N., Portmann, M.: NetFlow datasets for machine learning-based network intrusion detection systems. In: Big Data Technologies and Applications: 10th EAI International Conference, BDTA 2020, and 13th EAI International Conference on Wireless Internet, WiCON 2020, Virtual Event, December 11, 2020, Proceedings 10, pp. 117–135. Springer (2021). https:// doi.org/10.1007/978-3-030-72802-1_9
14. Sarhan, M., Layeghy, S., Portmann, M.: Towards a standard feature set for network intrusion detection system datasets. Mobile Netw. Appl. 1–14 (2022)
15. Shareena, J., Ramdas, A., AP, H., et al.: Intrusion detection system for IOT botnet attacks using deep learning. SN Comput. Sci. **2**(3), 1–8 (2021)
16. Yavanoglu, O., Aydos, M.: A review on cyber security datasets for machine learning algorithms. In: 2017 IEEE International Conference on Big Data (Big Data), pp. 2186–2193. IEEE (2017)

# MBMD-LoRa Scalable LoRaWAN for Internet of Things: A Multi-band Multi-data Rate Approach

Mkrm Almuhaya(✉), Tawfik Al-Hadhrami, Omparakash Kaiwartya, and David.J. Brown

Nottingham Trent University, Nottingham, UK
almohia82@yahoo.com

**Abstract.** Internet of things based applications are increasingly adopting low-power wide area network (LPWAN) technologies because they provide extensive coverage to many battery-powered devices. Due to its physical layer architecture and regulatory benefits, long-range wide area network (LoRaWAN) has become the most extensively adopted LPWAN solution which enables a new multiband technology. LoRaWAN unequivocally employs the ALOHA medium access control (MAC) protocol, resulting in a substantial reduction in the packet delivery rate, particularly in high-density networks where end devices (EDs) access the network randomly. This significantly and adversely affects the overall network performance. However, its scalability performance with simultaneous impact of multi-bandwidth approach has not yet been adequately investigated. This paper proposes a Multi-Band Multi-Data Rate (MBMD-LoRa) framework for enabling an scalabe LoRaWAN IoT use cases. Firstly, a system model for LoRaWAN is presented focusing on link, propagation and simulation scenarios of the framework. Secondly, the slim data rate, MBMD algorithm, and its zone-based implementation are presented detailing the framework. The comparative performance evaluation of the proposed framework attests to potential benefits considering several metrics related to the scalability of LoRAWAN for emerging IoT use cases.

**Keywords:** Internet of Things · LPWAN · LoRa · LoRaWAN · ADR

## 1 Introduction

The Internet of Things (IoT) is a network where electronic devices, machines, and ordinary items can detect their surroundings and exchange data with other devices and humans over the Internet. This will lead to the establishment of an IoT business that will generate a revenue of more than \$4 trillion [1]. Moreover,

Supported by Nottingham Trent University, Nottingham, UK.

there is an increasing demand for some applications. Examples include the implementation of intelligent urban environments (smart cities) and advanced domestic systems (smart homes), agricultural practices (smart agriculture) and driving the development of industrial processes (Industry 4.0) [2]. An obstacle in the field of the Internet of Things is to provide assistance to applications that require a large quantity of durable battery-operated devices that can communicate over distances and are inexpensive. Certain applications necessitate the utilisation of short-range radio communication technologies, such as IEEE 802.15.4/ZigBee [3]. Conversely, applications like precision agriculture or smart city initiatives may demand communication solutions capable of spanning distances exceeding 10km in rural regions. Long Range LoRa meets all previous application requirements; LoRa operates at the physical layer [4], provides various options for carrier frequency (CF), bandwidth (BW), coding rate (CR), and spreading factors (SFs). The symbol rate/chirp rate ratio-is one of the main Data Rate factors. The Spreading factor is linked to communication range and inversely to bit rate [5]. It can be set from 7 to 12, and the increment of one step leads to a decrement in the data rate and extends the communication range. The choice of communication technology for connecting numerous metering devices in urban regions depends on their capacity and scalability [6]. LoRaWAN's "star-of-stars" network architecture and simple medium access mechanism can support these applications. Nodes in LoRaWAN cellular networks communicate with gateways using single-hop LoRa links. This connection uses grant-free pure ALOHA medium access. This protocol lets several nodes send reports without handshaking. ALOHA is collision-prone because it lacks a contention mechanism [7].

Recent wireless communications advances have created various frequency bands and standards to increase mobility and throughput. Semtech created the third-generation ultra-low power LoRa transceiver LR1121. This device supports connection via satellite S-Bands, LoRa, and LR-FHSS in sub-GHz and 2.4 GHz ISM bands [8]. LoRa Alliance's LoRaWAN physical layer requirements are followed by the LR1121. This versatile design meets the demands of various applications and proprietary protocols [9]. The increase in LoRa symbol spectrum bandwidth $B$ doubles data rate but reduces the communication range. Bandwidth $B$ from 500 kHz to 125. The frequency band can be divided into bands with adjustable bandwidths [10,11]. In spite of all the improvements in the physical layer of LoRa, ALOHA utilization in the LoRaWAN protocol still has a negative effect, which decreases the network's performance in high-density deployment. Many studies investigate how to improve LoRaWAN scalability, but they do not pay more enough investigation and potential to address this issue by utilizing the entire bandwidth supported by LoRaWAN. Previous research focused on two control parameters, SF and transmission power, which the Adaptive Data Rate algorithm manipulates to achieve reliable and energy-efficient communication across intermediate distances. Our study added bandwidth (BW) to node settings for long-distance communication to increase reliability and energy efficiency. We use the link between SF and transmission power to find the ideal BW and SF configuration to minimise ToA, improve LoRa scalability, and reduce power consumption.

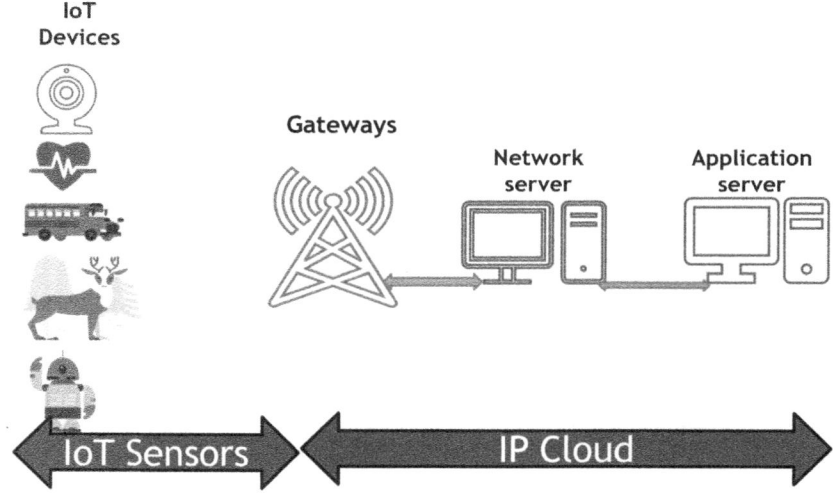

**Fig. 1.** LoRa Network Architecture

This work aims to optimise and improve the performance of LoRaWAN technology through an increase in data rate to minimise the Packet transmission time, as evidenced by the decrease in ToA. On the other hand, the presence of diversity inside a single cell in the context of SF (spreading factor) has a significant influence on mitigating packet collisions. This can be attributed to three main factors.

**Firstly** slim data rate decreases the time required for packet transmission (ToA) and then less transmission power.

**Secondly.** The small packet size due to using Slim data rate reduced collision occurrence between traffic and saved transmission power from retransmissions.

**Thirdly.** One notable contribution is the inclusion of diverse spreading factors within the cellular network. This incorporation serves the purpose of preventing node collisions, hence leading to energy conservation by eliminating the need for packet retransmission.

This article's remaining sections follow. Section 2 background of LoRa Sect. 3 system model of LoRaWANincludes the propagation, collision, and simulation system models. Section 4 describes our MBMD-LoRa and Z-MBMD-LoRa approaches. Section 5 evaluates our methodology performance relative to typical research based on results, followed by a conclusion in Sect. 6.

## 2    Background and Related Work

The architecture of LoRa network, as shown in [12], and illustrated in Fig. 1. The LoRa network is implemented using a star-of-stars topology, where the exchange of packets between End-Device and Gateways is facilitated by the LoRa communication protocol. On the right-hand side of the figure, the LoRa nodes are multifunction sensors that cater to various applications, such as industrial automation, smart cities, smart healthcare, smart agriculture, etc. On the other hand, the Gateways refer to devices that provide the function of demodulating LoRa traffic and facilitating the transmission of data between the network server and end-devices. The Network Server (NS) is linked to the gateways using conventional Internet Protocol (IP) connections, such as WiFi, Ethernet, or 3G/4G. Specifically, it is important to note that ED systems are not linked to a single Gateway but rather to a NS. All Gateways that receive data from an ED subsequently transmit it to the network server. The network server serves as the primary backend component of a LoRaWAN system, responsible for aggregating the data sent by all end-devices inside the network and then relaying this data to an application server (AS). AS is responsible for taking action, which the system is built for based on the collected data from the end devices. This sort of network may include several end devices. The network server may be accessed by the end devices via the gateways, using either the LoRa modulation scheme or FSK modulation. Several of studies present some solutions for LoRa to be scalable such as in metropolitan areas, the selection of communication technology for connecting a large number of metering devices is contingent upon the capacity to support the devices and their scalability [6].

A scalable IoT ecosystem is explored in [13] using LoRaWAN LPWAN technology without considering the packet delivery ratio. The use of LoRaWAN networks for extensive smart city applications is exemplified by M. Loriot and colleagues [14]. All tests conducted for this study took place at the Scientific Campus of the University of Lille, France. The experiments were conducted to cover distances of up to one hundred kilometers. Other research [15] introduces a method for placing gateways based on graphs. In order to address the difficulty of scalability when installing gateways in very large networks, this study demonstrates that the technique minimises the collision probability and the needed number of gateways while performing similarly to state-of-the-art related work in the worst scenario. The primary Data Rate parameters encompass the spreading factor, which represents the ratio between the symbol rate and the chirp rate. The SF is directly associated with the communication range and inversely related to the bit rate [5]. The increase in SF leads to a half reduction in bit rate. On the other hand, the increase in bandwidth results in an exact halving of the data rate. Additionally, the bandwidth $B$ is another significant parameter, referring to the spectrum occupied by a symbol. The spreading factor $f$ is adjustable within a range of 7 to 12, while the bandwidths $B$ can be adjusted within a range of 500 kHz to 125 KHz. This allows for the adaptation of both the communication range and data rate. Lastly, the Coding Rate denotes the rate at which coding is applied [16]. LoRa networks with different Multiband Multi data rate (MBMD) parameters are modelled in this study. We wish to scale LoRaWAN to an appropriate packet deliver ratio probability for each frequency band and deployment distance. Our technique optimises the data rate for each spreading factor and bandwidth to optimise node configuration. So, as a result, we obtain the following: First, it maximises airtime, improving data transfer. Second, it boosts the communication channel's capacity, allowing more data to be transferred in a given period. By taking advantage of the Slim Data rate, we obtained a short data packet short packet, which reduced the network overhead and increased network performance.

## 3   System Model for LoRaWAN

The Bit Rate (BR) holds significant importance in the network systems as it plays a decisive role in determining the pace at which data is transferred. Additionally, it directly impacts network performance, ensuring that application requirements are met, optimising the utilisation of available bandwidth, expanding the network's capacity, and facilitating technical improvements. The LoRa network bit rate can be calculated using Eq. (1) [17].

$$R_b = f * \frac{b}{2^f} * cr \quad [bits/s] \tag{1}$$

where $f$ is the spreading factor and $b$ is the channel bandwidth and the $cr$ is the coding rate. The spreading factor in LoRa modulation pertains to the quantity of chips employed for encoding each symbol, and it has a direct impact on both the data rate and communication range. The duration of each symbol transmission and the sensitivity to noise are determined by the $f$.

### 3.1   Propagation Model

The link budget of a wireless system or network refers to the comprehensive assessment of the total gains and losses incurred during the transmission process, encompassing the transmitter, propagation channel, and the intended receiver. The gains and losses encompass several factors such as system gains and losses related to the antenna, matching networks, and other components, as well as losses linked with the propagation channel itself, which can be determined using either modelling or observed data. In general, when considering channel mechanisms that exhibit random variations, such as multipath and Doppler fading, it is customary to incorporate supplementary margin based on the expected severity. The link budget for a wireless network link can be mathematically represented in Eq. (2) as follows:

$$P_{rx}(dBm) = P_{tx}(dBm) + G(dB) + PL(dB) \tag{2}$$

where $P_{rx}$ is receiving power, the $P_{tx}$ is transmission power, $G(dB)$ is the antenna gain of transmitter and receiver, and the $PL(dB)$ is the path loss for more explain the all gain can be written as Eq. (3):

$$G(dB) = G_{tx}(dB) + G_{rx}(dB) \tag{3}$$

while the all-path loss is represented in Eq. (4):

$$PL_{Total}(dB) = PL_{Env} + PL_{tx}(dB) + PL_{rx}(db) - X_\sigma \tag{4}$$

With $G_{tx}(dB)$, $PL_{tx}(dB)$, $G_{tx}(dB)$ and $PL_{rx}(dB)$ plus minus set to zero, $X_\sigma$ is the fading margin while $PL_{Env}(dB)$ is dictated by the communication environment. Different surroundings (urban, suburban) affect route loss in several models.

### 3.2   Simulation Model

There are multiple factors that affect whether a receiver can decode one or two packets, or none at all, when two LoRa signals collide. These factors include Carrier Frequency (CF), Spreading Factor, voltage, and duration. The collision between packets $p_1$ and $p_2$ happens only when all the conditions defined in Eqs. (5) are met:

$$C_{pckt}(p_1, p_2) = \begin{cases} 1 & if & (O(p_1, p_2)C_{fr}(p_1,p_2)C_f(p_1,p_2) \\ & & C_{pw}(p_1,p_2)C_t(p_1,p_2)) \\ 0 & else \end{cases} \tag{5}$$

The situation in which two transmissions collide on $C_{fr}(p_1, p_2)$ may be defined by considering the centre frequencies of transmission $(p_1, p_2)$, denoted as $fr_1$ and $fr_2$ respectively.

## 4   Multi-band Muti-data Rate for LoRaWAN

### 4.1   Slim Data Rate

We assume that N nodes are evenly distributed in a D-radius region centred on a gateway. This research assumes the gateway enables simultaneous multi-bandwidth owing to the latest LoRa version, which includes additional spreading factors [9]. Suppose the gateway has three fundamental channels with distinct bandwidth and sub-channels. For data collection, each node must send data within a time frame, defined as $T_t$. Our study calculates the mean packet reception probability for each spreading factor and bandwidth combination. The spreading factor-based LoRa MAC layer is an ALOHA MAC protocol without acknowledgements. Nodes are expected to send packets independently of each other and their locations. The percentage of nodes with spreading factor $f$ is shown in [18].

$$\sum_{i=7}^{12} \beta_f = 1 \quad \forall f \in SFs \tag{6}$$

where $f$ from 7 to 12 according to the proportion of nodes configured with bandwidth $b$ designed as the following equation.

$$\sum_{i=1}^{3} \alpha_b = 1 \quad \forall b \in BWs \tag{7}$$

where $b$ from 1 to 3 as 500,250 and 125 kHz. A packet production at the deployment area of nodes follows a Poisson distribution with rate $B_f N$ in all zones of spreading factors $zone_f$. Now, let us suppose a node located at a distance $di$ from the gateway, which is transmitting messages $\mu$ with a spreading factor of $f$. Considering the phenomenon of the capture effect, the successful transmission of a packet by a node can be determined by two conditions: (a) the absence of any other packet with the same spreading factor overlapping with the current packet within the same receiving time $t_r$, or (b) the power level at the gateway of any other packet with the same SF surpassing the power level of the current packet by a minimum threshold value $PW_{thld}$ [19]. Given the assumption of uniform transmission parameters across all nodes, the potential sources of interference can be identified based on the path-loss characteristics of the signal. Specifically, all nodes located at a distance of di from the gateway, where maximum distance is defined as the following equation:

$$D = d0 * 10^{\frac{PL_{env}}{10\lambda}} \tag{8}$$

where $\lambda$ represents the path loss exponent, which can be considered as potential interferers. $\beta f N \frac{(min(d_i, D)^2}{D^2}$ where $D$ is the range. The probability of successful transmission $P_s(d)$ Ensuring that no potential interfering nodes initiate a transmission during a vulnerability period of 2Tf is crucial to maintaining a secure and reliable system.

$$P_s(d) = e^{-2T_f \mu \beta_f N (\frac{(min(d_i, D))^2}{D^2})} \tag{9}$$

The slim data rates are derived from the Eq. (1). The likelihood of success data rate is derived from equation [18].

$$Pc_f = \frac{f}{2^f} / \sum_{i=7}^{12} \frac{i}{2^i} \quad \forall f \in SFs \tag{10}$$

Unfortunately, Eq. (10) does not consider bandwidth $b$ and coding rate (CR) variables and (7) does not consider SF. In order to accurately predict the success probability of our approach, it's important to take into account the effect of two crucial factors: bandwidth and spreading factor. By accounting for the impact of BW and SF on the success probability, we can ensure that our approach is optimized for maximum efficiency and efficacy. And from Eq. (6) and (7) the success probability $P_s(d)$ and throughput $Tht$ will be:

$$P_s(d) = e^{-2T_f \mu \beta_f \alpha_b (N \frac{(min(d_i, D))^2}{D^2})} \tag{11}$$

$$Tht = \frac{P_s * P_L * N}{\tau} \qquad (12)$$

where $P_L$ is the data payload and $N$ is the number of EDs while the collision probability $P_{f,b}$ will be

$$P_{f,b} = \frac{p_f * b}{\sum_{i \in BWs} i} \quad \forall f \in SFs \& b \in BWs \qquad (13)$$

Moreover, considering multiple channel frequencies, the Eq. 13 will be (Fig. 2):

$$P_{f,b,cr} = \frac{p_{f,b} * ch}{\sum_{i \in CHs} i} \forall f \in SFs \& b \in BWs \& ch \in CHs, \qquad (14)$$

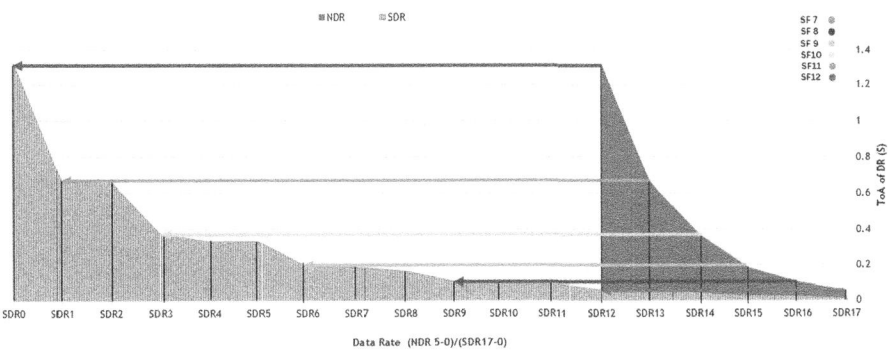

**Fig. 2.** Normal Data Rate vs Slim Data Rate

## 4.2   Zone-Based MBMD Implementation (MBMZ-LoRa)

The MBMZ-LoRa method uses MBMD-LoRa to find the best route and rank for a revolutionary agile data rate to build a stable and efficient connection. The network server assesses nodes' Received Signal Strength Indicator (RSSI) and compares it to gateway sensitivity, then assigns them the configuration that fits their needs and assigns them to the proposed zone by increasing $k$ in zone $Bf$. If the node's RSSI is less than the gateway sensitivity, the algorithm boosts its transmission power as in steps 15 to 19, then reassesses its link budget as in row 6 to find the appropriate setting $set_i$, and so on. Different transmission factors like the spreading factor and bandwidth affect data rates and airtime. Due to airtime allocation differences, collision probability vary, resulting in an unequal distribution of resources across nodes in a zone. The MBMZ-LoRa method employs $\beta f$ and $\alpha b$ to ensure fair distribution and improve the packet delivery ratio. The multiplication Eqs. (6) and (7) yield the number of slim data rates ($\zeta \in rows[0, 17]$).

The procedure ensures that the number of nodes in Zone $Z_f$ allocated to spreading factor $f$ is less than $\beta_f$, else it moves to the next zone. Use the following equations to calculate $Z_{f+1}$:

$$\sum_{i=7}^{12} Zone_f = 1 \quad \forall f \in SFs \tag{15}$$

$$\sum_{i=7}^{12} \beta_f \sum_{i=1}^{3} \alpha_b = 1 \quad \forall f \in SFs, \forall b \in BWs \tag{16}$$

---

**Algorithm 1:** Z-MBMD-algorithm

---

    **Input**   : n List of N nodes corresponding distance D by Ploss, RSSI, Z deployment zones.

    **Output**: $Ch, TP, SF, BW, CR$ settings for each $i$ of the N nodes$ToA_i$ and k for each zone Processed list.

1  $PtxLevel = [\ ], SF = [\ ], B = [\ ], Ch = [\ ], Zone = [\ ], Se = [\ , \ ], Set[SF , B, Cr ]$

2  /* Assign settings to each node in the list of N nodes.        */

3  **while** $i \leq N$ **do**

4     **if** $RSSI[i] > MinSens$ **then**

5        **for** $j \leftarrow 0$ **to** $MDR$ **do**

6           /* assign parameters setting to set[i] .        */

7           **if** $RSSI[i] > Sen[j]$ **then**

8               **if** $Z_f > \beta_f$ **then**

9                  $SFi \leftarrow fj$

10                 $BWi \leftarrow bj$

11                 $Chi \leftarrow Chz$

12                 $ToAi \leftarrow ToA(Sj, Bk, CR)\ Z_f.append([k] + 1)$   // count n in each zone.

13               **else**

14                 $Z_f = Z_f + 1$

15                 $\beta_f = \beta_f + 1$

16               **end if**

17           **end if**

18        **end for**

19     **else**

20        // Update Node's Transmission power

21        $TP[i] = TPwLevel[]+ = 1\ RSSI[i] = TPwLevel[] - PLoss + GN$

22        $Go\ to\ step\ 6\ Eng[i] \leftarrow CalculateEnergy$

23     **end if**

24     $TotalEng = Sum(Eng[i])$

25     $i+ = 1$

26  **end while**

27  **return** *Setting of Each Node and The Average of Energy Consumption and Node's zone*

Our strategies MBMD-LoRa and MBMZ-LoRa are for improving LoRaWAN regarding the data rate doubling is intended to reduce the packet transmission time which leads to fewer collisions and less transmission power. On the other hand, the degree of diversity in SF within a single cell significantly influences the ability of *Pckts* to avoid collisions. Our method will achieve greater energy efficiency initially due to the fact that an increase in data rate results in a reduction in *Pckts* time, in addition to a decrease in collisions caused by *Pckt* size, which reduces transmission power and retransmission for *Pckt* (Fig. 3).

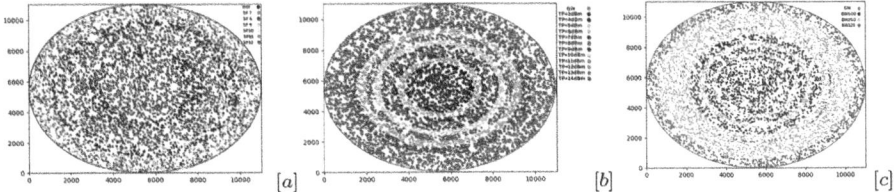

**Fig. 3.** LoRa Network Deployment on MBMD and Z-MBMD

## 5    Performance Evaluation

The effectiveness of our proposed resource allocation solutions, which are based on the multi-band multi-data rate (MBMD) algorithms, is illustrated in this section. We evaluate their performance using a model based on LoRaSim [19], a discrete-event simulator developed by Bor et al. We use the Simpy library to analyze scalability and collision issues in LoRa networks. Python 3.9 is utilized to construct the simulations. The allocation of SFs differs between two approaches: MBMD-LoRa, which uses a random zone without $\beta_f$ distribution, and Z-MBMD-LoRa, which uses six distributed zones $Z_f$. Our methodologies are built for large-scale dense networks hence the simulations have 500 to 6000 nodes. The nodes were randomly distributed throughout an 11 km$^2$ region. Assume a LoRa network with one gateway in the region's center. Each node sends 50-byte packets. The mean time between packet arrivals is 600 s, following an exponential distribution. Similar to [20], the LoRa physical layer follows European regional requirements, including a 1% duty cycle for nodes and gateways. Table 1 summarizes simulation parameters. We evaluate our method against BE-LoRa and LoRaWAN protocol. The evaluation considers collision likelihood, throughput, and energy consumption.

Table 2 shows that Data Rate DR0 changes slightly, but DR1-DR5 increases. DR1 data rate increased by 48 bits per second. In contrast, DR2 has four times the data rate of DR1. From DR2 to DR5, data rate doubles. The drop in Time on Air indicates that this data rate increase reduces packet size on the air and transmission time. However, diversity within a zone and spreading effects reduce

**Table 1.** System parameters used in evaluation [21]

| Parameter | Value | Comments |
|---|---|---|
| N | 500–6000 | Network Size |
| $f$ | 7 to 12 | Spreading factors |
| $d0$ | 1000 m | initial distance |
| $\lambda$ | 2 2.32 dBm | PLoss exponent |
| $PL_{Env}(d0)$ | 128.95 | Ploss of initial distance |
| $TPLevel$ | 2 dBm to 14 dBm | Transmission Power |
| $cr$ | 2 4/5 | Coding Rate |
| $b$ | [125, 250, 500]kHz | Bandwidth |
| MD(R) | 5500 m | Field radius |
| CF | [860, 864, 868] | Carrier Frequency (MHz) |
| T(s) | One day | Simulation time |
| $\tau$ | 10 min | Round time |

packet collisions. We improved energy efficiency with our technique. Due to the slim data rate, we can achieve a short packet transmission time, minimizing collisions, reducing transmission power, and eliminating the need for retransmissions. The inclusion of various SFs in the cellular network is noteworthy. This prevents node collisions, saving energy by avoiding packet resend. The algorithm was assessed by comparing it to the standard solution and an alternative technique from previous research [19,22]. Integration of simulations and testing was used in this review. Multiple urban and suburban networks' packet delivery ratios were assessed using a single ED sending data to a GW coupled to an NS. Network bandwidth, energy efficiency, SF/TP usage. We tested MBMD-LoRa and Z-MBMD-LoRa, assuming the LoRaWAN Gateway can handle various bands. Scalability and range were evaluated on two NS installations.

**Table 2.** The Enhancement in LoRaWAN Data Rate For EU 862–872

| DR | SDR | BW (kHz) | SF | Bit Rate [bit/s] | Increment [bit/s] |
|---|---|---|---|---|---|
| 0 | 0 | 125 | 12 | 293 | 0 |
| 1 | 1,2 | 125→250 | 11,12 | 537–585 | 48 |
| 2 | 3,4,5 | 125→500 | 10,11,12 | 976–1172 | 196 |
| 3 | 6,7,8 | 125→500 | 9,10,11 | 1757–2148 | 391 |
| 4 | 9,10,11 | 125→500 | 8,9,10 | 3125–3906 | 781 |
| 5 | 11→17 | 125→500 | 7,8,9 | 5468–7031 | 1563 |
| 6 | 6,7 | 125→500 | 7,8,9 | 5468–7031 | 1563 |

## 5.1 Network Throughput

Figure 4.a shows how End-device count affects network performance of network throughput. Equation (12) shows that $Tht$ rises proportionately with Nodes for small values. Increasing the number of EDs improves the proposed solutions over the usual ones. First, both iterations of our technology transport data faster than LoRaWAN. Once 2000 nodes are added, LoRaWAN throughput stabilizes with a slight increment. In other words, LoRaWAN throughput peaked when the number of nodes was 6000. So, our technique is more scalable than LoRaWAN. Our suggested protocol using MBMZ-LoRa has twice the throughput of LoRaWAN with 6000 nodes.

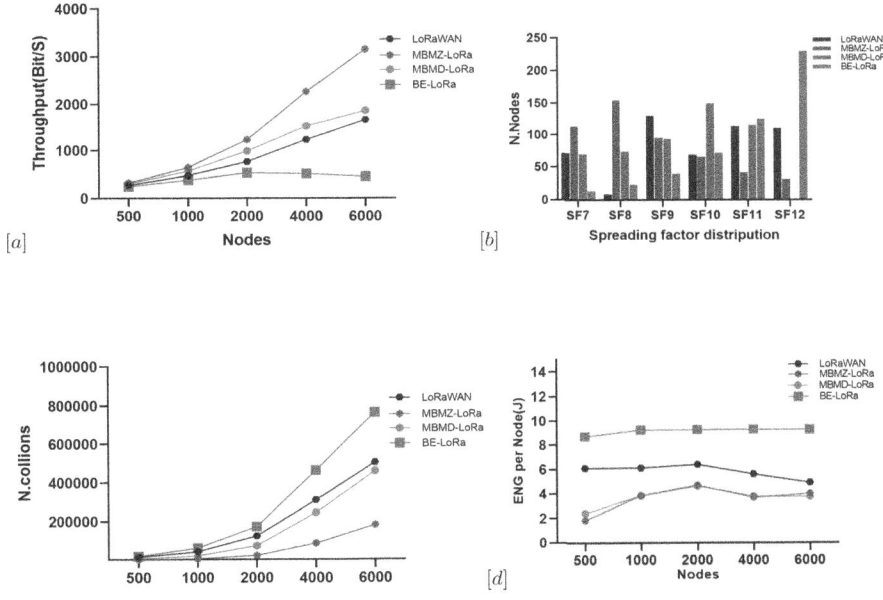

**Fig. 4.** Performance Metrics

## 5.2 Energy Consumption

Figure (4.d) shows both systems' energy utilisation per successful transmission. Divide the total energy expended by all LoRa nodes by the number of packets received by the network server to compute energy consumption. The MBMZ-LoRa algorithms improved energy efficiency significantly, with both algorithms performing identically. Without SF12, the MBMZ-LoRa technique consumes the same amount of energy due to the substantial use of SF10 and SF11 in nonBf. The $B_f$ method uses SF7 and SF8, but seldom SF12. Our methods use less energy than LoRaWAN and BE-LoRa due to spreading factor distribution. The spreading factor usage in each $f$ is compared in Fig. (4.b). While in LoRaWAN all nodes are near in spreading factor distribution except for a tiny proportion in sf8, BE-LoRa's greater use of SF12 while moving from lower to higher SF increases energy consumption. LoRaWAN's random distribution prevents it from exploiting all levels between the same spreading factors, unlike our technique, which leverages all levels inside one spreading factor. These data show how the proposed strategies might boost performance.

## 6    Conclusion

This article examines the Multi_Band Multi_Data Rate as a means to achieve scalable communication in long-range IoT networks. The proposed MBMZ_LoRa algorithm is a collaboration between the Bandwidth Allocation and Spreading Factor (BASF). These methods match LoRa node activity to network server data. This paper proposed a LoRa network size solution that optimises node configuration, network throughput, and energy efficiency. This method based on Slim Data Rate technique proved that an optimal solution exists and is unique for each Data Rate level. MBMZ-LoRa, based on MBMD-LoRa and six zones for all LoRa nodes, has also been shown to work. The MBMZ-LoRa algorithm has increased the data rate ratio, especially in data rate levels 1 to 6, whereas DR level 0 has not changed. This data rate doubling reduces time-on-air packet size and initially increases energy efficiency because it reduces *Pckts* time (ToA) and collisions caused by *Pckt* size, which reduces transmission power and retransmission. The cell's SF diversity is the most important enhancement since it eliminates node collisions and saves energy by not retransmitting *Pckt*. The simulations showed that the suggested technique outperforms BE-LoR and LoRaWAN in terms of network throughput and energy use.

**Acknowledgements.** The authors would like to thank the Computer Science Department, Faculty of Science and Technology, Nottingham Trent University, Nottingham, UK.

## References

1. Raza, U., Kulkarni, P., Sooriyabandara, M.: Low power wide area networks: an overview. IEEE Commun. Surv. Tutor. **19**, 855–873 (2017)
2. Marini, R., Mikhaylov, K., Pasolini, G., Buratti, C.: LoRaWANSim: a flexible simulator for LoRaWAN networks. Sensors **21**(3), 695 (2021)
3. Palattella, M.R., et al.: Standardized protocol stack for the internet of (important) things. IEEE Commun. Surv. Tutor. **15**(3), 1389–1406 (2012)
4. Gkotsiopoulos, P., Zorbas, D., Douligeris, C.: Performance determinants in LoRa networks: a literature review. IEEE Commun. Surv. Tutor. **23**(3), 1721–1758 (2021)
5. Altayeb, M., Zennaro, M., Pietrosemoli, E., Manzoni, P.: TurboLoRa: enhancing LoRaWAN data rate via device synchronization. In: 2021 IEEE 18th Annual Consumer Communications & Networking Conference (CCNC), pp. 1–4

6. N. S. et al.: LoRa⒭ and LoRaWAN⒭: a technical overview, technical paper (2019)
7. Xanthopoulos, A., Valkanis, A., Beletsioti, G., Papadimitriou, G.I., Nicopolitidis, P.: On the use of backoff algorithms in slotted aloha LoRaWAN networks. IEEE
8. Boquet, G., Tuset-Peiró, P., Adelantado, F., Watteyne, T., Vilajosana, X.: LR-FHSS: overview and performance analysis. IEEE Commun. Mag. **59**(3), 30–36 (2021)
9. S.P. Confidential: long range, low power, multi-band LoRa⒭ transceiver (2023)
10. Hashemi, H., Hajimiri, A.: Concurrent multiband low-noise amplifiers-theory, design, and applications. IEEE Trans. Microwave Theory Tech. **50**, 288–301 (2002)
11. Yegin, D.K.S.A., (Actility), Seller O. (Semtech): LoRaWAN⒭ regional parameters RP002-1.0.4 (2022)
12. Yegin, O.A.: LoRaWAN specication v1.0.4 (2020). https://www.lora-alliance.org/technology
13. Dumitru, M.C., Pietraru, R.N., Moisescu, M.A.: LoRaWAN as open scalable IoT ecosystem. In: 2023 13th International Symposium on Advanced Topics in Electrical Engineering (ATEE), pp. 1–6 (2023)
14. Loriot, M., Aljer, A., Shahrour, I.: Analysis of the use of lorawan technology in a large-scale smart city demonstrator. In: 2017 Sensors Networks Smart and Emerging Technologies (SENSET) (2017)
15. Loh, F., Mehling, N., Geißler, S., Hoßfeld, T.: Efficient graph-based gateway placement for large-scale LoRaWAN deployments. Comput. Commun. **204**, 11–23 (2023)
16. Semtech: An1200.22 LoRa™ modulation basics available on Semtech corporation and LoRa Allaince (2015)
17. Gao, H., Huang, Z., Zhang, X., Huang, L.: Design of LoRa communication protocol for image transmission. In: 2023 8th International Conference on Intelligent Computing and Signal Processing (ICSP), pp. 2142–2146 (2023)
18. Reynders, B., Meert, W., Pollin, S.: Power and spreading factor control in low power wide area networks. IEEE
19. Bor, M., Roedig, U., Voigt, T., Alonso, J.: Do LoRa Low-Power Wide-Area Networks Scale? (2016)
20. De Jesus, G.G.M., Souza, R.D., Montez, C., Hoeller, A.: LoRaWAN adaptive data rate with flexible link margin. IEEE Internet Things J. **8**(7), 6053–6061 (2021)
21. Slabicki, M., Premsankar, G., Francesco, M.D.: Adaptive configuration of LoRa networks for dense IoT deployments. In: NOMS 2018 - 2018 IEEE/IFIP Network Operations and Management Symposium, pp. 1–9 (2018)
22. Al-Gumaei, Y.A., Aslam, N., Chen, X., Raza, M., Ansari, R.I.: Optimizing power allocation in LoRaWAN IoT applications. IEEE Internet Things J. **9**(5), 3429–3442 (2022)

# M2M Interface for IoT Traffic Light with Computer Vision and AnyLogic PLE

Madina Mansurova[1]([✉]) [ID], Baurzhan Belgibayev[1], Sanzhar Abdrakhim[2], Assiya Boltaboyeva[1] [ID], Zhanel Baigarayeva[1] [ID], and Talshyn Sarsembayeva[1] [ID]

[1] Al-Farabi Kazakh National University, Almaty, Kazakhstan
madina.mansurova@kaznu.edu.kz
[2] The Hong Kong Polytechnic University, Hong Kong, China

**Abstract.** This paper proposes an advanced traffic light control system using IoT devices and computer vision, integrated through M2M interactions and modeled with AnyLogic PLE. The key contribution is the combination of IoT and computer vision for real-time, adaptive traffic light control. The study highlights the practical value of M2M technology, facilitating seamless interaction between web camera-equipped traffic lights and personal computers, overcoming the complexity of traditional wired methods like Siemens microcontrollers. Using a socket library for communication between Windows and Linux-based Raspberry Pi, the system implements interactive Wi-Fi information exchange for video monitoring and real-time road situation recognition. These data inputs control traffic lights via computer vision, enabling automated, adaptive traffic management. The prototype demonstrates real-time animated simulation managed by a dispatcher, enhancing the efficiency of traffic systems. The integration of M2M, IoT, and computer vision marks a significant advancement in intelligent transportation systems.

**Keywords:** M2M Interaction · Client-Server Interaction · IoT Traffic Light · Raspberry Pi 4 · YOLOv8n Neural Network · AnyLogic PLE

## 1 Introduction

Machine-to-Machine (M2M) interaction of IoT devices enables automatic data exchange between devices without human intervention, using wired or wireless technologies. This is particularly relevant for traffic management systems, where M2M interactions allow traffic lights to share information, adapt cycles, and coordinate with neighboring lights to improve road throughput and reduce delays [1]. IoT devices, such as motion sensors and cameras, collect real-time data on traffic, road, and weather conditions for centralized analysis and decision-making. For instance, sensor data can signal accidents or traffic jams, prompting the system to adjust routes and traffic light cycles automatically. Computer vision technologies analyze video streams to identify vehicles, pedestrians, and emergency situations in real-time, allowing cameras to detect traffic jams or accidents and relay this information to the management system for traffic light adjustments [2].

Traditional tools like Siemens WinCC for creating SCADA systems do not offer the advanced and visually appealing animations of AnyLogic PLE for online monitoring and

© The Author(s), under exclusive license to Springer Nature Switzerland AG 2024
N.-T. Nguyen et al. (Eds.): ICCCI 2024, CCIS 2166, pp. 68–80, 2024.
https://doi.org/10.1007/978-3-031-70259-4_6

dispatching [3]. AnyLogic PLE is widely used for visual control of discrete-event simulation models, including vehicle movement at signal-regulated intersections. However, standard options for coupling model parameters with technical control object actuators are not detailed in technical documentation.

In Kazakhstan, specifically Almaty, cameras at intersections transmit road conditions to a dispatch center where control devices adjust traffic light durations. The dispatcher manually adjusts traffic light phases based on analog camera images [4]. The city aims to transition to wireless digitalization and intelligent traffic management processes [5].

Modeling in AnyLogic PLE allows vehicles to move along defined routes with customizable properties and traffic laws represented by probabilistic Excel tables based on experimental research. Traffic lights are modeled with dynamic phase changes in Excel, unlike fixed cycles in most works. Optimization of adjacent traffic lights can be done visually on the simulation map [6]. AnyLogic PLE's advanced interfaces visualize road dynamics in 3D, exchange traffic light states with PyCharm via Excel, and support Python programming on Raspberry Pi 4 for automatic, computer vision-based traffic light adjustment. This setup enhances intelligent traffic flow management, addressing real-time adaptability and psychological aspects like drivers yielding at intersections.

The use of computer vision traffic lights with server communication is a current challenge. Autonomous operation of traffic lights with webcams has been discussed in existing works [7]. These principles form a multi-level intelligent control system in Python, aimed at reducing congestion through adaptive traffic lights and optimized green light cycles, showing significant improvements in traffic flow. Theoretical works [8] emphasize the necessity of constructing new interchanges and implementing adaptive smart traffic lights to regulate green light allowance and improve traffic flow. Studies indicate that a 30–40 s green light cycle reduces travel time from the city center to the suburbs.

Numerous techniques were investigated, including the use of IoT and artificial intelligence for traffic signal administration [9]. The authors used IoT approaches and MARL to improve traffic flow across six intersections, showing significant efficiency gains over traditional systems. Another study [10] addressed urban gridlock using IoT sensors and AI algorithms, particularly tree-based regressors like LGBM, to enhance short-term traffic predictions. A study [11] highlighted the application and challenges of using drones for video surveillance in smart cities, emphasizing real-time video monitoring's role in traffic management and public safety. In [12], a novel multi-hop secure LTE-D2D communication protocol for IoT was proposed, crucial for M2M interfaces in IoT traffic signals, ensuring reliable and secure communication. A smart visual sensor network for urban mobility was presented [13], using visual data from IoT traffic signals with computer vision to enhance traffic analysis and control. Another study [14] introduced an IoT-based autonomous vehicle with computer vision for traffic signal detection and automation, providing insights into autonomous decision-making for smart traffic signal control systems.

The proposed solution integrates M2M interactions and IoT devices with computer vision, significantly enhancing the intellectualization of traffic management systems. Unlike existing systems that rely on labor-intensive wired connections and incompatible software-hardware interfaces, this solution uses a client-server architecture over Wi-Fi, facilitating real-time data exchange and automation. Existing approaches often struggle with the complexity of visualizing and animating traffic control using traditional tools like Siemens WinCC. By leveraging the socket library for communication between Windows 10 computers and Linux-based Raspberry Pi mini-computers, the proposed solution overcomes compatibility issues with AnyLogic PLE. This practical automation streamlines the management process, making it more efficient and less labor-intensive.

AnyLogic PLE is a discrete-event simulation software that allows you to create complex models of systems and processes. In this context, AnyLogic PLE is used to create a virtual environment where road intersections involving traffic lights and cars are simulated. Raspberry Pi 4 a mini-computer that can perform various tasks, from data processing to device control. In this case, Raspberry Pi 4 is used to collect data from IoT devices and execute computer vision algorithms to detect cars at intersections. The interaction between the Raspberry Pi and AnyLogic PLE is done using Python to write scripts that control, providing traffic flow data and making traffic light adjustment decisions based on analyzing camera data.

The proposed system enables interactive Wi-Fi information exchange for traffic monitoring and road situation recognition via webcams. Real-time data is used for traffic light control with computer vision, providing dynamic animations of the simulation model on the server. This improves upon static visualizations in current systems. The prototype showcases automated traffic management, combining computer vision and discrete-event simulation models, allowing adaptive traffic light control based on real-time conditions, enhancing efficiency and responsiveness compared to traditional methods.

Existing traffic light control methods such as schedule program, traffic sensors and centralized control have limitations in adapting to changing traffic conditions and optimizing traffic flow in real time. In contrast, the proposed solution utilizes state-of-the-art technologies such as Raspberry Pi, AnyLogic PLE and computer vision algorithms to adaptively and intelligently control traffic lights. This allows the system to respond quickly to changes in traffic flow and optimize traffic lights in real time, which provides more efficient traffic management and improves road safety. Thus, the novelty of the proposed system lies in its ability to adaptively and intelligently adjust traffic lights based on the analysis of real traffic data, resulting in more efficient traffic flow management.

## 2 Materials and Methods

### 2.1 Tasks Related to Integrating Traffic Lights with Computer Vision and a Server Based on M2M (Machine-to-Machine) Communication

To integrate the Raspberry Pi 4 mini-computer as an intelligent traffic light sensor with computer vision with a server based on a high-performance computer using M2M technologies [15], we first identified and mapped interface masks and ports for consistent data exchange between the server and the Raspberry Pi 4 minicomputer with computer vision. Secondly, we solve the problem of client-server connection, which is solved at the system level of both platforms, experimentally determining the possibility of adequate data exchange. The next step was the exchange of data on the Internet, monitoring and notification on social networks about the state of the traffic light phase, server failure and the possibility of autonomous operation of the traffic light based on commands from the Raspberry Pi 4 control mini-computer, for this we solve the problem of providing compact uninterruptible power supply, lighting and alarm on the Internet indicating usage and charge level batteries for traffic lights with computer vision. It should also be taken into account that there should be adaptability of climate control through cooling and thermal insulation at high and low temperatures, that is, to ensure the optimal processor temperature for a mini-computer within 45–55 °C. Also, the next step is to create a universal embedded enclosure for a Raspberry Pi 4 mini computer with a video camera using three-dimensional printing methods, housed in a vandalism-proof, waterproof and dust-proof enclosure that can be easily integrated or combined with existing types of industrial traffic lights. After development, the main step in the implementation of this system is the certification and protection of intellectual property for a semi-industrial prototype of a hardware and software complex and the creation of an experimental industrial small-scale production with marketing, technical support and product modernization systems.

At the initial stage, as reflected in the already published articles [16], the problem of adequate application of the advanced yolov8n neural network for the Raspberry Pi 4 mini-computer with limited RAM and video memory was solved. This neural network is known for its use in autopilots of 4th and 5th generation electric and gadget mobiles equipped with powerful web microcontrollers. For example, the appearance of a new single-board control mini-computer Orange Pi 5 with 16 GB of RAM and an advanced graphics card creates prerequisites for further modernization of an intelligent video device for the Internet of Things. This mini-computer has higher technical characteristics compared to the Raspberry Pi 4, used in operation with 8 GB of RAM. However, it is not yet available for sale in Kazakhstan, and its price is two to three times higher. It should be noted that there is full compatibility with the Debian operating system for the Raspberry Pi 4 mini-PC, which creates prerequisites for the modernization of our hardware and software complex.

## 2.2  The Design Stage

Let's consider the design stages of a multi-level intelligent system with the Raspberry Pi 4 mini-computer as a client and a server based on a high-performance computer. Figure 1 illustrates the functional features of the hardware and software complex of the proposed compact and wireless intelligent control system with computer vision for managing and dispatching the automated traffic management system in Almaty.

As seen in Fig. 1, there are two independent blocks connected via wireless data exchange channels. A traffic patrol service officer monitors the current situation by comparing the calculated animation with the real video image at the interchange. On the left side, there is a standard computer server with a minimum of 16 GB of RAM, enhanced disk memory, and a graphics card. On the right side of Fig. 1, a single-board Raspberry Pi 4 mini-computer is presented, connected to it is an autonomous uninterrupted power supply of 5 V and a current of at least 3 A, a USB webcam, and a relay block, all housed in a casing and attached to standard traffic lights. The dashed-line outlined blocks with a USB keyboard and an HDMI monitor are connected only for debugging and testing purposes.

Therefore, the application of an expert mobile dispatching method allows for the automation of traffic light phase management through the use of a digital PLC with computer vision. The neural network can autonomously assess traffic intensity, and the PLC automatically adjusts the traffic light phases. During peak hours, to prevent traffic congestion, the on-duty road patrol service employee compares the real traffic situation with its predictive model on the server's simulation model. Through wireless data exchange channels with the intelligent attachment with computer vision, they adjust the operation of the PLC traffic light control system. The proposed method significantly reduces the manual traffic management portion in the metropolis.

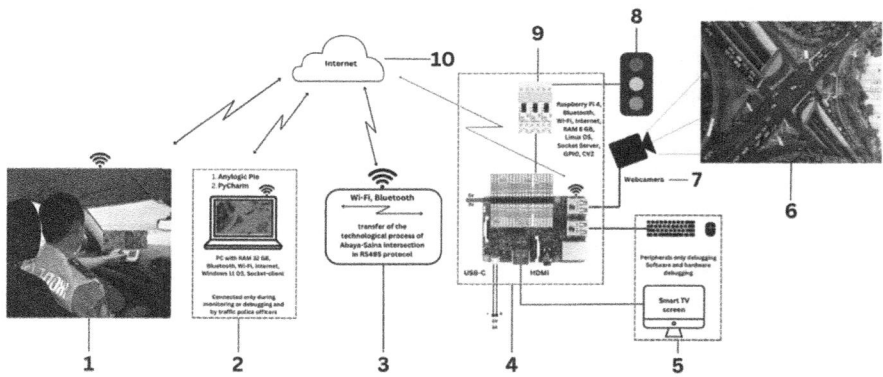

**Fig. 1.** The schematic interaction of a road patrol service employee with elements of the hardware-software complex of a traffic light with computer

In Fig. 1 the Traffic Patrol Service 1 constantly monitors traffic and detects any violations. They transmit the received information, i.e. the real picture of traffic 7 to the dispatcher. The dispatcher, having received information from the traffic police, analyzes the situation and decides on the necessary measures. It can send commands through the traffic management system. Special devices are installed on the ground, such as smart set-top boxes with video vision 5 and traffic lights 6, which control and regulate traffic on the road. They are connected to the control center 2. The control Center 2 receives commands from the dispatcher 1 and controls the operation of these devices in the field. It can also receive information from other sources, such as webcams with a self-focusing and pointing system 8. Webcams 8 installed on roads constantly monitor the situation on the road and transmit information to the Raspberry Pi 4 control microcomputer. Cameras capture real-time video footage of traffic. Once the data is collected, it is processed and analyzed using computer vision algorithms Yolo version 8N neural network algorithm to extract information from this footage, such as the number and type of vehicles, pedestrians and cyclists, as well as their speed and direction of travel. The algorithms track the movement of objects over time, which can be used to determine their speed and direction of travel. All traffic data, commands and decisions are transmitted over 3 wireless communication channels between the system devices. The unit with keyboard and display 9 is used to control and debug the system on site. Internet access 10 allows you to obtain additional information and resources to optimize traffic management and safety.

## 2.3 Implementation Stage

To construct a real prototype of an intelligent attachment to the traffic light, it is necessary to explore the capabilities of existing neural networks for adaptation in a compact control mini-computer. The relatively large amount of RAM allows for the application of the advanced Yolov8 neural network in the mini-computer [17]. Additionally, there are constraints on adjusting the aspect ratio and standard length of the video image, which is equal to 640. Furthermore, the color recognition pixels were limited to 255. These settings optimize the video memory of the Raspberry Pi mini-computer and enable real-time recognition. The results of test experiments are shown in Fig. 2. YOLOv8 provides fast and accurate processing of video streams and real-time capability allows traffic light control to be adapted to changing traffic conditions, ensuring efficient use of road infrastructure. Image analysis in the program is conducted at a frequency of 1 min, allowing for the assessment of traffic intensity and adjustment of the duration of the green light phase of the traffic light.

**Fig. 2.** The results of tests for recognizing the intensity of traffic of cars using the YOLOv8n nano neural network.

YOLOv8 is used to detect and count the number of vehicles at an intersection. Based on this information, the system analyzes the traffic and makes automatic decisions to adjust the duration of the traffic light green light (see Fig. 3).

The code above is a program fragment that automatically adjusts the duration of the green light phase of a traffic light based on the detected number of cars. If the number of cars is more than 4, the duration of phase 4 is increased to 30 s, otherwise it is 20 s. This code demonstrates the decision making process based on camera data and implements an automated traffic light control system.

The construction of an animated scene in the AnyLogic PLE environment is well-documented [18]. This technology overlays graphical primitives onto satellite photos of road intersections, using drawing tools for roads with color options for 3D animation graphics, creating vehicle types such as Car and Bus with bus stop simulation, probabilistic process diagrams for modeling vehicle movement throughout the day, and adding traffic lights to regulate traffic at simulated intersections.

```
print ('Recognising...')
numberOfCars = getCarsFromCam()
if numberOfCars > 4:
phase4 = 30
else:
phase4 = 20
traffic_Sain_N.phase_times = [phase1, phase2, phase3 - 3, 3, phase4 - 3,3]
print(f'Recognised cars: {numberOfCars}')
starting = time.time()
```

**Fig. 3.** A snippet of code from the PLC that automatically adjusts the duration of the green light phase of the traffic light.

## 2.4   The Testing Stage

In Fig. 4, a graphical model in AnyLogic PLE and a layout of traffic organization at the Abay-Sain intersection in Almaty city are presented [19]. It should be noted that the graphical model contains all the features of the actual intersection: stop lines, traffic lanes in all directions, a tunnel along Sain Street, and side roads for maneuvers for left and right turns.

**Fig. 4.** The graphical model and the schematic principle of organizing road traffic (ORT) at the Abay-Sain intersection.

The control of traffic animation with traffic light regulation [20] is carried out using a graphical programming language (see Fig. 5). In our case, the program code and the animated picture of the road traffic organization have the following form:

Let's describe the features of programming in the AnyLogic PLE simulation environment. Firstly, the movement of traffic and traffic light control is implemented using standard tools and is documented in technical literature. Secondly, the use of commands (see Fig. 5) allows for direct and reverse exchange of Excel tables with buffer data of the animation model. This opens up possibilities for integrating animation in AnyLogic with automated control systems for a wide range of discrete-event tasks with Markov processes. The work proposes an example of such automated traffic light phase control with neural vision for the side street of Sain-North.

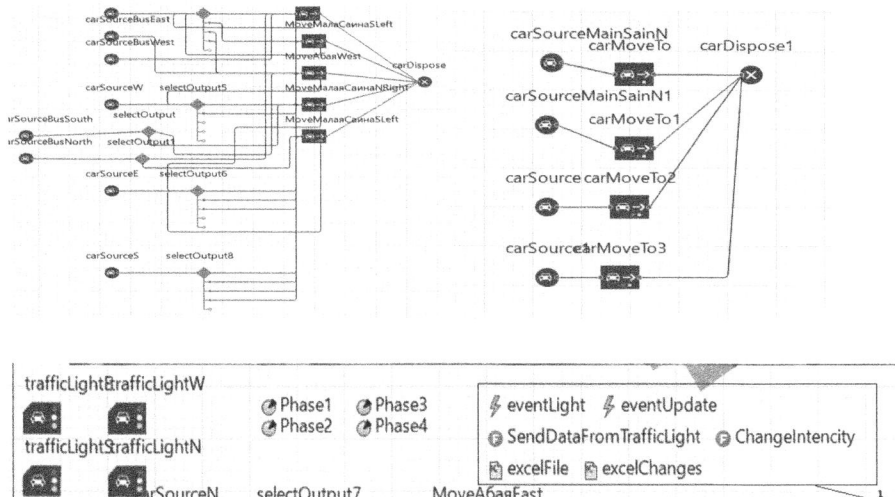

**Fig. 5.** The program code for the animation of traffic movement at the Abay-Sain intersection with traffic light regulation and variable data exchange with a buffer.

These settings allow for interactive changes in the operation time of traffic light phases (Phase4), which in the test version has adaptive computer vision and can adjust the duration of the green signal from 20 s to 30 s depending on the intensity of road traffic.

In the AnyLogic simulation environment, created vehicle types such as Car and Bus. To do this, created new agent classes for each vehicle type (e.g., Car and Bus), defining parameters such as speed and payload. Set up traffic parameters and defined routes and intersection behavior rules for each vehicle type. Used the AnyLogic Traffic library to create roads, intersections, and traffic lights, adding the created agents to them. Created and linked animated shapes to the agents to visualize their movement in the model. Finally, tested the model and calibrated parameters based on real data to ensure modeling accuracy.

Below is a fragment of the program (see Table 1) installed on the computer server, which transmits data from an Excel table to the program installed on the Raspberry Pi 4 mini-computer via Bluetooth channels. The computer connection is established by default within the Bluetooth channel visibility range of the intelligent traffic light accessory. The use of the Bluetooth channel allows for free local collaborative work between the server-dispatcher and the client-traffic light. The internet option requires editorial editing of the program text and does not pose fundamental algorithmic difficulties. In case of no connection with the server, the traffic light control unit operates autonomously based on data from the video camera.

Let's consider the features of the client application fragment of the Intelligent Traffic Light Control System (ITLCS) with computer vision. The block of the program, which receives control signals from the dispatcher and transmits situational data to the monitor of the traffic police officer, is presented (see Table 2). It is worth noting that matching

**Table 1.** Fragment of the server-side of the client-server application for connecting AnyLogic with the client-traffic light with a web camera

| | | |
|---|---|---|
| import socket<br>import time<br>import pandas as pd<br>import keyboard<br>hostMACAddress =<br>'D8:F8:83:7C:70:53'<br>port = 4<br>backlog = 1<br>s =<br>socket.socket(socket.AF_BL<br>UETOOTH,<br>socket.SOCK_STREAM,<br>socket.BTPROTO_RFCOM<br>M) | s.bind((hostMAC<br>Address, port))<br>s.listen(backlog)<br>s.settimeout(10)<br>start =<br>time.time()<br>qt = False<br>while not qt:<br>try:<br>print('Connecting<br>')<br>client, address =<br>s.accept()<br>except Exception<br>as e:<br>continue<br>while not qt: | if time.time() - start > 0.2:<br>start = time.time()<br>dataframe =<br>pd.read_excel('C:/Users/sanzh/One<br>Drive/Рабочий<br>стол/modelcvetofor/TestModel/dat<br>a.xlsx')<br>text = ''<br>slAbaiW =<br>dataframe['Data'].iloc[0]<br>slAbaiE =<br>dataframe['Data'].iloc[1]<br>slSainN =<br>dataframe['Data'].iloc[2]<br>slSainS =<br>dataframe['Data'].iloc[3]<br>text = str(slAbaiW) + '$' +<br>str(slAbaiE) + '$' + str(slSainN) +<br>'$' + str(slSainS)<br>text = text.replace('"', '')<br>print(text) |

the serverMACAddress for both the client and the server is a mandatory requirement and can only be found using system commands.

**Table 2.** The block of data exchange between the server and the client - the traffic light with computer vision.

| | |
|---|---|
| serverMACAddress =<br>'D8:F8:83:7C:70:53'<br>port = 4<br>size = 2048<br>listener = keyboard.Lis-<br>tener(on_press=on_press)<br>listener.start()<br>starting = time.time()<br>startForConn = time.time()<br>print('Starting')<br>s = None<br>offline = False | while not qt:<br>if not offline:<br>try:<br>if s is None:<br>s = socket.socket(socket.AF_BLUETOOTH,<br>socket.SOCK_STREAM,<br>socket.BTPROTO_RFCOMM)<br>s.connect((serverMACAddress, port))<br>s.settimeout(2)<br>work_online()<br>except Exception as e:<br>print(e)<br>offline = True<br>else:<br>work_offline()<br>print("Closing socket") |

As a result of the conducted circuit design and hardware-software work, screenshots of the prototype operation of the client-server complex are displayed (see Fig. 6), allowing the collective operation of the traffic light's intelligent control system with a local server with a dispatcher - TPS employee.

green traffic light is on

**Fig. 6.** Animation of the traffic simulation model on Abay-Saina (left) and a view of the prototype of the traffic light's intelligent control system with computer vision (right).

## 3    Discussion

In Fig. 6, the simultaneous appearance of the green light at the stop line of the auxiliary road Saina-North and the green bulb on the mounting board is shown. This indicates that the informational signal has been converted into a 5-V voltage on the bulb, which can be relayed to produce the power voltage for the traffic light's executive device at 24 V DC or 220 V AC. Pre-congestion processes characteristic of the Abay-Saina type junction are visible on the left image. Predictive animation models indicate the existence of a critical intensity threshold for stable traffic flow at the road junctions in the city.

Thus, a prototype of a multi-level automated traffic light control system with computer vision has been developed. This prototype offers several advantages compared to existing traffic management systems in the city of Almaty and provides opportunities for expanded practical application to digitize traffic management in the city.

In the discussion of this topic, it's essential to highlight the significance of the findings and their implications for traffic management and automation systems. The analysis conducted on classical multi-level automation systems revealed notable challenges in implementing animated simulation modeling for complex traffic management processes. This underscores the need for innovative approaches, such as the integration of computer vision and IoT technologies, as demonstrated in the study.

The utilization of simulation modeling in the AnyLogic environment proved to be highly beneficial for assessing traffic dynamics in urban areas across different countries. The ability to visualize and forecast traffic patterns serves as a valuable resource for urban planners and traffic management authorities, enabling them to make informed decisions to optimize traffic flow and mitigate congestion.

Moreover, the successful animation of simulation models for traffic and traffic light control on complex intersections using client-server applications highlights the potential for real-time monitoring and management of traffic systems. The integration of Python-based sockets as intermediaries facilitates seamless data exchange between the server and the traffic light control systems, enhancing the overall efficiency of traffic management operations.

The proposed traffic light control system is highly scalable, but requires significant investments in hardware (cameras, sensors, mini-computers, servers), software and network infrastructure. Operating costs include hardware and software maintenance and upgrades. The investment in the proposed system can be justified by significant resource savings in the future. Improved traffic management will reduce congestion, wait time and fuel consumption, resulting in cost savings and reduced emissions. Traffic optimization will also reduce wear and tear on infrastructure and vehicles, which will reduce repair and maintenance costs.

The development of a prototype for collective automated interaction between the server, traffic management simulation model, and real-time video presents promising opportunities for proactive traffic congestion prevention, particularly during peak hours. By autonomously adjusting the duration of green light signals based on traffic intensity, the system demonstrates adaptive and responsive behavior, further optimizing traffic flow and reducing delays.

Looking ahead, future research efforts will concentrate on enhancing the real-time capabilities of the system and refining the interaction between the server and traffic light control systems. This includes exploring advanced algorithms for predictive traffic modeling, integrating machine learning techniques for adaptive traffic signal control, and addressing scalability and interoperability challenges for widespread deployment of such systems in urban environments.

## 4   Conclusion

Based on the conducted research, the following conclusions and recommendations can be made: Analysis of existing classical multi-level automation systems revealed difficulties in implementing animated simulation modeling of complex traffic management processes. Simulation modeling in AnyLogic environment finds wide application in assessing traffic in large cities of various countries. The obtained results are used for visualization and forecasting of traffic as an informational resource. Animation of simulation models of traffic and traffic light control on complex intersections in AnyLogic environment can be achieved by coupling a client-server application with sockets, written in Python 3, acting as an intermediary between the computer server and the client traffic light control system with computer vision for direct and reverse data exchange regarding traffic light operation phases. The created prototype of collective automated interaction between the server with the visualization of traffic management simulation model and real-time video allows the DPS patrol car dispatcher to prevent traffic congestion during peak hours. In normal situations, the traffic light control system with computer vision can autonomously and automatically adjust the duration of the green light signal depending on the intensity of vehicle traffic.

Future works will focus on enhancing the real-time capabilities of the system and further optimizing the interaction between the server and the traffic light control system for improved traffic management efficiency.

**Acknowledgments.** This study was funded by Committee of Science of Republic of Kazakhstan AP19678998 "Neurocomputer Vision of Smart Traffic Lights in Megacities of the Country" (2021-2023).

**Disclosure of Interests.** The authors have no competing interests to declare that are relevant to the content of this article.

# References

1. Smith, J.R.: Simulation Modeling with AnyLogic, vol. 8, p. 150. Springer, New York (2020)
2. Johnson, A., Brown, M., Wilson, L.: Real-time object tracking on raspberry Pi CM 4 platform. In: Proceedings of the International Conference on Computer Vision (ICCV), Vancouver, Canada (2023)
3. Williams, R., Jones, S.: Exploring SCADA Systems: A Practical Guide to Siemens WinCC, p. 200. Routledge, London (2020)
4. Tengri News. https://tengrinews.kz/kazakhstan_news/poveliteli-almatinskih-svetoforov-zac hem-abaya-dolgo-gorit-351601/. Accessed 10 Mar 2024
5. Mansurova, M., Belgibaev, B., Zhamangarin, D., Zholdas, N.: Adaptive urban traffic lights as collective internet of things devices. In: Dolinina, O., et al. (eds.) AIES 2022. SSDC, vol. 457, pp. 697–712. Springer, Cham (2023). https://doi.org/10.1007/978-3-031-22938-1_48
6. Zhamangarin, D.S.: Integrated intelligent systems based on IoT technologies for urban traffic management. In: ADAL KITAP, Almaty, p. 242 (2023)
7. Belgibaev, B., Mansurova, M., Abdrakhim, S., Ormanbekova, A.: Smart traffic lights with video vision based on a control minicomputer in Kazakhstani megacities. In: 14th International Conference on Emerging Ubiquitous Systems and Pervasive Networks (EUSPN), Almaty (2023)
8. Duyssembayeva, L., Belgibaev, B., Mansurova, M., Abdrakhim, S.: Neural computer visualization of smart programs in megacities of the country. Certificate of Authorship of the Republic of Kazakhstan No. 39772 dated 19 October 2023
9. Jensen, M.B., Philipsen, M.P.: Vision for looking at traffic lights: issues, survey, and perspectives. IEEE Trans. Intell. Transp. Syst. (2019). https://doi.org/10.1109/TITS.2019.250 9509
10. Tsalikidis, N., et al.: Urban traffic congestion prediction: a multi-step approach utilizing sensor data and weather information. Smart Cities **7**, 233–253 (2024). https://doi.org/10.3390/sma rtcities7010010
11. Manjunathan, A., Kumar Suresh, A., Udhayanan, S., Thirumarai Selvi, C., Alexander Stonier, A.: Design of autonomous vehicle control using IoT. IOP Conf. Ser. Mater. Sci. Eng. 1055 (2021). https://doi.org/10.1088/1757-899X/1055/1/012008
12. Hossai, M.R.T., Shahjalal, M.A., Nuri, N.F.: Design of an IoT based autonomous vehicle with the aid of computer vision. In: 2017 International Conference on Electrical, Computer and Communication Engineering (ECCE), pp. 752–756 (2019). https://doi.org/10.1109/ECACE. 2017.7913003
13. OFDM Simulation in MATLAB. https://www.semanticscholar.org/paper/c380d9561973654 31f7f091bc721762b26a7e8ca

14. Leone, G., et al.: An intelligent cooperative visual sensor network for urban mobility. Sensors (Basel, Switzerland) **17** (2021). https://doi.org/10.3390/s17112588
15. Shen, T., Jiang, Y., Zhang, Y., Wang, X.: Traffic light recognition based on deep learning. In: 2018 IEEE Intelligent Vehicles Symposium (IV). IEEE (2018)
16. TensorFlow Official Documentation on Image Classification. https://www.tensorflow.org/tut orials/images/classification
17. PyImageSearch article on Traffic Light Detection. https://www.pyimagesearch.com/2021/05/24/traffic-light-detection-and-classification-with-yolov5/
18. Aimsun Live – Key Benefits. https://www.aimsun.com/aimsun-live
19. Pedestrian, M.: Traffic and vehicle interaction model using SUMO and OMNeT++. arXiv (2020)
20. Szeliski, R.: Computer Vision: Algorithms and Applications. Springer, Cham (2019). https://doi.org/10.1007/978-3-030-34372-9

# An Automated and Verbose Approach for Detecting Anomalies in Cloud Computing Platform Using Logs

Arthur Vervaet[1], Yousra Chabchoub[1], Mar Callau-Zori[2(✉)], and Raja Chiky[3]

[1] ISEP - Institut Supérieur d'Électronique de Paris 10 rue de Vanves, 92130
Issy-les-Moulineaux, France
**yousra.chabchoub@isep.fr**
[2] 3DS OUTSCALE, 1 rue Royale, 92064 Saint-Cloud, France
**mar.callau-zori@outscale.com**
[3] 3il Ingénieurs, Limoges, France
**chiky@3il.fr**

**Abstract.** Logs represent a valuable source of runtime information, serving various purposes such as monitoring, diagnosing issues, assessing performance, and facilitating maintenance. However, the task of automating real-time anomaly detection based on logs remains highly challenging. The process of parsing log messages is intricate and prone to errors. Furthermore, identifying relationships between logs is often unfeasible, particularly in complex systems like cloud computing platforms. We propose here Monilog, an automated system designed for log-based anomaly detection in cloud computing environments. Monilog introduces a novel approach to parsing and encoding logs, enabling the creation of appropriate inputs for traffic forecasting models. By generating comprehensive summaries of detected anomalies, Monilog empowers practitioners to develop intelligent alerting systems on top of it. Our evaluation, conducted using real-life log data obtained from Kernel Based Virtual Machine at the scale of a cloud region, highlights the effectiveness of Monilog in accurately forecasting server failures and providing meaningful insights into reported anomalies.

**Keywords:** Cloud Computing · Data Mining · Anomaly Detection · Monitoring · Log Analysis

## 1 Introduction

Cloud computing platforms are complex systems composed of a multitude of physical components (e.g., routers, servers, hard drives) geographically distributed amongst multiple data centers. Monitoring the health of such infrastructure is a challenging task for the operational teams [18]. Log files record runtime information about a system [1], and have already proved their effectiveness for a wide range of mining tasks such as identifying the origin of a failure [26],

N.-T. Nguyen et al. (Eds.): ICCCI 2024, CCIS 2166, pp. 81–93, 2024.
https://doi.org/10.1007/978-3-031-70259-4_7

analyzing performance issues [14], or detecting attacks [16]. With the increasing amount of generated log lines, many automated log-based approaches have been proposed [8] to exploit the diversity of information they contain.

For large-scale online systems and services expected to work without interruption such as cloud platforms, detecting anomalies in real-time allows monitoring teams to mitigate their consequences. Thus alleviating the pain of dealing with events that can impact millions of users at the same time [9,12]. This need for reliability drives the development of automated Anomaly Detection (AD) methods [3,20]. Different log-based AD approaches have been proposed within the previous years based on neural approaches such as LSTM [15,25] or BERT structures [17] with strong claims regarding their performances. Those methods use parsing algorithms to identify log underlying patterns and then seek for anomalous sequences. In practice, within a cloud environment, knowing or inferring log relationships is a challenging and often not possible task. This is due to logs most of the time not having a unique operation identifier. Also, inferring related events is hard due to the multiplicity of logging devices, the involved volumetry and the temporal proximity of operations.

In this paper, we present Monilog, an automated log-based AD system able to detect anomalous events at cloud scale. Monilog can produce comprehensive reports regarding the reported anomaly candidates including the concerned system and applications. It does not require any knowledge regarding the relationships between logs and is able to work in a streaming fashion allowing close to real-time alerting. Presented research work was conducted in partnership with OUTSCALE, a cloud-providing company. We implemented and tested our proposal at Cloud scale using all the Syslog messages [6] generated by the Kernel-based Virtual Machines (KVMs) of the company European cloud region during 11 consecutive days. Our precision evaluation displays the relevance of the reported abnormal events and the ability of the system to predict server crashes.

The paper is organized as follows. Section 2 presents existing log-based monitoring methods related to our work. Section 3 details Monilog architecture. Detailed information regarding the experimental setup as well as the selected model and metrics can be found in Sect. 4. Evaluation results regarding our implementation in a live cloud computing context are presented under Sect. 5. Section 6 and, Sect. 7 sums up the article and present future work leads.

## 2    Background and Related Work

To the best of our knowledge, the first reported log-based AD system was brought up by Liang et al. in 2007 [10]. The proposed system was based on the Support Vector Machines set of classifiers and aimed to detect failures affecting the IBM BlueGene/L supercomputer. In the following years, multiple machine learning-based approaches were published [11,13,23] to detect system runtime problems. These methods need to perform several passes on a log dataset to extract anomalies which is a problem in a context with a lot of data and real-time requirements.

In 2017, Du et al. proposed Deeplog [5], a log-based AD method based on LSTM neural network. Deeplog models logs as natural language sequences, it is able to process them in a streaming manner and its experimental evaluation displayed a higher accuracy than the existing non-neural methods. The Deeplog approach was extended later on by LogRobust [25] and LogAnomaly [15], both are based on bidirectional LSTM neural networks and aim to detect sequential log anomalies. Those methods also proposed new embedding mechanisms to alleviate the impact of log concept drift on the model. More recently, Yang et al. introduced PleLog [24] based on Convolutional Neural Network coupled with a clustering step to train more efficiently within unsupervised contexts.

All of the presented neural-based methods follow a common workflow. First, they use a log parsing algorithm to mine logs underlying templates and then they use a previously trained deep-learning model to predict the next element of a log template sequence. An anomaly is raised if the predicted probability of the verified next log template of a given sequence is under a determined threshold. When trying to work with existing methods in a cloud computing context, we run into several shortcomings. We found that reconstructing log sequences is not always possible due to the volumetry, the lack of unique identifiers and, the different layers of components. Logs are implemented by developers and contain a free text part, this semi-structured format makes it more difficult to exploit them in an automated way. Also, the aforementioned log-based AD solutions rely on a log parsing step to be able to work with stable input.

Logs are generated by log statements inside the code, and a log line can be broadly divided into two parts, a header and a message. As it is common to embed information describing the current state of the system inside the message, log parsers aim to retrieve those when you do not know the log statement. Log parsing is an active research area and multiple methods [26] have been proposed since SLCT by R. Vaarandi in 2003 [21]. More recent methods such as Drain [7] and USTEP [22] use parsing trees to process logs in a streaming fashion and infer new log templates on the fly; Spell [4] uses the longest common subsequence algorithm to regroup logs based on their similarities. Despite all the recent progress, no parsing solutions achieve perfect accuracy and, parsing errors can impact negatively downstream solutions. One of our studies [22] based on Drain parsing algorithm and Deeplog model showed that going from 1.0 parsing accuracy to 0.8 parsing accuracy decreases Deeplog AD accuracy by 0.75.

With Monilog, we propose a way to detect anomalous events within complex environments such as a cloud-computing platform. It does not require any knowledge about the relationship between logs and does not rely on a parsing step. Monilog details the concerned host and applications of the reported anomalous candidates. Those candidates can therefore be correlated and aggregated using domain-specific models to produce relevant alerts from a reliability engineer point of view.

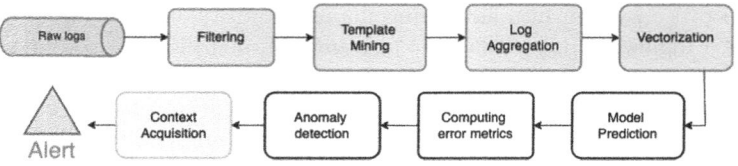

**Fig. 1.** Monilog Architecture and Workflow

## 3   Monilog Approach

Within this section, we detail the architecture of Monilog. Monilog workflow (Fig. 1) is divided into three parts: (Sect. 3.1) First (in blue), Monilog infers log templates and vectorizes them; (Sect. 3.2) It seeks anomalous events and generates alert candidates (in white); (Sect. 3.3) (in yellow) it acquires the context of an anomaly and produces a report.

### 3.1   Vectorizing Logs

**Log Filtering:** Logs are time-stamped and generated by machines at runtime. The input of Monilog is an infinite and temporally ordered sequence of logs. A preliminary filtering step to retain only the logs linked to one or a set of systems.

**Template Mining:** The template of a log message is set to be the combination of its application name and its severity (e.g., kernel_6). As it uses only header-embedded information, this step is not affected by log parsing errors, and the number of different possible templates is bound by the number of running applications times the number of severity levels.

**Log Aggregation:** Logs with the same host and template are aggregated by $w$ nanoseconds windows. We define $L_{h,e}(t)$ as the count of logs associated with template $e$ for host $h$ at timestamp $t$. We define the log count window as a time series $C_{h,e}(w)$.

$$C_{h,e}(t) = \sum_{i \in [t,t+w]} L_{h,e}(i) \tag{1}$$

**Vectorization:** Every unique log template within the training set is associated with a unique id. We define $\mathcal{T}$ as the ensemble of unique log templates within the model training set, $f_e : \mathcal{T} \mapsto 1, 2, .., |\mathcal{T}|$ is a bijective function. The id $e_i$ of the log template is defined by the following equation:

$$e_i = \begin{cases} f_e(i) & \text{if } i \in \mathcal{T} \\ |\mathcal{T}| + 1 & \text{otherwise} \end{cases} \tag{2}$$

The event count vector $V(t)$ is constructed as follows:

$$V(t)(i) = \sum_{e_i = i} C_{h,e}(t) \qquad (3)$$

$$|V(t)| = |\mathcal{T}| + 1 \qquad (4)$$

## 3.2  Detecting Anomalies

**Model Predictions:** Models aim to forecast the log traffic. They are provided a matrix regrouping the r previous observations and output a forecast of the next event count vector. Each model can be assimilated to a prediction function P:

$$P(M(t)) = \hat{V}(t) \qquad (5)$$

With M(t) a matrix composed of the r last count vectors at time t. M(t) is formally defined as follows:

$$M(t)(j) = V(t - (w \times (r - j))), j \in [1, .., r] \qquad (6)$$

**Compute Forecasting Error:** The previous step model prediction is compared to the actual observation $V(t)$ using an error metrics (Sect. 4.4). This metric evaluates the forecasting accuracy of the models for the elapsed time step.

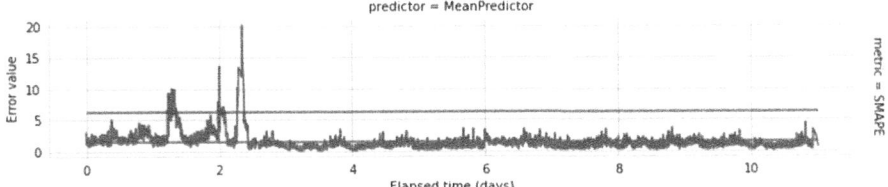

**Fig. 2.** An Example of Error Time Series

**Alert Candidate Generation:** The error time series are used to generate alert candidates. Points that are considered abnormal regarding the three sigma rules based on the weekly mean error are labeled as alert candidates. Figure 2 is an error-time series extracted from our experimental evaluation (Sect. 4). It plots the error value over time, the green line represents the mean error value, and the red one the 3-sigma threshold. Any point above this threshold will be considered an alert candidate.

### 3.3    Consolidation and Alerting

```
{"Timestamp": "2022-11-02 10:32:00",          1
  "Name": "Host-13",                           2
  "Anomaly Score": 4.5,                        3
  "Applications": [{                           4
    "Name": "App-13"                           5
    "Score Weight": "92.4%"}]}                 6
```

**Listing 1.1.** An Alert Report

**Context Acquisition:** Forecasting error metric is applied over a forecast vector and an observation vector. Every record inside those vectors is linked to a template. It is therefore possible to quantify the influence of a given template over the final error. This information is used to retrieve the most contributing templates to the total error. As templates are forged using the application name and the severity, reverting this process returns the application and the severity of the top error-contributing templates. Listing 1.1 is an example of an anomaly report outputted by Monilog.

**Alert Consolidation:** False positive or irrelevant alerting are the main obstacles to the adoption of an anomaly detection system by the operational teams [2]. For each alert candidate, the system returned the host as well as the top contributing applications to the alert. Operators can define additional rules to reduce the number of alerts. For instance, alerts linked to similar equipment inside a data center can be grouped, or alerting scenarios can be defined based on the involved applications.

## 4    Experimental Setup

We implemented 9 different versions of Monilog based on different model/metric pairs and runs it at a Cloud region scale.

### 4.1    Dataset

For 11 consecutive days, our Monilog implementations were reading a Kafka stream filled with syslog messages issued from all the Kernel-Based Virtual Machines (KVMs) linked to the OUTSCALE European region. In total, over several billion lines of logs were consumed during the monitored period at the average rate of 238 logs per host per minute.

### 4.2    Anomalies

To evaluate the relevance of Monilog architecture as well as the performance of each model/metric combination, we ran a manual inspection of all the reported abnormal periods to assign them a label. We used three distinct values during our labeling: *Probable Anomaly* (PA) for the possible legitimate abnormal events;

*False Positive* (FP) for the identified events that are not likely to represent anomalies, and *Unknown* for the remaining ones.

We assign to the *Probable Anomaly* category events such as a high concentration of kernel errors, dropped log messages due to excessive rates or network-related errors. Unusual but not abnormal events such as an unusually large number of virtual machines being deployed simultaneously, booting operations, or admin operations such as network configurations changes were labeled as *False Positive*. Events inside the *Unknown* are linked to log events referring to external systems or metrics. Assigning a clear label to such events is a difficult task as it requires correlating the information from multiple systems. In our case, due to the high number of concerned events, we choose to set them aside as it was reasonably impossible to inspect them all in detail.

## 4.3   Models

We selected and implemented three different kinds of forecasting models. By comparing the performance of those models coupled with error metrics described in the next section, we want to give industrial insight of what to expect from the different approaches in an industrial cloud computing context. We set $r = 24$, the number of last count vectors forwarded to the model. As each vector represents a five-minute time window, this is the equivalent of asking the model to predict the next five minutes based on the last two hours of logs.

**Repetitor:** This model always predicts the most recent count vector as the next output. With this model:

$$\hat{V}(t) = V(t) \tag{7}$$

The repetitor model is a very naïve way of forecasting the log traffic by assuming it remains constant over time. However, as it only considers the last log count vector, it is less influenced by trailing noises.

**Mean Channel Value:** This model returns the channel wise mean of the last r observations. For this model:

$$\hat{V}(t)(i) = \frac{\sum_{j=1}^{r} M(j)(i)}{r} \tag{8}$$

The mean channel value model is another simple way of forecasting the log traffic. It aims to mitigate the influence of chaotic changes such as wide variation affecting one channel by using its overtime mean.

**LSTM Autoencoder:** LSTM is a type of recurrent neural network that retains long-term dependencies between data at a given time from many time steps before. They have been used for many sequence learning tasks including handwriting recognition, speech recognition, and sentiment analysis. An LSTM-based encoder is used to map an input sequence to a vector representation of fixed dimensionality. The decoder is another LSTM network that uses this vector representation to produce the target sequence.

We train an LSTM autoencoder to reconstruct instances of normal time series. Training log data were collected during the two weeks prior to our 11 days train dataset, and are issued from the same KVMs. A linear layer on top of the LSTM decoder layer is used to predict the next log count vector.

**Table 1.** Predictor Alerting Precision per Selected Error Metric

| | Repetitor | | | MeanPredictor | | | LSTM-AE | | |
|---|---|---|---|---|---|---|---|---|---|
| | RMSE | SMAPE | LAR | RMSE | SMAPE | LAR | RMSE | SMAPE | LAR |
| # Relevant alerts | 281 | 840 | 453 | 37 | 825 | 352 | 1 | 10 458 | 4 085 |
| # False positive | 577 | 103 | 98 | 60 | 497 | 363 | 0 | 745 | 865 |
| # Unknown | 1 911 | 2 | 3 | 1 274 | 818 | 849 | 11 315 | 2 016 | 5 842 |
| **Total** | 2 769 | 945 | 554 | 1 371 | 2 140 | 1 563 | 11 316 | 13 219 | 10 792 |
| **Precision** | 0.327 | 0.891 | 0.822 | 0.381 | 0.624 | 0.493 | N.A | **0.933** | 0.825 |

### 4.4 Error Metrics

The workload can be very different between two identical devices and be subject to vast variation over time. The workload of a server can operate between 0 and 100% of the equipment computing capacity, and quickly evolve within this range of values without being abnormal. We considered three different metrics for our experiment evaluation. Each metric is given a forecast vector $\hat{V}$ and an observation vector $V$ and returns a numeric value. We assume here that vectors have the same dimension $|\mathcal{T}| + 1$.

**Root Mean Square Error:** The Root Mean Square Error (RMSE) is a commonly used metric in statistics. It is defined as the square root of the mean square error between a forecast vector $\hat{V}$ and an observation vector $V$:

$$RMSE = \sqrt{\frac{\sum_{i=1}^{|\mathcal{T}|+1}(\hat{V}(i) - V(i))^2}{|\mathcal{T}| + 1}} \qquad (9)$$

As it does not involve volumetric comparisons, this metric is heavily influenced by the most represented log templates. Two unpredicted logs related to rare logging events will have the same, weight in the final error than two unpredicted logs related to a frequent logging event.

**Symmetric Mean Absolute Percentage Error:** The Symmetric Mean Absolute Percentage Error (SMAPE) is also a commonly used metric in statistics, it expresses the accuracy as a ratio defined by the formula:

$$SMAPE = \frac{100}{|\mathcal{T}| + 1} \sum_{i=1}^{|\mathcal{T}|+1} \frac{|\hat{V}(i) - V(i)|}{|V(i)| + |\hat{V}(i)|} \qquad (10)$$

The lower the SMAPE value of a forecast, the higher the model accuracy. This error includes volumetric comparisons between the expected and obtained number of each logging template. With this metric, rare unpredicted events will have a stronger impact on the final value and fluctuation of the number of common events will be less impacting. One limitation of SMAPE is that if the actual value or forecast value is 0, the value of error will approach for the concerned channel 100%.

**Log Accuracy Ratio:** Mean Absolute Percentage Error metric systematically promotes methods with lower predictions. The Log Accuracy Ratio (LAR) was introduced by Tofallis in 2015 [19] to avoid this bias. LAR is defined as the square logarithm of the ratio observation over the forecast:

$$LAR = \frac{1}{|\mathcal{T}| + 1} \sum_{i=1}^{|\mathcal{T}|+1} ln(\frac{V(i)}{\hat{V}(i)})^2 \tag{11}$$

## 5    Evaluation

### 5.1    Precision Evaluation

A significant proportion of false positive can hide the relevant alerts and, in the long run, leads the system to be abandoned. To be accepted by technical teams, it is more important for an alerting model to be precise than to be accurate. Computing the accuracy of each model/metric pair would require knowing all the abnormal events that occurred during the monitored period. As previously discussed, this is impracticable due to the number of logs inside the dataset. However, we spent a consequent amount of time inspecting and labeling all the abnormal sequences raised by the different model/metric pairs to assign them a label. The precision for each model/metric combination is computed based on those labels using the following equation:

$$Precision = \frac{relevant\_alerts}{relevant\_alerts + false\_positives} \tag{12}$$

It is the ratio of relevant alerts over the total number of alerts. Note, that we left aside all the *Unknown* labeled events, as we weren't able to assign them a clear label we chose to ignore them when computing the precision. Table 1 regroups the obtained results for each of the nine model/metric combinations.

SMAPE is the precision-wise best-performing metric. Assigning a label to anomalies reported by the LSTM-AE/RMSE pair was a complicated task and we end up assigning the Unknown label to almost all of them (11315 over 11316). We considered this to be a strong sign that the reported anomalous period will not be easily exploitable. Regarding this, we set aside the precision result for this combination. The most precise model/metric pair in our evaluation is therefore the LSTM-AE coupled with the SMAPE error. This combination achieves a precision of 0.933 while also being the most verbose one with 13219 anomaly candidates, 11203 if we remove the unknown ones.

RMSE on its side is the worst-performing metric for each model. We explain this by its sensibility to volume change regarding the common templates. Most of the false positives raised by the model/RMSE pairs are linked to a massive deployment of virtual machines, automated CRON jobs or maintenance operations. All of those operations tend to generate a usually high number of log messages and trigger an alert by model/RMSE combinations. The model/LAR pairs produce in-between results, Tofallis [19] concern regarding metrics favoring under-forecasting models does not seem to be harmful in our case. Regards to the previously stated results, SMAPE appears to be the go-for choice to optimize the precision regarding generated anomaly candidates.

Regarding the models, the repetitor achieves surprisingly good precision despite its simplicity. It obtained a precision of 0.891 for the Repetitor/SMAPE pair and a precision of 0.822 for the Repetitor/LAR pair. We believe this result to be driven by the chaotic aspect of logs. A repetitor model with a proportional metric such as SMAPE or LAR is good to detect short-term major evolution in logging behaviors. Also, the 5-minute duration of the window is huge compared to cloud operations periods that are usually measured in milliseconds. However, the total number of relevant alerts (840 for Repetitor/SMAPE) is way lower compared to the 10458 reported by LSTM-AE/SMAPE.

## 5.2   Critical Events Forecasting

During the 11 days experimental period, three server crashes were reported by operational teams. As each crash affected a different KVM and occurred on a different day, we assumed here that they are independent anomalies. Forecasting such events would help the monitoring team mitigate their impact on the hosted client's virtual machines. For each crash, we reported the model/metric pairs that detected the crash (i.e., crash time occurs during one of the reported anomaly candidate periods), and if applicable the forecast. The forecast is computed using the time difference between the reported crash time and the beginning of the associated abnormal period.

The MeanPredictor/LAR detects two crashes and the MeanPredictor/SMAPE detects all of the three reported crashes. For the five detected crashes, the forecasting time is superior to 80 min. Given the low number of alleged crashes, we are working with, it is difficult to reach any hard conclusion. However, obtained results support our hypothesis that KVM log files can be used to predict server crashes.

The MeanPredictor model was the only one able to forecast some of the reported crashes. We believe this to be linked to this model's sensitivity to the accumulation of small anomalies over time. When diving into logs before the crash, we observed a behavioral-drift within the hours before the crash. The MeanPredictor/SMAPE pair appears to be especially efficient regarding the detection as well as the long-time alerting regarding those drifts. Figure 2 displays the overtime error of the SMAPE error metric over MeanPredictor forecasts. The crash occurred on Day 3, the crash error time is visible as the associate reported error value is the highest for the considered KVM during the experimental period.

Repetitor and LSTM-AE models quickly adapt to new behaviors despite their abnormal nature. Repetitor and LSTM-AE models coupled with SMAPE or LAR metrics raised anomaly at the same time as the MeanPredictor/SMAPE and MeanPredictor/LAR models for the alleged crashes. However, they stop raising alerts after a short period, around 30 min for Repetitor/metric pairs and after one abnormal period (5 min) for LSTM-AE/metric pairs. Regarding this, we could debate the ability of such models to forecast crashes because to a certain extent, they detect their premises. When looking at the alert candidate detailed report raised by repetitor and LSTM-AE models, it was obvious that it was an alleged anomaly but hard to predict a crash. On the other hand, MeanPredictor-generated reports give us more clues.

## 6   Conclusion

In this paper, we presented Monilog a log-based anomaly detection system designed for the cloud computing context. It is able to detect anomalous events in environments where log relationships are not identifiable. Monilog uses a model to forecast the log traffic and it evaluates the forecasting accuracy using error metrics. A period is concerned abnormal when the associated error takes an unusual value. Monilog way of parsing and embedding log events allows it to generate detailed events regarding the reported anomaly candidate including the concerned system and the applications whose behavior led to raising an alert.

We conducted an evaluation of our proposal at industrial scale using 11 consecutive days of logs issued from the KVMs of the European cloud region operated by 3DS OUTSCALE. For the evaluation we considered three forecasting models: Repetitor, MeanPredictor and LSTM-AE and three error metrics: RMSE, SMAPE, LAR. We were interested in the capacity of these 9 model/metric pairs to raise relevant alerts. Regarding the precision of each of the considered pairs, the SMAPE error metric appears to be the best-performing error metric. Regarding the models, LSTM-AE is the most effective model to detect abnormal periods. During the monitored period, three KVM crashes were reported, and the MeanPredictor model was able to predict all of them with at least 80 min of forecasts.

## 7   Lesson Learned and Future Work

In the future, we plan to extend Monilog monitoring scope by experimenting with a broader range of cloud computing equipment and software. Reducing the multiplicity of alerts is an underlying concern for our work, by enlarging the monitoring scope we expect abnormal events to raise multiple alerts candidates as they will probably be linked to more devices. In order to mitigate this phenomenon, we would like to experiment combining Monilog-reported anomaly candidates with other existing KPIs and metrics to produce fewer alerts and with an extended context.

**Acknowledgments.** The work described in this paper is supported by the French National Research and Technology Association (CIFRE program $N°2020/0289$).

**Disclosure of Interests.** The authors have no competing interests to declare that are relevant to the content of this article.

# References

1. Chen, B., Jiang, Z.M.: A survey of software log instrumentation. ACM Comput. Surv. (CSUR) **54**(4), 1–34 (2021)
2. Debnath, B., et al.: Loglens: a real-time log analysis system. In: 2018 IEEE 38th International Conference on Distributed Computing Systems (ICDCS). IEEE (2018)
3. Decker, L., Leite, D., Giommi, L., Bonacorsi, D.: Real-time anomaly detection in data centers for log-based predictive maintenance using an evolving fuzzy-rule-based approach. In: 2020 IEEE International Conference on Fuzzy Systems (FUZZ-IEEE), pp. 1–8. IEEE (2020)
4. Du, M., Li, F.: Spell: streaming parsing of system event logs. In: 2016 IEEE 16th International Conference on Data Mining (ICDM), pp. 859–864. IEEE (2016)
5. Du, M., Li, F., Zheng, G., Srikumar, V.: Deeplog: anomaly detection and diagnosis from system logs through deep learning. In: Proceedings of the 2017 ACM SIGSAC Conference on Computer and Communications Security, pp. 1285–1298 (2017)
6. Gerhards, R.: The Syslog Protocol. RFC 5424 (Proposed Standard), March 2009. http://www.ietf.org/rfc/rfc5424.txt
7. He, P., Zhu, J., Zheng, Z., Lyu, M.R.: Drain: an online log parsing approach with fixed depth tree. In: 2017 IEEE International Conference on Web Services (ICWS), pp. 33–40. IEEE (2017)
8. He, S., He, P., Chen, Z., Yang, T., Su, Y., Lyu, M.R.: A survey on automated log analysis for reliability engineering. ACM Comput. Surv. (CSUR) **54**(6), 1–37 (2021)
9. Liang, H., Song, L., Wang, J., Guo, L., Li, X., Liang, J.: Robust unsupervised anomaly detection via multi-time scale DCGANs with forgetting mechanism for industrial multivariate time series. Neurocomputing **423**, 444–462 (2021)
10. Liang, Y., Zhang, Y., Xiong, H., Sahoo, R.: Failure prediction in IBM Bluegene/l event logs. In: Seventh IEEE International Conference on Data Mining (ICDM 2007), pp. 583–588. IEEE (2007)
11. Lin, Q., Zhang, H., Lou, J.G., Zhang, Y., Chen, X.: Log clustering based problem identification for online service systems. In: 2016 IEEE/ACM 38th International Conference on Software Engineering Companion (ICSE-C), pp. 102–111. IEEE (2016)
12. Liu, P., et al.: Fluxrank: a widely-deployable framework to automatically localizing root cause machines for software service failure mitigation. In: 2019 IEEE 30th International Symposium on Software Reliability Engineering (ISSRE), pp. 35–46. IEEE (2019)
13. Lou, J.G., Fu, Q., Yang, S., Xu, Y., Li, J.: Mining invariants from console logs for system problem detection. In: 2010 USENIX Annual Technical Conference (USENIX ATC 10) (2010)

14. Lu, J., Liu, C., Li, F., Li, L., Feng, X., Xue, J.: Cloudraid: detecting distributed concurrency bugs via log-mining and enhancement. IEEE Trans. Softw. Eng. (2020)
15. Meng, W., et al.: Loganomaly: unsupervised detection of sequential and quantitative anomalies in unstructured logs. In: IJCAI, vol. 19, pp. 4739–4745 (2019)
16. Moh, M., Pininti, S., Doddapaneni, S., Moh, T.S.: Detecting web attacks using multi-stage log analysis. In: 2016 IEEE 6th International Conference on Advanced Computing (IACC), pp. 733–738. IEEE (2016)
17. Nedelkoski, S., Bogatinovski, J., Acker, A., Cardoso, J., Kao, O.: Self-attentive classification-based anomaly detection in unstructured logs. In: 2020 IEEE International Conference on Data Mining (ICDM), pp. 1196–1201. IEEE (2020)
18. Pourmajidi, W., Zhang, L., Miranskyy, A., Steinbacher, J., Godwin, D., Erwin, T.: The challenging landscape of cloud monitoring. Knowl. Manag. Dev. Data-Intens. Syst. (2021)
19. Tofallis, C.: A better measure of relative prediction accuracy for model selection and model estimation. J. Oper. Res. Soc. **66**(8), 1352–1362 (2015)
20. Ullah, W., Ullah, A., Haq, I.U., Muhammad, K., Sajjad, M., Baik, S.W.: CNN features with bi-directional LSTM for real-time anomaly detection in surveillance networks. Multimedia Tools Appl. **80**(11), 16979–16995 (2021)
21. Vaarandi, R.: A data clustering algorithm for mining patterns from event logs. In: Proceedings of the 3rd IEEE Workshop on IP Operations and Management (IPOM 2003) (IEEE Cat. No. 03EX764). IEEE (2003)
22. Vervaet, A., Chiky, R., Callau-Zori, M.: Ustep: unfixed search tree for efficient log parsing. In: 2021 IEEE International Conference on Data Mining (ICDM), pp. 659–668. IEEE (2021)
23. Xu, W., Huang, L., Fox, A., Patterson, D., Jordan, M.: Largescale system problem detection by mining console logs. In: Proceedings of SOSP'09 (2009)
24. Yang, L., et al.: Semi-supervised log-based anomaly detection via probabilistic label estimation. In: 2021 IEEE/ACM 43rd International Conference on Software Engineering (ICSE), pp. 1448–1460. IEEE (2021)
25. Zhang, X., et al.: Robust log-based anomaly detection on unstable log data. In: Proceedings of the 2019 27th ACM Joint Meeting on European Software Engineering Conference and Symposium on the Foundations of Software Engineering, pp. 807–817 (2019)
26. Zhu, J., et al.: Tools and benchmarks for automated log parsing. In: 2019 IEEE/ACM 41st International Conference on Software Engineering: Software Engineering in Practice (ICSE-SEIP), pp. 121–130. IEEE (2019)

# Cooperative Strategies for Decision Making and Optimization

# Efficiency of Specialized Genetic Operators in Non-dominated Tournament Genetic Algorithm (NTGA2) Applied to Multi-objective Multi-skill Resource Constrained Project Scheduling Problem

Michał Antkiewicz[1], Paweł Borys Myszkowski[1]([⊠]), Konrad Gmyrek[1]([⊠]), Adam Krzeminski[1,2], and Jose Luis Calvo-Rolle[2]

[1] Faculty of Information and Communication Technology, Wrocław University of Science and Technology, Wrocław, Poland
{michal.antkiewicz,pawel.myszkowski,konrad.gmyrek}@pwr.edu.pl
[2] Department of Industrial Engineering, University of A Coruña, Coruña, Spain
jlcalvo@udc.es

**Abstract.** The Multi-Objective Multi-Skill Resource Constrained Project Scheduling Problem (MS-RCPSP) is an NP-hard real-world problem that can be solved by metaheuristics like the Non-Dominated Tournament Genetic Algorithm (NTGA2). NTGA2 method is effective as a generic black-box metaheuristic. In the paper, we present experiments to examine how effective NTGA2 is in multi-objective optimization when the black-box rule is omitted, and specialized operators are used: Cheaper Resource Crossover, Less Assignment Crossover, and Resource-Leveling Mutation. Experimental results show that specialized operators have extra computational costs, but finally, the NTGA2 method is the most effective. Results are based on the benchmark iMOPSE library, compared to state-of-the-art methods, and statistically verified.

**Keywords:** multi-objective optimisation · evolutionary computation · specialized operators · scheduling

## 1 Introduction

The Multi-Skill Resource-Constrained Project Scheduling Problem (MS-RCPSP) represents an extension of the classical Resource-Constrained Project Scheduling Problem. It includes highly constrained solution space and interwoven opposite objectives, therefore we resolve it in a multi-objective manner. The multi-objective optimization (MOO) presents a paradigm shift from traditional single-objective optimization (SOO) by addressing the optimization of conflicting objectives simultaneously. Unlike SOO, which seeks a single optimal solution, MOO aims to identify a set of solutions (Pareto Front Approximation,

N.-T. Nguyen et al. (Eds.): ICCCI 2024, CCIS 2166, pp. 97–110, 2024.
https://doi.org/10.1007/978-3-031-70259-4_8

PFA) representing trade-offs between objectives. Specifically, bi-objective optimization considers exactly two objectives, while many-objective optimization extends this to scenarios with multiple objectives. This paper focuses on the bi-objective optimization of the MS-RCPSP, considering duration and cost.

Conventional optimization methodologies for combinatorial problems, both single- and multi-objective, adhere to the black-box principle, where solutions are optimized based on the sole feedback information provided by the objective/fitness function. This is a desired approach in the literature concerning methodological research, as it keeps the method problem-agnostic, enhancing its potential for generalization and application across diverse domains. However, in real-world scenarios, where the specific problem is considered, deviating from this principle can be advantageous. Domain knowledge can be leveraged through specialized operators, which can be incorporated into metaheuristic algorithms such as evolutionary methods, replacing generic mutation or crossover operators.

In this paper, we apply three specialized operators for the bi-objective MS-RCPSP, where domain-knowledge-guided operators aim to lead to potentially superior solutions. Each operator focuses on either cost or makespan minimization. Rather than focusing on operator construction itself, the ultimate goal is to explore how different single-criteria-guided operators shift the exploration in multi-objective space. The proposed operators possess the potential to enhance effectiveness in multi-objective optimization if given efficient integration.

The rest of the paper is structured as follows. Section 2 presents a brief related work review, focusing on the specialized operators. The investigated MS-RCPSP problem is shortly defined in Sect. 3. The applied method and specialized operators are described in Sect. 4. Section 5 includes results for the experiments conducted to answer the research questions. Lastly, the paper is concluded, and future work directions are given in Sect. 6.

## 2    Related Work

The specialization of genetic operators spans a spectrum from universally applicable modifications to those tailored for exact application using domain knowledge. Such specialization can be based on various considerations, like gene value boundaries, encoding type, instance size, knowledge of the objective function, or domain-specific insights. Operators can be 'lightly' (semi-) specialized, so they are applied to specific types of problems, e.g., permutation problems, as described in Subsect. 2.1. The problem-specialized operators can often exploit the knowledge deeper and use the expert metrics and indicators as enumerated in Subsect. 2.2.

### 2.1   Semi-specialized Operators

While not always explicitly termed as specialization in literature, survey [2] enumerates permutation operators found in the literature, such as Order Crossover

(OX), Partially Matched Crossover (PMX), or Inversion (INV). These semi-specialized operators are commonly applied in genetic algorithms (and similar) for solving diverse optimization problems. For instance: PMX, OX, Reverse One-Point, Reverse PMX, Reverse OX, Linear-Order Crossover (LOX), C1, NABEL, Swap, Insertion (INS), Inversion (INV), and Reverse (REV) can be applied in the Flowshop Scheduling [17], while using permutation encoding.

Authors of the [10] introduce the Very Greedy Crossover (VGX) operator and apply it to the Genetic Algorithm solving the TSP. It prioritizes selecting the shortest edges from parental tours and improves it further by selecting a starting city randomly and iteratively appending edges based on their lengths. Similarly, in the [6], authors propose the Greedy Single Point Crossover (GSPC) operator. Each selected gene is checked one by one, and if there is an allele that has the shortest path, it is exchanged. This approach aims to prioritize the selection of genes that contribute to shorter paths, potentially leading to improved solutions. The processing of genes in this manner adds computational overhead compared to traditional crossover methods, but the test results indicate superior fitness outcomes.

## 2.2 Problem Specialized Operators

Traveling Salesman Problem with Job-Times (TSPJ) introduces additional complexity by integrating job-time constraints into the TSP framework. Researchers solving this problem utilized the domain knowledge in both initialization and perturbation operators as presented in [8]. Among the mechanisms employed to enhance solution efficiency is the Local Search for Job Assignment (LSJA). It identifies the node (city) with the highest current completion time and subsequently engages in iterative job exchanges between the identified node and others, aiming to minimize completion time.

The authors of the [12] proposed a novel approach, termed the IF-Aware operator for routing services in Optical Transport Networks (OTN). This operator is designed to leverage domain-specific knowledge to enhance the convergence and effectiveness of the genetic algorithm. They state that the knowledge-based implementation of the genetic algorithm presents on average a 59% increase in the Hypervolume rate when compared to the purely random evolutionary algorithm for the same number of generations, and more so decreasing the computation time. The proposed IF-Aware crossover introduces a strategy to improve convergence by intelligently selecting parent solutions biasing towards minimizing the total number of interfaces.

The authors of the [7] applied domain-specific knowledge in developing the Greedy Mutation (GM) operator to optimize energy management in residential scenarios. This operator targets specific solution characteristics, focusing either on direct cost reduction or comfort improvement. The cost-greedy operators prioritize regions with lower total prices for shiftable loads and EV charging, while the comfort-greedy operator selects based on minimizing time slot penalties for shiftable loads. This utilization of domain-specific knowledge facilitates addressing the issue of conflicting objectives.

The authors of the [15] address the Travelling Thief Problem (TTP) (an extension of TSP) by proposing a Multi-Objective Evolution by Variation algorithm (MOEV), employing variation operators. The Greedy Fill (GF), operates on a partially filled packing plan, adding the most profitable items until capacity is reached. The Weight Improver (WI) operator seeks to improve both objectives simultaneously by identifying lighter items. Similarly, the Profit Improver (PI) operator focuses on profit improvement by sorting items according to profit.

The domain-specific knowledge has been leveraged in [9] in addressing the Flowshop Sequence-Dependent Group Scheduling Problem (FSDGSP) to minimize the makespan, total flow time, and total energy consumption, simultaneously. The authors introduced a Greedy Cooperative Co-evolutionary Algorithm (GCCEA) designed to delve into the solution space. The key utilization of the domain-specific knowledge is the indicator value used during the job sequencing within families.

The authors of [18] implemented an additional greedy operator, following mutation and crossover operations. This approach aimed to optimize Virtual Machine (VM) task scheduling algorithms by minimizing total execution time and balancing workload across VMs. Tasks assigned to a VM with the maximum total execution time will transfer to another VM with the minimum total execution time, to reduce the total execution time, thus improving the efficiency of the task scheduling algorithm.

## 3   Multi-skill Resource Constrained Project Scheduling Problem

The classical Resource Constrained Project Scheduling Problem (RCPSP) focuses on scheduling activities within a project, ensuring that the project is completed in the shortest possible time while adhering to constraints such as resource limitations and precedence relationships between activities. The MS-RCPSP extends the classical problem by introducing the concept of multiple skills (MS), where resources possess different skills at varying proficiency levels. This adds a layer of complexity, requiring the assigned resources to have the requisite skills to perform the task.

The MS-RCPSP encompasses two interrelated sub-problems: task sequencing, and resource assignments. The objective of the MS-RCPSP is to determine a feasible schedule. This involves assigning available resources to tasks and arranging the tasks on a timeline. To be considered feasible, a schedule, denoted as $PS$, must adhere to a predetermined set of **constraints**.

Each resource is connected to its salary $r_{salary}$, no salary can be negative. The set of skills $S_r$ possessed by the resource $r$ cannot be empty. Each task's duration $d_t$ and finish time $F_t$ are not negative. Tasks are constrained by a precedence relation – all task's predecessors must be finished before work on it can be started. Each task must have assigned exactly one resource.

The skill extension of the MS-RCPSP is described in Eq. 1. The resource must have the skill at the required level or higher if assigned to a task.

$$\forall_{t \in T^r} \exists_{s_r \in S^r} \; h_{s_t} = h_{s_r} \wedge l_{s_t} \leq l_{s_r} \tag{1}$$

where $T^r$ is a set of tasks assigned to a resource $r$, $s_t$ is the skill required by the task $t$, $S^r$ is the set of skills possessed by the resource $r$, $h$ and $l$ are the type and level of the skill respectively.

The latest definition of the MS-RCPSP is a many-objective optimization problem with five objectives [13]. The original two objectives – schedule duration (makespan) and cost – can be defined by Eq. 2 and Eq. 3. Further MS-RCPSP objectives tackle specific project scheduling aspects: average cash flow, skill overuse, and the average use of resources. The **Makespan $f_\tau(\mathbf{PS})$** of the project schedule $PS$ is given as Eq. 2.

$$f_\tau(PS) = \max_{t \in T} t_{finish} \tag{2}$$

where $T$ is a set of all tasks, $t_{finish}$ is the finish time of the task $t$. The **Cost** of the schedule is $\mathbf{f_C(PS)}$ defined as Eq. 3.

$$f_C(PS) = \sum_{i=1}^{n} R_i^{salary} * T_i^{duration} \tag{3}$$

where $n$ is the number of all task-resource assignments, $R_i^{salary}$ is the salary of a resource of the i'th assignment, $T_i^{duration}$ is the duration of the task of the i'th assignment.

The MS-RCPSP originally optimises 2-objectives $\mathbf{f_\tau(PS)}$ and $\mathbf{f_C(PS)}$, where all objectives must be minimized:

$$\min f(PS) = \min \left[ f_\tau(PS), f_C(PS) \right] \tag{4}$$

## 4   Method

In this section, we present and briefly describe the Non-Dominated Tournament Genetic Algorithm 2 method, as well as three different mutations providing a deeper understanding of the discussed operators.

### 4.1   Non-dominated Tournament Genetic Algorithm

Non-Dominated Tournament Genetic Algorithm 2 (NTGA2) [13] (see Pseudocode 1) is a metaheuristic designed for multi-objective optimization, it introduces the Gap Selection (GS) operator to further promote diversity in PFA by targeting the least explored areas of the solution space.

In the original implementation of NTGA2 [13], the representation of the MS-RCPSP is defined by task-resource assignments, accounting for skill constraints by using a list of capable resources for each task. This representation ensures that only valid assignments are considered, with the genotype indicating the specific capable resource assigned to each task. The actual solution also requires

**Algorithm 1.** Pseudocode for the NTGA2[13]

---

$archive \leftarrow \{\}$
$P_{current} \leftarrow generateInitialPopulation()$
$evaluate(P_{current})$
$updateArchive(P_{current})$
**for** $i \leftarrow 0$ to $genenrationLimit$ **do**
    $P_{next} \leftarrow \{\}$
    **while** $|P_{next}| < |P_{current}|$ **do**
        **if** $i$ mod $(2 \cdot gsGenerations) < gsGenerations$ **then**
            $parents \leftarrow select_{tour}(P_{current})$
        **else**
            $parents \leftarrow select_{GS}(P_{current} \cup archive)$
        $children \leftarrow crossover(parents)$
        $children \leftarrow mutate(children)$
        **while** $P_{next}$ contains $children$ **do**
            $children \leftarrow mutate(children)$
        $evaluate(children)$
        $P_{next} \leftarrow P_{next} \cup children$
        $updateArchive(children)$
    $P_{current} \leftarrow P_{next}$

---

placing tasks on a timeline, achieved through a **greedy Schedule Builder** that assigns tasks to their earliest possible starting times.

For genetic operations within NTGA2, the initial population for MS-RCPSP is generated randomly. A single-point crossover (SX) method is employed, where a random slice-point is chosen, and genes from one parent are combined with genes from another. The mutation is executed randomly, altering one gene to a different value within its domain, ensuring variability and exploration within the search space.

## 4.2    Specialized Genetic Operators for MS-RCPSP

In this section, we describe different genetic operators investigated in the experiments section to provide a deeper understanding of the influence they have on the evolution process.

**4.2.1    Cheaper Resource Crossover** (CRX) [14] operator aims to achieve lower-cost resource allocations to tasks from both parents. It compares the resource assignments of the parents for each task and selects the one with the lower resource cost for the offspring's genotype (listing 2). If the costs are equal, it compares the number of assignments for each resource and selects the one with fewer assignments. If equal, the gene is randomly selected from the parents. This approach ensures that the offspring will not have a higher cost than either parent and in most cases, will be significantly cheaper. However, this method may lead to more conflicts, potentially increasing project duration.

---

**Algorithm 2.** Pseudocode for the Cheaper Resource Crossover Operator [14]

---

```
 1: Input: Parent Genotypes P1, P2
 2: Output: Offspring Genotypes O1
 3: for i from 1 to length(P1) - 1 do
 4:     if P1(i).cost < P2(i).cost then
 5:         O1(i) = P1(i)
 6:     else
 7:         if P1(i).cost > P2(i).cost then
 8:             O1(i) = P2(i)
 9:         else
10:             if P1(i).assignments < P2(i).assignments then
11:                 O1(i) = P1(i)
12:             else
13:                 O1(i) = P2(i)
14: return O1
```

---

**4.2.2   Less Assignment Crossover** (LAX) [14] operator aims to reduce the project duration by assigning tasks to resources with fewer total assignments, thereby potentially reducing conflicts and balancing the workload among resources (see Pseudocode 3). When the number of assignments is equal, resources with lower cost are preferred. If equal, the selection is made randomly. This method promotes a more balanced distribution of tasks among resources, potentially leading to fewer conflicts and a more efficient project schedule. However, it has a limitation: if tasks in the parent genotypes are associated with a single resource, LAX cannot redistribute the task load.

**4.2.3   Resource-Leveling Mutation** (RLM) [16] operator aims to balance the workload among resources. It seeks to adjust task allocations to resources by considering their constraints and efficient use. As shown in the pseudocode below, it operates by selecting a task through roulette wheel selection and then reassigning it to the least loaded resource capable of performing it. This task-to-resource reassignment process is repeated until the number of mutations reaches a predefined parameter $\epsilon$. Ideally, this leads to better resource utilization and potentially shorter project schedules while maintaining feasibility.

**Algorithm 3.** Pseudocode for the Less Assignment Crossover[14]

---

1: **Input:** *Parent Genotypes P1, P2*
2: **Output:** *Offspring Genotype O1*
3: **for** *i from 1 to length(P1) - 1* **do**
4:     **if** $P1(i).assignments < P2(i).assignments$ **then**
5:         $O1(i) = P1(i)$
6:     **else**
7:         **if** $P1(i).assignments > P2(i).assignments$ **then**
8:             $O1(i) = P2(i)$
9:         **else**
10:            **if** $P1(i).cost < P2(i).cost$ **then**
11:                $O1(i) = P1(i)$
12:            **else**
13:                $O1(i) = P2(i)$
14: **return** O1

---

**Algorithm 4.** Pseudocode for the Resource-Leveling Mutation [16]

---

1: **Input:** *Individual Genotype IG*
2: **Output:** *Mutated Individual Genotype MG*
3: **Parameter:** *Resource level scale $\epsilon$*
4: **for** *i from 1 to $\epsilon$* **do**
5:     *Select a task T by roulette wheel selection*
6:     *Select a capable resource R, which has the least assignments*
7:     *Reassign task T to R*

---

## 5  Experiments

To investigate the effectiveness of specialised operators in NTGA2 applied to MS-RCPSP, the following **Research Questions** have been developed: (Q1). Which specialized genetic operators, when examined individually, enhance solution quality? (Q2). When applied together in mutation and crossover, which specialized genetic operators yield improved outcomes? (Q3). Which combination of genetic operators yields optimal performance metrics? (Q4). What is the computational overhead associated with utilizing specialized genetic operators regarding execution time and resource consumption? And (Q5). How do the outcomes achieved with specialized genetic operators compare with those employing conventional operators? How effective are the proposed methodologies in comparison to state-of-the-art methods in the literature? Mainly, this section contains research details and answers to the above questions.

### 5.1  Experimental Setup

In experiments, benchmark iMOPSE [1] dataset is used, which contains 36 MS-RCPSP instances created using real-world scheduling problems. All instances have varying tasks, resources, and skills to define a range of problems. The dataset contains two main parts: 100 and 200 tasks. The number of resources varies between 5 and 20 for the first part and 10 and 40 for the second part. The number of skills varies between 9 and 15. Due to the non-deterministic nature of metaheuristics, all runs have been repeated 30 times, and results have been averaged and statistically verified. Moreover, to verify the quality of resulted PFAs and get broader context as state-of-the-art have been used, such as the basic version of NTGA2 and state-of-the-art methods: $\theta$-DEA, NSGA-II, and U-NSGA-II [13]. All methods are configured as follows: population size $popSize = 200$, probability of mutation $P_m = 0.005$, probability of crossover $Px = 0.9$, and tournament size $tSize = 2$. The best-found configuration of NTGA2 is population size $popSize = 50$, probability of mutation $P_m = 0.01$, probability of crossover $Px = 0.6$, tournament size $tSize = 6$ and $gsGenerations = 50$. NTGA2 that uses specialised operators needs more frequent mutation $P_m = 0.6$ and crossover $P_m = 0.9$. The computational budget for all examined methods is $Generations = 2000$. All experiments are developed using Java programming language and a computer with AMD EPYC 7H12 64-Core equipped with Ubuntu 22.04.

## 5.2    Results

In this paper, we want to discover how specialized operators in NTGA2 effectively solve MO MS-RCPSP. Thus, a combination of genetic operators (mutation and crossovers) is compared to the basic version of NTGA2. As a result, each MOO method gives PFA, which cannot be evaluated so simply as in SOO. Thus, to compare results, standard quality measures for MO have been used. HyperVolume ($HV$ – see in Tab. 1) – quantifies the volume of objective space dominated by solutions. The higher value indicates a better coverage of the objective space. Inverted Generative Distance ($IGD$) measures an average distance from each *True Pareto Front approximation* (TPFa)(approximated by gathering all found solutions) point to the closest point in PFA. Lower IGD values signify that solutions are closer to the ideal Pareto front. *Purity* (see Tab. 2) – interpreted as the part of TPFa that the given method provided (where 1 means that the method found all of the points in TPFa). Each table contains the average and standard deviation of QMs for every method. At the bottom of each table, the complete average of the method and Wilcoxon signed-rank test results on whether the proposed method has improved the results can be found.

The results presented in Tab. 1 show that the most effective approaches are NTGA2_CRX_RLM (in 18 cases gained the best results) and NTGA2_LAX_RLM (20 the best results) as the combinations that link specialized mutation (RLM) and specialised crossover (CRX or LAX). However, in 8 cases, these two methods in the context of $HV$ measure give the same results, i.e., are statistically unimportant. Thus, the resulting PFAs should be measured by the other MOO measure, like *Purity* (see Tab. 2), to get a dominance relation between resulted PFAs.

The results presented in Tab. 2 confirm that NTGA2 with specialised operators (NTGA2_CRX_RLM and NTGA2_LAX_RLM) are more effective than

**Table 1.** Results ($HV$) for each spec. operator: (R)LM, (C)RX and (L)AX.

| instance | NTGA2 avg | std | NTGA2_C avg | std | NTGA2_L avg | std | NTGA2_R avg | std | NTGA2_C_R avg | std | NTGA2_L_R avg | std | p-value | |
|---|---|---|---|---|---|---|---|---|---|---|---|---|---|---|
| 100_5_20_9_D3 | 0.3911 | 0.3549 | 0.3744 | 0.3744 | 0.3752 | 0.3752 | 0.5607 | 0.1901 | 0.7396 | 0.0036 | **0.7437** | 0.0025 | <0.0001 | ++ |
| 100_5_22_15 | **0.6407** | 0.1738 | 0.5784 | 0.0003 | 0.578 | 0.0009 | 0.5783 | 0.0082 | 0.586 | 0.0447 | 0.576 | 0.009 | <0.0001 | ++ |
| 100_5_46_15 | **0.5923** | 0.1615 | 0.5674 | 0.0023 | 0.5871 | 0.1027 | 0.5545 | 0.019 | 0.5807 | 0.1049 | 0.566 | 0.0021 | <0.0002 | ++ |
| 100_5_48_9 | **0.5356** | 0.1375 | 0.4808 | 0.0116 | 0.483 | 0.0006 | 0.5116 | 0.1756 | 0.4781 | 0.0031 | 0.4793 | 0.0124 | 0.0015 | ++ |
| 100_5_64_15 | 0.6136 | 0.0939 | 0.6206 | 0.001 | **0.6217** | 0.0008 | 0.6118 | 0.0286 | 0.6193 | 0.0046 | **0.621** | 0.0615 | 0.3041 | == |
| 100_5_64_9 | 0.5666 | 0.0347 | 0.5711 | 0.0007 | **0.5713** | 0.0005 | 0.556 | 0.0455 | 0.5678 | 0.0031 | 0.5691 | 0.0021 | 0.0098 | ++ |
| 100_10_26_15 | 0.6614 | 0.0297 | **0.6714** | 0.0013 | 0.6699 | 0.0016 | 0.6649 | 0.0024 | 0.6634 | 0.0081 | 0.6697 | 0.0024 | 0.0023 | ++ |
| 100_10_27_9_D2 | 0.7738 | 0.0017 | **0.7782** | 0.0016 | 0.776 | 0.0027 | 0.767 | 0.0337 | 0.7628 | 0.0568 | 0.7685 | 0.0227 | 0.0004 | ++ |
| 100_10_47_9 | 0.6758 | 0.0704 | 0.6864 | 0.0019 | 0.6718 | 0.0053 | **0.6896** | 0.0294 | 0.6886 | 0.0026 | **0.6897** | 0.0028 | 0.6547 | == |
| 100_10_48_15 | 0.7022 | 0.0285 | 0.7076 | 0.0009 | 0.7069 | 0.0009 | 0.7047 | 0.0108 | **0.7097** | 0.0034 | 0.702 | 0.0159 | 0.0027 | ++ |
| 100_10_64_9 | 0.7593 | 0.012 | **0.7774** | 0.0021 | 0.771 | 0.0028 | 0.7626 | 0.0354 | 0.7738 | 0.0042 | **0.7764** | 0.003 | 0.3632 | == |
| 100_10_65_15 | 0.6149 | 0.0717 | 0.6101 | 0.0011 | 0.6105 | 0.0075 | 0.6099 | 0.0306 | **0.6175** | 0.0021 | 0.6077 | 0.004 | <0.0001 | ++ |
| 100_20_22_15 | 0.8192 | 0.0132 | 0.8351 | 0.0019 | 0.826 | 0.0029 | 0.8331 | 0.0027 | 0.8344 | 0.0024 | **0.836** | 0.0025 | 0.0002 | ++ |
| 100_20_23_9_D1 | 0.4984 | 0.2599 | 0.4272 | 0.327 | 0.4209 | 0.3348 | 0.6831 | 0.1012 | **0.7525** | 0.0068 | **0.7519** | 0.0015 | 0.1678 | == |
| 100_20_46_15 | 0.7197 | 0.0029 | 0.7385 | 0.0023 | 0.7225 | 0.0039 | 0.734 | 0.0034 | 0.7371 | 0.0036 | **0.74** | 0.0025 | 0.0034 | ++ |
| 100_20_47_9 | 0.8167 | 0.0285 | 0.8392 | 0.0027 | 0.8199 | 0.0051 | 0.8432 | 0.0026 | 0.8432 | 0.0032 | **0.8482** | 0.0028 | <0.0001 | ++ |
| 100_20_65_15 | 0.757 | 0.0025 | 0.7647 | 0.001 | 0.7603 | 0.0012 | 0.7657 | 0.0021 | **0.7666** | 0.0013 | **0.7665** | 0.001 | 0.9117 | == |
| 100_20_65_9 | 0.7293 | 0.035 | 0.7605 | 0.0025 | 0.7432 | 0.0056 | 0.7635 | 0.0053 | **0.7648** | 0.0033 | 0.7607 | 0.0265 | <0.0001 | ++ |
| 200_10_128_15 | 0.7052 | 0.0017 | 0.7161 | 0.0011 | 0.7105 | 0.0019 | 0.7135 | 0.0034 | 0.7154 | 0.0023 | **0.717** | 0.0019 | 0.0065 | ++ |
| 200_10_50_15 | 0.8211 | 0.046 | 0.8371 | 0.0016 | 0.8289 | 0.0023 | 0.8313 | 0.0033 | 0.8325 | 0.0037 | **0.8387** | 0.0025 | 0.0008 | ++ |
| 200_10_50_9 | 0.7179 | 0.0636 | 0.7599 | 0.0016 | 0.7414 | 0.0043 | 0.7569 | 0.0035 | 0.7542 | 0.0541 | **0.7691** | 0.0016 | 0.0003 | ++ |
| 200_10_84_9 | 0.7639 | 0.0038 | 0.8141 | 0.0014 | 0.7947 | 0.0042 | 0.8112 | 0.0392 | **0.8252** | 0.052 | 0.8206 | 0.0022 | <0.0001 | ++ |
| 200_10_85_15 | 0.5157 | 0.0288 | 0.5511 | 0.0013 | 0.5372 | 0.0028 | 0.5561 | 0.001 | 0.5557 | 0.0017 | **0.5577** | 0.0015 | <0.0001 | ++ |
| 200_10_135_9_D6 | 0.7256 | 0.1669 | 0.683 | 0.2073 | 0.6747 | 0.2153 | 0.8841 | 0.0038 | 0.8855 | 0.0023 | **0.8862** | 0.0015 | <0.0001 | ++ |
| 200_20_145_15 | 0.7001 | 0.0377 | 0.7614 | 0.0022 | 0.7337 | 0.0065 | 0.7652 | 0.0022 | **0.7707** | 0.0014 | **0.7685** | 0.0026 | 0.6204 | == |
| 200_20_54_15 | 0.6776 | 0.0044 | 0.735 | 0.0032 | 0.7165 | 0.0045 | 0.7339 | 0.0026 | **0.741** | 0.0018 | **0.745** | 0.0017 | 0.0965 | == |
| 200_20_55_9 | 0.7917 | 0.0047 | 0.8805 | 0.0027 | 0.8582 | 0.0062 | 0.8855 | 0.0023 | **0.8958** | 0.002 | 0.8945 | 0.0019 | <0.0003 | ++ |
| 200_20_97_15 | 0.6717 | 0.0055 | 0.7324 | 0.0021 | 0.7022 | 0.0059 | 0.7376 | 0.0022 | **0.7394** | 0.0024 | **0.7437** | 0.0007 | 0.1224 | == |
| 200_20_97_9 | 0.6962 | 0.0059 | 0.8061 | 0.0032 | 0.7643 | 0.0062 | 0.8139 | 0.0031 | **0.8231** | 0.0017 | 0.8207 | 0.0025 | <0.0004 | ++ |
| 200_20_150_9_D5 | 0.7188 | 0.1549 | 0.6794 | 0.1917 | 0.6739 | 0.2001 | 0.7524 | 0.0771 | **0.8681** | 0.0034 | 0.8668 | 0.0028 | <0.0002 | ++ |
| 200_40_130_9_D4 | 0.6878 | 0.0071 | 0.7923 | 0.0038 | 0.7396 | 0.01 | 0.8157 | 0.0077 | **0.8201** | 0.0047 | **0.8185** | 0.005 | 0.2643 | == |
| 200_40_133_15 | 0.787 | 0.006 | 0.9058 | 0.0037 | 0.8615 | 0.0074 | 0.9179 | 0.0022 | **0.9244** | 0.0015 | **0.9241** | 0.0017 | 0.7393 | == |
| 200_40_45_15 | 0.7502 | 0.0056 | 0.8522 | 0.0028 | 0.82 | 0.0073 | 0.8597 | 0.0025 | **0.8709** | 0.0015 | 0.868 | 0.0023 | <0.0008 | ++ |
| 200_40_45_9 | 0.7733 | 0.005 | 0.8767 | 0.0033 | 0.8215 | 0.0064 | 0.8857 | 0.0032 | **0.8994** | 0.0018 | 0.8939 | 0.002 | <0.0002 | ++ |
| 200_40_90_9 | 0.7515 | 0.0061 | 0.8815 | 0.0047 | 0.827 | 0.0137 | 0.8893 | 0.0026 | **0.9084** | 0.002 | 0.9002 | 0.003 | <0.0004 | ++ |
| 200_40_91_15 | 0.7696 | 0.0055 | 0.8629 | 0.0028 | 0.8179 | 0.0058 | 0.8739 | 0.0025 | **0.8787** | 0.0014 | **0.8784** | 0.0019 | 0.501 | == |
| Avg | 0.7107 | 0.048 | 0.7423 | 0.0024 | 0.7248 | 0.0076 | 0.7442 | 0.0162 | **0.7481** | 0.0064 | **0.7473** | 0.0122 | n.u. | n.u. |
| Wins | 3 | | 3 | | 2 | | 1 | | 18 | | **20** | | | |

other NTGA2 variants. The *Purity* measure points out as the winner of the NTGA2_CRX_RLM, as it resulted in the best PFA in 16 cases and near 54% of the average APF generated by this method. The second place gets to NTGA2_LAX_RLM, which wins in 12 cases and on average 39% non-dominated points are generated by this method. Additionally, it could be concluded that these two methods, on average, generate 95% of APF. But there is a cost to this efficiency – the average computational time (see Tab. 2) of NTGA2 is nearly 20 s, while the specialized version of NTGA2 needs more than 200 s.

The results presented in Tab. 1 and Tab. 2 show that potentially not in each case specialised operators get the best results – instance 100_5_22_15 is the

**Table 2.** Results (*Purity*) for each spec. operator: (R)LM, (C)RX and (L)AX.

| instance | NTGA2 | NTGA2_C | NTGA2_L | NTGA2_R | NTGA2_C_R | NTGA2_L_R |
|---|---|---|---|---|---|---|
| 100_5_20_9_D3 | 0.0231 | 0.0244 | 0.0436 | 0.2256 | 0.5705 | **0.6346** |
| 100_5_22_15 | 0.3413 | 0.3234 | 0.4012 | 0.5629 | **0.6527** | 0.5269 |
| 100_5_46_15 | 0.3894 | 0.375 | 0.3798 | **0.6707** | 0.5841 | 0.637 |
| 100_5_48_9 | 0.0529 | 0.0824 | 0.1667 | **0.6824** | 0.5765 | 0.6333 |
| 100_5_64_15 | 0.2607 | 0.2636 | 0.2579 | **0.616** | 0.5673 | 0.1719 |
| 100_5_64_9 | 0.0553 | 0.0308 | 0.018 | **0.7712** | 0.6954 | 0.7661 |
| 100_10_26_15 | 0.0072 | 0.0108 | 0.0072 | 0.0719 | 0.518 | **0.6583** |
| 100_10_27_9_D2 | 0.0021 | 0.0064 | 0.0 | **0.6497** | 0.4862 | 0.552 |
| 100_10_47_9 | 0.0049 | 0.0016 | 0.0016 | **0.5916** | 0.4911 | 0.5397 |
| 100_10_48_15 | 0.0309 | 0.0077 | 0.0077 | 0.3398 | **0.4363** | 0.2239 |
| 100_10_64_9 | 0.0022 | 0.0022 | 0.0022 | 0.4771 | 0.4553 | **0.5229** |
| 100_10_65_15 | 0.0128 | 0.0213 | 0.0128 | 0.5489 | 0.0298 | **0.6894** |
| 100_20_22_15 | 0.0159 | 0.0079 | 0.004 | 0.4286 | **0.4643** | 0.2778 |
| 100_20_23_9_D1 | 0.0371 | 0.0609 | 0.0305 | 0.6675 | 0.5325 | **0.6715** |
| 100_20_46_15 | 0.043 | 0.0161 | 0.0 | **0.5269** | 0.3548 | 0.0591 |
| 100_20_47_9 | 0.0058 | 0.0038 | 0.0096 | 0.1173 | **0.8308** | 0.0327 |
| 100_20_65_15 | 0.0042 | 0.0 | 0.0084 | 0.0 | **0.7616** | 0.2257 |
| 100_20_65_9 | 0.0 | 0.0 | 0.0 | 0.2379 | **0.559** | 0.2484 |
| 200_10_128_15 | 0.0025 | 0.0076 | 0.0 | 0.5178 | 0.264 | **0.6827** |
| 200_10_50_15 | 0.0221 | 0.0 | 0.0083 | 0.2569 | 0.2735 | **0.4392** |
| 200_10_50_9 | 0.0026 | 0.0 | 0.0009 | 0.1217 | 0.5004 | **0.5724** |
| 200_10_84_9 | 0.0039 | 0.0031 | 0.0 | 0.1048 | **0.8144** | 0.0738 |
| 200_10_85_15 | 0.0036 | 0.0012 | 0.0 | 0.4837 | **0.7521** | 0.5284 |
| 200_10_135_9_D6 | 0.0282 | 0.0411 | 0.0231 | 0.516 | **0.6367** | 0.4069 |
| 200_20_145_15 | 0.0105 | 0.0 | 0.0021 | 0.1572 | **0.6289** | 0.2013 |
| 200_20_54_15 | 0.0035 | 0.0 | 0.0 | 0.2695 | **0.4007** | 0.3262 |
| 200_20_55_9 | 0.0095 | 0.0 | 0.0 | 0.1394 | **0.6631** | 0.1881 |
| 200_20_97_15 | 0.0 | 0.0 | 0.0 | 0.2412 | 0.4475 | **0.5272** |
| 200_20_97_9 | 0.0022 | 0.0045 | 0.0 | 0.0247 | **0.6494** | 0.3191 |
| 200_20_150_9_D5 | 0.0472 | 0.0354 | 0.0446 | 0.1404 | 0.6037 | **0.7533** |
| 200_40_130_9_D4 | 0.0 | 0.0 | 0.0 | 0.0892 | **0.831** | 0.0798 |
| 200_40_133_15 | 0.0407 | 0.0 | 0.0 | 0.126 | 0.3537 | **0.4797** |
| 200_40_45_15 | 0.0068 | 0.0 | 0.0 | 0.0544 | **0.534** | 0.4048 |
| 200_40_45_9 | 0.0 | 0.0 | 0.0 | 0.0852 | **0.5742** | 0.3407 |
| 200_40_90_9 | 0.0032 | 0.0064 | 0.0 | 0.1051 | 0.4108 | **0.4745** |
| 200_40_91_15 | 0.0 | 0.0 | 0.0 | 0.0328 | **0.772** | 0.1952 |
| avg | 0.0419 | 0.0367 | 0.0403 | 0.316 | **0.5417** | 0.3937 |
| wins | 0 | 0 | 0 | 7 | **16** | 12 |
| comp. time [s] | **20.22**± 9.46 | 131.46 ± 38.05 | 182.98 ± 51.71 | 79.07 ± 19.6 | 204.34 ± 45.59 | 212.98 ± 52.55 |

most interesting example (see Fig. 1a), where the PFA of all NTGA2 variants are very similar, and the basic NTGA2 gets the best *HV* value. However, the more detailed analysis of this case shows that NTGA2_CRX_RLM gets near *Purity* = 65%. The more evident case (200_40_133_15, see Fig. 1b) shows the dominance of the NTGA2_LAX_RLM method, where it generates an average of about 48% of PFA.

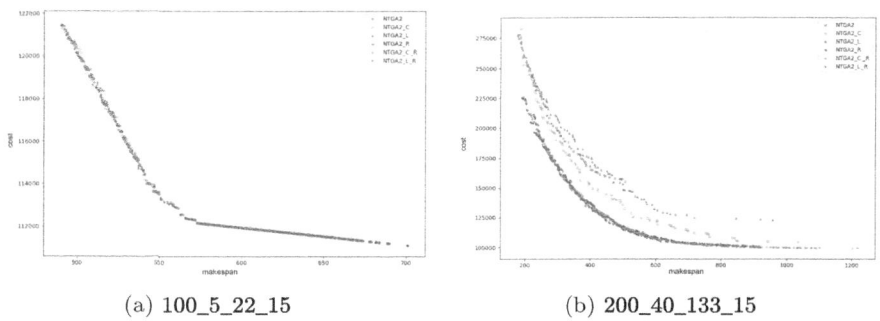

(a) 100_5_22_15                    (b) 200_40_133_15

**Fig. 1.** Comparison of PFAs for investigated NTGA2 variants.

To verify the results presented in Tab. 1 another MOO measure has been used – see *IGD* values presented in Tab. 3. The minimal (preferable) *IGD* values get the NTGA2_CRX_RLM method, which is statistically significant. Additionally, to get a broader context of experimental research in Tab. 3, the summarised results for several variants of NTGA2 have been compared to stat-of-the-art methods, like $\theta$-DEA, NSGA-II, and U-NSGA-II. The final winner is the NTGA2_CRX_RLM method, which gets the best results in *Purity* and *IGD* measures, and wins *ex aequo* with NTGA2_LAX_RLM.

**Table 3.** Summary of MO metrics for investigated methods

| | HV↑ | | IGD↓ | | Purity↑ | |
|---|---|---|---|---|---|---|
| | avg | std | avg | std | avg | std |
| NSGA-II | 0.6847 | 0.0039 | 0.0193 | 0.0079 | 0.0305 | 0.0527 |
| UNSGA-III | 0.6881 | 0.0059 | 0.0177 | 0.0064 | 0.0403 | 0.066 |
| $\theta$-DEA | 0.7011 | 0.0047 | 0.0188 | 0.0069 | 0.0466 | 0.0623 |
| NTGA2 | 0.7107 | 0.048 | 0.0123 | 0.0052 | 0.0507 | 0.1755 |
| NTGA2_C | 0.7423 | 0.0024 | 0.0052 | 0.0074 | 0.0494 | 0.0607 |
| NTGA2_L | 0.7248 | 0.0076 | 0.0095 | 0.0076 | 0.0541 | 0.0696 |
| NTGA2_R | 0.7442 | 0.0162 | 0.0054 | 0.0049 | 0.2002 | 0.0404 |
| NTGA2_C_R | **0.7481** | 0.0064 | **0.0037** | 0.0021 | **0.483** | 0.0864 |
| NTGA2_L_R | **0.7473** | 0.0122 | 0.0043 | 0.0035 | 0.3421 | 0.0581 |
| *p-value* | 0.2546 | == | 0.0021 | ++ | 0.0067 | ++ |

It's worth noticing that *Purity* values presented in two tables (Tab. 2 and Tab. 3) are not the same. The reason is that *Purity* measures how the resulted PFAs of a given method dominate other compared methods. As in Tab. 2, only NTGA2 variants are compared; in Tab. 3 in comparison also used stat-of-the-art methods, the values are not the same.

The experimental results presented in this section show that the NTGA2 method could be effectively specialised if it works in tandem with a specialized mutation and crossover operator. Using specialised operators has the extra computational cost, but it could be profitable if a *decision maker* needs a high-quality decision. Moreover, NTGA2, as metaheuristics in its nature, could be easily parallelized, which reduces the time to get a final decision.

## 6   Summary and Future Work

In this paper, we considered the impact of specialized crossover and mutation operators in a MOO. We evaluated two crossover operators and one mutation operator dedicated to the MS-RCPSP problem and investigated their effectiveness and efficiency. Based on the obtained results, we observed that individual operators dedicated to specific criteria indeed improve the method's effectiveness in the expected direction. The basic combination of crossover and mutation operators additionally improves these results. However, it should be noted that this is at the expense of the method's efficiency - handling specialized logic increases the execution time of operators. These results are consistent with the literature, where defining specialized operators regularly improves method effectiveness. Still, the impact on efficiency varies - in some cases, execution time is also increased, while in others, simple mechanisms do not affect efficiency.

What is more, some of the methods in the literature combine multiple operators [8] [4] and use computation-heavy operators sparingly [3], which pushes exploration from the local optima, yet does not contribute to the general computation time. In the presented Pareto front graphs, it can be observed that each operator affects the results in different areas and different ways depending on the instance. The multi-operator approach, together with the process of operator selection and composition is a very promising direction that requires further research. More advanced approaches are also applicable, which adaptively [5] [17] [15] decide which operators should be selected at a particular stage of optimization, even learning the selection of operators using the Q-learning algorithm [11]. In our future work, we want to focus on the method of combining and adaptively selecting specialized operators. This includes how to measure their effectiveness, which is a significant challenge in MOO, and whether this effectiveness should be measured globally or locally for a specific region of the PFA.

## References

1. imopse library. https://imopse.ii.pwr.edu.pl
2. Cicirello, V.A.: A survey and analysis of evolutionary operators for permutations (2023). https://doi.org/10.5220/0012204900003595
3. Contreras-Bolton, C., Gatica, G., Barra, C.R., Parada, V.: A multi-operator genetic algorithm for the generalized minimum spanning tree problem. Expert Syst. Appl. **50**, 1–8 (2016)

4. Contreras-Bolton, C., Parada, V.: Automatic combination of operators in a genetic algorithm to solve the traveling salesman problem. PLoS ONE **10**(9), e0137724 (2015)

5. Dong, L., Lin, Q., Zhou, Y., Jiang, J.: Adaptive operator selection with test-and-apply structure for decomposition-based multi-objective optimization. Swarm Evol. Comput. **68**, 101013 (2022)

6. Fernandez, B., Fanggidae, A., Pandie, E., Mauko, A.: Travelling salesman problem: greedy single point crossover in ordinal representation. In: Journal of Physics: Conference Series, vol. 2017, pp. 012012. IOP Publishing (2021)

7. Gonçalves, I., Gomes, Á., Antunes, C.H.: Optimizing residential energy resources with an improved multi-objective genetic algorithm based on greedy mutations. In: Proceedings of the Genetic and Evolutionary Computation Conference, pp. 1246–1253 (2018)

8. Gutiérrez-Aguirre, P., Contreras-Bolton, C.: A multioperator genetic algorithm for the traveling salesman problem with job-times. Expert Syst. Appl. **240**, 122472 (2024)

9. He, X., Pan, Q.k., Gao, L., Wang, L., Suganthan, P.N.: A greedy cooperative co-evolution ary algorithm with problem-specific knowledge for multi-objective flow-shop group scheduling problems. IEEE Trans. Evol. Comput. **27**(3), 430–444 (2021)

10. Julstrom, B.A.: Very greedy crossover in a genetic algorithm for the traveling salesman problem. In: Proceedings of the 1995 ACM Symposium on Applied Computing, pp. 324–328 (1995)

11. Karimi-Mamaghan, M., Mohammadi, M., Pasdeloup, B., Meyer, P.: Learning to select operators in meta-heuristics: an integration of q-learning into the iterated greedy algorithm for the permutation flowshop scheduling problem. Eur. J. Oper. Res. **304**(3), 1296–1330 (2023)

12. Moniz, D., Pedro, J., Horta, N., Pires, J.: Multi-objective framework for cost-effective OTN switch placement using NSGA-II with embedded domain knowledge. Appl. Soft Comput. **83**, 105608 (2019)

13. Myszkowski, P.B., Laszczyk, M.: Diversity based selection for many-objective evolutionary optimisation problems with constraints. Inf. Sci. **546**, 665–700 (2021)

14. Myszkowski, P.B., Skowroński, M.: Specialized genetic operators for multi-skill resource-constrained project scheduling problem. In: 19th International Conference on Soft Computing MENDEL 2013, Brno, Czech Republic, pp. 57–62 (2013)

15. Santana, R., Shakya, S.: Evolutionary approaches with adaptive operators for the bi-objective TTP. In: 2022 IEEE Symposium Series on Computational Intelligence (SSCI), pp. 1202–1209. IEEE (2022)

16. Tian, Y., Xiong, T., Liu, Z., Mei, Y., Wan, L.: Multi-objective multi-skill resource-constrained project scheduling problem with skill switches: model and evolutionary approaches. Comput. Ind. Eng. **167**, 107897 (2022)

17. Zhang, L., Wang, L., Zheng, D.Z.: An adaptive genetic algorithm with multiple operators for flowshop scheduling. Int. J. Adv. Manuf. Technol. **27**, 580–587 (2006)

18. Zhou, Z., Li, F., Zhu, H., Xie, H., Abawajy, J.H., Chowdhury, M.U.: An improved genetic algorithm using greedy strategy toward task scheduling optimization in cloud environments. Neural Comput. Appl. **32**, 1531–1541 (2020)

# Maximum Entropy Model of Synonym Selection in Post-editing Machine Translation into Kazakh Language

Assem Shormakova[✉] [iD] and Ualsher Tukeyev[iD]

Al-Farabi Kazakh National University, Almaty, Kazakhstan
shormakovaassem@gmail.com

**Abstract.** The work presents a model, algorithm, and experimental studies for selecting synonyms of incorrectly translated words from English into Kazakh within the framework of machine translation post-editing technology. As a model for choosing synonyms, a maximum entropy model has been developed, the distinctive feature of which is the consideration of contextual words located at any distance from the translated word in a sentence (non-consecutive collocations), which takes into account the peculiarities of the Kazakh language. A feature of the proposed solution algorithm for this model is the use of the semantic cube model proposed by the authors. The developed maximum entropy model was learned on the 250 0000 sentences parallel Kazakh-English corpus, and a test set containing 25,000 sentences was conducted. Experiments on post-editing of machine translation of the Kazakh language compared with machine translation of Google Translate showed an improvement in the BLEU metric by 6 positions.

**Keywords:** Maximum entropy model · synonym selection · post-editing · machine translation · Kazakh

## 1 Introduction

The main advantage of machine translation (MA) after the post-editing process is increased productivity. It's a solution that helps customers produce content in multiple languages faster and more cost-effectively. A machine translation tool edited by a professional linguist is very useful for large-scale projects. However, it should be remembered that machine translation is still not equal to human translation [1, 2]. MT post-editing is a process of improving the translation quality of a text translated from one language to another. However, the quality of the translation depends on many parameters, such as speed and time, so it is difficult to achieve a professional translation immediately. To facilitate the work, translators (users) are recommended to use computer-aided translation (CAT) tools that help ensure the speed of quality materials [3]. The newest generation of MT is Neural Machine Translation (NMT). In many scientific works, NMT began to produce professional translations well, taking into account the quality and accuracy of translation results. In addition, neural MT is a good tool for faster translation of

© The Author(s), under exclusive license to Springer Nature Switzerland AG 2024
N.-T. Nguyen et al. (Eds.): ICCCI 2024, CCIS 2166, pp. 111–123, 2024.
https://doi.org/10.1007/978-3-031-70259-4_9

large texts. However, correcting errors in neural MT results also requires the help of a professional translator. This process is called post-edited MT (MTPE- Machine Translation of Post-editing). This type of service is usually provided by linguists, editors, and proofreaders who are well-versed in the required language. Many works have used post-editing approaches based on neural MA. However, during testing, problems may occur if proper nouns of different terminologies (idioms in the language) are encountered in the database. Different aspects are taken into account before applying neural MA: the clarity of the text to be translated; and working with certain specific sectors (legal, medical, etc.). Moreover, this approach requires very large corpora [4].

## 2  Related works

Every year at the MT WMT conference, the task of automatic post-editing is discussed [5]. Jacob Mundt [6] in his article "Learning to Automatically Post-Edit Dropped Words in MT" found that automatic post-editing (APE) was able to improve the accuracy of MT results through the re-entry process. But it is very important to consider the place of entering words. In this paper, a probabilistic approach to learning re-entry rules for specific languages and MT systems is proposed. The method of synthesis of ready data from translations was also described. Validated input logic for Chinese-to-English and Arabic-to-English MT systems. As a result, the adequacy index between three words in the APE system adapted by the neural MT approach reached 73% in the Arabic-English MT text, and 67% in the Chinese-English MT text. Compared to a rule-based input approach, an improved output in terms of automated sufficiency metrics was presented. Specific aspects of the NMT problem and the effectiveness of its application to machine learning solutions are considered.

Also, Santanu Pal et al. [7] proposed an automatic post-editing (APE) system based on a neural network to improve the result of MA. Their APE (NNAPE) neural model was based on a bidirectional recurrent neural network (RNN) model and consisted of an encoder that encodes the result of the MT into a fixed-length vector. Its decoder was used for post-edited full-sentence translation.

In the article "Log-linear Combinations of Monolingual and Bilingual Neural Machine Translation Models for Automatic Post-Editing" Marcin Junczys-Dowmunt [8] a group of scientists from Adam Mickiewicz University about the work of automatic post-editing described. They investigated the use of neural translation models. The results of the study stated that the evaluation metrics achieved better results than the baseline with 3.2% TER and + 5.5% BLEU, better than any other system.

Another concept closely related to the integration of MT and IS is the development of interactive MT systems. Torregrosa et al. [9] believed that an approach to interactive MT allows for the almost continuous addition of new sources of bilingual information.

Chris Hokamp et al. [10] in their work "Ensembling Factored Neural Machine Translation Models for Automatic Post-Editing and Quality Estimation" used the tools of specialized neural MT for automatic post-editing (APE) and proposed a new approach to word quality evaluation (QE).

Negri et al. [11] seamlessly integrated systems using the OL [online learning] process into APE systems to produce NMT output. For neural MA, the utility of the proposed systems was highlighted to eliminate the need to fine-tune or retrain MT systems from scratch.

In 2018, the APE task consisted of two automatic post-editing subtasks, one for Phrase-Based Statistical MT (PBSMT) and the other for Neural Machine Translation (NMT) for English-German and other languages. Chatterjee et al. (Chatterjee et al. 2018) [12] research team conducted this MT test and focused on two important aspects of any APE system: fast learning and review of source text.

The work of Negri's group [13] focuses on the development of neural systems operating in networks for continuous user development through stepwise learning. This includes quite a bit of other work on neural network-based post-editing approaches.

Antonio's work [14] studied user-requested translations using a non-monotonic automatic post-editing approach. The resulting system from their approach was compared to a system trained in left-to-right and random order during post-editing. In addition to the proposed approach, a Transformer-based model pre-trained using the BERT tool, which learns to process automatic post-editing of translations, is also proposed.

An approach pioneered by Félix's group [15] compared post-editing based on four editing tasks to translation with an autocomplete function.

The MT collection shows the results of many international tasks related to APE. For example, Chatterjee et al. [16] presented the results of the sixth round of the WMT APE competition. Here, participants attempted to use APE to correct the output of a black-box MT system by learning user corrections within different sentences.

In another case, Yang et al. [17] tried to show that the method can be improved by increasing the performance of MT when tuning ready-made NMT models with a small amount of APE corpus.

Sharma et al. [18] adapted neural machine translation for automatic post-editing. Automatic post-editing patterns were used to correct the output of the MT system by learning the user's post-editing patterns.

Many articles describe the work of English-German and other language pairs, not English-Kazakh. Post-editing work for the English-Kazakh pair was not considered until 2021. Rakhimova's scientific group considers machine learning in the article "The Development of the Light Post-editing Module for English-Kazakh Translation" [19].

## 3   Method

### 3.1   Semantic Cube Model and Algorithm Used to Correct Wrong Words

The article [20] used the "semantic cube" method for the realization of the maximum entropy model. Our contribution to the current paper is the use of non-consecutive collocations of the considered sentence and applying the maximum entropy model with non-consecutive collocations for the optimal choice of synonym for the wrong translated word in Kazakh.

The maximum entropy classification formula is used:

$$\hat{c} = \underset{c \in C}{\arg \max}\, P(c|x) \tag{3.1}$$

In the maximum entropy model [21, 22], the probability of a certain class c is estimated:

$$P(c|x) = \frac{1}{Z} \exp \sum_j \lambda_j f_j \qquad (3.2)$$

Here Z is the normalizing factor.

The equation that calculates the probability of given x in class c in maximum entropy model (Eq. 3.3):

$$P_i^e(c|x) = \frac{\exp(\sum_{j=1}^{N^e} \lambda_{ij}^e f_{ij}^e(c,x))}{\sum_{c' \in S^e} \exp(\sum_{j=1}^{N} \lambda_{ij}^e f_{ij}^e(c',x))} \qquad (3.3)$$

Here,

$$f_{ij}^e(c,x) = \begin{cases} 1, & \text{if } x = z_j^e \ \& \ c = s_i^e \ (x \text{ is a non - consecutive colleaction of } c) \\ 0, & \text{other case} \end{cases}$$

$\omega^e$ - polysemous word, c- class of synonyms, $z_j^e$ - jth colocation word of class c ($s_i^e$), x – observed word, $N^e$ - number of features for $\omega^e$, $\lambda_{ij}^e$ - the weight of the feature $f_{ij}^e$, $S^e$ - set of synonyms for $\omega^e$.

**Table 1.** $\omega^e$ word frequency table $\omega^l$

| $\omega^e$ | $z_1^e$ | $z_2^e$ | $z_3^e$ | $z_4^e$ | $z_5^e$ | ... | $z_N^e$ |
|---|---|---|---|---|---|---|---|
| 1 | 2 | 3 | 4 | 5 | 6 | ... | N |
| $s_1 f_{1j}$ | 0 | 1 | 0 | 1 | 0 | ... | 0 |
| $s_1 g_{1j}$ | 0 | 5 | 0 | 6 | 0 | | 0 |
| $s_2 f_{2j}$ | 1 | 0 | 0 | 0 | 1 | ... | 0 |
| $s_2 g_{2j}$ | 1 | 0 | 0 | 0 | 6 | | 0 |
| | | | | | | ... | |

Here $g_{ij}$ – frequency $z_j^e$ for $s_i^e$. For this, the data in Table 1 was used:

$$\lambda_2 = \frac{g_{12} f_{12}}{\sum_{j=1}^{N^e} g_{1j} f_{1j}} = 5/11 = 0,45$$

$$\lambda_{14} = \frac{s_1 f_4}{\sum_{j=1}^{N^e} g_{1j} f_{1j}} = 6/11 = 0,54$$

here, $g_{ij}$ - $z_j$ frequency for $s_i$;

the calculation of weights $\lambda^e$ for $s_2{}^e$ is as follows:

$$\lambda_{21} = \frac{g_{21}f_{21}}{\sum\limits_{j=1}^{N^e} g_{2j}f_{2j}} = 1/7 = 0,14$$

$$\lambda_{25} = \frac{g_{25}f_{25}}{\sum\limits_{j=1}^{N^e} g_{2j}f_{2j}} = 6/7 = 0,85$$

The full result can be seen in Table 2.

**Table 2.** The semantic cube table for the word $w^e$

| $\omega^e$ | $z_1$ | $z_2$ | $z_3$ | $z_4$ | $z_5$ | ... | $z_N^e$ |
|---|---|---|---|---|---|---|---|
| 1 | 2 | 3 | 4 | 5 | 6 | ... | N |
| $s_1$ $f_{1j}$ | 0 | 1 | 0 | 1 | 0 | ... | 0 |
| $s_1$ $\lambda_{1j}$ | 0 | 0,45 | 0 | 0,54 | 0 | | 0 |
| $s_2$ $f_{2j}$ | 1 | 0 | 0 | 0 | 1 | ... | 0 |
| $s_2$ $\lambda_{2j}$ | 0,14 | 0 | 0 | 0 | 0,85 | | 0 |
| | | | ... | | | | |

Calculation of the probability of the class of synonyms according to the formula (3.3) is as follows:

$$P(s_1|x) = \frac{e^{0,45} * e^{0,54}}{e^{0,45} * e^{0,54} + e^{0,14} * e^{0,85}} = \frac{0,243}{0,243 + 0,119} = \frac{0,243}{0,362} \approx 0.67$$

$$P(s_2|x) = \frac{e^{0,14} * e^{0,85}}{e^{0,45} * e^{0,54} + e^{0,14} * e^{0,85}} = \frac{0,119}{0,243 + 0,119} = \frac{0,119}{0,362} \approx 0.33$$

Then the classification formula (formula 3.1) is used. That is, by maximum $P(s_i|x)$ the synonym class $s_i$ is selected. For example, the chosen synonym is $s_1$, because, $P(s_1|x)$ value is the largest (maximum).

**Algorithm. Algorithm for choosing a high-probability synonym for a wrongly translated Kazakh word**

The description of the algorithm, which is implemented by selecting a word with a high probability related to the text after identifying the wrongly translated word, is as follows:

1. Input data: list of synonyms of wrongly translated Kazakh words, the corpus of English-Kazakh pairs.

2. Synonyms in the Kazakh language are the formation of the semantic cube. (Teaching the corpus in the Kazakh language). Constructing a semantic cube of synonym probabilities for mistranslated words.

3. $\omega_j^{каз}$ choose the highest possible (max) synonym of the wrongly translated Kazakh word.

4. $k_l$ the probability of the incorrectly translated Kazakh word in the sentence is to replace it with the synonym value max.

5. Produce a post-edited sentence.

6. End.

As a result of this third task, by using the semantic cube method, a correct full sentence was found the correct complete sentence was found and the closest version of the wrongly translated word was found in terms of meaning.

The basic post-editing stage includes the following steps:

• First of all, it is necessary to find the wrongly translated word from English to Kazakh.
• Searching and creating a catalog of synonyms in the second place; if a mistranslated word is found, the word is searched in the catalog to find synonyms.
• In the third step, the system uses the semantic cube method, selects a suitable synonym, and updates the mistranslated word with this appropriate synonym.

The post-editing technology of the proposed scientific work has implemented the above-mentioned steps and has shown good results. Information about this is discussed in more detail in the next section on various calculation indicators and quality assessment indicators.

### 3.2  An Example of Choosing a High-Probability Synonym for a Wrongly Translated Kazakh Word

Practical calculations of the application of the semantic cube method based on maximum entropy can be seen in the following tables and descriptions [64–66]. Table 3 shows the wrongly translated word with synonyms in English and Kazakh languages.

**Table 3.** The example of the wrong translation of the word *"baby"* with synonyms in English and Kazakh languages

| $\omega^e$, wrong translated word | Synonym1 | Synonym2 | Synonym3 | Synonym4 . | Synonym5 |
|---|---|---|---|---|---|
| baby | сәби (a baby) | ергежейлі (a dwarf) | кішкентай (little) | кіші (junior) | шағын (small) |

This list consists of translations based on parallel English-Kazakh words. Words taken from the context, wrongly translated words (synonyms), and their frequencies are recorded in the appropriate table. Possible translations are defined using Table 4.

Context is a sentence introduced at the beginning. It uses the frequency of words found in the corpus and found in context and synonyms that have been mistranslated.

**Table 4.** Variants of Kazakh *"baby" translation in context*

| z, from the corpus, already taken words | $\omega^e$ and $S^e$, wrong translated say a word and synonyms | g, their frequencies |
|---|---|---|
| Мен (I) | сәби (a baby) | 201 |
| Көйлек (shirt) | ергежейлі (a dwarf) | 20 |
| Мінез (character) | кішкентай (little) | 189 |
| ... | ... | ... |
| Бойжеткен (a girl) | кіші (junior) | 31 |

The list of lemmas of mistranslated words in both languages was taken from sentences found in parallel English-Kazakh corpora and stored in memory. After updating the catalog of wrong words, the frequency of use was calculated for the entire parallel corpus. To calculate the frequency of mistranslated words in the corpus, the intersection of English and Kazakh sentences is considered. If only one side (direction, English, or Kazakh) is used, the necessary words may not be found. The wrongly translated words in the Kazakh and English corpora may not appear in either Kazakh or English sentences. If only one version is taken, then the meaning of the sentence may be different. Then, using the semantic cube method, tables consisting of incorrect synonyms are automatically created from the Kazakh language versions of the words.

As explained earlier, the wrongly translated words are found in the target sentence. Then the alternatives (synonyms) of these words are searched in the catalog. If the catalog contains this mistranslated word, the table associated with that line is opened and the probability of occurrence of only the words in the input sentence is calculated. If the necessary wrong word is not found in the catalog, it goes to the algorithm of adding new wrongly translated words to the new list of synonyms (catalog). Thus, the catalog will continue to grow. The next consideration is the semantic cube creation algorithm, a probabilistic model and a corpus were used to select the correct alternative in the target sentence. The volume of the parallel (English-Kazakh) corpus is 250,000 sentences. To improve the application of this semantic cube method, the following steps were taken:

First, find the lemmas of all the sentences in the corpus.

Second, synonyms were used for each mistranslated Kazakh word from the catalog of mistranslated synonyms for polysemous words, and a table was compiled with all its synonyms and the frequency of occurrence of these synonyms in the context (from the corpus).

A semantic cube is created for each mistranslated synonym. The main steps of the semantic cube creation algorithm are as follows.

First, mistranslated words previously found in the English corpus were searched together with their synonyms. Kazakh translations were collected using English synonyms. These synonyms were searched from a parallel corpus of 250,000 sentences. The lemmas of the words were found by Apertium. The probability of occurrence of synonyms in the context was calculated. Probabilities of their contexts (in the corpus) were calculated for the required words in the tables of identified wrong synonyms. For this, the number of rows (number) of synonyms found in the catalog was equaled to the number of the table associated with these synonyms (Table 5). The table shows the frequency of occurrence of all words found in the corpus and incorrectly translated words.

**Table 5.** Part of the frequency table of the mistranslated word *"baby" with synonyms and words found in the corpus*

| $z$, words in the corpus | $\omega^e$ and $S^e$, mistranslated words and synonyms |
| --- | --- |
| тіл (language) | baby: 42 |
| үзінді (excerpt) | baby: 1 |
| мен (i) | baby: 201 |
| үст (on) | baby: 38 |
| ойын (game) | baby: 1 |
| Кеш (late) | baby: 84 |
| ана (mother) | baby: 113 |
| мен(I) | dwarf: 12 |
| бой (height) | Junior: 29 |
| із (trace) | Junior: 19 |
| біз (we) | junior: 45 |
| дым(wet) | junior: 63 |
| Ал (take) | junior: 65 |
| оқырман (reader) | junior: 5 |
| бір (one) | child: 16 |
| Түсін (understand) | child : 3 |
| Берә (since ) | child: 5 |
| Не (what) | child: 37 |
| Бер (give ) | child: 22 |
| Үй (house) | child: 26 |
| Ай (moon) | child: 61 |
| ... | ... |

If this wrong translation is in the catalog, the system takes the necessary alternatives from the catalog, chooses the correct version of the wrongly translated words, and produces a corrected sentence. If it is not in the catalog, the catalog is supplemented, and the system creates a table of synonyms (catalog) and a table of the frequency of occurrence of those synonyms in the corpus. Only the parameters of the word *"baby"* in the original sentence *I have a nice baby* are shown. That is, the update sequence of the semantic cube is explained and described, showing the wrongly translated word given as an example.

From the frequency Table 6, we get only the synonyms of the words that appear in the target sentence *T ( мен, сүйкімді, бар )*. As mentioned earlier, only the corpus frequencies of words and mistranslated synonyms in a given sentence are used. Details can be found in Table 6.

**Table 6.** Frequency table of words appearing in the target sentence, the mistranslated word *"baby"* and its synonyms

| $\omega^e$ and $S^e$ $z$, context last word | сәби (a baby) | Ергежейлі (a dwarf) | Кішкентай (little) | Кіші (junior) | Бала (child) |
|---|---|---|---|---|---|
| Мен(I) | 201 | 12 | 104 | 153 | 145 |
| сүйкімді (cute) | 258 | 15 | 40 | 136 | 23 |
| бар(have) | 233 | 136 | 199 | 144 | 89 |

*My cute boy* is there is calculated using the formulas (5.1) and (5.2): 250000 sentences are used and Kazakh sentences from the bilingual corpus are used. That is, the synonym class $P(s_i|x)$ with the maximum value $s_i$ is selected. Therefore, the probability of the identified wrong word and its synonyms was calculated.

$$babysynonym : P(s_1|x) = (201 + 258 + 233)/250000 = 0.0027$$

$$dwarf\,synonym : P(s_2|x) = (12 + 15 + 136)/250000 == 0.00065$$

$$smallsynonym : P(s_3|x) = (104 + 40 + 199)/250000 == 0.001372$$

$$minorsynonym : P(s_4|x) = (153 + 136 + 144)/250000 == 0.001732$$

Probability of the synonym *child*: $P(s_5|x) = (145 + 23 + 89)/250000 == 0.001028$

(3.3) As shown in the formula, the highest value was selected from among the found probabilities, and $P(s_1|x)$ the synonym of the wrongly translated word "baby" with the maximum value was selected.

only the words in $T$ target sentences is that the tables may contain many words from the corpus. That is, the total frequency of words is not important, only the frequency of words and synonyms in a given sentence is important. Table 6 shows only wrong words with equivalents related to the entered initial sentence.

Finally, a short list of corrected sentences as a result of the proposed post-editing technology is presented in Table 7.

**Table 7.** List of sentences corrected as a result of post-editing technology

| Source sentence | Sentence translated by Google Translate | A sentence produced using PE-LC technology | Synonyms probability index of mistranslated words |
|---|---|---|---|
| You are the most *beautiful* woman I have ever seen in my life. | Сіз менің өмірімде көрген ең *әдемі* әйелсіз. (You are the most beautiful woman I have ever seen.) | Сіз менің өмірімде көрген ең *сұлу* әйелсіз. (You are the most beautiful woman I have ever seen.) | $P(s_1\|x)$=0.00132 (nice) $P(s_2\|x)$=0.00471 (beautiful) $P(s_3\|x)$=0.00087 (attractive) $P(s_4\|x)$=0.00014 (beautiful) $P(s_5\|x)$=0.0014 (elegant) $P(s_6\|x)$=0.0008 (cute) $P(s_7\|x)$=0.00017 (slender) |
| Algorithms and *data* structures are central to computer science. | Алгоритмдер мен *деректер* құрылымдары информатика үшін орталық болып табылады. (Algorithms and data structures are central to computer science.) | Алгоритмдер мен *мәліметтер* құрылымдары информатика үшін орталық болып табылады. (Algorithms and data structures are central to computer science.) | $P(s_1\|x)$=0.00198 (data) $P(s_2\|x)$= 0.00243 (data) $P(s_3\|x)$=0.00187 (material) $P(s_4\|x)$=0.00063 (info) |
| This year has become a year of comprehensive transformations and *genuine* renewal. | Биылғы жыл жан-жақты қайта құрулар мен *шынайы* жаңару жылы болды. (This year was a year of comprehensive reconstructions and true renewal.) | Биылғы жыл жан-жақты түрлендірулер мен *нақты* жаңару жылы болды. (This year was a year of comprehensive transformations and real renewal.) | $P(s_1\|x) = 0.00143$ (true) $P(s_2\|x) = 0.00281$ (exact) $P(s_3\|x) = 0.00017$ (legal) |

## 4 Experiment

To evaluate the proposed post-editing technology, the learning of the Maximum entropy model was made on the 250 0000 sentences parallel Kazakh-English corpus, and a test set containing 25,000 sentences was conducted. Since the goal of the work is to translate the text from English to Kazakh, only English sentences were used as input sentences. The sentences were taken from various English-language news portals, grammar sites, and literary sources. Sentences were collected from sites about fairy tales, government, news, statistics, history, and law. Various types of sentences have been added to complete the catalog in identifying suitable alternatives. These are not ready-made bilingual sources, as it is difficult to obtain translations of any sentence with correct clear references. Therefore, only English sentences were taken and wrongly translated words were found using the proposed method.

The content of the test included news from different fields, different sentences with synonyms, and fairy tales to calculate WER (Word Error Rate) [23], TER (Translation Error Rate) [24] and BLEU [25] (4.1 - table). The texts used during testing can be viewed at this link. https://github.com/assem7shormak/Data-set.

Table 8 shows the comparison of the proposed post-editing model and Google Translate for 100 sentences. In these experiments, 100 English sentences translated by Google Translate into Kazakh. Received Kazakh sentences processed by the proposed post-editing model.

**Table 8.** The comparison of Google Translate and the proposed post-editing model for 100 sentences.

| BLEU ($\mu \pm 95\%$ CI) | TER ($\mu \pm 95\%$ CI) | chrF2 ($\mu \pm 95\%$ CI) |
|---|---|---|
| Google Translate - 41.0103 (40.9405 $\pm$ 6.0153) | Google Translate - 40.6349 (40.6292 $\pm$ 4.9062) | Google Translate - 70.3059 (70.2557 $\pm$ 3.4295) |
| Proposed post-editing model- 47.4809 (47.4171 $\pm$ 5.8691) | Proposed post-editing model - 35.9788 (35.9826 $\pm$ 4.8104) | Proposed post-editing model -75.6458 (75.6220 $\pm$ 2.9849) |

## 5   Conclusion and future works

According to the experimental results, BLEU's quality index of post-editing technology with the Maximum Entropy Model of synonym selection in post-editing machine translation into Kazakh language showed +6.47% better results than Google's MT. The quality indicators of the proposed post-editing technology showed a lower TER of − 4.66% and WER of −4.28% than Google MT, which means that the proposed technology has obtained relatively better results. In the future, it is planned to conduct other research using the proposed technology.

**Acknowledgments.** This research was performed and financed by the grant Project IRN AP 19677835 of the Ministry of Science and Higher Education of the Republic of Kazakhstan.

# References

1. Koponen, M.: Machine translation post-editing and effort: empirical studies on the post-editing process. Doctoral thesis, University of Helsinki, Helsinki, Finland (2016). http://hdl.handle.net/10138/160256. Accessed 18 Feb 2022
2. Shormakova, A.N., Tukeyev, U.A.: Machine translation technology with teaching English into Kazakh. In: Materials of the International Conference of Students and Young Scientists "The World of Science", 23–26 April 2012, p. 154. Almaty: Kazakh University (2012)
3. Papineni, K., Roukos, S., Ward, T., Henderson, J., Reeder, F.: Corpus-based comprehensive and diagnostic MT evaluation: initial Arabic, Chinese, French, and Spanish results. In: Proceedings of Human Language Technology 2002, San Diego, CA (2002)
4. Barrault, L., et al.: Findings of the 2019 conference on machine translation (WMT19). In: Proceedings of the Fourth Conference on Machine Translation (Volume 2: Shared Task Papers, Day 1), pp. 1–61. Association for Computational Linguistics, Florence, Italy (2019). https://doi.org/10.18653/v1/W19-5301
5. Carmo, F., et al.: A review of the state-of-the-art in automatic post-editing. Mach. Transl. **35**, 101–143 (2021). https://doi.org/10.1007/s10590-020-09252-y.18.02.22
6. Mundt, J.: Learning to automatically post-edit dropped words in MT. In: Association for Machine Translation in the Americas (AMTA 2012), Columbia San Diego, California, USA (2012). https://aclanthology.org/2012.amta-wptp.5/. Accessed 18 Feb 2022
7. Pal, S., Naskar, S.K., Vela, M., Genabith, J.: A neural network-based approach to automatic post-editing. In: Proceedings of the 54th Annual Meeting of the Association for Computational Linguistics, (Berlin, Germany, 7–12 August 2016, pp. 281–286 (2016)
8. Junczys-Dowmunt, M., Grundkiewicz, R.: Log-linear combinations of monolingual and bilingual neural machine translation models for automatic post-editing. In: Proceedings of the First Conference on Machine Translation, pp. 751–758. Association for Computational Linguistics, Berlin, Germany (2016). http://www.aclweb.org/anthology/W16-2378. Accessed 18 Feb 2022
9. Pérez-Ortiz, J.A., Torregrosa, D., Forcada, M.: Black-box integration of heterogeneous bilingual resources into an interactive translation system. In: EACL 2014 Workshop on Humans and Computer-assisted Translation (Gothenburg, 26 April 2014)
10. Hokamp, C.: Ensembling factored neural machine translation models for automatic post-editing and quality estimation. Computation and Language (cs.CL), arXiv preprint arXiv: 1706.05083 (2017). Accessed 18 Feb 2022
11. Negri, M., Turchi, M., Bertoldi, N., Federico, M.: Online neural automatic post-editing for neural machine translation. In: Cabrio, E., Mazzei, A., Tamburini, F. (eds.) Proceedings of the Fifth Italian Conference on Computational Linguistics (CLiC-it 2018), Torino, Italy, 10–12 December 2018. CEURWS.org, CEUR Workshop Proceedings, vol. 2253 (2018). http://ceur-ws.org/Vol-2253/paper63.pdf. Accessed 18 Feb 2022
12. Chatterjee, R., Negri, M., Rubino, R., Turchi, M.: Findings of the WMT 2018 shared task on automatic post-editing. In: Proceedings of the Third Conference on Machine Translation, Volume 2: Shared Task Papers, Brussels, Belgium. Association for Computational Linguistics (2018)
13. Negri, M., Turchi, M., Chatterjee, R., Bertoldi, N.: eSCAPE: a large-scale synthetic corpus for automatic post-editing. In: Proceedings of the Eleventh International Conference on Language Resources and Evaluation (LREC 2018), Miyazaki, Japan, 7–12 May 2018. CoRR abs/1803.07274 (2018)
14. Gois, A., Cho, K., Martins, A.: Learning non-monotonic automatic post-editing of translations from human orderings. In: Proceedings of the 22nd Annual Conference of the European Association for Machine Translation (EAMT 2020), pp. 205–214, Lisboa, Portugal, November 2020

15. do Carmo, Félix., et al.: A review of the state-of-the-art in automatic post-editing. Mach. Transl. **35**(2), 101–143 (2020). https://doi.org/10.1007/s10590-020-09252-y

16. Chatterjee, R., Freitag, M., Negri, M., Turchi, M.: Findings of the WMT 2020 shared task on automatic post-editing. In: Proceedings of the WMT 2020 Automatic Post-Editing Shared Task (2020). https://www.statmt.org/wmt20/pdf/2020.wmt-1.75.pdf. Accessed 18 Feb 2022

17. Yang, H., et al.: HW-TSC's participation at WMT 2020 automatic post editing shared task. In: Proceedings of the WMT 2020 Automatic Post Editing Shared Task (2020). https://aclanthology.org/2020.wmt-1.85.pdf. Accessed 18 Feb 2022

18. Sharma, A., Gupta, P., Nelakanti, A.: Adapting neural machine translation for automatic post-editing. In: Proceedings of the WMT 2020 Automatic Post Editing Shared Task (2021). https://aclanthology.org/2021.wmt-1.35.pdf. Accessed 18 Feb 2022

19. Rakhimova, D., Karyukin, V., Karibayeva, A., Turarbek, A., Turganbayeva, A.: The development of the light post-editing module for English-Kazakh translation. In: Proceeding: The 7th International Conference on Engineering and MIS 2021, pp. 1–5 (2021)

20. Tukeyev, U., Amirova, D., Karibayeva, A., Sundetova, A., Abduali, B.: Combined technology of lexical selection in rule-based machine translation. In: Nguyen, N., Papadopoulos, G., Jędrzejowicz, P., Trawiński, B., Vossen, G. (eds.) ICCCI 2017, Part II. LNCS (LNAI), vol. 10449, pp. 491–500. Springer, Cham (2017). https://doi.org/10.1007/978-3-319-67077-5_47

21. Berger, A.L., Pietra, S.A.D., Pietra, V.J.D.: A maximum entropy approach to natural language processing. Comput. Linguist. **22**, 39–71 (1996)

22. Jurafsky, D., Martin, J.H.: Automatic Speech Recognition. Speech and Language Processing: An Introduction to Natural Language Processing, Computational Linguistics, and Speech Recognition, pp. 213–220. Prentice Hall, Upper Saddle River

23. A Study of Translation Edit Rate with Targeted Human Annotation (TER). https://www.cs.umd.edu/~snover/pub/amta06/ter_amta.pdf. Accessed 18 May 2022

24. Papineni, K., Roukos, S., Ward, T., Zhu, W.J.: BLEU: a method for automatic evaluation of machine translation. In: Proceedings of the 40th Annual Meeting of Association for Computational Linguistics (ACL 2002), pp. 311–318, Association for Computational Linguistics, Philadelphia, Pennsylvania, USA (2002)

25. Post, M.: A call for clarity in reporting BLEU scores. In: Proceedings of the Third Conference on Machine Translation: Research Papers, pp. 186–191. Association for Computational Linguistics, Brussels, Belgium (2018).

# Cross-Domain Abbreviation Disambiguation on Vietnamese Clinical Texts in Online Processing

Chau Vo[✉] [ORCID] and Hua Phung Nguyen[✉] [ORCID]

Ho Chi Minh City University of Technology, Vietnam National University – Ho Chi Minh City, Ho Chi Minh City, Vietnam
{chauvtn,nhphung}@hcmut.edu.vn

**Abstract.** Readability of clinical texts in electronic medical records (EMRs) is more significant when EMRs are shared and required to be understood by both human users and computer programs. This feature is seldom reached successfully in the real world because of noises in clinical texts such as spelling errors, abbreviations, synonyms, and sentence incompleteness. Among noises, abbreviations are ubiquitous in many various short forms for many different long forms. Therefore, abbreviation disambiguation on clinical texts has been well researched worldwide for a long time. Many languages like English, German, Korean, Swedish, etc. have been considered. Recently Vietnamese clinical text analytics has been emerging due to the more popularity of EMRs. However, few works have been dedicated to abbreviation disambiguation on Vietnamese clinical texts. As one of the first works for abbreviation disambiguation on Vietnamese clinical texts, our work aims at an effective novel solution. Different from the existing works, our solution supports a cross-domain context where data shortage and imbalance exist simultaneously in online processing. It defines Nonparametric Self-Training, a parameter-free semisupervised learning algorithm, to get rid of the mismatch between the source and target domains. It also enhances the labeled dataset of one domain by adding more unlabeled data of another domain in favour of data imbalance. As a result, the proposed solution can tackle the task challenges well and outperform the others with both Accuracy and AUC of higher 90% in most experiments on the real Vietnamese clinical texts of 4 different note types in 2 different hospitals. Cleaned clinical texts can be further processed better for more readability and sharability.

**Keywords:** Clinical Text · Abbreviation Disambiguation · Vietnamese Natural Language Processing · Semisupervised Learning · Data Imbalance

## 1 Introduction

Healthcare is always more and more important to everyone and every country in the world. Electronic medical records (EMRs) are thus required more worldwide so that medical services can be better provided. In EMRs, clinical texts belong to a helpful part

N.-T. Nguyen et al. (Eds.): ICCCI 2024, CCIS 2166, pp. 124–140, 2024.
https://doi.org/10.1007/978-3-031-70259-4_10

to give details about many different aspects of patients and other stakeholders. However, they are full of noises such as spelling errors, abbreviations, synonyms, and sentence incompleteness. Among noises, abbreviations are ubiquitous in many various short forms for many different long forms because they are intended to be used for record simplification in a working environment under high presure with time shortage. Such ubiquitiy of abbreviations in clinical texts make EMRs less readable, sharable, even misinterpreted and confused [5, 25], especially as abbreviations are context-dependent [21]. To make the most of EMRs, abbreviation disambiguation is thus one of the important clinical text processing tasks that need to be solved soon.

Indeed, abbreviation disambiguation on clinical texts is part of abbreviation resolution to decide the most relevant sense for each ambiguous abbreviation associated with more than one (full) long form, called sense. This task was handled in many works [1, 2, 6, 9–12, 14–20, 22, 23, 26–28, 31–34]. In most works, English clinical texts were supported. Other languages like French [6], German [22], Korean [14], and Swedish [9, 28] have recently been considered. Except [29], none of the existing works was dedicated to Vietnamese, leading to a necessary solution to the task.

Moreover, among these works, few of them examined the task in a practical real-time context. In [32], the authors took this context into account for English notes from Vanderbilt Medical Center, using Support Vector Machines (SVM) and the 10-fold cross-validation evaluation scheme with Accuracy. Similar to [32] and different from the others, our work focuses on online processing for the task on Vietnamese clinical texts. In contrast to [32], our work experiments with semisupervised learning on the data from two different domains in the real-world cross-domain context of Vietnamese EMRs. This context results in more challenges like data shortage and imbalance which might not have existed with the context in [32] on English EMRs.

Regarding the approaches, we beware of three main groups: statistical in [2, 11, 12, 18, 32], unsupervised [1, 4, 9, 14, 18, 22, 28], and supervised [10, 15, 23, 26, 31, 32, 34]. As studied in [29], supervised learning was the most promising. Furthermore, an approach based on dictionary and rules in [6], GPT-3 edit in [2], unsupervised ensemble learning in [18, 19], and more deep learning in [1, 2, 12, 17, 27] were proposed. Although interesting, these approaches are nontrivially adapted to Vietnamese clinical texts because few labeled Vietnamese data resources exist. Therefore, our work addresses this task with a semisupervised learning approach as tackling cross-domain abbreviation disambiguation on Vietnamese clinical texts in online processing.

As one of the first works for abbreviation disambiguation on Vietnamese clinical texts, our work aims at an effective novel solution to this task as Vietnamese clinical text analytics is emerging with the more popularity of EMRs. Different from the existing works, our solution supports a cross-domain context where data shortage and imbalance exist simultaneously in online processing. It defines Nonparametric Self-Training, a parameter-free semisupervised learning algorithm, to get rid of the mismatch between the source and target domains. It also enhances the labeled dataset of one domain by adding more unlabeled data of another domain in favour of imbalance. Besides, its optimization can ensure the performance improvement or at least maintenance for the base supervised learning algorithm.

Via the experimental results from the real Vietnamese clinical texts of 4 different note types in 2 different hospitals, our solution has been proved effective with both Accuracy and AUC of higher 90% consistently for almost all the cases. Such results confirm our significant contribution to resolving abbreviation disambiguation on Vietnamese clinical texts in a practical online processing cross-domain context. Cleaned clinical texts can be further processed better for readability and sharability.

## 2   A Cross-Domain Abbreviation Disambiguation Task on Vietnamese Clinical Texts in Online Processing

First of all, we reintroduce the basic notions from [29] for more clarity and the readers may read more related details in [10]. An abbreviation is a short form of a word or phrase which is a full form of the abbreviation. Each full form is a sense of an abbreviation. One abbreviation is associated with one or many different full forms. One full form has one or many different abbreviations. Abbreviation disambiguation is to determine a relevant sense for the abbreviation linked to more than one sense.

In [29], the abbreviation disambiguation task on Vietnamese clinical texts was defined. As an initial study, [29] examined the task in batch processing with the cross-validation evaluation scheme when the full context of each abbreviation was processed via its surrounding words on both left and right sides of the abbreviation. In contrast to [29], this work considers the task in online processing so that cleaning clinical texts with abbreviations can be performed on the fly. Abbreviation disambiguation in online processing is abbreviation disambiguation with only the left context of each abbreviation, i.e. surrounding words on the left side. Such online processing is more practical as soon as clinical texts are generated and captured. It's also more challenging with data shortage from the left context of each abbreviation. Figure 1 sketches our task in online processing where the abbreviation is located at the end of the text.

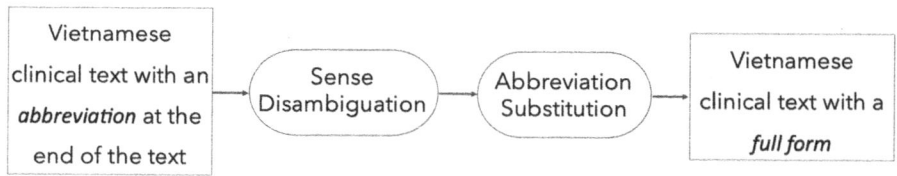

**Fig. 1.** Abbreviation disambiguation in Vietnamese clinical texts in online processing.

Moreover, different from [29], our work focuses on a cross-domain context where there is no available resource for the input note type. This context is realistic for Vietnamese clinical texts when Vietnamese clinical text analytics has been emerging and the resources like UMLS, MIMIC, PubMed, etc. don't exist in Vietnamese. Utilizing the resources in one domain to support the task in another domain is thus necessarily taken into account. In our work, domain means the note type of clinical texts and further the hospital (medical center) where the texts are originally generated. Some typical note types are nursing note, treatment note, and discharge summary. In particular, we formally define a cross-domain abbreviation disambiguation task on Vietnamese clinical texts in online processing as follows.

Let *type*1 and *type*2 be two different note types corresponding to the domains where *type*1 is associated with the available resources and *type*2 is not, $D = \{t_i \mid i = 1..n\}$ be a collection of $n$ clinical notes where $t_i$ is a note composed of unstructured clinical texts, and $A = \{a_j \mid j = 1..m\}$ be a set of $m$ distinct abbreviations. Each abbreviation $a_j$ is associated with a set of $q$ distinct senses: $S_j = \{s_{j,k} \mid k = 1..q\}$ where $j = 1..m$.

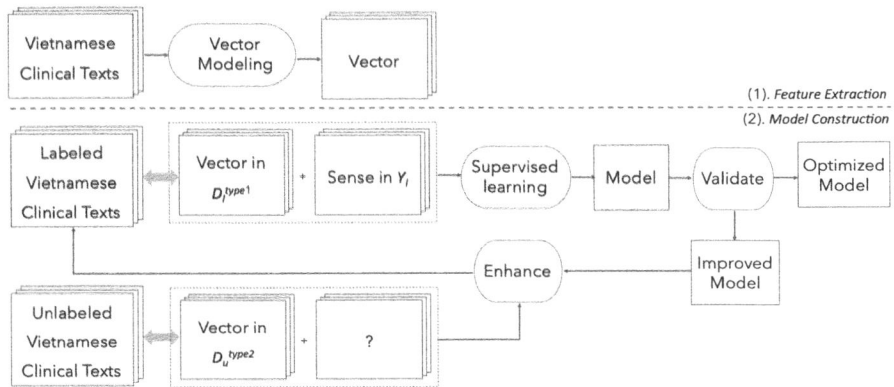

**Fig. 2.** The Proposed Solution

Let $D_l^{type1}$ and $D_u^{type2}$ in D be datasets of *type*1 and *type*2, respectively. As a given resource of *type*1, $D_l^{type1}$ is a labeled dataset of $n1$ labeled texts where each abbreviation is already associated with one corresponding sense. Different from $D_l^{type1}$, $D_u^{type2}$ is an unlabeled dataset of $n2$ new texts where abbreviations need to be disambiguated.

The task is now procedurally written as follows:

```
1. for each note tᵢ in Du^type2
2.     for each abbreviation aⱼ in tᵢ
3.         sⱼ,ₖ ← abbreviation_disambiguation(tᵢ, aⱼ, Sⱼ, Dₗ^type1)
```

## 3 The Proposed Solution

To resolve the aforementioned task, supervised learning was popularly used in many works [10, 12, 15, 17, 23, 26, 27, 31, 32, 34]. However, supervised learning needs labeled datasets and sometimes additional resources for the learning process. As a result, it is suitable for the well-studied language like English. On the other hand, unsupervised learning is practical with no requirement of labeled datasets. This approach was examined in many works for some various languages like English [1, 14, 18, 20], German [22], Korean [14], and Swedish [9, 28].

As for Vietnamese clinical texts, from the study in [29], supervised learning is promising with the highest performance in many experiments as compared to the statistical and unsupervised learning approaches. Nonetheless, its achievement needs to be validated in the new context of the task defined in our work. In addition, [16] introduced self-training in unsupervised resolution of acronyms and abbreviations in nursing notes based on document-context language models. This shows the potential of semisupervised learning because semisupervised learning exploits unlabeled data to enhance the labeled dataset and further improve the learning process. Different from [16], our work aims at a more practical cross-domain online processing context.

In short, we propose a novel solution to tackle the cross-domain abbreviation disambiguation task on Vietnamese clinical texts in online processing. Our solution includes two phases, (1). Feature Extraction and (2). Model Construction, described in Fig. 2. The first phase transforms texts into computable vectors in a vector space model while the second one conducts semisupervised learning to construct a model for abbreviation disambiguation. The solution is algorithmically written as follows:

(i) Feature Extraction:

$$V_l^{type1} \leftarrow vector\_modeling(D_l^{type1})$$
$$V_u^{type2} \leftarrow vector\_modeling(D_u^{type2})$$

(ii) Model Construction:

$$H \leftarrow semisupervised\_learning(V_l^{type1}, Y_l, V_u^{type2})$$

### 3.1  Solution Details

#### 3.1.1  Feature Extraction

Feature extraction is our feature engineering to do vector modeling for each abbreviation and its surrounding context in clinical texts. As effectively used in [29], we redefine the feature set defined in [32, 34] along with those added by [29] to finalize a feature set in our new online processing context.

In particular, based on [32, 34], we define a context window of 11 words and then extract word features, word features with direction, and position features for all the context words on the left side of each abbreviation. Based on [29], we reuse the following features: *Position, leftFirstCase, leftCaseCount, leftSpecial, leftNumber*, and *leftConsonant*. We remove the right-side features because each abbreviation being considered is now located at the end of the text and no context word on the right side of each abbreviation exists in online processing. Furthermore, to make feature sets computable, indexing is constructed according to their alphabetical order local to the value set of each feature. As a result, each abbreviation is now represented as a numeric vector in a 21-dimension vector space. This degree is stable for any text of each abbreviation regardless of text lengths. Such feature engineering also helps us avoid sparseness in case of short sentences that often occur in medical records.

Compared to representation learning in an end-to-end solution, feature extraction in our work is simpler but feasible for Vietnamese clinical texts. This is because no extra resource is required in our vector modeling while representation learning needs large datasets. On the other hand, pretrained models like BERT [7] might be taken into consideration in our feature extraction. This option is promising; however, not straightforwardly adapted to our work because Vietnamese clinical text sets are not widely published and none of the pretrained models has been built on Vietnamese clinical texts.

### 3.1.2  Model Construction

Our model construction is accomplished with semisupervised learning as this learning paradigm has been proved effective for cross-note abbreviation detection in Vietnamese clinical texts [30]. On the other hand, [29] has selected supervised learning with Support Vector Machines using Sequential Minimal Optimization [13, 24], where multiclass learning with the 1-vs-1 scheme in [8] was employed. Combining the semisupervised learning process in [30] and the supervised learning algorithm in [29] would result in an effective model for our new context.

From the theoretical perspectives, we expect to enhance the supervised learning process by incorporating more data characterisics of the unlabeled dataset. This decision is significant in our context, where no labeled data is available. Even more for a cross-domain task, instance-based transfer learning can be performed iteratively in semisupervised learning and both source and target domains can get aligned accordingly. The learning process is then generalized of the target domain.

Compared to the existing ones proposed for Vietnamese clinical texts in [30], our model construction does not require any user's setting when we define Nonparametric Self-Training as a parameter-free semisupervised learning algorithm. This design simplifies our solution and makes it applicable to the cross-domain context where different domains can be taken into account. Furthermore, it can be combined with other techniques to resolve other task challenges like data imbalance.

### 3.2  The Proposed Nonparametric Self-training Method

In this subsection, we propose Nonparametric Self-Training for our model construction. Nonparametric Self-Training is based on the Self-Training's learning style. It is also similar to Tri-Training [35] as utilizing the ensemble's principles and doesn't require any parameter settings. The following is its formal definition.

<u>Input</u>: $V_l^{type1}$, $Y_l$, $V_u^{type2}$
<u>Output</u>: a model $H$
<u>Process</u>:

```
1.    uset ← Vᵤ^type2;
2.    lset ← bootstrap_sampling (V₁^type1, Y₁);
3.    cur_H ← base_supervised_learning (lset);
4.    cur_Recall ← Recall (iH(V₁^type1), Y₁);
5.    repeat
6.        prev_H ← cur_H;
7.        prev_Recall ← cur_Recall;
8.        for k=1..3
9.            lset[k] ← bootstrap_sampling (lset);
10.       for k=1..3
11.           eH[k] ← base_supervised_learning (lset);
12.       for each instance i in uset
13.           for k=1..3
14.               label[k] ← eH1[k](i);
15.           if label[1]=label[2] || label[1]=label[3] then
16.               yᵢ ← label[1]
17.           else if label[2]=label[3] then
18.               yᵢ ← label[2]
19.           if yᵢ is not null then
20.               lset  ← lset ∪ {(i, yᵢ)};
21.               uset  ← uset - {i};
22.       cur_H ← base_supervised_learning (lset);
23.       cur_Recall ← Recall (cur_H(V₁^type1), Y₁);
24.   until prev_Recall > cur_Recall || uset is empty;
25.   H ← prev_H;
26.   return H;
```

Similar to Self-Training and other existing semisupervised learning algorithms, Non-parametric Self-Training is flexibly designed. As a general semisupervised learning algorithm, it can wrap any base supervised learning algorithm and then improve the performance of this algorithm by enhancing its training dataset. As a specific semisupervised learning algorithm for the proposed task, it is coupled with Support Vector Machines using Sequential Minimal Optimization to form an effective solution because this base algorithm has been proved effective in [29].

By contrast, our algorithm is novel by defining a threshold-free instance selection scheme to enhance a labeled training dataset with more correctly predicted instances from an unlabeled dataset in a random manner. In particular, on lines 12–21, each unlabeled instance is selected according to the labeling agreement of two over three models different from the one being built. Using the principles of an ensemble model, a predicted label is wrong if more than half components predict it wrongly. The probability of being correct of a predicted label also increases along with the lower correlation of the components. In our work, all the components are built in each loop using bootstrap samples from the enhanced labeled dataset as shown on lines 8–11. As unlabeled instances keep being added into the labeled dataset, bootstrap samples get possibly different from the labeled dataset and a higher chance to be different from each other. As a result, more different models of lower correlation are built.

On the other hand, randomness is injected into our model via bootstrap sampling to make it more generalized. It is done at the beginning when the starting training set of our model is generated from the given labeled dataset on line 2. In each loop, on lines 8–9, it is repeated to obtain other bootstrap samples for the component models for the threshold-free instance selection scheme. Besides, such randomness helps us maintain the difference between our training and test datasets in each loop to evaluate the objective function in our learning process. Nevertheless, the randomness does not have any impact on the deterministics of our algorithm because of two following reasons with the termination condition of our algorithm on line 24:

(1) The unchange of the objective function is desired.
(2) The maximum number of selected instances is constrained to be the number of the unlabeled instances. This constraint always enforces the deterministics of our algorithm regardless of any infinite loop posibility.

Finally, we use Recall as our objective function to favor the retrieval of true positives. F-measure can be used as our objective function alternatively like the one in [30]. By doing that, our learning process pays more attention to the instances of the minor classes in each loop. When the learning process stops, our objective function reaches its maximum Recall value. The returned model is now an optimized model as depicted in Fig. 2. It is also important to admit that the optimization of our semisupervised learning process is locally converged. However, for the task, the optimization of the nested supervised learning model, Support Vector Machines, is globally converged. This partially ensures the effectiveness of our final model.

### 3.3   The Characteristics of the Proposed Work

In this subsection, the main characteristics of our work are discussed.

Firstly, the task defined in our work is a novel challenging task on Vietnamese clinical texts in online processing. In this task, data shortage and imbalance problems are inherent when a practical cross-domain context is examined. For these problems, [12] has considered data imbalance in supervised learning with a cross-validation evaluation scheme on English clinical notes. Different from [12], our work tackles the task on Vietnamese clinical texts of two various domains in online processing from many different perspectives for these problems with data enhancement and objective-based optimization of the semisupervised learning process.

Secondly, we define Nonparametric Self-Training as the first solution to the aforementioned task. The algorithm is generally utilized for any multiclass classification task and straightwardly for resolving the task because of no user-defined parameter setting in the algorithm. It is also expected to just improve the performance of the base supervised learning algorithm because of its performance maintenance for the base supervised learning algorithm in the semisupervised learning process. Whenever the current performance of the base supervised learning algorithm is impaired, early stopping proceeds. Nonetheless, the current version of the proposed algorithm supports only one performance measure in its objective function. More measures can be included in its optimization process for more effectiveness.

In summary, our work contributes greatly to Vietnamese clinical text analytics as dealing with ambiguous abbreviations which are ubiquitous in clinical texts of EMRs.

# 4  An Empirical Evaluation

## 4.1  Research Questions

In this section, we raise three following research questions to evaluate our work. Question 1 is asked for the significance of the task in practice while Questions 2 and 3 are examined for the effectiveness contribution of Nonparametric Self-Training.

- Question 1: Is cross-domain abbreviation disambiguation on Vietnamese clinical texts in online processing challenging?
- Question 2: Does Nonparametric Self-Training really resolve this task?
- Question 3: Does Nonparametric Self-Training outperform some existing semisupervised learning methods for the aforementioned task?

## 4.2  Experiment Settings

In this work, we reused the datasets prepared for abbreviation disambiguation in [29]. These datasets were from Vietnamese EMRs in two hospitals in Vietnam. Due to privacy protection, their details are not disclosed. They are separated into two domains corresponding to different note types in two hospitals: treatment and care notes (TCN) and discharge summaries (DS). They contain Vietnamese clinical texts from care progress notes (CPN), treatment progress notes (TPN), and care order notes (CON) in the first domain and from discharge summaries (DS) in the second domain, written by different nurses and physicians for different purposes (care, treatment, delivery). However, all of them have a lot of abbreviations and at least one abbreviation was used on average in each record. Table 1 describes these datasets while Tables 2 and 3 show more details about the abbreviations in our experiments.

In [29], twenty abbreviations and their senses were processed while six abbreviations were further discussed due to their challenging characteristics. By contrast, our current work is resolving an even more challenging task where a cross-domain context in online processing is considered. We took into account three abbreviations including HC, M, and T as presented in Tables 2 and 3. The reasons for choosing them are: (i). they are available in both domains, (ii). they are very highly imbalanced, (iii). they are of varying size and complexity in terms of the number of samples and senses. To illustrate this new context, we use the datasets in the first domain (TCN) as training datasets while those in the second one (DS) as test datasets. In addition, for M's case, we also conducted the experiments using the dataset in DS as a training dataset and the one in TCN as a test dataset. The rationales behind these training and test datasets are to ensure data sufficiency for the task and to study the impacts of the solution characteristics on how effectively the task is solved.

In the experiments, for the base supervised learning algorithm, we used SMO [13, 24], which supports multiclass learning with the 1-vs-1 scheme in [8] and

**Table 1.** Dataset Descriptions with Disambiguated Abbreviations.

| Domain | Dataset | Record# | Abbreviation Occurrence# | | Sense Occurrence# | |
|---|---|---|---|---|---|---|
| | | | All | Examined | All | Examined |
| Treatment and care notes | TPN | 3,846 | 7,976 | 4,559 | 18,351 | 9,669 |
| | CON | 10,247 | 7,922 | 145 | 45,036 | 13,279 |
| | CPN | 11,255 | 13,946 | 6,159 | 23,077 | 9,969 |
| Discharge summaries | DS | 1,117 | 3,243 | 1,934 | 25,483 | 10,075 |
| **All** | | **26,465** | **33,087** | **12,797** | **111,947** | **42,992** |

**Table 2.** Examined Abbreviations and their Senses.

| Abbreviation | Senses |
|---|---|
| HC | *hố chậu* (pelvis), *hội chẩn* (consult), *hội chứng* (syndrome), *hồng cầu* (red cell) |
| M | *mạch* (pulse), *mặt* (face), *miệng* (mouth), *mông* (buttock), *mũi* (nose), *mét* (metter) |
| T | *nhiệt độ* (temperature), *tay* (hand), *theo* (follow), *thứ* (day), *tiêm* (inject), *tim* (heart), *trái* (left) |

**Table 3.** Examined Abbreviation Details.

| Abbreviation | Sense# | Samples# | Imbalance Ratio | Sample# in TCN | Sample# in DS |
|---|---|---|---|---|---|
| HC | 4 | 168 | 0.0857 | 132 | 36 |
| M | 5 | 5,515 | 0.0244 | 3,446 | 2,069 |
| T | 7 | 10,638 | 0.0671 | 9,061 | 1,577 |

for the base semisupervised learning algorithms, we used Self-Training with a probability threshold of 0.75 and Tri-Training [35]. These algorithms were chosen due to their best performances for the abbreviation-related tasks on Vietnamese clinical texts in [29, 30]. For data imbalance, we considered Resample, SMOTE [3], and SpreadSubSample with default settings. Resample is used to generate bootstrap datasets by random oversampling with replacement and SpreadSubSample creates datasets by adjusting the difference in class frequencies, while SMOTE performs oversampling the minority classes with synthetic instances using the $k$-nearest neighbors approach. These methods are used to analyze the challenges of the task and highlight the contributions of our proposed algorithm for dealing with data imbalance.

Recording the experimental results, we observe Accuracy in [0, 100] and Area under the ROC curve (AUC) in [0, 1.0]. Accuracy reflects the overall performance of each solution while AUC is used for data imbalance handling in our work. Their higher values are expected for better solutions while their best ones are shown in bold.

### 4.3  Experimental Results and Discussions

For the first question, we conducted several experiments with two evaluation schemes, 10-fold cross-validation and hold-out, and two processing styles, online and offline. Their results are displayed in Table 4.

In our experiments, the cross-validation evaluation scheme is used for the in-domain context while the hold-out one for the cross-domain context. In all cases, the cross-domain context is much more challenging because it is non-trivial to determine a relevant sense for each abbreviation correctly. Indeed, the difference between these two contexts in terms of both Accuracy and AUC is very high. Each case in the in-domain context can be correctly resolved easily. This context has Accuracy of higher than 95% and AUC of higher than 97%. In contrast, the second cross-domain context receives 30%–60% for Accuracy and 52%–70% for AUC. Different writing styles, data sizes, and imbalance strongly influence the performance in the second context.

For processing types, the full in-domain context in offline processing has more details about abbreviations, leading to better Accuracy and AUC as compared to the partial in-domain context in online processing. In particular, the difference between these two cases in Accuracy is 1%–5% and that in AUC is 1%–2.5%. So, online processing certainly is one of the factors that have an impact on the task performance.

Putting them altogether, we conclude that our cross-domain abbreviation disambiguation task on Vietnamese clinical texts in online processing is challenging. The first examination on the challenges of the task helps us figure out the suitable solution to the task presented next.

**Table 4.** Cross-validation vs. Hold-out Evaluation Schemes for Cross-domain Challenges in Online Processing.

| Abbreviation | Processing | Evaluation scheme | Accuracy | AUC |
|---|---|---|---|---|
| HC in all notes | Offline | Cross-validation | **95.833** | **0.974** |
| HC in all notes | Online | Cross-validation | 91.667 | 0.951 |
| HC in discharge summaries | Online | Hold-out | 52.778 | 0.593 |
| M in all notes | Offline | Cross-validation | **98.640** | **0.992** |
| M in all notes | Online | Cross-validation | 97.697 | 0.985 |
| M in treatment and care notes | Online | Hold-out | 34.010 | 0.683 |
| M in discharge summaries | Online | Hold-out | 31.609 | 0.521 |
| T in all notes | Offline | Cross-validation | **99.445** | **0.998** |
| T in all notes | Online | Cross-validation | 95.159 | 0.980 |
| T in discharge summaries | Online | Hold-out | 57.451 | 0.574 |

**Table 5.** Rebalance for the Task.

| Abbreviation | Rebalance | Accuracy | AUC |
|---|---|---|---|
| HC in discharge summaries | Original | **52.778** | 0.593 |
| | Resample | 47.222 | 0.566 |
| | SMOTE | **52.778** | **0.611** |
| | SpreadSubSample | **52.778** | 0.593 |
| M in treatment and care notes | Original | 34.010 | 0.683 |
| | Resample | 27.539 | 0.638 |
| | SMOTE | **34.620** | **0.684** |
| | SpreadSubSample | 33.372 | 0.682 |
| M in discharge summaries | Original | 31.609 | 0.521 |
| | Resample | **41.179** | **0.675** |
| | SMOTE | 37.216 | 0.524 |
| | SpreadSubSample | 31.513 | 0.521 |
| T in discharge summaries | Original | 57.451 | 0.574 |
| | Resample | **57.641** | **0.618** |
| | SMOTE | 57.451 | 0.574 |
| | SpreadSubSample | 57.451 | 0.575 |

For Question 2, our solution with Nonparametric Self-Training really solves this task from some different perspectives via the experimental results in Tables 5 and 6. In particular, we first do rebalancing datasets for the task and then associate them with our Nonparametric Self-Training. Table 5 shows that both Accuracy and AUC are slightly improved with rebalancing with either Resample or SMOTE. Such results show the influence of data imbalance and signal a chance to overcome data imbalance. Little improvement can be achieved with rebalancing because data rebalancing just took into account the data characteristics of the source domain and had no information about those of the target domain. As a result, such preprocessing has not yet helped the solution solve the task challenges much.

Different from Table 5, Table 6 shows a great difference between the base supervised learning and our Nonparametric Self-Training in terms of Accuracy and AUC in almost all the cases. Regardless of rebalancing, our solution can achieve much higher Accuracy. However, with rebalancing, our solution can resolve the task more effectively with both higher Accuracy and AUC. For each abbreviation, we always find the solution with both Accuracy and AUC of higher than 90%. Such results reflect the fact that the challenges of the task have been resolved. Semisupervised learning helps us determine more correct relevant senses for the abbreviations. A mismatch between two domains has been lessened and data enhancement can be exploited successfully. In addition, imbalance between the senses of each abbreviation has been handled so that both frequent and rare senses can be correctly decided for their corresponding abbreviations.

**Table 6.** Semisupervised Learning and Rebalance for the Task.

| Abbreviation | Learning method | Rebalance | Accuracy | AUC |
|---|---|---|---|---|
| HC in discharge summaries | SMO | Original | 52.778 | 0.593 |
| | Nonparametric Self-Training | | **94.444** | **0.901** |
| | SMO | Resample | 47.222 | 0.566 |
| | Nonparametric Self-Training | | **77.777** | **0.848** |
| | SMO | SMOTE | 52.778 | 0.611 |
| | Nonparametric Self-Training | | **94.444** | **0.943** |
| | SMO | SpreadSubSample | 52.778 | 0.593 |
| | Nonparametric Self-Training | | **88.889** | **0.915** |
| M in treatment and care notes | SMO | Original | 34.010 | **0.683** |
| | Nonparametric Self-Training | | **41.788** | 0.643 |
| | SMO | Resample | 27.539 | 0.638 |
| | Nonparametric Self-Training | | **76.001** | **0.821** |
| | SMO | SMOTE | 34.620 | 0.684 |
| | Nonparametric Self-Training | | **88.886** | **0.845** |
| | SMO | SpreadSubSample | 33.372 | 0.682 |
| | Nonparametric Self-Training | | **91.207** | **0.933** |
| M in discharge summaries | SMO | Original | 31.609 | 0.521 |
| | Nonparametric Self-Training | | **98.695** | **0.990** |
| | SMO | Resample | 41.179 | 0.675 |
| | Nonparametric Self-Training | | **98.647** | **0.990** |
| | SMO | SMOTE | 37.216 | 0.524 |
| | Nonparametric Self-Training | | **79.217** | **0.929** |
| | SMO | SpreadSubSample | 31.513 | 0.521 |
| | Nonparametric Self-Training | | **73.030** | **0.839** |
| T in discharge summaries | SMO | Original | 57.451 | 0.574 |
| | Nonparametric Self-Training | | **92.327** | **0.654** |
| | SMO | Resample | 57.641 | 0.618 |
| | Nonparametric Self-Training | | **98.098** | **0.974** |
| | SMO | SMOTE | 57.451 | 0.574 |
| | Nonparametric Self-Training | | **97.464** | **0.950** |
| | SMO | SpreadSubSample | 57.451 | 0.575 |
| | Nonparametric Self-Training | | **97.844** | **0.831** |

**Table 7.** Accuracy-based Evaluation of Nonparametric Self-Training with its Base Methods.

| Abbreviation | Learning method | Rebalance | | | |
|---|---|---|---|---|---|
| | | Original | Resample | SMOTE | SpreadSubSample |
| HC in discharge summaries | SMO | 52.778 | 47.222 | 52.778 | 52.778 |
| | Self-Training | 47.222 | 47.222 | 47.222 | 55.556 |
| | Tri-Training | 52.778 | 58.333 | 58.333 | 50.000 |
| | Nonparametric Self-Training | **94.444** | **77.777** | **94.444** | **88.889** |
| M in treatment and care notes | SMO | 34.010 | 27.539 | 34.620 | 33.372 |
| | Self-Training | 31.573 | 36.680 | 38.595 | 47.911 |
| | Tri-Training | 33.691 | 27.597 | 32.443 | 35.519 |
| | Nonparametric Self-Training | **41.788** | **76.001** | **88.886** | **91.207** |
| M in discharge summaries | SMO | 31.609 | 41.179 | 37.216 | 31.513 |
| | Self-Training | 34.896 | 40.696 | 41.421 | 49.009 |
| | Tri-Training | 41.324 | 41.179 | 40.358 | 39.053 |
| | Nonparametric Self-Training | **98.695** | **98.647** | **79.217** | **73.030** |
| T in discharge summaries | SMO | 57.451 | 57.641 | 57.451 | 57.451 |
| | Self-Training | 59.670 | 57.451 | 56.944 | 58.782 |
| | Tri-Training | 57.451 | 57.578 | 57.070 | 57.451 |
| | Nonparametric Self-Training | **92.327** | **98.098** | **97.464** | **97.844** |

As for Question 3, the results in Table 7 confirm the appropriate design of our Nonparametric Self-Training for the task. In all cases, it outperforms its corresponding existing base supervised and semisupervised learning algorithms. The difference between the other algorithms and ours in Accuracy is very high from about 10% to 60% regardless of data rebalancing. Such difference stems from the enhancement scheme of our algorithm compared to the others. Its optimization process also contributes to this performance. Furthermore, they confirm the hypothesis we made on the design of our algorithm as discussed earlier. On the other hand, we beware of the importance of the supporting data in the source domain in the case of abbreviation M in treatment and care notes. Compared to other cases, this M's case achieved smaller differences. Those can be explained in such a way that M has a more severe imbalance and the number of samples in the source domain is less than that in the target domain. As a result, more rare senses in the source domain exist but cannot be recognized in the target domain. Misrecognition makes them unselected in data enhancement and then there are not enough samples of these senses in the enhanced training dataset for building a model to recognize their corresponding abbreviations in the target domain. Nonetheless, our semisupervised learning mechanism proposed in this work is effective when more different models of lower correlation were used.

In short, this evaluation study has proved the effectiveness of our proposed solution to the abbreviation disambiguation task on real-world Vietnamese clinical texts in the online processing cross-domain context. The experimental results have highlighted the better performance of the solution in this context from several aspects.

## 5  Conclusion

In this paper, abbreviation disambiguation in Vietnamese clinical texts has been discussed in a practical context where Vietnamese electronic medical records are managed on the fly. As soon as an abbreviation is detected, our work will check the ambiguity of the abbreviation and for an ambiguous abbreviation, a multiclass classification model is executed to determine its relevant sense. Within such a context, this is the first time the aforementioned task is defined for Vietnamese clinical texts. As a result, we obtain more readable cleaned texts helpful for clinical text analytics like medical term extraction and treatment plan recommendation.

Another significant contribution of our work is Nonparametric Self-Training, which plays a role of semisupervised learning to build the multiclass classification models for the task. This solution is not only practical but also effective for imbalanced Vietnamese clinical texts when a cross-domain context is supported. Via the experimental results on Vietnamese clinical texts of four various note types from two different hospitals, our solution can handle the task effectively by aligning different domains in semisupervised learning with and without data rebancing.

In the future, we first evaluate the generality of Nonparametric Self-Training with more note types, i.e. more various domains. We also plan to examine more end-to-end solutions as more resources can be prepared for Vietnamese clinical text processing. Above all, Vietnamese clinical text mining applications of our work are being considered to reflect the effectiveness of our solution in practice.

**Acknowledgment.** This research is funded by Vietnam National University – Ho Chi Minh City (VNU-HCM) under grant number C2022-20-11.

In addition, our sincere thanks go to Dr. Nguyen Thi Minh Huyen and her team at University of Science, Vietnam National University, Hanoi, Vietnam, for the helpful resources. We also thank the providers of the Vietnamese electronic medical records very much.

## References

1. Adams, G., Ketenci, M., Bhave, S., Perotte, A., Elhadad, N.: Zero-Shot clinical acronym expansion via Latent Meaning Cells. Proc. Mach. Learn. Res. **136**, 12–40 (2020)
2. Argawal, M., Hegselmann, S., Lang, H., Kim, Y., Sontag, D.: Large language models are Zero-Shot clinical information extractors. arXiv:2205.12689v1 [cs.CL] 25 May 2022, pp. 1–31 (2022)
3. Chawla, N.V., Bowyer, K.W., Hall, L.O., Kegelmeyer, W.P.: SMOTE: synthetic minority over-sampling technique. J. Artif. Intell. Res. **16**, 321–357 (2002)

4. Chondrogiannis, E., Karanastasis, E., Andronikou, V., Varvarigou, T.: Building a repository for inferring the meaning of abbreviations used in clinical studies. J. Comput. **12**(1), 76–88 (2017)

5. Collard, B., Royal, A.: The use of abbreviations in surgical note keeping. Ann. Med. Surg. **4**, 100–102 (2015)

6. Cossin, S., Jolly, M., Larrouture, I., Griffier, R., Jouhet, V.: Semi-automatic extraction of abbreviations and their senses from electronic health records. In: Proc of IA & Santé 2021, pp. 1–13 (2021)

7. Devlin, J., Chang, M., Lee, K., Toutanova, K.: BERT: pre-training of deep bidirectional transformers for language understanding. In: NAACL-HLT (2018)

8. Hastie, T., Tibshirani, R.: Classification by pairwise coupling. In: Advances in Neural Information Processing Systems (1998)

9. Henriksson, A., Moen, H., Skeppstedt, M., Daudaravicius, V., Duneld, M.: Synonym extraction and abbreviation expansion with ensembles of semantic spaces. J. Biomed. Semant. **5**(6), 1–25 (2014)

10. Jaber, A., Martinez, P.: Disambiguating clinical abbreviations using pre-trained word embeddings. In: Proceedings of the 14th International Joint Conference on Biomedical Engineering Systems and Technologies (BIOSTEC 2021) 5 (HEALTHINF), pp. 501–508 (2021)

11. Jin, Q., Liu, J., Lu, X.: Deep contextualized biomedical abbreviation expansion. arXiv:1906.03360v1 [cs.CL] 8 Jun 2019, pp. 1–9 (2019)

12. Joopudi, V., Dandala, B., Devarakonda, M.: A convolutional route to abbreviation disambiguation in clinical text. J. Biomed. Inform. **86**, 71–78 (2018)

13. Keerthi, S.S., Shevade, S.K., Bhattacharyya, C., Murthy, K.R.K.: Improvements to Platt's SMO algorithm for SVM classifier design. Neural Comput. **13**(3), 637–649 (2001)

14. Kim, J.-B., Oh, H.-S., Nam, S.-S., Myaeng, S.-H.: Using candidate exploration and ranking for abbreviation resolution in clinical documents. In: Proceedings of the 2013 IEEE International Conference on Healthcare Informatics, pp. 317–326 (2013)

15. Kim, Y., Hurdle, J., Meystre, S.M.: Using UMLS lexical resources to disambiguate abbreviations in clinical text. In: Proceedings of the AMIA Symposium, pp. 715–722 (2011)

16. Kirchhoff, K., Turner, A.M.: Unsupervised resolution of acronyms and abbreviations in nursing notes using document-level context models. In: Proceedings of the Seventh International Workshop on Health Text Mining and Information Analysis (LOUHI), pp. 52–60 (2016)

17. Li, I., et al.: A neural topic-attention model for medical term abbreviation disambiguation. arXiv:1910.14076v1 [cs.CL] 30 Oct 2019, pp. 1–9 (2019)

18. Link, N.B., et al.: Acronym disambiguation in clinical notes from electronic health records. medRxiv, pp. 1–24 (2020)

19. Link, N.B., et al.: Binary acronym disambiguation in clinical notes from electronic health records with an application in computational phenotyping. Int. J. Med. Inform. **162**, 104753 (2022)

20. Liu, Y., Ge, T., Mathews, K.S., Ji, H., McGuinness, D.L.: Exploiting task-oriented resources to learn word embeddings for clinical abbreviation expansion. arXiv:1804.04225v1 [cs.CL] 11 Apr 2018, pp. 1–6 (2018)

21. Long, W.J.: Parsing free text nursing notes. In: Proceedings of AMIA Annual Symposium, p. 917 (2003)

22. Oleynik, M., Kreuzthaler, M., Schulz, S.: Unsupervised abbreviation expansion in clinical narratives. In: MEDINFO 2017: Precision Healthcare Through Informatics, pp. 539–543 (2017)

23. Peng, M., Quan, H.: Clinical abbreviation disambiguation using deep contextualized representation. Digit. Personal. Health Med. 88–92 (2020)

24. Platt, J.: Fast training of support vector machines using sequential minimal optimization. In: Schoelkopf, B., Burges, C., Smola, A. (eds.) Advances in Kernel Methods - Support Vector Learning (1998)
25. Shilo, L., Shilo, G.: Analysis of abbreviations used by residents in admission notes and discharge summaries. QJM: Int. J. Med. **111**(3), 179–183 (2018)
26. Skreta, M., Arbabi, A., Wang, J., Bruno, M.: Training without training data: Improving the generalizability of automated medical abbreviation disambiguation. Proc. Mach. Learn. Res. **116**, 233–245 (2020)
27. Skreta, M., et al.: Automatically disambiguating medical acronyms with ontology-aware deep learning. Nat. Commun. **12**(5319), 1–10 (2021)
28. Tengstrand, L., Megyesi, B., Henriksson, A., Duneld, M., Kvist, M.: EACL - expansion of abbreviations in clinical text. In: Proceedings of the 3rd Workshop on Predicting and Improving Text Readability for Target Reader Populations (PITR) @ EACL 2014, pp. 94–103 (2014)
29. Vo, T.N.C., Nguyen, H.P.: An initial study of abbreviation disambiguation in Vietnamese clinical texts. In: Proceedings of the 18th International Conference on Ubiquitous Information Management and Communication (IMCOM 2024), pp. 1–8 (2024)
30. Vo, T.N.C., Nguyen, H.P.: Nested semisupervised learning for cross-note abbreviation detection in Vietnamese clinical texts. In: Proceedings of the 15th Asian Conference on Intelligent Information and Database Systems (ACIIDS 2023), pp. 1–13 (2023)
31. Wen, Z., Lu, X.H., Reddy, S.: MeDAL: medical abbreviation disambiguation dataset for natural language understanding pertraining. arXiv:2012.13978v1 [cs.CL] 27 Dec 2020, pp. 1–8 (2020)
32. Wu, Y., et al.: A preliminary study of clinical abbreviation disambiguation in real time. Appl. Clin. Inform. **6**, 364–374 (2015)
33. Wu, Y., et al.: A long journey to short abbreviations: developing an open-source framework for clinical abbreviation recognition and disambiguation (CARD). J. Am. Med. Inform. Assoc. **24**(e1), e79–e86 (2017)
34. Wu, Y., Xu, J., Zhang, Y., Xu, H.: Clinical abbreviation disambiguation using neural word embeddings. In: Proceedings of the 2015 Workshop on Biomedical Natural Language Processing (BioNLP 2015), pp. 171–176 (2015)
35. Zhou, Z.H., Li, M.: Tri-training: exploiting unlabeled data using three classifiers. IEEE Trans. Knowl. Data Eng. **17**(11), 1529–1541 (2005)

# Exploring the Potential of Generative Models in Promoting Local Language News Consumption

Md Mahbub Ul Alam and Doina Logofătu$^{(\boxtimes)}$

Department of Computer Science and Engineering, Frankfurt University of Applied Sciences, Frankfurt am Main, Germany
md.ulalam@stud.fra-uas.de, logofatu@fb2.fra-uas.de

**Abstract.** In the present era of digital platforms, there exists a notable disparity in the accessibility of news content across various languages, resulting in a disadvantage for individuals who speak Swahili. This research seeks to rectify this disparity by employing advanced generative artificial intelligence models, specifically tailored iterations of the BERT algorithm, to facilitate the production of Swahili news articles. Our methodology entails enhancing and refining these models to interpret and generate linguistically diverse information effectively. The study's findings illustrate the effectiveness of AI-driven techniques in generating accurate and culturally suitable Swahili news content, thereby bridging the linguistic divide in the digital domain. Furthermore, this study emphasizes the vital significance of innovative AI applications in advancing inclusion and diversity on digital news platforms, setting a benchmark for future initiatives to improve language representation in the digital domain.

**Keywords:** Generative AI · Swahili · Local Language News · and Modified BERT

## 1 Introduction

The worldwide news environment has changed as a result of the unprecedented availability of information brought about by the digital era. Nevertheless, not all language groups have benefited equally from this digital revolution, which has resulted in a discernible difference in news consumption among populations that do not speak English. The language that is most impacted by this discrepancy is Swahili, which is spoken by millions of people in East Africa. Swahili is widely spoken, yet its material is notably underrepresented in digital news outlets, leaving its speakers with a large knowledge vacuum [1]. The objective of this project, Harnessing Generative Models for Enhancing Local Language News Access, is to close this gap by utilizing generative models' artificial intelligence (AI) capabilities.

© The Author(s), under exclusive license to Springer Nature Switzerland AG 2024
N.-T. Nguyen et al. (Eds.): ICCCI 2024, CCIS 2166, pp. 141–154, 2024.
https://doi.org/10.1007/978-3-031-70259-4_11

Local dialects play a major role in the preservation of culture and individuality. To energize worldwide differences within the advanced circle, it is required to ensure their perceivability on advanced platforms [2]. Improving the availability and accessibility of Swahili news substance on the web is significant for promoting dialect differences within the advanced news field, as the analysis emphasizes.

The problem of limited substance availability can be effectively and modernly solved with generative models of imitation information, such as modified versions of the Bidirectional Encoder Representations from Transformers (BERT) demonstrated for generative assignments [3]. These models can produce authentic and distinctive fabric, which is especially helpful for dialects with limited amounts of advanced resources. Our research suggests using these models to create excellent Swahili news articles. This will provide a broad range of phonetic content, helping to close the technological gap and improve the online news landscape.

This study goes beyond certainly addressing the urgent requirement for extra Swahili information material; it also targets researching the broader outcomes of utilizing generative AI to promote language inclusivity. The examine aims to establish a widespread for comparable packages targeting different underrepresented languages by employing a radical methodology that consists of collecting information, schooling models, and evaluating effects.

However, there are difficulties to overcome when applying generative models for producing news in local languages. These relate to the shortage of information, the present biases in models, and the complex complexities included in dialect generation. This introduction provides an outline of the tactics used to overcome such barriers, with a focus on technological improvements.

The project envisions a time when generative AI, by visualizing digital news platforms that are as different as the people they serve worldwide, would contribute to the promotion of linguistic variety. The aim of the project is to lay the groundwork for generative AI research and development in the future. Its objective is to spur more development in this area so that generative AI may be used in a wider variety of languages. This will democratize the availability of information by promoting equitable access to it in all languages.

## 2   Problem Description

In this modern era, having access to news in the native languages is essential to having informed citizens. Even in nations where the majority of people speak languages other than English, there is still a notable lack of local language news coverage, despite the abundance of digital news sources. This discrepancy results from a combination of factors, including the restricted resources available for producing and distributing news in regional languages as well as the digital divide. The challenge is particularly apparent for millions of people in Eastern Africa who speak Swahili, a widely spoken language, since the demand for news in the native tongue far outweighs the availability of resources. [4].

The main issue stems from the scarcity of Swahili news information, exacerbated by the intricacies of language processing for generative models. Conventional news platforms frequently give more importance to content in global languages, while local language news is often considered less significant. This scenario leads to a dearth of varied, easily obtainable, and promptly delivered news material for individuals who speak Swahili, resulting in discrepancies in knowledge that impact social, political, and economic engagement.

**Fig. 1.** Train text data frequency visualisation by category.

Figure 1 illustrates the text frequency train data sheet. Here words in Swahili grouped by category in different sizes and orientations.

Furthermore, the technological domain is included in the challenge. As demonstrated that that machine learning and natural language processing can automate content creation and classification. Furthermore, not much study has been done on the potential use of these technologies to promote the consumption of local news. One of the challenges in building and training models that can effectively generate and classify news material is the absence of large datasets in local languages. Thus, creative approaches based on generative models are desperately needed to overcome this shortcoming and enable the production of relevant, reliable, and varied news pieces in Swahili.

A comprehensive strategy that goes beyond simple technological remedies is needed to address this problem. It requires the establishment of sophisticated generative models specifically designed for Swahili. The compilation of large datasets to train these models and a strategy framework that takes into account the language and cultural subtleties of the intended audience. The project aims to solve these challenges in order to improve information accessibility for a large audience and create an inclusive digital news ecosystem that values local language representation.

## 3   The Evolutionary Trajectory of Generative Models

A significant phase in the development of artificial intelligence has been started by the exploration of generative models. Beginning with the first use of simple statistical models to simulate data distributions and culminating in the development of neural networks that are able to learn and produce complex data representations.

**Earlier stages:** Focus on the development of patterns and simple probabilistic models.

**The Evolution of Neural Networks:** The extensive use of neural networks to increasingly difficult data-generating jobs [5].

**Generative Adversarial Networks (GANs):** Generative Adversarial Networks have revolutionized the discipline by introducing their remarkable capabilities. This generates outputs that are highly realistic [5] [6] [7].

**Variational Autoencoders (VAEs) and Beyond:** The use of models such as Variational Autoencoders (VAEs) allows for the improvement and broadening of generation by refining and diversifying the output in terms of both quality and diversity [5] [8] [9].

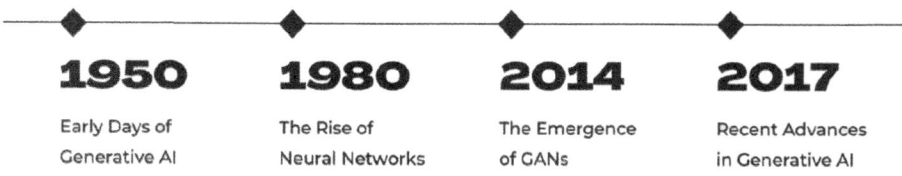

**Fig. 2.** Evolution of Generative AI Through Time [10].

Figure 2 [10] demonstrate This graphical timeline reference point for our chronological progression.

**Current Innovations:** Continual progress in the design of model structures, methods for training, and utilization in diverse fields.

This graphical timeline in Fig. 2 [10] is a visual reference point for our chronological progression. Incorporating the crucial moments that have shaped the field of generative modeling.

## 4    Related Work

Significant advancements in computational linguistics, machine learning (ML), and natural language processing (NLP) have been aimed at improving news consumption in various languages. However, these advancements have predominantly focused on widely spoken languages, creating a gap for languages like Swahili. Researchers such as Bird et al. have emphasized the importance of developing NLP tools for resource-scarce languages to preserve linguistic diversity. Although Swahili is widely spoken language, it has only recently started to gain attention in computational linguistics, evidenced by the work of Owuor and Mwangi in developing a Swahili news classification system [11].

Deep learning architectures, such as generative models like GPT and BERT, have revolutionized content creation by their ability to generate grammatically correct text. While English has benefited most from these advancements, new initiatives aim to extend their advantages to indigenous languages. For instance, Kumar's framework for Hindi content creation suggests a potential strategy for similar initiatives in Swahili [12].

Advancements in news content categorization and summarization using machine learning have enabled the extraction and presentation of concise news summaries [13]. However, integrating these technologies with regional languages remains a significant challenge due to the lack of extensive datasets. Efforts to generate resources for languages like Urdu highlight the potential for similar developments for Swahili, facilitating the application of generative models and NLP techniques to create and classify local language news.

Research on the subject highlights the potential that has yet to be discovered when using advanced computational techniques to promote news consumption in local languages. While there are still important obstacles, like the need for better datasets and language complexity, the rapidly developing fields of natural language processing (NLP) and generative modeling offer exciting possibilities for research and innovation to further improve the usability and attractiveness of local language news.

## 5    Data Analysis and Preprocessing

In this section, the data analysis and preprocessing steps taken to prepare the Swahili news articles dataset for training and validation of the generative BERT model. The dataset was sourced from the "ChatGPT ML Challenge: Swahili News Classification" on Zindi, comprising over 5151 entries in the "train.csv" file, each with a unique ID, content, and category.

**Dataset Division:** The dataset was split into training and testing subsets to facilitate model training and validation.

**Exploratory Data Analysis:** Figure 3 represents training data analysis that examined the categorical distribution of the news articles, revealing the dominance of specific news genres over others.

The news item length distribution graph is crucial for figuring out the preprocessing stages. It reveals that there is a skewness towards shorter articles,

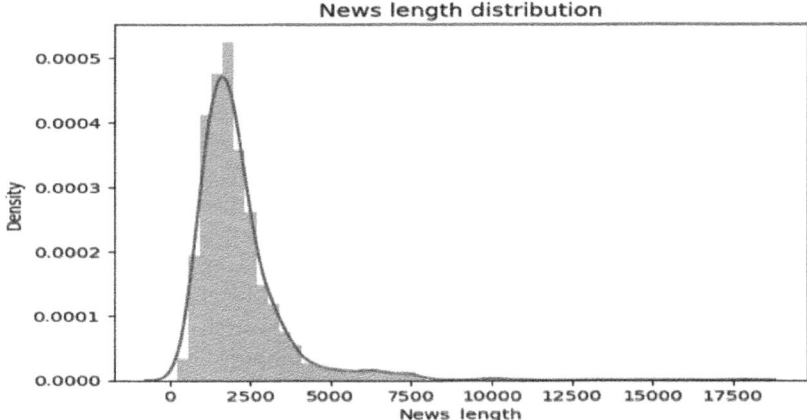

**Fig. 3.** News length distribution of Train data set.

with a height that shows a common duration among the various majority of articles. This height, followed by an extended tail, suggests that even though most articles are concise, some are notably longer. The perception received from this distribution informs the technique for padding or truncating articles during preprocessing, making sure to have a uniform input period without dropping crucial content.

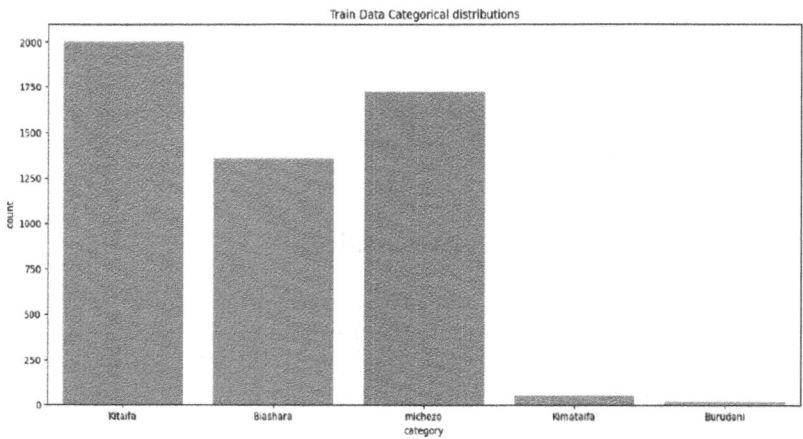

**Fig. 4.** Training data categorical distributions.

Figure 4 shows training data's distribution based on the category. The categorization information highlights disparities in representation that the model has to address, suggesting potential issues with class imbalance.

**Sample Data Inspection:** The raw text and associated categories are included in the dataset sample, demonstrating the complex and diverse nature of the news material that will be fed into the generative model.

| | id | content | category |
|---|---|---|---|
| 0 | SW0 | SERIKALI imesema haitakuwa tayari kuona amani... | Kitaifa |
| 1 | SW1 | Mkuu wa Mkoa wa Tabora, Aggrey Mwanri amesiti... | Biashara |
| 2 | SW10 | SERIKALI imetoa miezi sita kwa taasisi zote z... | Kitaifa |
| 3 | SW100 | KAMPUNI ya mchezo wa kubahatisha ya M-bet ime... | michezo |
| 4 | SW1000 | WATANZANIA wamekumbushwa kusherehekea sikukuu... | Kitaifa |

**Fig. 5.** Sample data from Training dataset.

**Preprocessing Steps:** Text Normalization: Converted all text to lowercase and used regular expressions to remove extraneous characters, standardizing the text to its most basic components.

Tokenization: The text was tokenized, or separated into discrete parts, so that the model could examine and learn from the linguistic structure.

Label Encoding: Following text preprocessing, label encoding was used to transform the categorical news categories into a numerical structure that is suitable for machine learning models.

**BERT-Specific Preprocessing:** A tokenizer has been designed to satisfy the unique requirements of BERT in order to preprocess the data for the BERT model. Important tasks, including inserting special tokens, trimming or padding sequences to maintain a uniform length, and creating attention masks to distinguish important content from padding.

**Data Splitting:** A stratified division approach function was used to separate the dataset into sets for training and validation. With this method, a sample of the full dataset is guaranteed to be represented in each set. Making a plan for preparation guarantees that the data used for model validation and training is well organized and of the highest caliber. This is necessary to develop generative models that are specifically tailored to the Swahili language in an efficient manner.

These comprehensive preprocessing steps are crucial for developing an effective generative model tailored to the Swahili language, ensuring the model can accurately interpret and generate Swahili news content.

## 6   Methods and Algorithms

Major developments the study involves training a modified BERT model using advanced machine learning techniques so that it can produce news items in

Swahili [14]. To facilitate the creation of coherent text sequences, generative layers are added to the model once it has been initialized using the multilingual version of BERT. The methodology encompasses several key steps:

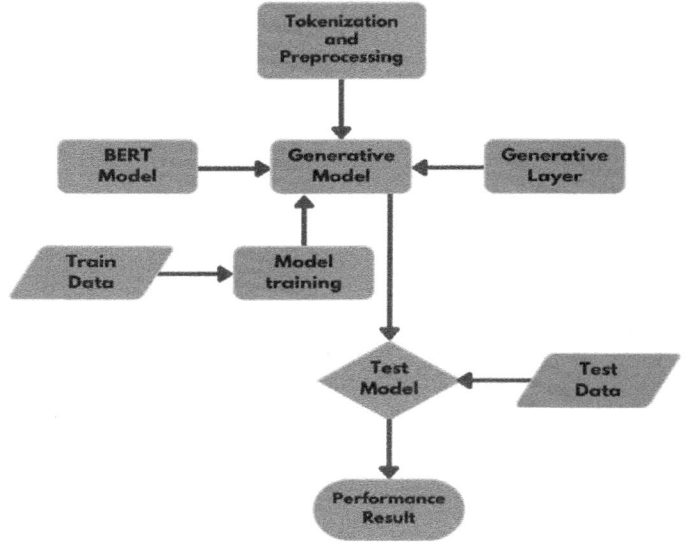

**Fig. 6.** Generative Model building approach.

Figure 6 illustrates the flow of the model building. The procedures for using a BERT model as a generative model

**Tokenization and Data Preprocessing:** The preparation of the data for model training starts with these procedures. The Swahili news text data is first tokenized to aid with machine comprehension, then preparation is carried out to guarantee data uniformity

**Model Initialization:** The BERT model is initialized using a multilingual version pre-trained on diverse text corpora. This provides a robust foundation for further fine-tuning on Swahili news data.

**Generative Layer Integration:** A generative layer is added to the BERT model to enable coherent text generation. This layer leverages contextual embeddings produced by the BERT encoder to generate new text sequences.

**Training Process:** The model is trained using a combination of advanced machine learning techniques, including early stopping and batch training. The training data is split into training and validation sets to monitor performance and prevent overfitting.

**Performance Evaluation:** The model's performance is evaluated using various metrics, including accuracy, precision, recall, and F1-score. These metrics provide insights into the model's ability to generate accurate and culturally relevant Swahili news content.

This detailed workflow ensures the production of high-quality Swahili news articles, addressing the linguistic and cultural nuances of the target audience.

# 7    Data Modelling and Training

## Model Training Approach

The training program of the generative BERT model is carefully designed to guarantee high efficiency in producing news items in Swahili.

**Training Setup:** Tokenized and preprocessed Swahili news items are fed into the model in batches, and this is the input data used to train the model. There is an Early Stopping callback to monitor the model's performance and minimize overfitting. It stops training if the validation loss does not decrease for three consecutive epochs. The training process is also recorded using the CSVLogger callback, gathering metrics that will be examined subsequently to comprehend the model's learning behavior over time.

**Training Execution:** A total of 16 batches are trained over a period of 20 epochs. In order to continually assess performance and generalize ability outside of the training dataset, the model's fitting procedure makes use of the training data while verifying against an independent validation set.

**Model Optimization::** To ensure a smooth and successful training process through the incorporation of callback functions like EarlyStopping and CSVLogger that constantly adjust to the model's learning curve and prevent overtraining,

## Model Output and Generative Capabilities

After training, the model is capable of producing new text sequences. The BERT model's generative layer has been refined to capture the subtleties of Swahili news syntax and semantics, allowing for the production of contextually rich and linguistically coherent news items.

# 8    Performance Measurement

The empirical assessment of the generative BERT model for Swahili news article production provided measurable insights into its performance. Conventional criteria was employed to assess the model's categorization prowess.

## Training and Validation Performance

The learning process of the model is demonstrated by its accuracy, which is calculated at each epoch using the given equation [15]:

$$\text{Accuracy} = \frac{NumberofCorrectPredictions}{TotalNumberofPredictions} \tag{1}$$

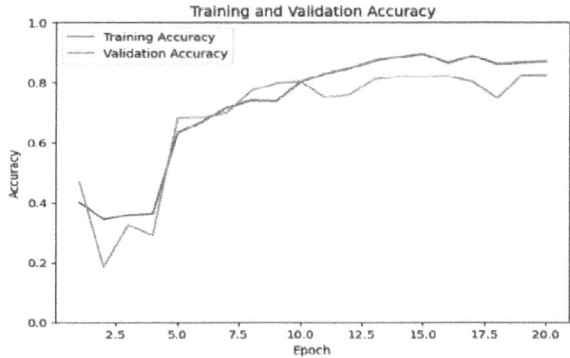

**Fig. 7.** Training and Validation Accuracy Over Epochs.

The training phase demonstrated a steady increase in accuracy, eventually leveling off, which suggests that the model has reached its maximum learning potential on the training dataset.

Figure 7 shows the graph of training and validation accuracy throughout epochs. Validation accuracy remains lower, while training accuracy somewhat rises.

The validation accuracy, together with the training accuracy, is essential to guarantee the model's capacity to generalize(Fig. 7). The loss for both the training and validation sets dropped gradually over time(Fig. 8), as determined by the cross-entropy loss function used for multi-class classification.

$$\text{Loss} = -\sum_{c=1}^{M} y_{o,c} \log(p_{o,c}) \tag{2}$$

where $M$ is the total number of classes, $y$ is the binary indicator (0 or 1) of whether class label $c$ is the correct classification for analysis $o$, and $p$ is the predicted probability that analysis $o$ is of class $c$ [16].

Figure 8 illustrates the model's training and validation losses over epochs are depicted in the graph. Training loss goes down as the number of epochs goes up, while validation loss goes up.

**Classification Performance**

The categorization report provided metrics such as precision, recall, and F1-score. These metrics are defined as follows [17]:

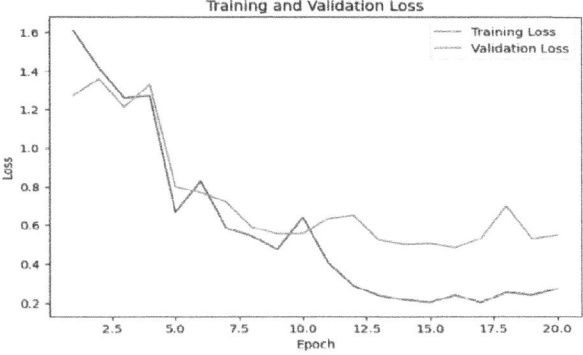

**Fig. 8.** Training and Validation Loss Trends Over Epochs.

$$Precision = \frac{TruePositives}{TruePositives + FalsePositives}$$

$$Recall = \frac{TruePositives}{TruePositives + FalseNegatives}$$

$$F1\text{-score} = 2 \times \frac{Precision \times Recall}{Precision + Recall}$$

Precision quantifies the model's level of accuracy in correctly classifying an item into a specific category, whereas recall evaluates the model's capability to correctly identify all articles belonging to that category. The F1-score achieves an equilibrium between precision and recall, providing a unified metric for assessing the model's accuracy and resilience.

The 'michezo' type had an effective ability to offer sports activity-associated information, as visible with the aid of its exceptional F1-rating and good excessive accuracy and recall. On the other hand, several categories displayed areas for development, suggesting feasible avenues for extra-schooling of models and data enhancement.

Table 1 represents the model's performance across a number of categories and offers a comprehensive categorization report. The report provides weighted average, macro-average, and overall model accuracy ratings in addition to precision, recall, and F1-score data for each category. Support provides an idea of the dataset's distribution by displaying the number of true cases for each category.

**Macro and Weighted Averages**

The macro-average F1-score is a useful metric for datasets with an uneven magnificence distribution. It is calculated by taking the average of the F1-scores for each character class

**Table 1.** Model classification report

| Class | Precision | Recall | F1-score | Support |
|---|---|---|---|---|
| biashara | 0.76 | 0.80 | 0.78 | 680 |
| burudani | 0.67 | 0.44 | 0.53 | 9 |
| kimataifa | 0.55 | 0.22 | 0.32 | 27 |
| kitaifa | 0.83 | 0.79 | 0.81 | 1000 |
| michezo | 0.95 | 0.97 | 0.96 | 860 |
| **Accuracy** | | | 0.85 | 2576 |
| **Macro avg** | 0.75 | 0.65 | 0.68 | 2576 |
| **Weighted avg** | 0.85 | 0.85 | 0.85 | 2576 |

$$\text{Macro-average F1-score} = \frac{1}{N} \sum_{i=1}^{N} F1_i \qquad (3)$$

where $N$ is the number of classes and $F1_i$ is the F1-score of the $i^{th}$ class [17].

The weighted common F1-rating contains the support, that's the variety of true cases for every label. This metric appropriately represents the contribution of each magnificence to the whole performance [17].

$$\text{Weighted-average F1-score} = \sum_{i=1}^{N} \left( \frac{Support_i}{Total Support} \times F1_i \right) \qquad (4)$$

The model's effectiveness is proven by its overall accuracy and weighted-common F1-score of 0.85, which demonstrate its strong overall performance across all categories. This shows that the generative version performs properly regardless of the version in terms of the quantity of samples for each category.

All things considered, these measures show how well the generative BERT model performs, lending credence to the notion that it may be useful for producing accurate and varied Swahili news material. Continued improvement and adjustment of the model can increase its accuracy and flexibility. This will help get the ultimate goal of promoting the spread of news in local languages.

## 9    Conclusion and Future Work

This study investigated the disparity in virtual language usage in news intake within the Swahili-speaking network. Using generative AI models has made large improvements in generating excellent and various information memories in Swahili, efficiently lowering factual gaps. Despite going through challenges including limited record availability and inherent biases, the efforts represent numerous progress in leveraging AI to foster linguistic diversity.

In the future, there are expect several potential paths for advancing this research. Further investigations ought to update the generative models by enhancing them, correcting biases, and improving capacity to create more complex information. In order to archive this goal, it is important to reinforce the size of the datasets for Swahili and other languages that require progressed illustration. Furthermore, studying the usage of comparable models in different local languages may want to extend the effect of the findings, promoting international linguistic diversity within the virtual domain.

The wider impacts of work move beyond simple technological achievements; they dive into the core of a protective way of life and encourage international inclusivity. Generative AI has the capacity to enhance groups, ensure equitable dissemination of information, and promote a numerous and better digital environment by making information more convenient in surrounding languages.

# References

1. Masua, B., Masasi, N.: Enhancing text pre-processing for Swahili language: Datasets for common Swahili stop-words, slangs and typos with equivalent proper words. Data Brief **33**, 106517 (2020). https://doi.org/10.1016/j.dib.2020.106517
2. Hidayat, D., Rahmasari, G., Wibawa, D.: the inhibition and communication approaches of local languages learning among millennials. Int. J. Lang. Educ. **5**(3), 165–179 (2021)
3. Oh, S., et al.: Deep generative design: integration of topology optimization and generative models. J. Mech. Des. **141**(11), 111405 (2019)
4. Lugoba, E.M.: The case for Kiswahili as a regional broadcasting language in East Africa. J. Pan Afr. Stud. **2**(8), 1–19 (2012)
5. Cao, H., et al.: A Survey on generative diffusion models. IEEE Trans. Knowl. Data Eng. (2024).https://doi.org/10.1109/TKDE.2024.3361474
6. Goodfellow, I., et al.: Generative adversarial networks. Commun. ACM **63**(11), 139–144 (2020). https://doi.org/10.1145/3422622. Accessed 14 Mar 2024
7. Creswell, A., White, T., Dumoulin, V., Arulkumaran, K., Sengupta, B., Bharath, A.A.: Generative adversarial networks: an overview. IEEE Signal Process. Mag. **35**(1), 53–65 (2018). https://doi.org/10.1109/MSP.2017.2765202
8. Kingma, D.P., Welling, M., et al.: An introduction to variational autoencoders. Found. Trends® Mach. Learn. **12**(4), 307–392 (2019). (document)
9. Oussidi, A., Elhassouny, A.: Deep generative models: Survey. In: 2018 International Conference on Intelligent Systems and Computer Vision (ISCV), pp. 1-8. Fez, Morocco (2018). https://doi.org/10.1109/ISACV.2018.8354080
10. Deep Tech Stars. "Unveiling the Evolution of Generative AI: A Journey Through Time." LinkedIn, 28 Dec. 2023, https://www.linkedin.com/pulse/unveiling-volution-generative-ai-journey-through-time-fmtfc/.Accessed 14 Mar 2024
11. Gesare, O.G.: Literary Strategies in Yvonne Owuor's Dust and Meja Mwangi's The Last Plague (2018)
12. Ahmad, T., et al.: News article summarization: analysis and experiments on basic extractive algorithms. Int. J. Grid. Distrib. Comput. **13**(2), 2366–2379 (2020)
13. Maurya, K.K., et al.: ZmBART: an unsupervised cross-lingual transfer framework for language generation. arXiv preprint arXiv:2106.01597 (2021)
14. Martin, G., et al.: Swahbert: language model of swahili. In: Proceedings of the 2022 Conference of the North American Chapter of the Association for Computational Linguistics: Human Language Technologies (2022)

15. Bressler, N.: How to check the accuracy of your machine learning model. Deepchecks, 23 Nov 2022. https://deepchecks.com/how-to-check-the-accuracy-of-your-machine-learning-model/. Accessed 14 Mar 2024

16. Draelos, R.: Connections: Log Likelihood, Cross Entropy, KL Divergence, Logistic Regression, and Neural Networks. Glass Box Medicine, 7 Dec 2019. https://tinyurl.com/Glass-Box-Medicine. Accessed 14 Mar 2024

17. Kundu, R.: F1 Score in Machine Learning: Intro & Calculation. V7 Labs 16 (2022). https://www.v7labs.com/blog/f1-score-guide. Accessed 14 Mar 2024

# New Methodology for Attack Patterns Classification in Deep Brain Stimulation

Jihen Fourati[1,2,7](✉) 📵, Mohamed Othmani[3,4], and Hela Ltifi[5,6]

[1] National Engineering School of Sfax, University of Sfax,
1173 Sfax, Tunisia
jihen.fourati@enis.u-sfax.tn
[2] Research Lab: Technology, Energy, and Innovative Materials Lab,
Faculty of Sciences of Gafsa, University of Gafsa, Gafsa, Tunisia
[3] Faculty of Sciences of Gafsa, University of Gafsa,
2100 Gafsa, Tunisia
[4] ATES: Advanced Technologies on Environment and Smart City,
Gafsa, Tunisia
[5] Faculty of Sciences and Techniques of Sidi Bouzid, University of Kairouan,
Kairouan, Tunisia
Hela.ltifi@ieee.org
[6] Research Groups in Intelligent Machines Lab,
3038 Sfax, Tunisia
[7] Unit of Scientific Research, Applied College, Qassim University,
Buraydah, Saudi Arabia
j.fourati@qu.edu.sa

**Abstract.** Deep brain stimulation is a surgical treatment using implantable devices to generate electrical impulses and alleviate motor symptoms in Parkinson's disease, primarily the resting tremor. Recently, deep learning (DL)–based approaches have provided incredible benefits for a wide range of problems, including classification, segmentation, and feature engineering. Although DL-based techniques performed very well in the classification tasks, they remain challenged in handling time series data (i.e., rest tremor velocity signals recorded from patients suffering from PD). Multiple issues persevere with time-series data, including difficulties in extracting relevant features, heavily weighted data, etc. This research aims to develop a hybrid model for classifying different attack types in deep brain implants using a convolutional neural network (CNN) and bidirectional long short-term memory (BiLSTM). Our model autonomously extracts features from raw data with minimal preprocessing. Because of the advantages of CNN and BiLSTM, they can learn both high-level features and long-term dependencies in sequential data. Various model architectures were explored to construct the optimal structure. A comprehensive experimental study has been made in this research, which shows that the proposed approach offers the most efficient tool for accurate classification and ranks at the top of the list of recently published algorithms on the Physionet dataset. Ten-fold cross-validation is carried out. The proposed model achieved accuracy, precision, recall, and an F1-score of 97.69%, 99,62%, 99,62%, 99.5%, and 97.63%, respectively. The proposed model provides a robust tool for the classification of different attack types for deep brain implants.

© The Author(s), under exclusive license to Springer Nature Switzerland AG 2024
N.-T. Nguyen et al. (Eds.): ICCCI 2024, CCIS 2166, pp. 155–167, 2024.
https://doi.org/10.1007/978-3-031-70259-4_12

**Keywords:** Deep brain stimulation · Multi-class classification · Convolutional Neural Network · Bidirectional long-short term memory · k-fold cross-validation

# 1   Introduction

Deep brain stimulation (DBS) is a popular therapeutic option for decreasing the movement symptoms of advanced Parkinson's disease. It is a type of neurosurgery that enables targeted circuit-based neuromodulation. Integration of modern technology into the neurological system of humans has considerable advantages. However, as with all medical therapy, DBS has side effects and risks. Several of these adverse effects have garnered significant ethical scrutiny elsewhere, in light of the concerns they raise for personal identity and autonomy. Indeed, there are several ways to attack implantable therapeutic equipment. For instance, a potential opponent may interfere with, halt, or cause unnecessary shocks in the cerebral cortex through the generation of false signals. The outcome of attacks can be life-threatening for patients because these sudden changes can have an impact on their lives. Furthermore, the attacking systems have attained a point where they can manipulate political and social institutions on a worldwide scale to establish new types of social control. These prospect attacks could involve inducing severe pain in the recipient, whereas altering the pulse width could have unintended side effects and potentially damage brain tissue. The idea behind coupling neuronal circuitry into the software directly through digital devices could bear unpredictable risks. Based on statistical analysis, [21] has been focused on devoting a thorough assessment of the literature on the security and privacy problems with the Internet of medical things and their solutions utilizing machine learning approaches. In this work, the authors mentioned that 37% of the research publications were concerned with detecting anomalies in sensors or medical equipment. It was clear from this that in previous years, the main emphasis had been on detecting medical device intrusions including fake measurement injection, resource exhaustion attacks, and psychological and physical attacks. This can be related to the seriousness of these attacks and the major health problems they cause. For instance, attacks on implantable medical equipment may lead to death. In addition, 35% of the reported research (7% for malware and 28% for other assaults) received considerable attention on intrusion and malware attack detection. Because of the scarcity of research in this field (i.e. brainjacking), continuous discussions and investigations on these issues are essential to reduce dangers and prevent unexpected consequences. Deep learning methods [1–6] have proven efficient in addressing various challenges involved in accurate classification problems in recent years. For the current study, a DL-based model is suggested, which employs both raw data and generated data with minimal data preprocessing. In this research, we present a CNN-BiLSTM neural network model with different candidate structures for the classification of attack types for deep brain implants in PD patients using RT data. The suggested model merges the advantages of both CNNs and BiLSTMs. The BiLSTMs can handle long-term dependencies, whereas the convolutional neural networks are good at extracting inherent characteristics. BiLSTMs -in contrast to baseline LSTMs- make use of both future information along with past information when the entire sequence of time series data is accessible. As a result of the added context, the network can now make more accurate predictions. Several experiments were performed on the Physionet Parkinsonian tremor dataset to examine the architecture of our suggested approach. The experimental findings illustrated that the hybrid deep learning-based approach to attack pattern classification could lead to hitting rates higher than 97.69%. Moreover, it can be of competitive performance with the prior state-of-the-art studies.

The current study contributions can be recapped in the next points: Firstly, emulating different types of attack patterns by adjusting the stimulus parameters including pulse rate, the intensity of the pulse, pulse width, stimulation mode, etc. Secondly, three CNN-BiLSTM neural network model candidates with various architectures were offered to construct the best model structure for the suggested attack patterns classifier. Next, the model makes use of CNNs and BiLSTMs to capture inherent features and long-term dependencies in sequential data. Then, a new hybrid model was involved in coupling the CNN and BiLSTM benefits to anticipate various attack stimulations in DBSs and classify different types of attacks for deep brain implants and evaluate the performance compared to other prediction methods.

The remainder of this research paper is organized as follows. The literature that is pertinent to the current work is discussed in Sect. 2 after the introduction. In Sect. 3, the general approach to the problem and the procedures followed encompass are elucidated. The performance and evaluation of our system are covered in Sect. 4. Finally, Sect. 5 specifies the scope of additional future work in more detail to complete the research.

## 2   Recent Works

Several machine learning methods have been employed in numerous studies to classify the resting tremor of patients with Parkinson's disease. For instance, Seung-BoLee et al. [17] proposed a machine-learning approach to improve the diagnostic accuracy of PD. The proposed gradient boosting decision tree model has reached a high classification accuracy of 89.3%.

In [8], the authors have presented two predictive models for classifying resting tremors of PD between high and low frequencies, which correspond to the intensity of Parkinson's motor symptoms. A high classification accuracy of 92.8% was attained by these models. Besides this, Pedrosa et al. [7] examined the effect of using gait characteristics and tremors for monitoring and early identification of Parkinson's disease using statistical analysis and machine learning approaches, with an accuracy of no more than 86.9% but this method is limited to a binary classification issue where only the presence of Parkinson's disease is recognized and its stages have not been determined.

Most of the reported work to date has focused on detecting and quantifying rest tremor velocity (RT). Nonetheless, a few studies have attempted to categorize different types of attacks for brain implants.

For example, Abdaoui et al. [18] created a monitoring system for distinguishing genuine alarms from fake ones based on deep learning on the Raspberry Pi 3. It was found that their system can learn and predict fake signals with an accuracy of roughly 97%. In a recent paper, In [9], the authors proposed an algorithm for third-party prediction of various attacks on DBSs. They analyzed RT values for designing and training the neural network, which can be used to distinguish between multiple attacks in the DBS context. Nonetheless, the shortcoming of this suggestion was that it was not lightweight and did not provide accuracy. In [22], the authors proposed a deep learning algorithm that determines the individual diagnosis of patients with PD by focusing on a limited cross-sectional brain structural Magnetic resonance imaging scan to predict fake and natural stimulation.

# 3    Methodology

This section provides details about the research methodology for the classification of attack types for deep brain implants. The proposed model is provided in three phases. Firstly, we exploited some signal preprocessing techniques. Secondly, we conducted data generation of different attack patterns by modifying the stimulus parameters before we fed data to the network. Then, we thoroughly detailed the proposed CNN-BiLSTM-based approach and its architecture to classify the attack patterns based on resting tremor signals.

## 3.1    Signal Preprocessing

Normalization of the input is carried out before applying it to the neural network architecture for classification. The RT recordings need some manipulation. First, to guarantee data consistency, it was determined that the first second of every subject's recording was discrepant; the tremor would reach extremely high levels, leading to inaccuracies in the graph and algorithm. The first "second of the minute" record was thus eliminated in every instance. Besides, the tremor velocity units in the files of the dataset [13] differ amongst patients. The data were reported for some patients in millimeters per second and others in meters per second. The data must be normalized for the subsequent processing and outcomes to be consistent across all patient data. All data were normalized in millimeters per second using a straightforward technique. The attacker can employ a variety of attack strategies, as stated in [9]. The different attacks on data might be defined as excessive compared to what is considered to be normal behavior [10]. We have generated different attack patterns by modifying the stimulus parameters [9].

## 3.2   Convolutional Neural Network

The convolutional neural network is a very useful neural network proposed from the human nervous system and achieves high performance across a variety of applications compared to traditional methods [11]. A typical CNN is a mathematical construct that is composed of multiple building blocks [11]: the convolution layer and the subsampling layer, which are two computational layers, and an ultimate classification throughout a fully connected layer. The convolutional layer computes the convolution operation of input data with a set of kernel filters to extract high-level features. It contains a set of filters, the parameters of which are to be learned throughout the training. Each kernel is convolved with the input data to perform an activation feature map. To generate the convolutional layer's output volume, the activation maps of all filters are stacked along the depth dimension [11]. The subsampling layer provides typical downsampling of the output size dimension by averaging or maximizing the pooling over the feature maps in the convolutional layer [11]. It is used to reduce both overfitting as well as the computational burden. The fully connected layer carries out a high-level logical operation by assembling the features from the pooling layer. It produces the output data for the CNN model [11].

## 3.3   Bidirectional Long Short-Term Memory Network

BiLSTM is a particular type of recurrent neural network that combines forward and backward hidden layers, in opposite directions to the same output, to capture the future context in addition to the past context. As a consequence, more temporal information flows in both directions, and the network learns more efficiently [12]. We adopt the BiLSTM network for attack pattern classification because it is efficient for identifying long- and short-term dependencies both in the forward and the reverse directions and because it enables end-to-end fine-tuning with a straightforward structure [12].

The input is $x_t = (x_1, x_2, \ldots, x_n) \in \mathbb{R}^{n \times d}$ for a given timestep $t$, where $d$ is the dimension of the hidden state. The forward and backward hidden states are $\overrightarrow{h} \in \mathbb{R}^{n \times d}$ and $\overleftarrow{h} \in \mathbb{R}^{n \times d}$. respectively [12]. The forward and backward hidden state has been carried on based on the equations given below [12]:

$$\overrightarrow{h} = \sigma(W_{\overrightarrow{h}} x_t + W_{\overrightarrow{h}\,\overrightarrow{h}} \overrightarrow{h}_{t-1} + b_{\overrightarrow{h}}) \tag{1}$$

$$\overleftarrow{h} = \sigma(W_{\overleftarrow{h}} x_t + W_{\overleftarrow{h}\,\overleftarrow{h}} \overleftarrow{h}_{t-1} + b_{\overleftarrow{h}}) \tag{2}$$

$$y_t = \sigma(W_{\overrightarrow{h}\,\overrightarrow{y}} h_t + W_{\overleftarrow{h}\,\overrightarrow{y}} h_t + b_y) \tag{3}$$

The output is a function of both forward and backward layers. The BiLSTM produces the sequence of the hidden states as follows [12]:

$$y_t = [\overrightarrow{h}, \overleftarrow{h}] \tag{4}$$

The $\sigma$ function is employed to combine the two output sequences. The forward- and backward-layer outputs are combined and then the final hidden state $y_t$ represents the entire sentence and can be computed based on both LSTMs, where $h_t = [\overrightarrow{h}_t, \overleftarrow{h}_t]$ [12].

### 3.4   Convolutional Neural Network and Bidirectional Long Short-Term Memory for Different Attack Patterns Classification

In this research, the convolutional neural network is used to handle nonlinear features of raw data and extract high-level characteristics. The BiLSTM network is then chosen to receive abstracted data provided by the convolutional neural network, extract temporal information, and output classification findings via a fully connected network. These networks can cooperate, retain prior information, and enhance the ability to extract pertinent information from RT data.

The RT signals are fed to the network through a couple of convolution layers and a pooling layer in between. The data is then passed through a stack of multiple BiLSTM layers, one dropout layer, one fully connected layer, and one output layer.

The input layer is convolved with 128 filters and a kernel size of 2 to obtain layer 1. The max-pooling layer of size 4 is used for each feature map to minimize computational complexity by linking feature maps to a window with a pre-fixed dimension. Then, the feature maps from layer 2 are convolved with 64 filters and a kernel size of 2 to produce layer 3. In our architecture, we used the rectified linear unit activation function $f(x) = max(0, x)$ to increase the non-linearity of the feature maps. Then, after three layers of BiLSTM with 128, 128, and 64 output shapes (layers 4, 5, and 6), one hidden dropout layer (rate $= 0.5$) with one-hot number of length 8 is outputted through the fully connected layer to determine whether an attack is occurring, with a unique value for each type of attack. The final fully connected layer has a single unit layer with a softmax activation function that classifies each data point into multiple categories. The softmax activation function has the following mathematical computation:

$$\text{Softmax}(z_i) = \frac{\exp^{z_i}}{\sum_{j=1}^{N} \exp^{z_j}} \tag{5}$$

where $z$ is the vector of raw output values from a neural network, and $i$ denotes the entry in the softmax output vector. This can be thought of as the predicted probability of the test input belonging to class $i$.

The details of parameters in each layer of the proposed RT classification model are displayed in Table 2. The network weights in the fully connected layer are optimized during training using an adaptive moment estimation algorithm. The Adam optimizer and categorical cross-entropy loss function were employed with default parameter settings. The dropout layer was used with a value of 0.5, which could randomly remove part of the network to prevent overfitting [15]. For all the experiments, the model is trained with a batch size of 64 for 100 epochs.

## 4   Experimental Results

### 4.1   Dataset Description

The recordings of this database are of rest tremor velocity in the index finger of 16 Parkinsonian patients who receive electric deep brain stimulation at chronic

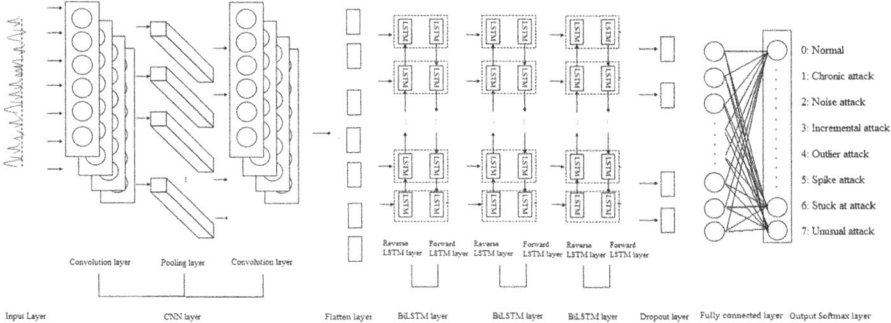

**Fig. 1.** The proposed CNN-BiLSTM model.

**Table 1.** 10-fold cross-validation results.

| Experimental results | Training accuracy | Training loss | Validation accuracy | Validation loss |
|---|---|---|---|---|
| fold 1 | 0.9861 | 0.0453 | 0.9954 | 0.0270 |
| fold 2 | 0.9948 | 0.0141 | 0.9861 | 0.0478 |
| fold 3 | 0.9876 | 0.0594 | 0.9861 | 0.0617 |
| fold 4 | 0.9938 | 0.0283 | 0.9769 | 0.0749 |
| fold 5 | 0.9700 | 0.1209 | 0.9861 | 0.0583 |
| fold 6 | 0.9881 | 0.0510 | 0.9907 | 0.0311 |
| fold 7 | 0.9954 | 0.0172 | 0.9861 | 0.0371 |
| fold 8 | 0.9933 | 0.0234 | 0.9954 | 0.0161 |
| fold 9 | 0.9923 | 0.0297 | 0.9954 | 0.0229 |
| fold 10 | 0.9948 | 0.0242 | 0.9861 | 0.0618 |
| **avg (± std)** | 0.98962 (±0.007666) | 0.031641 (±0.04135) | 0.98843 (±0.005878) | 0.04387 (±0.0198055) |

**Table 2.** The architecture of our classification model.

| Layers | Layer name | Number of filters | Kernel size | Stride | Output shape |
|---|---|---|---|---|---|
| 0 | Convolution | 128 | 4 | 1 | 4093 × 128 |
| 1 | Maxpooling | - | - | 4 | 1023 × 128 |
| 2 | Convolution | 64 | 2 | 1 | 1022 × 64 |
| 3 | Bidirectional LSTM | 128 | - | - | 1022 × 256 |
| 4 | Bidirectional LSTM | 128 | - | - | 1022 × 256 |
| 5 | Bidirectional LSTM | 64 | - | - | 128 |
| 6 | Dropout | - | 0.5 | - | 128 |
| 7 | Dense | - | - | - | 8 |

high frequency, uni, or bi-lateral in any of the three targets (ventral intermediate nucleus, globus pallidus internus, subthalamic nucleus). In Parkinson's disease patients undergoing high-frequency DBS, rest tremor was continually monitored as the deep brain stimulator was turned on and off at an effective frequency. For long-term stimulation at frequencies exceeding 100 Hz, an electrode is implanted in subcortical regions during this operation. It is not known how high-frequency DBS suppresses tremors and reduces other symptoms of Parkinson's disease.

### 4.2  Experiment Setup and Result Analysis

The proposed hybrid network was designed in Python language using Theano [20] backend with the Keras [19] framework. The general software platform for the training hardware was an Intel Pentium N3520 processor running Windows 10 64-bit. The approach was tested on the Physionet dataset. The Physionet dataset is composed of both the original resting tremor velocity signals as well as the ground truths for RT classification. The recordings were considered genuine measurements. Hence, we established a new dataset by generating misleading signals that correspond to the attack strategies along with modulating the learned stimulation pattern. In order to prevent the biasing phenomena, the ten-fold cross-validation strategy was employed. The training and testing processes were repeated ten times until each tremor dataset was tested. The ten model tests were then averaged for the evaluation of the model (as shown in Table 1). The results are denoted by mean (avg) and standard deviation (std) of the avg ± std cross-validation.

The mean and standard deviation values for training and validation accuracy after ten-fold cross-validation were 98.96% (±0.0.007666) and 98.84% (±0.005878), respectively.

In the main experiment, we emulated different types of attack patterns by modifying the stimulus parameters. We evaluated the performance of our proposed classifier compared with other classification methods. We first designed several CNN-BiLSTM model candidates and then evaluated them. The best-performing model was selected, and we evaluated its performance compared to other classification models.

Three candidate models were presented to construct a reasonable network structure for the CNN-BiLSTM model. The first model combined two convolutional layers and a pooling layer. The second model consists of a combination of two convolutional layers with one layer of BiLSTM. The third model combines three layers of BiLSTM with the entire convolutional (layers = 2) and pooling layers.

The convolutional layer is convolved with 128 filters and a kernel size of 4. Each feature map receives a max-pooling of size 4. The first and second BiLSTM layers each have 128 hidden nodes per layer per direction, while the third layer has 64 neurons.

Table 3 displays the experimental findings, and these results revealed that the proposed hybrid model I achieved an accuracy of 78.7% with a recall of 63.51%, precision of 88%, AUC of 0.9537, and kappa coefficient of 0.7099, and F1-score of 0.9377. Next, model II predicted accuracy of 91.20% with a recall of 87%, precision of 87%, AUC of 0.982, kappa coefficient of 0.857, and f1-score of 0.87. Besides, model III predicted an accuracy of 97.69% with a recall of 99.6%, precision of 99.62%, AUC of 0.998, kappa coefficient of 0.9952, and F1-score of 0.995.

**Table 3.** Experimental results for three CNN-BiLSTM hybrid models.

| Parameter | Model I | Model II | Model III |
|---|---|---|---|
| Accuracy | 0.787 | 0.912 | 0.976 |
| Precision | 0.88 | 0.87 | 0.996 |
| Recall | 0.635 | 0.87 | 0.996 |
| AUC | 0.953 | 0.982 | 0.998 |
| F-score | 0.737 | 0.87 | 0.995 |
| Kappa's coefficient | 0.709 | 0.857 | 0.99 |

The findings show that the rest tremor classification performance of model III, with an accuracy equal to 97.69%, was better than other models. From the above experimental findings, It can be concluded that model III is the optimal model for our attack patterns classifier. The performance evaluation was conducted with the same data in each model. It can be noted that increasing the number of hidden layers improves the model's performance. From the recall and precision values, it is clear that the proposed method has been effective in achieving high true positive and negative rates by properly classifying the actual positive and actual negative samples. Figure 2 depicts the confusion matrix for all models evaluated on the test set. From the confusion matrices, it has been found that model III predicted excellent results for all classes. However, the model I had somewhat good results for Class 0 (normal measurement), and from Class 3 (incremental attack) to Class 6 (stuck attack). However, it predicted low performance for Class 1 (chronic attack) and Class 7 (unusual attack). Further, model II had good results for the majority of classes except for classes 5 (spike attack) and 7 (unusual attack).

In this research, five evaluation indexes were adopted to evaluate the performance of the proposed attack pattern classifier: The accuracy index (ACC), recall index (REC), precision index (PR), F1-score (F1), and Cohen's kappa coefficient (K).

Based on the aforementioned findings, we considered that the CNN-BiLSTM model III was the best model for our RT classification. It should be mentioned that our proposed method earned the best accuracy, F1-score, and kappa score, i.e., 97.69%, 99.5%, and 99.04%, which attests to its exceptional capability for addressing the challenge. The receiver operating characteristic curve for each candidate model in the classification with resting tremor is shown in Fig. 3. It graphically displays the trade-off between sensitivity and specificity.

The area under the curve is another measure of test performance. It was achieved at 99.9%, demonstrating the algorithm's overall ability to distinguish between classes and providing a summary of the ROC curve.

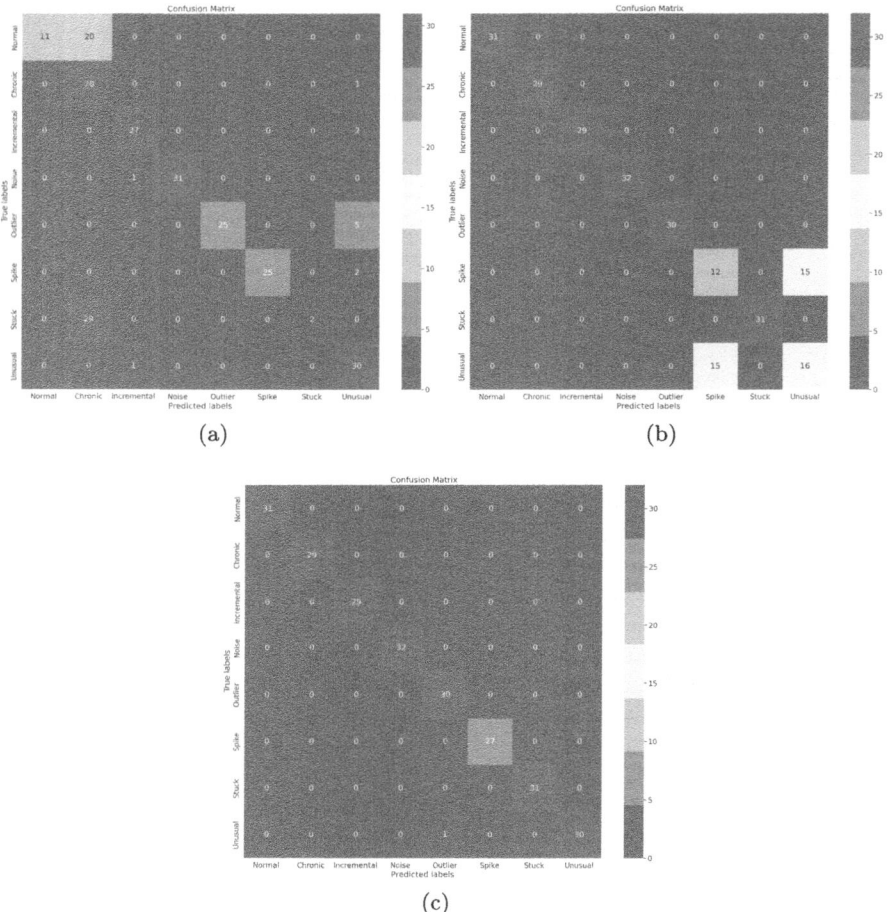

**Fig. 2.** Confusion matrix (a) Model I (b) Model II (c) Model III

Compared with the literature (for the resting tremor classification, as shown in Table 4), Our method demonstrated higher accuracy (97.69%) and a higher F1-score (99.5%). The CNN and BiLSTM network structures combined offer advantages to the signal abstraction process, allowing for more effective extraction of relevant features. The BiLSTMs can handle long-term dependencies, whereas the convolutional neural networks are good at extracting inherent characteristics. The novelty of this method is the introduction of three CNN-BiLSTM neural network model candidates with various structures to construct an optimal model structure. Indeed, no hand-picked features are required in this study. This significantly decreases the procedure of experimentation and selection of the best set of features for classification.

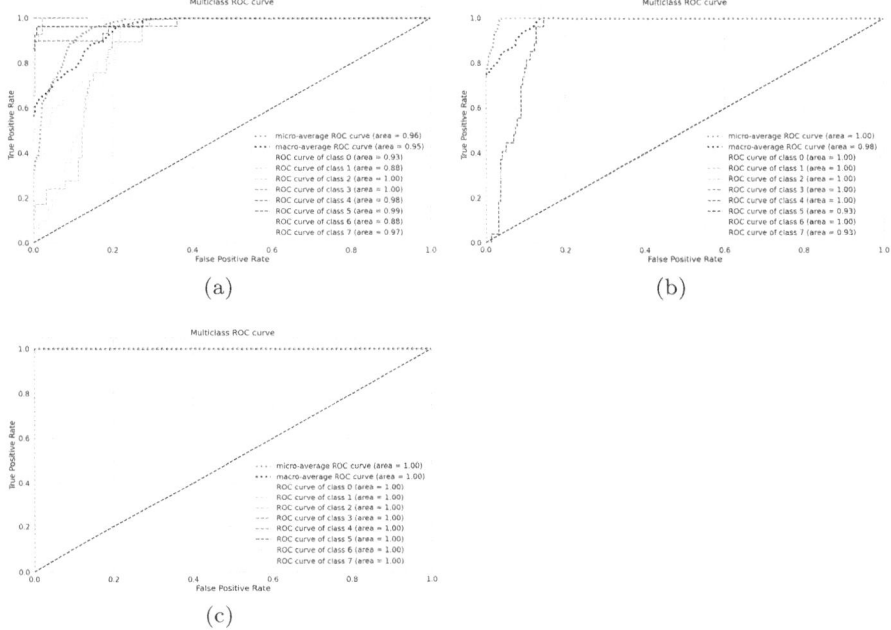

**Fig. 3.** ROC curve (a) Model I (b) Model II (c) Model III

**Table 4.** Proposed method classification accuracy rate with existing methods based on the same dataset.

| Authors | Year | Techniques | Accuracy | Precision | Recall | F1-score |
|---|---|---|---|---|---|---|
| Perumal et al. [8] | 2016 | LDA | | 0.869 | - | - |
| Pedrosa et al. [7] | 2018 | KNN+SVM | 0.928 | 1 | 0.833 | 0.909 |
| **Ours** | 2024 | model I | 0.7870 | 0.88 | 0.6351 | 0.7377 |
| **Ours** | 2024 | model II | 0.9120 | 0.87 | 0.87 | 0.87 |
| **Ours** | 2024 | model III | **0.9769** | 0.99625 | 0.99625 | **0.995** |

Note: preferred model for each metric in **bold**.

## 5   Conclusion

In this work, a CNN-BiLSTM model has been proposed to classify different attack patterns that operate on both raw data and generated data with minimal pre-processing. This model makes use of CNNs and BiLSTMs to capture inherent features and long-term dependencies in sequential data. In addition, three CNN-BiLSTM neural network models were offered as candidates to construct the optimal model structure for the proposed attack pattern classifier. The efficiency of the proposed model was evaluated on the Physionet dataset and achieved 97.69% of accuracy. The experimental findings indicate that the proposed CNN-

BiLSTM model performed better than the earlier suggested methods in the literature. This advancement holds significant promise within the research domain, highlighting the potential of the developed model to transform the classification landscape for various attack types associated with deep brain implants. As a follow-up of this research line, the authors aim to enhance further and evaluate the model's performance in clinical practice.

# References

1. Fourati, J., Othmani, M., Ltifi, H.: An improved approach for Parkinson's disease classification based on convolutional neural network. In: Nguyen, N.T., et al. (eds.) ICCCI 2023. CCIS, vol. 1864, pp. 123–135. Springer, Cham (2023). https://doi.org/10.1007/978-3-031-41774-0_10
2. Fourati, J., Othmani, M., Ltifi, H.: A hybrid model based on convolutional neural networks and long short-term memory for rest tremor classification. In: ICAART, vol. 3, pp. 75–82 (2022)
3. Ben Salah, K., Othmani, M., Fourati, J., Kherallah, M.: Advancing spatial mapping for satellite image road segmentation with multi-head attention. Vis. Comput. 1–11 (2024)
4. Salah, K.B., Othmani, M., Saida, S., Kherallah, M.: Improved approach for semantic segmentation of MBRSC aerial imagery based on transfer learning and modified UNet. In: 2023 International Conference on Cyberworlds (CW), pp. 46–53. IEEE (2023)
5. Othmani, M., Issaoui, B., El Khediri, S., Khan, R.U.: Hybrid active shape model and deep neural network approach for lung cancer detection. Int. J. Info. Technol. 1–12 (2024)
6. Othmani, M.: A vehicle detection and tracking method for traffic video based on faster R-CNN. Multimedia Tools Appl. $81(20)$, 28347–28365 (2022)
7. Pedrosa, T.Í., et al.: Machine learning application to quantify the tremor level for Parkinson's disease patients. Procedia Comput. Sci. $138$, 215–220 (2018)
8. Perumal, S.V., Sankar, R.: Gait and tremor assessment for patients with Parkinson's disease using wearable sensors. ICT Express $2(4)$, 168–174 (2016)
9. Rathore, H., et al.: A novel deep learning strategy for classifying different attack patterns for deep brain implants. IEEE Access $7$, 24154–24164 (2019)
10. Ni, K., et al.: Sensor network data fault types. ACM Trans. Sensor Netw. (TOSN) $5(3)$, 1–29 (2009)
11. Pak, U., Kim, C., Ryu, U., Sok, K., Pak, S.: A hybrid model based on convolutional neural networks and long short-term memory for ozone concentration prediction. Air Qual. Atmos. Health $11(8)$, 883–895 (2018)
12. Al Hamoud, A., Hoenig, A., Roy, K.: Sentence subjectivity analysis of a political and ideological debate dataset using LSTM and BiLSTM with attention and GRU models. J. King Saud Univ.-Comput. Inf. Sci. $34(10)$, 7974–7987 (2022)
13. Goldberger, A.l., Amaral, L.an., Glass, L., et al.: PhysioBank, PhysioToolkit, and PhysioNet: components of a new research resource for complex physiologic signals. Circulation $101(23)$, e215–e220 (2000)
14. Livieris, I.E., Pintelas, E., Pintelas, P.: A CNN-LSTM model for gold price time-series forecasting. Neural Comput. Appl. $32(23)$, 17351–17360 (2020)
15. Esteban, S., et al.: Deep bidirectional recurrent neural networks as end-To-end models for smoking status extraction from clinical notes in Spanish. bioRxiv: 320846 (2018)

16. Yuan, H., et al.: Detection and quantification of resting tremor in Parkinson's disease using long-term acceleration data. Math. Probl. Eng. (2021)
17. Lee, S.B., Kim, Y.J., Hwang, S., Son, H., Lee, S.K., Park, K.I., Kim, Y.G.: Predicting Parkinson's disease using gradient boosting decision tree models with electroencephalography signals. Parkinsonism Rel. Disord. **95**, 77–85 (2022)
18. Abdaoui, A., Al-Ali, A., Riahi, A., Mohamed, A., Du, X., Guizani, M.: Secure medical treatment with deep learning on embedded board. In: Energy Efficiency of Medical Devices and Healthcare Applications, pp. 131–151. Academic Press (2020)
19. Chollet, V.: Keras: theano-based deep learning library (2015). Code https://github.com/fchollet, Documentation http://keras.io
20. The Theano Development: A Python framework for fast computation of mathematical expressions. arXiv:1605.02688 (2016)
21. Hameed, S.S., et al.: A systematic review of security and privacy issues in the internet of medical things; the role of machine learning approaches. PeerJ Comput. Sci. **7**, e414 (2021)
22. Joseph, A.S., Lazar, A.J.P., Sharma, D.K., Maria, A.B., Ganesan, N., Sengan, S.: ConvNet-based deep brain stimulation for attack patterns. In: Agarwal, P., Khanna, K., Elngar, A.A., Obaid, A.J., Polkowski, Z. (eds.) Artificial Intelligence for Smart Healthcare. EAI/Springer Innovations in Communication and Computing, pp. 275–292. Springer, Cham (2023). https://doi.org/10.1007/978-3-031-23602-0_16

# An Evolutionary Algorithm and a Clustering Technique to Select Good Subsets of Test for Finite State Machines

Miguel Benito-Parejo$^{(\boxtimes)}$ ⑩, Manuel Méndez⑩, and Mercedes G. Merayo⑩

Design and Testing of Reliable Systems Research Group,
Universidad Complutense de Madrid, Madrid, Spain
{mibeni01,manumend}@ucm.es, mgmerayo@fdi.ucm.es

**Abstract.** Testing is the technique most widely used to validate the correct behaviour of systems. Essentially, a test consists of applying an input to the system and decide whether it returns the expected output. Unfortunately, budget and temporal constraints limit the amount of testing that can be applied to the system. Therefore, a *good* selection of tests will reduce the resources devoted to testing while keeping an effective validation process. In this paper, we tackle this problem by using mutation testing, which effectively simulates the possible faults that the system under test may have and suggest which tests are best in finding potential faults. In order to perform test selection, we use a multi-objective genetic algorithm that focuses on two targets: minimising the number of inputs the test suite has to perform and maximising the mutation score. We have performed several experiments and exhaustively compared our proposal with a Machine Learning method, specifically clustering, which groups the tests into classes, from which we select the most suitable test to be applied.

**Keywords:** Mutation Testing · Genetic Algorithms · Clustering · Test case selection

## 1 Introduction

Software testing is the most widely used technique to validate the correctness of a system, i.e. to check it for errors. One of the main drawbacks of software testing is that the number of tests needed to check all the system behaviours can be infinite. Therefore, there is a need to select a (finite) number of tests that can detect most of the errors based on a certain criterion. In order to obtain this subset, it is necessary to establish a metric that allows us to determine the quality of each test. The final goal is to filter out and keep only the *best*

This work has been supported by the Spanish MINECO/FEDER project AwESOMe (PID2021-122215NB-C31).

ones in the final subset. Therefore, we have to rely on test case prioritisation techniques, which allow us to sort tests based on a desirable property, such as their ability to detect errors [4]. In this line, *mutation testing* [8,9,12,13,22,23] has shown its effectiveness to evaluate the quality of a set of tests. For this purpose, systems similar to the system under test (SUT) are created by applying small modifications to the SUT, such as removing code fragments, swapping arithmetic operators or swapping relational operators. By applying the tests to the mutants and to the original system, it is possible to determine those with the highest error detection capability, that is, those tests that reveal the most differences in the behaviour of the mutants compared to that of the original system. The idea is that if a set of tests distinguishes the system under test from other systems similar to it but with faults, then it can be a *good* set for determining errors in that system. In this way, prioritisation of test cases can be carried out based on this error detection capability. Another aspect to take into account during the selection of the test cases is their length, that is, the number of inputs to be applied for their execution. The shorter the length of the tests, the faster their execution.

In this paper, we propose two techniques for selecting a subset of test from an initial set of test cases that optimise both criteria, their error detection capacity and their diversity when detecting errors, as well as the length of the test cases. The first one corresponds to a genetic algorithm, that applies an advanced multi-objective optimisation algorithm [5]. The second technique that we propose applies a clustering method. Clustering is a traditional unsupervised grouping technique that has been utilised across various fields [20,26]. However, this technique is not widely applied to test selection tasks. Some works in this area examine the effectiveness of test case prioritisation techniques through a requirements-based clustering approach that integrates traditional code analysis information [1], or apply spectral clustering to mutant reduction [25]. This work is an extension of a previous proposal for test case selection in the field of finite state machines [2].

The main objective of this work is to apply and combine mutation testing techniques with genetic algorithms and clustering for the design and implementation of a test case selection framework. This framework should guide the minimisation of the number of inputs to be applied while maximising the number of different mutants that the test cases kill. To this end, we will follow the next steps:

– We will generate a set of *mutants* by injecting small changes in the original SUT, with the goal of simulating faults. We will use the *mutation score* for determining the quality of the tests. This metric is based on the number of *mutants* a set of test cases is able to detect. The more mutants the test cases kill, the better this subset of test cases will be. We will use the *mutation score* together with the number of inputs to be applied by the subset of test cases to determine the quality of this subset.

- We will design a genetic algorithm in which we apply the multi-objective algorithm NSGA-II [5] during the replacement phase. This would allow us to order the partial solutions in each generation based on our criteria.
- We will apply a clustering method to group the tests depending on which mutants they kill, with the goal of selecting the shortest test from each cluster.
- Finally, we compare the effectiveness of these methods to obtain the best subset of test cases.

The rest of the paper is organised as follows. In Sect. 2 we present the framework and the main problem to solve with the formalism and objectives to optimise as well as the methods that we use to solve this problem. In Sect. 3 we present the results of our experiments. Finally, in Sect. 4 we present our conclusions and some lines for future work.

## 2    Description of the Problem and Methods

In this section, we present our framework. We consider a formalism able to represent systems and the mutants that we will deal with. We also introduce tests and how they interact with systems. Finally, we include an introduction to genetic algorithms, multi-objective optimisation and clustering.

### 2.1    The Formalism

In order to represent the SUT and mutants, we use FSMs. This formalism is simple enough to represent the basic behaviour of systems, but our work can be adapted to deal with other more complex structures or formalisms. This allows a more progressive extension of this work, while maintaining the clarity of the execution and the performance of the experiments.

**Definition 1.** *A* Finite State Machine *(FSM) is a tuple* $M = (S, I, O, Tr, s_{in})$ *where* $S$ *is a finite set of* states, $I$ *is the set of* input *actions,* $O$ *is the set of* output *actions,* $Tr$ *is the set of* transitions *and* $s_{in} \in S$ *is the* initial state. *A transition belonging to* $Tr$ *is a tuple* $(s, s', i, o)$ *where* $s, s' \in S$ *are the initial and final states of the transition,* $i \in I$ *is the input action and* $o \in O$ *is the output action. We say that* $M$ *is* input-enabled *if for each* $s \in S$ *and input* $i \in I$, *there exist* $s' \in S$ *and* $o \in O$ *such that* $(s, s', i, o) \in Tr$. *We say that* $M$ *is* deterministic *if for each* $s \in S$ *and* $i \in I$, *there exists at most one transition* $(s, s', i, o)$ *belonging to* $Tr$.

In this work, we consider that our system will be input-enabled and deterministic, representing the usual behaviour of most systems, where all the states of the FSM can receive any possible input and perform only one change of state (always the same).

*Example 1.* Let us consider Fig. 1 (left). This FSM is both input-enabled and deterministic. The FSM shown in Fig. 1 (center) is not deterministic due to the

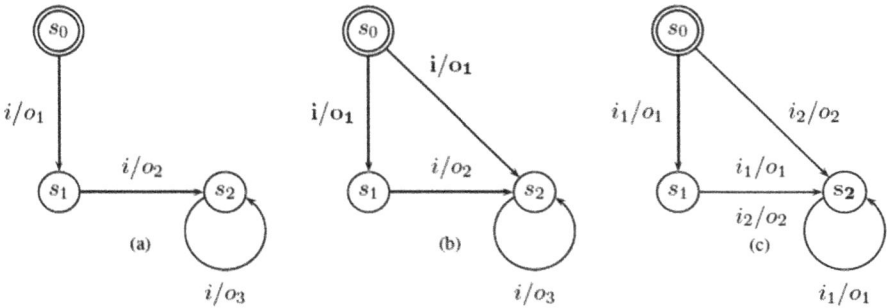

**Fig. 1.** FSMs with different properties

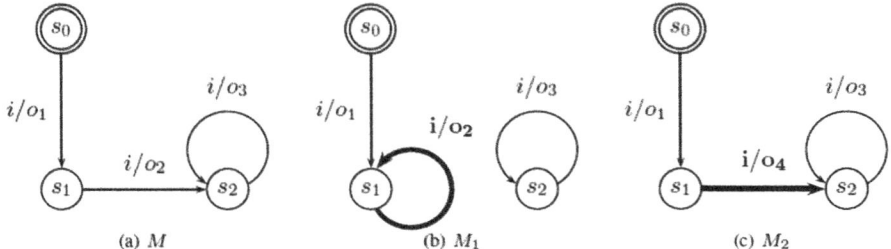

**Fig. 2.** System and two mutations

fact that there are two outgoing transitions labelled with the input $i$ from state $s_0$. The machine depicted in Fig. 1 (right) is not input-enabled since $s_2$ does not have any outgoing transition labelled by the input $i_2$.

**Definition 2.** *Let* $M = (S, I, O, Tr, s_{in})$ *be a FSM, a FSM* $M' = (S, I, O, Tr', s_{in})$ *is a* mutant *of* $M$ *if* $Tr'$ *differs from* $Tr$ *in only one transition. This mutation can be produced by choosing one transition* $(s, s', i, o) \in Tr$ *and replacing it by either* $(s, s', i, o') \in Tr'$, *where* $o' \in O$ *and* $o \neq o'$, *or* $(s, s'', i, o) \in Tr'$, *where* $s'' \in S$ *and* $s' \neq s''$. *A mutant is called* equivalent *when it is semantically identical to the original FSM, that is, for each sequence of inputs both produce the same sequence of outputs. Duplicated mutants are a special form of equivalent mutants. They are equivalent to each other, but not to the original FSM.*

It is important to note that the mutations we are performing do not change the existence nor the uniqueness of the transitions associated to each input. As such, the mutants that we obtain are deterministic and input-enabled.

*Example 2.* Consider the FSM given in Fig. 2 (left), being $s_0$ the initial state. Two possible mutants are shown in Fig. 2 (center) and 2 (right): the first one represents the change of the final state of a transition while the second one represents a change of an output.

**Table 1.** Tests killing mutants

|       | $m_1$ | $m_2$ | $m_3$ | $m_4$ |
|-------|-------|-------|-------|-------|
| $t_1$ | 1     | 1     | 1     | 1     |
| $t_2$ | 0     | 1     | 1     | 0     |
| $t_3$ | 0     | 0     | 1     | 1     |
| $t_4$ | 1     | 0     | 1     | 0     |

**Table 2.** Differences in mutation score

| test sets              | killed mutants          | ms   |
|------------------------|-------------------------|------|
| $T_{S_1} = \{t_4\}$           | $m_1, m_3$               | 0.5  |
| $T_{S_2} = \{t_3, t_4\}$      | $m_1, m_3, m_4$          | 0.75 |
| $T_{S_3} = \{t_1\}$           | $m_1, m_2, m_3, m_4$     | 1    |
| $T_{S_4} = \{t_2, t_3, t_4\}$ | $m_1, m_2, m_3, m_4$     | 1    |

**Definition 3.** *Let $M = (S, I, O, Tr, s_{in})$ be a FSM. A test for $M$ is a tuple $\sigma = (\sigma_{in}, \sigma_{out})$ where $\sigma_{in} \in I^*$ is a sequence of inputs, $\sigma_{out} \in O^*$ is the sequence of outputs that $M$ produces when applying $\sigma_{in}$.*

*Let $t = (\sigma_{in}, \sigma_{out})$ be a test for $M$ and $M'$ be a mutant of $M$. We say that $M'$ passes $t$ if the application of $\sigma_{in}$ to $M'$ produces $\sigma_{out}$; otherwise, we say that $M'$ fails $t$.*

*Example 3.* Let us consider again the mutants depicted in Fig. 2 and the tests $t_1 = (i, o_1)$, $t_2 = (ii, o_1 o_2)$ and $t_3 = (iii, o_1 o_2 o_3)$ for $M$. We have that $M_1$ passes $t_1$ and $t_2$ and fails $t_3$ while $M_2$ passes $t_1$ and fails $t_2$ and $t_3$.

Note that the set of input actions is fixed for a given FSM. In this work we want to minimise the number of inputs that the tests apply to the SUT. In other words, we want to reduce the length of the sequences of inputs (respectively outputs) of the tests that we select to be performed on the SUT. This is because the execution of each input of the tests is associated to a cost, either in terms of budget or time.

The classical approach to assess the goodness of a test suite is to compute its mutation score, which indicates the ratio of killed mutants.

**Definition 4.** *We define the mutation score of the set of tests $\mathcal{T}$ for the set of mutants $\mathcal{M}$ as*

$$ms(\mathcal{M}, \mathcal{T}) = \frac{\#\text{mutants killed}}{\#\text{nonequivalent mutants}}$$

It is important to note that groups of duplicated mutants are only considered once, or otherwise this metric would be biased towards tests that kill many duplicated mutants.

*Example 4.* Let us consider Table 1, where rows represent different tests and columns different mutants. Each cell indicates whether the test kills the mutant (1) or not (0). In this case, we have that tests $t_2$, $t_3$ and $t_4$ all kill two mutants. They all detect $m_3$ and then individually kill $m_2$, $m_4$ and $m_1$ respectively. Also, we have that test $t_1$ kills all four mutants.

Now, let us consider Table 2, which uses different subsets of tests from Table 1. We observe $T_{S_1}$ only kills two mutants, having $ms = 0.5$. If we look at $T_{S_4}$, we are able to kill all mutants, having $ms = 1$, but several tests are required to achieve that goal. We can also achieve $ms = 1$ with $T_{S_3}$, that only needs one test to detect all the mutants.

## 2.2   Multi-objective Genetic Algorithm

We address the problem of selecting a quality subset of tests focused on optimising two objectives: the total number of inputs to be applied and the mutation score of the subset corresponding to a set of mutants. Since these two objectives are interrelated in the sense that fewer inputs are likely to reduce the mutation score, we are faced with a classic multi-objective optimisation problem.

Genetic Algorithms (GAs) are stochastic methods for heuristic adaptive search that are based on the natural selection and in the genetic evolution in nature [6,11,24]. These methods are generally used when looking for a good enough approximation for problems where the optimal solution is unreachable or when it is too expensive to obtain it, generally with an exponential cost. The general idea is to have a population of individuals or chromosomes that evolve over a number of iterations, making changes that eventually yield a solution close to the optimum [16].

Figure 3 presents the flowchart of a GA: Initially, an initialisation of the population takes place, usually yielding a diverse spectrum of initial solutions that could easily find a good approximation. Each iteration of a GA is called *generation*, and consists of a *selection* of individuals, usually representing the best solutions so far, which are *crossed* with each other producing an offspring where some of the new individuals are *mutated* for further exploration of the solutions. Finally, during the replacement phase, the parents and offspring compete to continue into the next generation.

Usually, the use of a fitness function is required to measure how good a solution is. However, we are aiming to optimise two objectives at the same time, and it is often the case that two solutions are incomparable. That is, one is better on one objective but worse on the other. This introduces the concepts of domination and Pareto front, defined as follows.

**Definition 5.** *We say that a solution $x$ dominates another one $y$, denoted by $x \prec y$, if no objective of the latter improves with respect to the former and at least one objective of the former improves compared to the latter. A Pareto front, or front is a set of solutions that are not dominated among them.*

Although GAs have previously used for software testing [2,3,7,10,15,19], in our approach, we use a multi-objective genetic algorithm, the Non-dominated Sorting Genetic Algorithm II (NSGA-II) [5], that has not been considered yet for this specific problem. The idea is not to use a fitness function that guides the execution of the GA, indicating how good each solution is, but to compare and classify the solutions in the different *fronts* that select those that are non dominated [5,21]. Note that, in general, NSGA-II does not work well on many-objective problems. This is due to the fact that dominated solutions are rare when the number of objectives grows, where the front has a higher dimension and the dominated region is smaller. However, our previous experience with NSGA-II in two-objective problems has shown very good results because given a point, half of the solution space is dominated or being dominated by such point. The other half are incomparable points, as it can be seen in Fig. 4, where

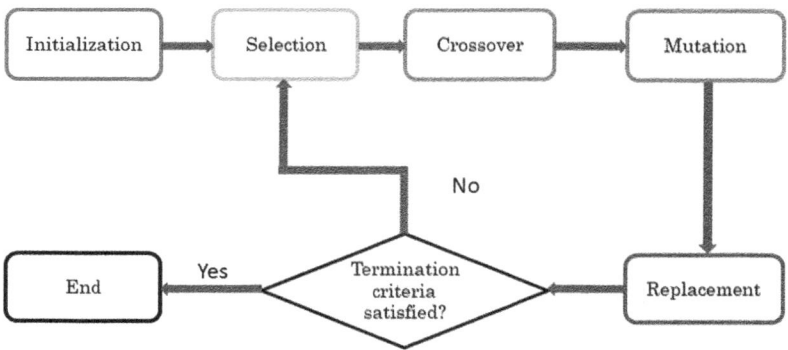

**Fig. 3.** Flowchart for a standard GA

region 1 is dominated by $x$, region 3 dominates $x$ and the regions 2 and 4 are not dominated and do not dominate $x$.

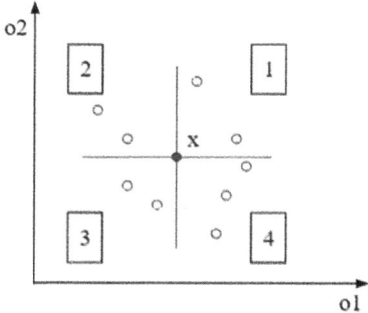

**Fig. 4.** Dominance regions with 2 objectives

It is worth noting that the mutation score $ms$ is a maximisation objective. Therefore, in order to adequate it for the standard use of the NSGA-II. Instead of using such value we will minimise $1 - ms$, as $ms$ takes values from 0 to 1.

Next, we will describe the specific structure of the phases of the GA and the methods we propose to solve the considered problem.

**Population.** The chromosomes or individuals represent a possible solution to the problem to solve. In our case, such problem is to obtain a good subset of tests. Thus, each chromosome is a subset of the initial set of tests, which we have implemented as a list.

**Initialisation.** We have used *random initialisation*, as there is not any special criterion to bias the initialisation, and a diverse population might help to achieve a better solution faster. The number of tests and inputs each solution may initially have should vary, having chromosomes that specialise in each of the objectives.

**Selection.** In order to obtain the best results at the end of each iteration, without exponentially increasing the size of the population, we have to determine which individuals are selected for the rest of the process. In this case, we focus on the individuals that are able to dominate in both objectives over the rest, forming *fronts*.

   Although there exist several methods to perform the selection, most of them require a percentage of relevance for each individual, highly related to a fitness function [11] (roulette wheel, stochastic universal). Since we are targeting this multi-objective problem, we focus on two methods that do not require such percentage. These methods are *truncation* and *tournament*:

- *Truncation* is an elitist method that chooses a percentage of the population that fits better the objectives we are minimising. The number of individuals that should be selected should correspond to a range varying from 10% to 50% of the actual population. Then, the selected individuals are duplicated as many times as required to maintain the size of the population. In our case, we consider a 25% of individuals, where each of them appears 4 times.
- *Tournament* is a selection technique that individually selects each chromosome from small groups of, usually, two or three participants. The way of deciding who wins each small tournament can vary. The deterministic approach selects the individual that dominates the others. Another option is a probabilistic tournament, where a random number between 0 and 1 is chosen. If the number is smaller than an initial parameter, then the first chromosome is the one that wins the tournament and is selected. Otherwise, the other participants have a chance to compete without this rejected competitor. In this way, we increase the diversity, and a high fixed parameter would still produce a high fitted population. In our case, we consider 3 participants with a 0.8 ratio of victory for the better suited participant.

**Crossover.** The crossover phase focuses on the exploitation of the search space, as it aims to find new solutions that are combinations of already existing ones. The crossover operators that we use exchange tests in several pairs of individuals. We have considered the following two methods:

- *Standard crossover*: we select a point in the list of tests of an individual that we will use to split it into two parts. Then, we produce two individuals with the first half of one of the parents and the second half of the other parent.
- *Continuous crossover*: for each test in each parent individual, there is a chance to be swapped for the respective test in the other parent's list. In this way, the children generated have a more mixed configuration of the tests.

**Mutation.** The mutation phase aims at fully explore the search space, adding new solutions that were not considered before, trying to increment the diversity of the population. Even if a bad solution arises from this method, the small ratio of mutation operations does not have a negative impact, as such solution would be discarded soon. However, if a promising solution is obtained, it is likely to stay and improve the overall set of solutions. The mutation usually makes small modifications on an individual to avoid local optima. Here we propose two methods:

- The *adding mutation* approach introduces a new test to the subset that was not being considered before. Since this increases the size of the solution, it is ideal for small solutions that are not complete.
- The *replacement mutation* approach substitutes a test that was being considered by a chromosome by a new one, where more changes could be produced. This mutation is oriented to big sets of tests, where there are redundant tests or partial solutions that do not generate a substantial increase in mutation score nor reduces the inputs to apply.

**Replacement.** The last step of a GA is to fix the population for the next generation. For that, we have to decide which parents and children continue, as the size of the population should stay constant. We have only considered a replacement method oriented to solve multi-objective problems.

In this NSGA-II replacement we combine the parent and children population, having twice the desired size. Then, we generate fronts of solutions that do not dominate each other, having different levels of fronts. This ensure elitism, producing an increase of the performance of the solution after each generation. Then, we include the best fronts in the new population until we reach the specified size. If at any point it is impossible to add a complete front, we add the chromosomes that are more spread in such front, as they are more likely to have important information rather than a group of solutions really close to each other.

### 2.3   Cluster Algorithm

In this section we briefly describe the fundamentals of the cluster technique that we use. First, it is important to note that clustering is a machine learning algorithm used to group similar objects based on certain criteria and features, without any prior knowledge of the groups. In our experiments, we use a method known as Agglomerative Clustering [17]. This clustering algorithm begins by considering each test case as a separate cluster. Then, the similarity between each pair of clusters is calculated using the Euclidean distance. Subsequently, the two most similar clusters are merged into a single cluster. At this stage, it is necessary to employ a linkage criterion to establish how similarity between clusters is measured. In this case, we use the Ward criterion [18]. The process of calculating the similarity between each pair of clusters and merging the two most similar clusters into a single cluster iterates until the number of clusters reaches a predetermined value established beforehand. For our experiments, we use clustering on the tests only knowing what mutants each of them kills. Also we perform this clustering to obtain from 2 to 30 clusters, each of them yielding

---

**Algorithm 1.** Agglomerative Clustering

---

    **Input:**   Dataset $\{t_1, t_2, \ldots, t_n\}$, distance metric $d$, number of clusters $T$
    **Output:** Final set of $T$ clusters

1: Initialise each test as a separate cluster: $C \leftarrow \{\{t_1\}, \{t_2\}, \ldots, \{t_n\}\}$
2: **while** $T < |C|$ **do**
3:     Compute pairwise distances between clusters in $C$ using distance metric $d$.
4:     Find the two closest clusters, $(C_i, C_j)$
5:     Merge clusters $C_i$ and $C_j$: $M \leftarrow C_i \cup C_j$
6:     $C \leftarrow (C \cup M) \setminus \{C_i, C_j\}$
7: **end while**
8: **return** Final set of $T$ clusters $C$

---

one solution. In particular, the subset of tests made of the shortest test of each cluster.

Algorithm 1 presents a generalised pseudo-code for the Agglomerative Clustering method. The required input is a set with $n$ tests and a predefined distance metric, $d$ (in our case, the Euclidean distance), along with the desired number of clusters, $T$. The output will be the final set containing $T$ clusters. In line 1, the algorithm assigns each test to a separate cluster. Then, the clusters are processed until the number of clusters equals to $T$. The algorithm computes the distances between all pairs of clusters (line 3). The two closest clusters are then identified (line 4), based on the computed distances, and merged into a single cluster, forming a new cluster (line 5). This new cluster, will replace the merged clusters (line 6).

## 3    Experiments

We have compared our two techniques with a total of 9 SUTs, ranging from smaller specifications that generated 400 mutants to bigger ones with 12800. The length of the sequence of inputs of our tests has a range that varies from 202 to 6428. We evaluate how good our methods are in these SUTs by obtaining representative results and we show how much we are able to improve the mutation score as the number of inputs grows. Additionally, we have included a random selection of solutions and a bad set of solutions to visualise the range of the possible results. This is important as no matter how you select a group of tests, at a given point adding more and increasing the sequence of inputs will result in detecting more faults, and thus the mutation score will improve as well. Because of that, the different approaches must have similar values near the extreme values, i.e. the empty set and the full set. In order to obtain the bad set of solutions[1], we used the GAs with opposite objectives. In this case, the target was to minimise the mutation score (maximising $1 - ms$) and to maximise the number of inputs at the same time. The implementation of the

---

[1] Represented in Figs. 6 and 7 as *Worst*. The *Random* set is chosen by randomly selecting tests to have different solutions with different sizes.

described methods, the SUTs, and all the results are available at https://github.com/miguelbpsg/ICCCI24.

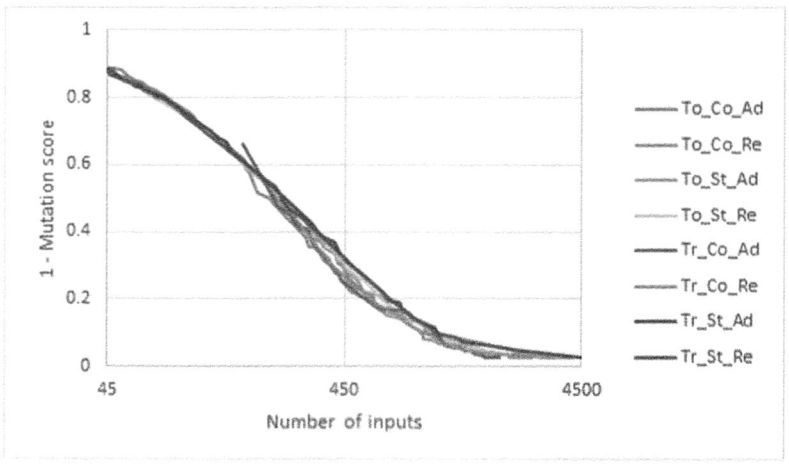

**Fig. 5.** Comparison of GAs

First, we have performed all 8 configurations of the GAs a total of 20 times. To control the random behaviour and reduce the threat of validity of our experiments we use the median of the 20 runs as the representative value for each configuration. This also means that the GAs may provide better solutions when executed once, but one individual run is not representative, and should be repeated more times. Figure 5 shows that the different configurations of the GA yield similar Pareto fronts. However, an important feature of these algorithms is diversity, that is distributing the solutions in an extensive manner. For that, the truncation method is too tight, as it focuses on a small number of solutions. Although the results of this selection method are comparable to the tournament method, no matter which crossover or mutation method is used, the range is smaller. Such range is skewed towards fewer inputs when using the standard crossover or the replacement mutation, and towards bigger inputs with the continuous crossover or the adding mutation (all the results can be checked in the files and graphs in github). Considering the tournament selection method, the solutions obtained are distributed through all the possible inputs. The four configurations for this selection are of an equivalent quality, but in particular the continuous crossover and replacement mutation yield solutions that are further from one another, thereby covering a wider solution space. For that reason, in order to compare the GAs with the clustering approach and to easily visualise the results, we will only show such configuration of the GAs corresponding to tournament selection, continuous crossover and replacement mutation.

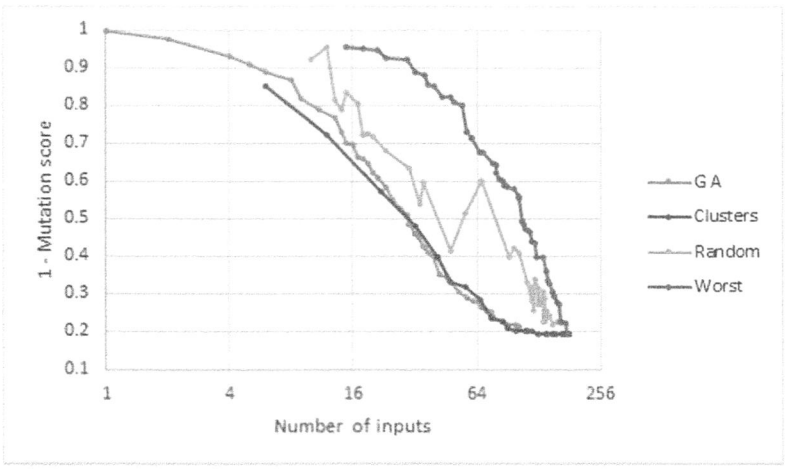

**Fig. 6.** Small experiment

As the Agglomerative Clustering method is deterministic, only one execution is needed to have a valid result which is later analysed and compared in Figs. 6 and 7. The former corresponds to the smallest experiment we perform with a set of tests with a total of 202 inputs and 400 mutants to detect. In the graph, we observe that the clustering method slightly improves the GA when it comes to selecting small subsets, that is when few clusters are being considered. However, the GA is able to give more uniform solutions, starting from one input to all of them. Besides, the lines for the GA and the clustering method are very similar, having closer solutions among the clustering method than among the GA. Both our methods always dominate considerably a random solution, that simulates selecting a subset without any criteria or knowledge but the number of inputs to perform. The latter corresponds to the biggest experiment we perform with a set of tests with a total of 6428 inputs and 12800 mutants to detect. Here the same trend is maintained. The clustering approach have a higher mutation score (lower $1 - ms$) than GAs for small subsets, but as more tests are selected and the number of inputs grows, both methods get closer while obtaining better solutions than any random one throughout the spectrum that the GA fully covers. It is important to note too that the scale for the number of inputs is logarithmic, so in order to increase the mutation score, the number of inputs has to substantially grow each time.

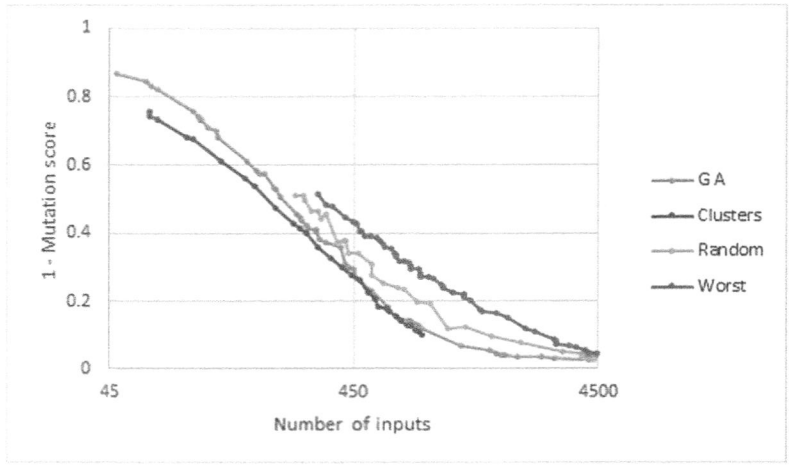

**Fig. 7.** Big experiment

From these experiments we determine that our proposed methods are able to obtain subsets of tests that minimise the cost while detecting many faults.

## 4    Conclusions and Future Work

In this paper we have presented a clustering method and a genetic algorithm to select a good subset of tests that detects a big number of mutants applying a small amount of inputs. We have performed experiments on several SUTs to obtain subsets that optimise these objectives, and compare them. We have concluded that both techniques yield great results, being the clustering slightly better but more specific, and the GA much more spread while consistently providing good solutions. Our proposal significantly simplifies the task of selecting *good* tests from a large test suite in order to validate a SUT and provides a diverse group of solutions depending on which objective is more important to prioritise.

Future lines of work include the use of other Machine Learning methods [14] to select good tests, extending the methods for the different phases of the GA, and specially combining the two techniques to obtain better solutions in a more spread space. Additionally, this work could be extended to other formalisms, or to systems coded in popular programming languages like Java, Python or C. Another approach involves considering different metrics to optimise. For example, some inputs may take more time than others to be executed in a real system, or some actions may be more critical and thus more expensive to be tested, changing the objective of minimising the number of inputs to minimise the *actual* cost or time of testing. These objectives do not necessarily have to replace the current ones, but could also be added to the problem, raising it to a many-objective one.

# References

1. Arafeen, Md.J., Do, H.: Test case prioritization using requirements-based clustering. In: 2013 IEEE Sixth International Conference on Software Testing, Verification and Validation, pp. 312–321 (2013)
2. Benito-Parejo, M., Merayo, M.G.: An evolutionary algorithm for selection of test cases. In: 2020 IEEE Congress on Evolutionary Computation (CEC), pp. 1–8 (2020)
3. Benito-Parejo, M., Merayo, M.G.: Using genetic algorithms to select test cases for finite state machines with timeouts. In: 2021 IEEE Congress on Evolutionary Computation (CEC), pp. 2403–2410 (2021)
4. Catal, C., Mishra, D.: Test case prioritization: a systematic mapping study. Softw. Qual. J. **21**(3), 445–478 (2013)
5. Deb, K., Pratap, A., Agarwal, S., Meyarivan, T.: A fast and elitist multiobjective genetic algorithm: NSGA-II. IEEE Trans. Evol. Comput. **6**(2), 182–197 (2002)
6. Goldberg, D.E.: Genetic Algorithms in Search, Optimisation and Machine Learning. Addison-Wesley, Boston (1989)
7. Harman, M., McMinn, P.: A theoretical and empirical study of search-based testing: local, global, and hybrid search. IEEE Trans. Softw. Eng. **36**(2), 226–247 (2010)
8. Hierons, R.M., Merayo, M.G., Núñez, M., Mutation testing. : Laplante, P.A. (ed.) Encyclopedia of Software Engineering, pp. 594–602. Taylor & Francis (2010)
9. Jia, Y., Harman, M.: An analysis and survey of the development of mutation testing. IEEE Trans. Softw. Eng. **37**(5), 649–678 (2011)
10. Jones, B.F., Eyres, D.E., Sthamer, H.-H.: A strategy for using genetic algorithms to automate branch and fault-based testing. Comput. J. **41**(2), 98–107 (1998)
11. Katoch, S., Chauhan, S.S., Kumar, V.: A review on genetic algorithm: past, present, and future. Multimedia Tools Appl. **80**, 8091–8126 (2021)
12. King, K.N., Offutt, A.J.: A Fortran language system for mutation-based software testing. Softw. Pract. Exp. **21**(7), 685–718 (1991)
13. Lou, Y., Hao, D., Zhang, L., Mutation-based test-case prioritization in software evolution. In: 26th International Symposium on Software Reliability Engineering. ISSRE'15, pp. 46–57. IEEE Computer Society (2015)
14. Méndez, M., Merayo, M.G., Núñez, M.: Machine learning algorithms to forecast air quality: a survey. Artif. Intell. Rev. **56**(9), 10031–10066 (2023)
15. Michael, C.C., McGraw, G., Schatz, M.A.: Generating software test data by evolution. IEEE Trans. Softw. Eng. **27**(12), 1085–1110 (2001)
16. Mirjalili, S.: Evolutionary Algorithms and Neural Networks. SCI, vol. 780. Springer, Cham (2019). https://doi.org/10.1007/978-3-319-93025-1
17. Müllner, D., Modern hierarchical, agglomerative clustering algorithms. arXiv, abs/1109.2378 (2011)
18. Murtagh, F., Legendre, P.: Ward's hierarchical agglomerative clustering method: which algorithms implement ward's criterion? J. Classif. **31**(3), 274–295 (2014)
19. Núñez, A., Merayo, M.G., Hierons, R.M., Núñez, M.: Using genetic algorithms to generate test sequences for complex timed systems. Soft. Comput. **17**(2), 301–315 (2013)
20. Gbeminiyi John Oyewole and George Alex Thopil: Data clustering: application and trends. Artif. Intell. Rev. **56**(7), 6439–6475 (2022)
21. Pareto, V.: Cours d'économie politique, vol. 1. Librairie Droz (1964)

22. Petrović, G., Ivanković, M., Fraser, G., Just, R.: Does mutation testing improve testing practices? In: 2021 IEEE/ACM 43rd International Conference on Software Engineering (ICSE), pp. 910–921 (2021)
23. Shin, D., Yoo, S., Papadakis, M., Bae, D.H.: Empirical evaluation of mutation-based test case prioritization techniques. Softw. Test. Verif. Reliab. **29**(1–2), e1695 (2019)
24. Sivanandam, S.N., Deepa, S.N.: Introduction to Genetic Algorithms. Springer, Heidelberg (2008). https://doi.org/10.1007/978-3-540-73190-0
25. Wei, C., Yao, X., Gong, D., Liu, H.: Spectral clustering based mutant reduction for mutation testing. Inf. Softw. Technol. **132**, 106502 (2021)
26. Xu, R., Wunsch, D.: Survey of clustering algorithms. IEEE Trans. Neural Netw. **16**(3), 645–678 (2005)

# Computational Intelligence for Digital Content Understanding

# Fusing Visual and Textual Representations via Multi-layer Fusing Transformers for Vietnamese Visual Question Answering

Cong Phu Nguyen[1,2] , Huy Tien Nguyen[1,2] , and Tung Le[1,2(✉)]

[1] Faculty of Information Technology, University of Science, Ho Chi Minh, Vietnam
21C11018@student.hcmus.edu.vn, {ntienhuy,lttung}@fit.hcmus.edu.vn
[2] Vietnam National University, Ho Chi Minh city, Vietnam

**Abstract.** In recent decades, artificial intelligence has made significant progress in understanding and interacting with images. One of the important applications of this technology is Visual Question Answering (VQA), a research field that requires computers to understand and answer questions about images in a natural manner. Despite extensive research and development in VQA for English, there have been very few similar efforts made for other languages, especially Vietnamese. This gap presents a significant challenge and opportunity for the advancement of VQA technology in the Vietnamese language context. By bridging this gap, the field of Vietnamese VQA not only enriches the diversity of research in artificial intelligence but also enables practical applications in various domains, such as education, healthcare, and entertainment, catering to Vietnamese-speaking populations worldwide. Thus, the exploration and development of Vietnamese VQA systems hold immense potential for advancing both research and practical applications in the intersection of computer vision and natural language processing. In this paper, we propose a Multi-layer Fusing Transformer model utilizing a cross attention module to combine multiple modality features of images and texts from different layers in an aggregated representation. Our architecture allows us extract information from low level to high level. Through detailed experiments and ablation studies, our model achieves promising results against the competitive baselines in ViVQA dataset for Vietnamese language.

**Keywords:** Cross Attention · Multilayer Fusing Attention · Visual Question Answering · ViVQA

# 1   Introduction

In everyday life, humans effortlessly receive and interpret information from multiple sources simultaneously, including visuals, sounds, and textual data. For machines to mirror this complex human capability, they must be adept at processing a variety of signal types concurrently. The field of multi-modal research, especially tasks that combine vision and language, has seen significant growth in recent years, reflecting the increasing interest in understanding and developing systems that can interpret and respond to combined data inputs. Among the myriad of tasks within the realm of image and text multi-modal processing, activities such as Visual Question Answering (VQA), image captioning, text-to-image synthesis, and cross-modal retrieval have emerged as key areas of focus. Notably, Visual Question Answering (VQA) represents a cutting-edge intersection of research and practical application, showcasing the potential of integrating visual and linguistic features. VQA systems are designed to analyze images and comprehend questions about them, synthesizing information from both domains to produce accurate and relevant answers. This ability not only highlights advancements in artificial intelligence (AI) but also opens up new pathways for human-computer interaction, making VQA a critical area of study in the broader field of AI research.

Vietnamese is a widely spoken language with a rich visual culture, making it an important target for developing language-enabled AI systems. However, the development of VQA systems for Vietnamese faces unique challenges due to linguistic nuances, cultural contexts, and the scarcity of annotated datasets in the Vietnamese language. Existing VQA models trained on English-centric datasets may not generalize well to Vietnamese due to language differences and semantic variations.

Therefore, the primary objective of this research is to design and develop a Vietnamese Visual Question Answering (VQA) system capable of understanding and accurately answering questions about visual content in the Vietnamese language. This system will leverage state-of-the-art techniques in computer vision and natural language processing, adapted and optimized for the Vietnamese language.

This study introduces a novel method designed to enhance the performance of Vietnamese Visual Question Answering (VQA) systems. By addressing the specific challenges of processing visual content and natural language in the Vietnamese language, our proposed method aims to significantly improve the accuracy and reliability of VQA systems tailored for Vietnamese-speaking users. This advancement not only contributes to bridging the gap in language-centric AI technologies but also opens up new possibilities for the development of inclusive and culturally relevant AI applications for Vietnamese communities.

In prior models within the realm of deep learning, the emphasis has predominantly been on employing architectures featuring multiple hidden layers designed to meticulously extract and articulate both visual and linguistic features. Traditionally, these models have harnessed solely the attributes derived from the terminal layer as semantically enriched inputs for ensuing predictive analyses.

The efficacy and robustness of such deep, multi-layered constructs are well-established and validated through extensive research across various domains. However, when addressing multi-modal challenges that necessitate the concurrent assimilation of both imagery and linguistic data, a sophisticated approach is not only requisite for the adept representation of features but also imperative for their harmonious integration. This integration aims to engender a more profound semantic synthesis. The conventional strategy of isolating features to the concluding layers of deep learning frameworks, thereby processing visual and textual streams in a segregated manner, inadvertently mitigates the potential for dynamic interaction between images and textual queries throughout the model's learning trajectory.

In light of these considerations, our research introduces the innovative Multi-layer Fusing Transformer (MFT) model. This model is specifically architected to not only represent but also intricately weave together visual and textual features across the entirety of its processing layers. By doing so, the MFT model facilitates a more integrated and continuous interaction between these two pivotal dimensions. This novel approach marks a significant advancement in the field of Vietnamese Visual Question Answering, promising to elevate the efficacy of models tasked with deciphering and responding to complex visual-textual inquiries. Through this cutting-edge methodology, our work sets a new benchmark in multi-modal AI research, particularly in enhancing the comprehension and interpretative capabilities of systems engaging with the rich tapestry of Vietnamese visual and textual content.

Section 2 reviews the background and relevant studies in Visual Question Answering (VQA) for Vietnamese, highlighting challenges and opportunities. Section 3 introduces our proposed VQA model for Vietnamese, detailing its architecture, components, and training methodology. Section 4 describes our experimental setup and evaluation, including the dataset, metrics, and comparative analysis with baseline models. Section 5 concludes with key findings, contributions to VQA research in Vietnamese, and future research directions.

## 2   Related Works

In recent years, interest in Visual Question Answering (VQA) in the Vietnamese language has significantly increased, with numerous studies exploring its potential and proposing new datasets as well as methods. In 2021, Tran and colleagues [11] introduced the first ViVQA dataset for VQA in Vietnamese, which has since become the standard in subsequent research.

With the rapid development of technology and computers in recent years, Deep Neural Networks (DNNs) have achieved significant success in various applications. In 2017, the paper "Attention is all you need" by the authors Vaswani et al. [14] from Google brought a new wave to the field of natural language processing research. The authors introduced the architecture of Transformers, which, instead of following traditional approaches to text representation, utilizes the self-attention mechanism to estimate the weight of relationships between words in a text and then aggregates the representative information of a word with the weighted information of all words in the text. This implies that in the encoding process, the Transformers architecture can simultaneously process the entire input text without the need for recurrent iterations. At the end of 2018, based on the Transformers architecture, the pre-trained model BERT [3] was announced. BERT was proposed with the ability to train using semi-supervised learning methods, making model training easier and more effective in terms of semantic representation of sentences, marking a new milestone in building transferable language models. With the success of the Transformers architecture in natural language processing, many models based on this architecture have been applied to the field of computer vision, such as Vision Transformer (ViT) [4] and BeiT [1], yielding promising results compared to models using CNNs.

Current approaches to the Vietnamese VQA problem utilize language and vision models trained on Vietnamese datasets. Notable examples include models like PAT [9], BARTPhoBEiT [12] and Multi-vision Contextual Attention (MCA) model [7], which are considered state-of-the-art (SOTA) for the VQA task on the ViVQA dataset [11]. While the PAT model utilizes Parallel Attention with a hierarchical feature extractor for questions, the BARTPhoBeiT model employs multiway transformers on the BardPho [13] and BeiT [1] models. Multi-vision Contextual Attention model simultaneously learns both global and local contextual visual features of the image and uses multi-branch contextual attention fusion that hierarchically integrates visual and textual features.

The findings from [5] confirm the intuitions researchers had about the hierarchical representations learned by deep architectures. Specifically, it reveals that the units in higher layers represent features that are significantly more complex and correspond to combinations of features from lower layers. And according to [2], different layers of BERT exhibit various specializations, suggesting that it could be more beneficial to integrate information from multiple layers rather than selecting a single layer based solely on its overall performance.

## 3   Our Model

We propose a new method utilizing cross-attention across different layers, called Multi-layer Fusing Transformer (MFT). Additionally, to ensure the model performs well on Vietnamese, we leverage language models that have been pre-trained on large Vietnamese datasets. The MFT model consists of 4 main components: the Language Embedding Module, the Visual Embedding Module, the Cross-attention Module, and the Answer Selector (see Fig. 2). The following section will describe in detail the architecture of the model.

### 3.1   Language Embedding Module

We use PhoBert [8], the pre-trained language model utilized as the Language Embedding Module in our Visual Question Answering (VQA) system. PhoBERT is built upon the Transformer architecture, a powerful neural network architecture renowned for its effectiveness in natural language processing tasks. The Transformer model consists of multiple layers of self-attention mechanisms and feedforward neural networks, allowing it to capture long-range dependencies and contextual information from input sequences. Specifically, PhoBERT employs a multi-layer bidirectional Transformer encoder architecture. Each layer of the encoder consists of self-attention mechanisms, which enable the model to weigh the importance of different words in the input sequence based on their contextual relevance. Additionally, each layer includes feedforward neural networks for feature transformation and activation functions for non-linear mapping. The architecture of PhoBERT allows it to generate contextualized embeddings for words and sentences, capturing the semantic relationships and nuances of the Vietnamese language effectively.

PhoBERT is pre-trained on a large corpus of Vietnamese text data using self-supervised learning techniques. The combination of the Transformer architecture and the self-supervised learning objectives enables PhoBERT to learn rich and meaningful representations of the Vietnamese language, making it well-suited for various downstream tasks, including Visual Question Answering.

### 3.2   Visual Embedding Module

The design of ViT draws inspiration from the success of the BERT [3] model in natural language processing. Just as BERT revolutionized NLP by capturing bidirectional context in text sequences, ViT leverages the Transformer architecture to capture global and local context in visual data. Specifically, ViT adapts the self-attention mechanism from BERT to process image patches, enabling it to model complex relationships among visual elements effectively. By treating images as sequences of patches and applying self-attention across these patches, ViT achieves state-of-the-art performance in various computer vision tasks (Fig. 1).

### 3.3   Cross-Attention Module

Taking inspiration from the VilBERT [6], we employ a cross-attentional transformer layer to effectively integrate information between the question and the image, yielding a representation vector used for answer classification. Each layer has two sub-layers. The initial layer comprises a multi-head cross-attention mechanism, while the subsequent layer consists of a straightforward, fully connected feed-forward network applied position-wise. In the first layer, one modality's representation is used as the Key and Value vectors, while the other modality's representation serves as the Query vector. Take the cross-attention of the text branch as an example, the query $\mathbf{Q_v}$ is generated from the image intermediate representations, while the key $\mathbf{K_t}$ and value $\mathbf{V_t}$ is generated from text intermediate representations, the computation can be formulated as:

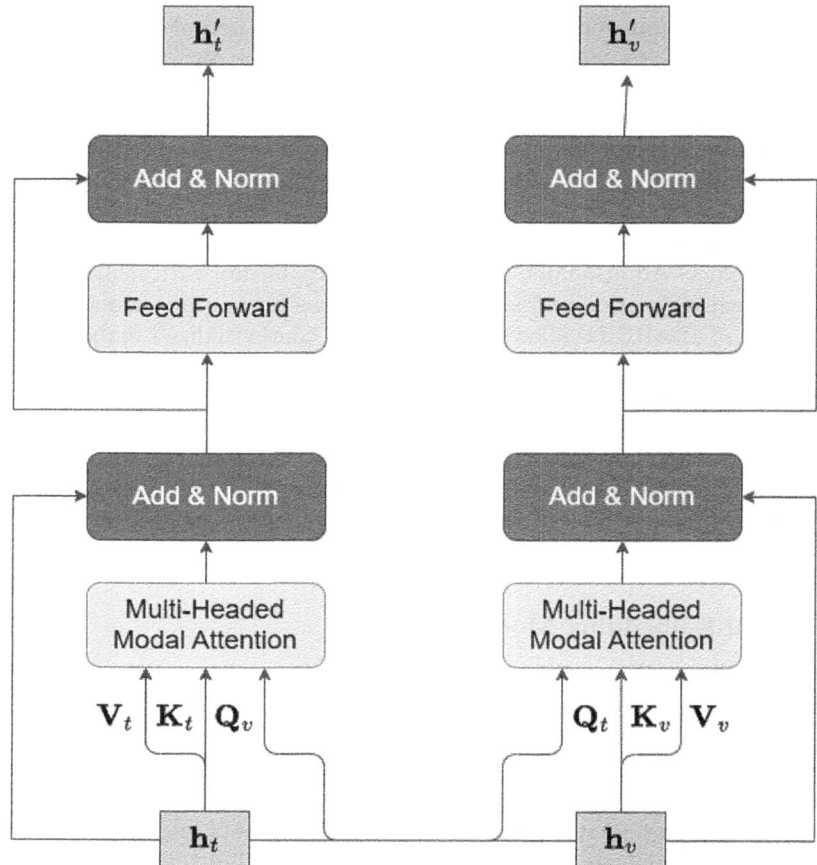

**Fig. 1.** Architecture of Cross-Attention

$$Attention(\mathbf{Q_v}, \mathbf{K_t}, \mathbf{V_t}) = Softmax\left(\frac{\mathbf{Q_v}\mathbf{K_t}^{\mathbf{T}}}{\sqrt{d_k}}\right)\mathbf{V_t}, \tag{1}$$

where the subscript t refers to text representations and v for visual representations, respectively. $\mathbf{K_t}^{T}$ is the transpose of the matrix $\mathbf{K_t}$. The $d_k$ represents the attention dimension. Using cross-attention fusion layers, both modalities can acquire pertinent and supplementary contextual representations from each other. We utilize a residual connection surrounding each of the two sub-layers, succeeded by layer normalization.

We perform cross-attention taking as input the outputs of language and image processing models at different layers. To integrate information from simple to complex from the two models, representation vectors at the beginning, middle, and end positions (corresponding to layers 0, 6, 12) are fed into the cross-attention layers. The representation vectors are averaged for both the text and vision parts, and the vector corresponding to the position of the [CLS] token of each part is concatenated together and passed through a pooling layer followed by a softmax operation to obtain probabilities for each answer.

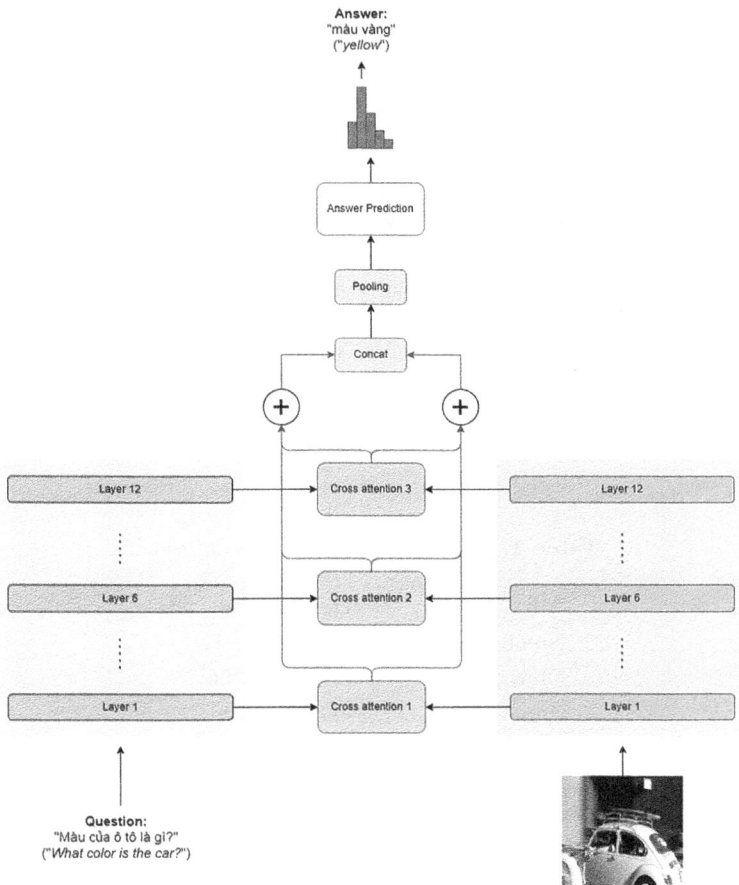

**Fig. 2.** Multi-layer Fusing Transformer: use cross-attention on different layers of pre-trained language model and vision model

## 3.4  Answer Selector

The Answer Selector module fuses visual features $h'_v$ and linguistic features $h'_t$ to produce fused features $h'_f$. These fused features are then projected into the vocabulary space, resulting in a probabilistic vector that indicates the most likely candidate answer.

## 4    Experiments

In this section, we conduct a series of experiments to evaluate the performance of our proposed MFT models alongside established state-of-the-art (SOTA) models. Our experiments are conducted on the ViVQA dataset [11], chosen for its relevance and suitability for assessing the efficacy of visual question answering systems. We provide comprehensive details regarding the implementation process and the selection of evaluation metrics, ensuring a fair and rigorous comparison between our models and existing benchmarks on the same dataset.

### 4.1    Dataset

The ViVQA dataset, created for Vietnamese Visual Question Answering (VQA), addresses the lack of resources in this domain despite the growing interest in VQA within natural language processing and computer vision communities. It consists of 10,328 images paired with 15,000 questions and answers in Vietnamese, facilitating the evaluation of VQA models tailored for the Vietnamese language. The dataset is divided randomly into training and test sets at an 8:2 ratio. These questions are categorized into types such as Object, Number, Color, and Location, and distributed across the train and test sets as shown in the Table 1

**Table 1.** Statistics of the ViVQA dataset

|                   | Training | Test   |
|-------------------|----------|--------|
| Number of samples | 11,999   | 3001   |
| Object            | 41.55%   | 41.56% |
| Number            | 14.81%   | 14.81% |
| Color             | 20.82%   | 20.82% |
| Location          | 22.82%   | 22.82% |

### 4.2    Evaluation Metrics

Following Rajpurkar et al. [10], we use two evaluation metrics: F1 score and Exact Match. The F1 score serves as a crucial metric in assessing the performance of visual question answering models. It quantifies the average overlap between predicted and ground truth answers, treating them as bags of tokens to compute their F1 score. Specifically, the F1-score for each answer is computed based on the tokens of both the gold answer (GA) and the predicted answer (PA), ensuring a comprehensive evaluation. This calculation is performed for each question within the dataset, with the overall F1 score being averaged across all questions to provide a comprehensive assessment of model performance. In the context of

Vietnamese VQA, F1 scores are calculated based on tokenized versions of the answers, facilitating precise evaluation and comparison across different models and datasets. Exact Match evaluates the proportion of predictions that exactly match any of the ground truth answers.

$$Precision\,(P) = \frac{GA \cap PA}{PA} \tag{2}$$

$$Recall\,(P) = \frac{GA \cap PA}{GA} \tag{3}$$

$$F1 = \frac{2 \times P \times R}{P + R} \tag{4}$$

### 4.3  Experimental Settings

In our model, we take advantage of pre-trained models for extracting features from images and questions. The name of initial version in our feature extraction modules are presented in Table 2. Besides, we also provide more details to make the reproducing process easier. For Multi-head Attention and Transformer components, we also show the range of our tuning process. The experiments were performed on a computing system with a CPU of Intel(R) Xeon(R) E5-2696 v3, a GPU of RTX3060, and a RAM of 48 GB.

**Table 2.** Detail of Component Setting

| Component | Value |
|---|---|
| Visual Embedding Module | google/vit-base-patch16-384 |
| Language Embedding Module | vinai/phobert-base-v2 |
| No. Connections | $i \in \{0, 0, 6, 6, 12, 12\}$ |
| Batch size | 32 |
| Optimizer | AdamW(lr = 1e$-$4, eps = 1e$-$8) |
| Learning rate Scheduler | OneCycleLR(lr = 1e-4, pct_start = 0.05) |

### 4.4  Results

Because there are few studies on Visual Question Answering (VQA) for Vietnamese, there are also few methods for this problem since the ViVQA dataset was published. In this section, we will compare the MFT model with recent SOTA models on the ViVQA dataset, including the Parallel Attention Transformer (PAT) [9], BARTPhoBeiT [12] and Multi-vision Contextual Attention (MCA) model [7]. For these models, the results are extracted from the publication of these models. The details of the comparison are presented in Table 3.

**Table 3.** The comparison of our model and competitive baselines (test-standard)

| Model | Precision | Recall | F1-score | Exact Match |
|---|---|---|---|---|
| $ViHieCoAtt - syllable_{100dims}$ [12] | 0.2271 | 0.2954 | 0.2234 | – |
| $ViHieCoAtt - word_{100dims}$ [12] | 0.2049 | 0.3044 | 0.2256 | – |
| $ViHieCoAtt - syllable_{300dims}$ [12] | 0.2185 | 0.3104 | 0.2277 | – |
| $ViHieCoAtt - word_{300dims}$ [12] | 0.2192 | 0.3236 | 0.2436 | – |
| $PAT$ [9] | – | – | – | 0.6055 |
| $MCA$ [7] | – | – | – | 0.6076 |
| $BARTPhoBeit - base_{syllable}$ [12] | 0.6684 | 0.6718 | 0.6610 | – |
| $BARTPhoBeit - base_{word}$ [12] | 0.6593 | 0.6646 | 0.6518 | – |
| $BARTPhoBeit - large_{syllable}$ [12] | 0.6931 | 0.6858 | 0.6777 | – |
| $BARTPhoBeit - large_{word}$ [12] | 0.6755 | 0.6749 | 0.6643 | – |
| $MFT(our)$ | **0.7057** | **0.7060** | **0.7044** | **0.6485** |

In comparison to existing models in the field, our proposed Multi-layer Fusing Transformer (MFT) model exhibits notable advantages across various performance metrics. While current state-of-the-art (SOTA) models have demonstrated commendable performance in Vietnamese Visual Question Answering (VQA) tasks, our MFT model surpasses them in terms of both F1-score and Exact metric. Specifically, when compared to the best-performing version of the BARTPhoBeit model, MFT showcases a significant improvement, achieving a 4% higher F1-score. Moreover, in terms of the Exact metric, our proposed model exhibits a similarly substantial lead of 4% over the PAT model. These results underscore the effectiveness and superiority of the MFT model in capturing intricate semantic relationships between visual and textual inputs, thereby enabling more accurate and nuanced responses in VQA tasks. Additionally, the innovative cross-attention mechanism integrated across different layers within the MFT architecture enhances its capability to effectively fuse information from both modalities, leading to superior performance outcomes. Overall, the comparative analysis highlights the MFT model's potential as a leading solution in advancing the state-of-the-art in Vietnamese VQA research.

## 4.5    Ablation Studies

In this section, we experiment with the ability to combine cross-attention modules at different layer positions as well as the number of cross-attention layers on the language model and image processing model (Table 4).

The findings indicate that integrating layers at non-matching positions does not yield favorable outcomes. Furthermore, exclusively combining layers either at the beginning, the middle, or at the end does not produce satisfactory results compared to distributing positions evenly at the beginning, middle, and end of the models (Table 5).

**Table 4.** The effect of using cross-attention on different layer pairs

| Connection layers | Precision | Recall | F1-score | Exact Match |
|---|---|---|---|---|
| (0, 0), (6, 6), (12, 12) | **0.7057** | **0.7060** | **0.7044** | **0.6485** |
| (0, 12), (6, 6), (12, 0) | 0.7007 | 0.7003 | 0.6990 | 0.6415 |
| (0, 6), (6, 12), (12, 0) | 0.6956 | 0.6954 | 0.6939 | 0.6351 |
| (0, 12), (6, 0), (12, 6) | 0.7012 | 0.7004 | 0.6992 | 0.6415 |
| (0, 0), (1, 1), (2, 2) | 0.6217 | 0.6207 | 0.6192 | 0.5548 |
| (5, 5), (6, 6), (7, 7) | 0.6602 | 0.6587 | 0.6578 | 0.5975 |
| (10, 10), (11, 11), (12, 12) | 0.6808 | 0.6820 | 0.6798 | 0.6198 |

**Table 5.** The effect of using cross-attention on number of cross-attention layers

| Num layers | Connection layers | Precision | Recall | F1-score | Exact Match |
|---|---|---|---|---|---|
| 2 | (0, 0), (12, 12) | 0.7033 | 0.7032 | 0.7017 | 0.6441 |
| 3 | (0, 0), (6, 6), (12, 12) | **0.7057** | **0.7060** | **0.7044** | **0.6485** |
| 4 | (0, 0), (4, 4), (8, 8), (12, 12) | 0.7009 | 0.7012 | 0.6996 | 0.6451 |
| 5 | (0, 0), (3, 3), (6, 6), (9, 9), (12, 12) | 0.7054 | 0.7053 | 0.7039 | 0.6468 |

Conducting experiments with varying numbers of layers follows a similar process as when there are 3 layers. The results in the table represent the best outcomes corresponding to different numbers of layers. From the results table, we can conclude that the model performs best when the number of layers is 3.

## 5    Conclusion

In conclusion, this study presents a novel approach aimed at enhancing the performance of Vietnamese Visual Question Answering (VQA) systems. By addressing the unique challenges associated with processing visual content and natural language in the Vietnamese language, our proposed method significantly improves the accuracy and reliability of VQA systems tailored for Vietnamese-speaking users. The Multi-layer Fusing Transformer (MFT) model, introduced in this study, outperforms existing models in the field across various performance metrics. Specifically, compared to state-of-the-art models such as BARTPhoBeit and PAT, our MFT model demonstrates a notable improvement, achieving a 4% higher F1-score and Exact metric. These results highlight the effectiveness of the MFT model in capturing complex semantic relationships between visual and textual inputs, leading to more accurate and nuanced responses in VQA tasks. The innovative cross-attention mechanism integrated across different layers within the MFT architecture further enhances its capability to fuse information from both modalities, resulting in superior performance outcomes. Overall, our findings underscore the potential of the MFT model as a leading solution in advancing Vietnamese VQA research, while also paving the way for the development of inclusive and culturally relevant AI applications for Vietnamese communities.

**Acknowledgment.** This research is funded by University of Science, VNU-HCM under grant number CNTT 2023-07.

# References

1. Bao, H., Dong, L., Piao, S., Wei, F.: Beit: Bert pre-training of image transformers. arXiv preprint arXiv:2106.08254 (2021)
2. De Vries, W., van Cranenburgh, A., Nissim, M.: What's so special about Bert's layers? a closer look at the NLP pipeline in monolingual and multilingual models. arXiv preprint arXiv:2004.06499 (2020)
3. Devlin, J., Chang, M.W., Lee, K., Toutanova, K.: BERT: pre-training of deep bidirectional transformers for language understanding. In: Proceedings of the 2019 Conference of the North American Chapter of the Association for Computational Linguistics: Human Language Technologies, Volume 1 (Long and Short Papers), Minneapolis, Minnesota, pp. 4171–4186. Association for Computational Linguistics (2019). https://doi.org/10.18653/v1/N19-1423, https://aclanthology.org/N19-1423
4. Dosovitskiy, A., et al.: An image is worth 16x16 words: transformers for image recognition at scale. In: International Conference on Learning Representations (2021)
5. Erhan, D., Bengio, Y., Courville, A., Vincent, P.: Visualizing higher-layer features of a deep network. Univ. Montreal **1341**(3), 1 (2009)
6. Lu, J., Batra, D., Parikh, D., Lee, S.: VILBert: pretraining task-agnostic visiolinguistic representations for vision-and-language tasks. In: Advances in Neural Information Processing Systems, vol. 32 (2019)
7. Nguyen, A.D., Le, T., Nguyen, H.T.: Combining multi-vision embedding in contextual attention for Vietnamese visual question answering. In: Wang, H., et al. (eds.) PSIVT 2022. LNCS, vol. 13763, pp. 172–185. Springer, Cham (2022). https://doi.org/10.1007/978-3-031-26431-3_14
8. Nguyen, D.Q., Nguyen, A.T.: PhoBERT: pre-trained language models for Vietnamese. In: Findings of the Association for Computational Linguistics: EMNLP 2020, pp. 1037–1042 (2020)
9. Nguyen, N.H., Van Nguyen, K.: PAT: parallel attention transformer for visual question answering in Vietnamese. In: 2023 International Conference on Multimedia Analysis and Pattern Recognition (MAPR), pp. 1–6. IEEE (2023)
10. Rajpurkar, P., Zhang, J., Lopyrev, K., Liang, P.: Squad: 100,000+ questions for machine comprehension of text. arXiv preprint arXiv:1606.05250 (2016)
11. Tran, K.Q., Nguyen, A.T., Le, A.T.H., Van Nguyen, K.: VIVQA: Vietnamese visual question answering. In: Proceedings of the 35th Pacific Asia Conference on Language, Information and Computation, pp. 683–691 (2021)
12. Tran, K.V., Van Nguyen, K., Nguyen, N.L.T.: Bartphobeit: pre-trained sequence-to-sequence and image transformers models for Vietnamese visual question answering. In: 2023 International Conference on Multimedia Analysis and Pattern Recognition (MAPR), pp. 1–6. IEEE (2023)
13. Tran, N.L., Le, D.M., Nguyen, D.Q.: BARTPHO: pre-trained sequence-to-sequence models for vietnamese. arXiv preprint arXiv:2109.09701 (2021)
14. Vaswani, A., et al.: Attention is all you need. In; NIPS'17, Red Hook, NY, USA, pp. 6000–6010. Curran Associates Inc. (2017)

# Analyzing the Publicization of Drought Debates in Arizona Newspapers

Anne Lise Boyer[1] , Brigitte Juanals[2] , and Jean-Luc Minel[3(✉)] 

[1] Environnement Ville Société, École Normale Supérieure de Lyon - CNRS,
Lyon, France
`annelise.boyer@ens-lyon.fr`
[2] Centre Norbert Elias, Aix Marseille University - CNRS - Avignon University,
Marseille, France
`brigitte.Juanals@univ-amu.fr`
[3] MoDyCo, CNRS - University Paris Nanterre, Nanterre, France
`jl.minel@orange.fr`

**Abstract.** The aim of this article is to analyze the publicization of debates on public policy orientations, socio-technical choices and alternative solutions put forward by stakeholders to deal with situations of drought and water shortage. The case study is the state of Arizona (USA), where these situations are multiplying and intensifying. The analysis combines knowledge gained from field surveys with the analysis of stakeholder networks and discourses collected in the two largest regional newspapers. It emerges that political and economic positions and decisions have a decisive impact on the distribution of water resources. The article shows how textual data represent an important source for assessing the spatio-temporal configurations of drought episodes, and for identifying the risks and measures implemented or to be implemented that preoccupy societies and are the subject of public debate.

**Keywords:** Drought · Media · Topic Modeling · Public debate

## 1 Introduction

In the current period of climate change, drought, whether anthropogenic or climatic in origin, is becoming a key issue on the environmental agenda and in public debate. On a global scale, the proportion of land affected by severe drought doubled between the 1970s and 2000, and five times as many territories could experience extreme drought by 2050 [21]. Meteorological, hydrological and ecological drought also affects human ecosystems (agricultural and socio-economic drought) [5].

The aim of this article is to analyze the publicization of debates on public policy orientations, socio-technical choices and alternative solutions put forward by players or groups of players, to deal with situations of drought and water shortage. It takes as a case study the state of Arizona, where these situations are multiplying and intensifying (cf. Fig. 1).

© The Author(s), under exclusive license to Springer Nature Switzerland AG 2024
N.-T. Nguyen et al. (Eds.): ICCCI 2024, CCIS 2166, pp. 197–209, 2024.
https://doi.org/10.1007/978-3-031-70259-4_15

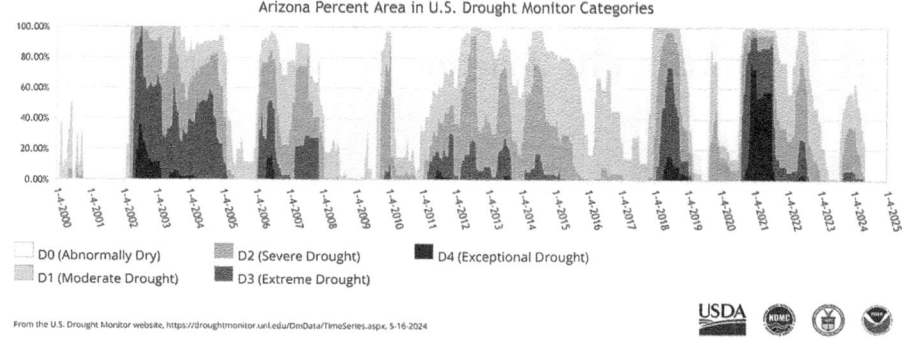

**Fig. 1.** Drought trends in Arizona between 2000 and 2023. Source: U.S. Drought Monitor.

Since the early 1990s, a third of Arizona's water resources have come from the Colorado River, through the Central Arizona Project (CAP) canal. This canal, over 500 kilometers long, provides 85% of the water used by the city of Tucson and 44% for the metropolis of Phoenix.

Until the early 2000s, Arizona was seen as proof that water scarcity could be resolved with substantial investment and strong technological mobilization. But the tense context of scarcity and uncertainties linked to climate change is challenging the system of large-scale hydraulic infrastructures, providing fertile ground for questioning the conditions for the emergence of new assemblages of public policies, developments and particular attitudes that characterize societies' relationship to water in situations of scarcity. In 2012, the US Bureau of Reclamation (USBR), the river's main manager, recognized for the first time two important facts: i) the demand for water in the Colorado basin is greater than the availability of the resource; ii) this imbalance is set to worsen as the river's reduced flow is due to a transition to more arid climatic conditions in the context of climate change.

Starting in 2018, when the levels of the main lake reservoirs (Lake Mead and Lake Powell) were at their lowest, the federal government asked water users to implement the Drought Contingency Plan. The aim of this plan was to delay the declaration of a shortage in the Colorado Basin. However, in 2021, the USBR announced that the Colorado watershed was now affected by a shortage, with progressive water cuts in the allocations reserved for each state: Colorado, Wyoming, Utah, New Mexico, California, Nevada, Arizona). Thus, from 2023, Arizona will lose 20% of the water it receives through the CAP. The areas affected by the water restrictions are primarily the urban expansion on the outskirts of the two metropolises of Phoenix and Tucson. These cities are in fact in a fragile situation due to their urban density: 75% of the population of the State of Arizona resides in the cities of Tucson and Phoenix (US Census, 2020). Secondly, intensive irrigated agriculture is in trouble, followed by industrial activities requiring high water resources, notably mining. In such a context of resource scarcity, marked by socio-environmental and economic choices to favor certain uses and not others, tensions and conflicts of use around the resource can therefore arise.

This interdisciplinary work combines the contributions of three researchers in communication and media, in environmental geography and in statistical processing of texts corpora to analyze, based on a field survey, the modes of presentation in the media of the problems that the drought represents. By crossing the analysis of corpora drawn from the local press, with field surveys (ethnographic observations and semi-directive interviews) we seek to understand how media discourses, along with the stakeholders who carry them, prioritize certain perspectives and downplay others. Our aim is to analyze how the construction of a dominant discourse on the management of water scarcity is associated with political and socio-economic choices, in our case the range of solutions available to mitigate or adapt to drought, implemented through public policies and particular institutional and industrial arrangements, but also emanating from local initiatives [3,5,18]. In the media, we question the forms of visibility and invisibility of the individuals and collectives - with their positions and discourses - involved in these debates. A wide range of stakeholders, including elected representatives, industrial stakeholders, scientific stakeholders, environmental associations (NGOs) and local associations, including citizens' initiatives, are developing interrelationships, strategies and communication actions. We relate their social positions and legitimacy [2] to the visibility they achieve in the local press. These stakeholders take a stand on water management policies, the choice of technical supply systems, alternative solutions and the modes of regulation implemented to organize access to and use of water resources.

The outline of this paper is the following. In Sect. 2, we set out the literature review and the methodology designed to analyze the publicization of drought debates in the media. The case study data, the textual corpora and their sources are presented, together with their modes of collection in Sect. 3. These data form the empirical basis for the analyses and interpretations presented in the Sect. 4. The conclusion places our contribution in the context of ongoing work on climate change and its impact on water use policies, and the scope of the work to be developed in this direction.

## 2   Literature Review and Methodology

### 2.1   Literature Review

In environmental geography, the phenomenon of drought is subject to complex definitions. It corresponds to a diffuse event in time and space that affects the meteorological, hydrological and ecological functioning of a territory, as well as the societies that inhabit it (agricultural and socio-economic drought). A drought can take the form of different levels of intensity (moderate, severe, extreme and exceptional) [23]. The analysis of the publicization of the problem of drought, and the water shortage that follows, leads us to take into account the many social science studies that describe water shortage as not only a consequence of drought, but just as much a social construction [15].

Numerous studies have been carried out using topic modeling (LDA) to identify topics or frames in the press. In recent years, these works focused mainly on press representations of COVID-19 and climate change. For example, [7] used LDA to analyze 7,655 climate change-related news articles published between 2010 and 2021 in three Pakistani English newspapers. They showed that climate change coverage in Pakistan has substantially increased over the years, and the focus has generally been on "climate politics", "climate governance and policy" and "climate change and society". A study carried out by [24] examines the major topics of net zero emission articles in South Korean newspapers. Latent Dirichlet allocation (LDA)-based topic modeling was applied to infer the major topics. Their findings suggest that newspapers can influence implementation of government policies in certain ways.

One of the few studies on drought is presented in [9]. This study rely on Structural topic modeling (STM). STM groups documents using FREX words (i.e., words that are occurring FRequently and EXclusively within a group). The corpus is composed of articles containing the word 'water' from 1990 to late 2017 in any of 37 local newspapers in the United States. Analysis of the corpus identified coherent topics on a variety of water resources issues and revealed temporal and spatial variations in coverage, in particular a topic on tribal issues showed coverage predominantly in the western newspapers.

## 2.2 An Instrumented Methodology

In order to make visible the treatment of water issues by stakeholders logics and their interactions, the analyses combine the knowledge provided by field surveys with the analysis of actor networks and discourses collected in the local press. This approach is in line with non-binary method proposed by [22]: *"non-binary methods are interested not only in exploiting the data generated by digital technology, but also and crucially in understanding the way in which these technologies make collective existence measurable and computable"*. The study of the media is based on work dedicated to the media treatment of information [17], combining quantitative and statistical processing of collected data, identification of stakeholders and their interactions, and analysis of media framing and actors' discourses in the press.

Stakeholder identification in the local press was based on the use of "named entity" search tools (graphic forms referencing stakeholders), supported by a classifier from the Spacy (https://v2.spacy.io/) suite that automatically identifies and categorizes these graphic forms. First, we kept the categorized entities "Person" (PER) and "Organization" (ORG), and used a Python script developed by [6] to identify morphological variants of person names in order to enrich classifier results. Then, we used the score SCP proposed by [6] to quantify the importance of a stakeholder in a corpus. This score computes the ratio N/Nall where N is the number of different articles in which the stakeholder is mentioned and Nall is the total number of articles. This score favors stakeholders who are mentioned in different articles and not stakeholders mentioned many times in the same article.

In the press, we look for manifestations of significant facts or events, to be associated with the contextualized lexicons and positions of the stakeholders. To achieve this, we identify media peaks, based on the number of articles in relation to average flows over the period studied. These media peaks refer to a set of discursive productions published in greater numbers in the media concerning real-world facts, in the sense of "a fact that becomes by and in the media an 'event'" [20]. These descriptive traits are matched with the field survey to assess whether they correspond to phenomena that are new in nature or scope, or to a trait of observed irreversibility.

To search for descriptive features in press articles, we choose the probabilistic Dirichlet Latent Analysis (LDA) model [1]. This model enables us to study the temporal evolution of topics, to identify the emergence or visibility of phenomena, and to compare the treatment of topics between different articles and between different sources. Several works have shown that LDA can be used to construct semantically coherent groups of topics in medium-length texts such as news articles, blogs and newspapers [1,13,14] or scientific articles [11]. Topic modeling algorithms are implemented in different languages (R, Python), we choose to use the R package topicmodels [10].

## 3   Corpora

The press corpus consists of articles published between January 2018 and December 2023 in the two largest regional newspapers, the Arizona Daily Star(ADS), which covers the Tucson area, and the Arizona Republic(AZR), which covers the Phoenix area. The articles were collected on the Proquest platform with the search equation "drought" OR "water". A total of 601 articles were collected. We point out that press corpus was lemmatized and tagged with part of speech categories; only adjective, adverb, noun, verb categories were kept before processing topic modeling step.

Field surveys covering the period 2018–2023 were carried out with water stakeholders in Arizona, on the one hand with institutions managing major infrastructures and water distribution, and on the other with environmentalist associations seeking to propose more local modes of water management (rainwater harvesting, reuse of treated water) and more participatory climate change adaptation strategies (watershed associations, environmental justice associations, environmental activists). This work is complemented by studies carried out on a local scale in Pima County, combined with various documentary sources produced by stakeholders (regulatory bodies, companies, associations). A total of 76 interviews were carried out. Table 1 shows the types of players interviewed and a detailed description of the interviews is provided in [4].

## 4   Analysis

The analysis and interpretation stage begins with the "crisis" situation experienced in Arizona as it enables us to situate the positions, recurrences and changes

**Table 1.** Categories of players interviewed, from [4]

| Categories of players | Number of interviews |
|---|---|
| Municipal water services | 18 |
| Regional water managers | 7 |
| Other public services in charge of adaptation measures | 12 |
| Environmental associations and activists | 39 |
| Total | 76 |

observed within debates in the press, with regard to water policies, management choices and access to water resources, as well as the network of stakeholders.

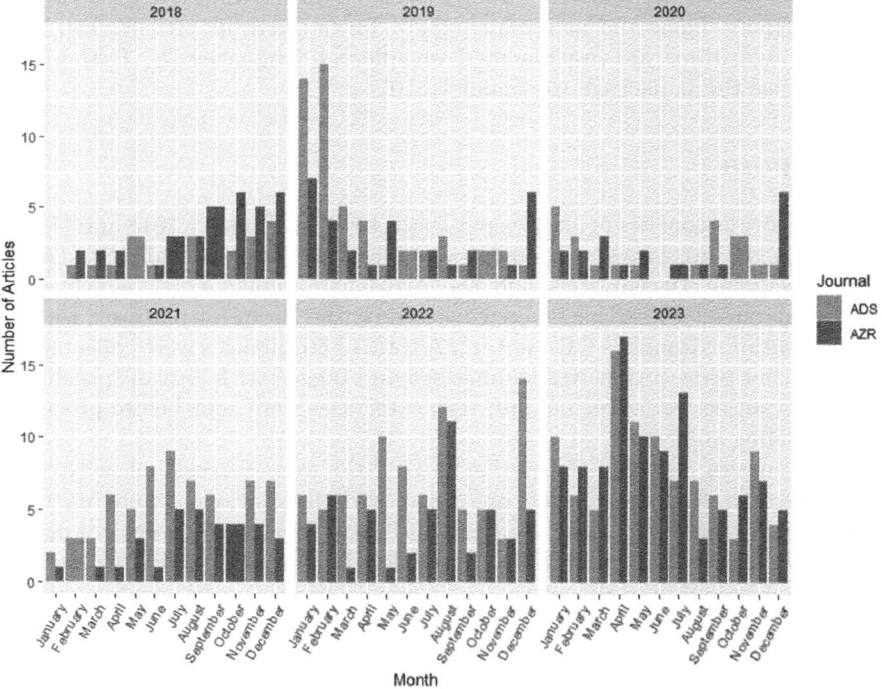

**Fig. 2.** Annual flow of articles in Arizona Daily Star (ADS) and Arizona Republic (AZR)

The graph of monthly flows (Fig. 2) for the Arizona Daily Star (ADS) and Arizona Republic (AZR) newspapers by year between 2018 and 2023 reveals a marked increase in monthly article flows from 2019. The diachronic distribution of annual flows in the press reveals the joint influence of climatic phenomena, political negotiations and social concerns. Several political and climatic factors may explain these media spikes. Firstly, the renegotiation of water drawing rights on the Central Arizona Project (CAP) by the signatory states of the Drought Contingency Plan before January 31, 2019 (cf. Introduction). On the other hand, the 2018, 2019 and 2020 electoral deadlines (Governor, Congress and Senate and President of the United States) led the candidates to take a stand on the means they intended to implement to deal with the drought. In correlation with the media peaks, the deterioration in drought levels observed and posted on monitoring agency websites (cf. Fig. 1), which are sources of information for journalists, contributed to increased media coverage on this subject (Fig. 2). The years 2019, 2021 and 2022 are classified as D4 ("exceptional Drought") by the US Drought Monitor. At the beginning of 2023, *"the U.S. Bureau of Reclamation, which manages the dams and the water, announced that the lakes had fallen to Tier 2A levels, triggering previously agreed-upon cuts to water supplied to states in the lower basin of the Colorado River, beginning on Jan. 1"* (AZR, January 4 2023). As a result, drought and its consequences has become a major headline in the press.

To analyze more precisely this point, we apply a topic modeling step, to emphasize the number of texts indexed by a topic and the words with the highest probability in each topic. Relying on [19] we choose to set the number of topics at 10. Table 2 shows the list of topics with the most significant words for each of them and the number of articles indexed by these topics.

**Table 2.** Topics in the press corpora

| Number | Themes of Topic | Most probable words | Articles |
|---|---|---|---|
| 01 | Tribe water rights | tribe, indian, bill, legislation, deal, Navajo Nation | 76 |
| 02 | Cities project | city, rate, utility, plant, conservation, infrastructure | 68 |
| 03 | Colorado issues | Colorado, CAP, shortage, cut, cutbacks, reservoir | 137 |
| 04 | Biodiversity | fish, wildlife, fire, species, saguaro, watershed | 67 |
| 05 | Mining | mines, permit, cooper, Navajo, Hopi, environmental | 25 |
| 06 | Aquifer | pump, crop, aquifer, agriculture, groundwater | 60 |
| 07 | Energy | energy, gas, emission, carbon, solar, heat | 43 |
| 08 | Foothills Conlict | Scottsdale, standpipe, unincorporated, cut off | 26 |
| 09 | Dam Glen Canyon | Glen Canyon, dam, hydropower, turbine, low level | 32 |
| 10 | Climate change | climate change, temperature, rain, record, increase | 66 |

Figure 3 shows annual evolution of these topics. There are two categories of topics. Topics that are covered on a recurring basis in the press, sometimes with an annual peak, such as "Cities project", "Colorado issues", "Biodiversity", "Mining", "Energy", "Climate change", and those that correspond to the coverage of a one-off event, such as "Aquifer", "Tribe water right", "Foothills conflict" and "Dam Glen Canyon".

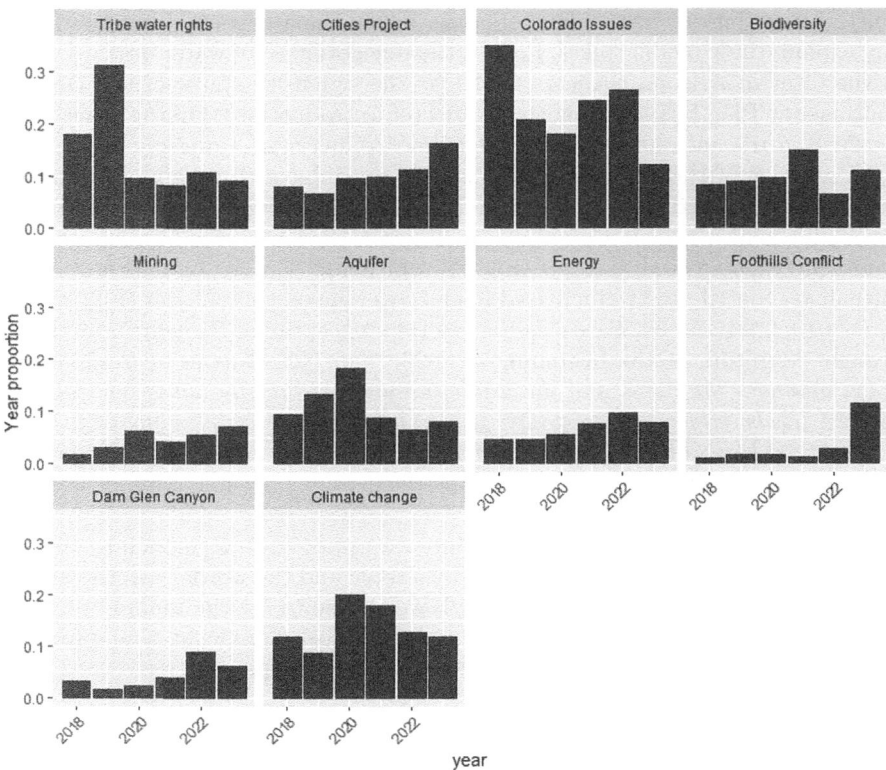

**Fig. 3.** Annual evolution of topics in the press

The press devotes the greatest number of articles to the "Colorado issues" topic with a first peak in 2018 and two others in 2021 and 2022: *Two new studies drive home the dramatic loss of water suffered by the Colorado River Basin this century, in which enough water disappeared from the river system to fill Lake Mead. One study, written by UCLA scientists, found that the impact of climate change on the river was severe enough that without it, the mandatory federal cutbacks in water deliveries that befell the Southwest in 2022 and 2023 wouldn't have been needed. Those cutbacks particularly affected Arizona, depriving the Central Arizona Project canal system serving Tucson and Phoenix of more than 20% of its supply last year and about one-third this year.* (ADS August 2023).

The second most important topic is "Tribe water rights" devoted to the conflict between Indian tribes and the State of Arizona. This conflict was settled by the Supreme Court: *The U.S. Supreme Court has sided with Arizona in its bid to keep the federal government from taking actions that could help the Navajo Nation get a larger share of Colorado River water. In a split decision Thursday, the justices accepted the state's arguments that the federal government has no legal obligation to help the tribal nation, which spans three states, secure the water it needs* (ADS June 2023).

The third most important topic is "Cities project", which has become increasingly important over the years. This topic deals with the solutions adopted by Tucson and Phoenix to deal with the consequences of the drought: *Phoenix wants to recycle wastewater into drinking water by the end of 2030 and share it with the Valley. The plan is to build an advanced water purification facility and treat, then reuse millions of gallons of wastewater that would have otherwise been discharged into the Salt River.* (AZR April 2023).

*The Tucson City Council unanimously approved a ban on non-functional grass planting in all new developments except single-family homes. The council also unanimously approved a new requirement Tuesday night for developers to install federally certified water-saving fixtures in all new development, including single-family homes. Both actions are part of a broader, ongoing effort by the City Council and city officials in general to strengthen Tucson's water conservation efforts in the face of ongoing drought and a water crisis on the Colorado River* (ADS June 2023).

The topic 'FootHills conflict' indexes a small number of articles (26), but is emblematic of conflicts that drought can generate. In January 2023, taps ran dry for a lot of homes in the unincorporated community of Rio Verde, located on the North of Scottsdale, because the city of Scottsdale cut off the accessibility of a water standpipe. *For weeks, the city[Scottsdale] has held that its decision to cut off Rio Verde Foothills residents from accessing a nearby city standpipe is on the right side of the law. ( . . . ) But doing so has plunged Rio Verde Foothills into a precarious situation. Private water haulers have still been able to provide some water to the roughly 1,000 people reliant on them through sources beyond Scottsdale. Still, those sources are unstable and can stop doing business with the haulers at any time* (AZR January 2023).

We searched for persons or organizations mentioned in the press by using a classifier; Fig. 4 shows the number of mentions in the press of people and organization with a score SCP higher than 0.03 over the four years (see Sect. 2.2)[1]. The most cited organization is the U.S. Bureau of Reclamation whose mission is "to manage, develop, and protect water and related resources in an environmentally and economically sound manner in the interest of the American public" (https://www.usbr.gov/). We note that the organizations most frequently mentioned in the press are large institutions at the Arizona state (Tucson Water, Salt River Project) or federal level (EPA, U.S. Bureau of Reclamation, Interior Department). The people most frequently mentioned fall into three categories: political figures, heads of local or federal agencies, and scientists.

These observations can be explained by the persistence of the bureaucratic model of the "a grand water-distribution machine" [8] in the southwestern United States as pointed out by the results of field survey. Indeed, decision-making

---

[1] The descriptions of people and organizations and their mentions is accessible in the form of a knowledge graph that visualizes their links with topics: https://shorturl.at/vxIMZ.

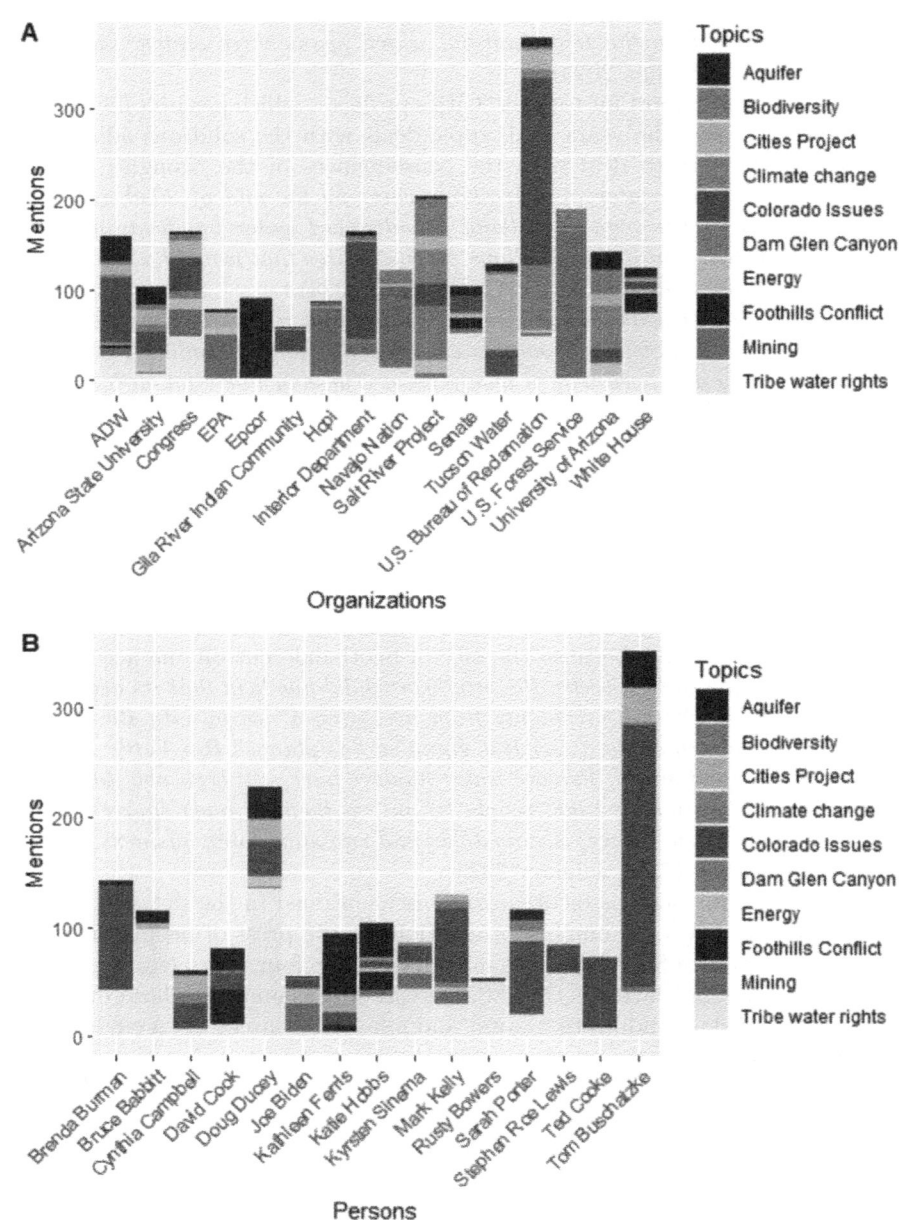

**Fig. 4.** Distribution of organizations and persons by topic

remains concentrated in the large institutions gradually created from the early 20th century, which have come to correspond to a highly complex framework operating at several scales (federal, regional, state) in which the role of experts (engineers, hydrologists) and techno-optimism remain dominant [3,16]. This leaves little room for alternative proposals, from actors who remain marginal. Indeed, while the field survey reports on water reuse practices of all kinds (individual, neighborhood or city-wide rainwater harvesting; grey water; recycled water) that are highly visible in local landscapes and discourses, these and the actors behind them (environmentalist associations, local figures, neighborhood resident collectives) appear only rarely in public discussion.

However, a political debate does exist between the path to be taken for water management in a situation of scarcity: on the one hand, increasing available water resources, and on the other, water conservation. In the Arizona case study, these two options are embodied by the figures of Bruce Babbitt, a fervent Democrat advocate of good water-saving and sustainable management practices, on the one hand, and Doug Ducey, Republican Party Governor of the State of Arizona from 2015 to 2023, particularly active in water management (water desalination, funding of new water supply projects for Arizona) on the other. The risk of these discourses focusing on dominant players and headline-grabbing technical projects, such as seawater desalination for a landlocked territory, obscures the range of alternatives, implemented on a local scale, to adapt the region and above all its cities to a warmer, water-depleted world.

Figure 4A shows also the central role played by universities in this area. Locally, the university functions as an "anchor institution" in the local institutional field. This is particularly the case in the cities of Phoenix-Tempe (Arizona State University, ASU) and Tucson (University of Arizona, UA), where UA is the leading employer, and is therefore able to initiate collaborations with local governments, private companies and civic associations on urban planning and local development projects. Both universities are mentioned in a large number of topics, demonstrating their active communication of scientific advice and mediation to the general public. The UA is home to fifteen departments specializing in environmental issues, which may explain the interest of a large proportion of Tucson's population in environmental conservation. ASU, for its part, stands out for its decision-making centers, which some of the biggest names in water management join as "Senior Research Fellows": such is the case of Kathleen Ferris at the Kyl Center for Water Policy.

Figure 4B shows that the theme of biodiversity has all but disappeared from the discourse associated with water management personalities, with the exception of Sarah Porter, Director of the Kyl Center for Water Policy. Indeed, the field survey showed that the issues at stake are primarily socio-economic, before taking into account the environmental consequences of the drought and the shrinking of the Colorado River, which no longer even reaches its mouth in Mexico's Sea of Cortez.

Analysis of the publicization of the drought issue in the press corpus, with a topic dedicated to "Cities project", reveals the importance of municipal actors. This can be explained by the fact that cities are recognized as being particularly vulnerable to the effects of climate change [12]. The high concentration of population, the high density of infrastructure and material goods, and the complexity of the organization of the urban system multiply the stakes in the face of drought or heat waves, which pose water supply problems as shown by the topic "FootHills conflict". The presence of municipal actors is all the more striking in the case of Arizona, where the first water cuts are likely to concern urban sprawl on the outskirts of the major metropolitan areas of Phoenix and Tucson. The field survey highlighted the proactive role played by municipalities in implementing water-saving campaigns and supporting local rainwater and greywater harvesting initiatives. Since 2020, they have been drawing up their own drought management plans.

## 5   Conclusion

It appears that the public discussion of hydrological drought in the media and by the various stakeholders is a complex subject, bringing to the fore issues of different kinds in Arizona. Indeed, while hydrological drought is linked to climate change and human activities, and has a decisive environmental dimension, water policy and management methods are just as much the result of political, economic, legal and technical-industrial choices. The synergy of field research and media analysis underscores the peril of media narratives fixating solely on major players large-scale, headline-grabbing technical projects, like seawater desalination. The methodology developed offers the possibility of deploying a cross-cutting approach enabling the implementation of a multi-sectoral impact study in connection with the multiplication and intensification of drought episodes. In this respect, this article shows a way to process textual data as an important source for accurately assessing the spatio-temporal configurations of drought episodes, on the one hand, and for determining, on the other, which risks and measures implemented or to be implemented are the subject of public debate.

**Acknowledgments.** This study was partially funded by the Labex DRIIHM (ANR-11-LABX-0010, OHM Pima County) and iGLOBES (UMI 3157).

## References

1. Blei, D., Y. Ng, A., Jordan M.: Latent Dirichlet allocation. J. Mach. Learn. Res. **3**, 993–1022 (2003)
2. Bourdieu, P.: Language and Symbolic Power. Polity Press, Cambridge (1991)
3. Boyer, A.L., Le Lay, Y.F., Marty, P.: Coping with scarcity: the construction of the water conservation imperative in newspapers (1999–2018). Glob. Environ. Chang. **71**, 102387 (2021). https://doi.org/10.1016/j.gloenvcha.2021.102387
4. Boyer, A.L., Le Lay, Y.F., and Marty, P. : Urban adaptation to water scarcity in Phoenix and Tucson (Arizona): a political ecology approach, Cybergeo: European Journal of Geography (2022). https://doi.org/10.4000/cybergeo.38002
5. Boyer, A.L., Juanals, B., Minel, J.L.: Debating Drought and Its Solutions. The Case of Water Shortages in Arizona's Semi-Arid Climate (2024). https://doi.org/10.60527/7m6q-1453

6. Brooks, C.F., Juanals, B., Minel, J.L.: Trends in media coverage and information diffusion over time: the case of the American earth systems research centre biosphere 2. J. Creative Commun. **17**, 88–107 (2022)
7. Ejaz, W., Ittefaq, M., Jamil, S.: Politics triumphs: a topic modeling approach of analyzing news media coverage of climate change in Pakistan. J. Sci. Commun. **22**(01), A02 (2023)
8. Fleck, J.: Water is for fighting over: and other myths about water in the west, Island Press (2016)
9. Gunda, T.: Evolution of Water Narratives in Local US Newspapers: A Case Study of Utah and Georgia. Technical report, US Dpt of Energy (2018). https://doi.org/10.2172/1467781.
10. Grün, B., Hornik, K.: Topicmodels: an R package for fitting topic models. J. Stat. Softw. **40**(13), 1–30 (2011)
11. Hannigan, T., et al.: Topic modeling in management research: rendering new theory from textual data. Acad. Manage. Ann. **13**(2), 586–632 (2019). https://doi.org/10.5465/annals.2017.0099
12. IPCC, Impacts, Adaptation and Vulnerability, Contribution of Working Group II to the Sixth Assessment Report of the Intergovernmental Panel on Climate Change, Cambridge University Press (2022)
13. Jacobi, C., et al.: Quantitative analysis of large amounts of journalistic texts using topic modeling. Digital Journalism, pp. 1–18 (2015)
14. Juanals, B., Minel, J,L.: A comparative analysis of long covid in the French Press and Twitter. Communications in Computer and Information Science, vol. 1864(1), pp. 379–392. Springer (2023)
15. Millington, N.: Producing water scarcity in São Paulo, Brazil: The 2014–2015 water crisis and the binding politics of infrastructure. Polit. Geogr. **65**, 26–34 (2018). https://doi.org/10.1016/j.polgeo.2018.04.007
16. O'Neill, B.F., Boyer, A.L.: Locking in' Desalination in the U.S.-Mexico borderlands: path dependency, techno-optimism and climate adaptation. Water Alternatives **16**(2), 480–508 (2023)
17. Opperhuizen, A., Schouten, K.: Dynamics and tipping point of issue attention in newspapers: quantitative and qualitative content analysis at sentence level in a longitudinal study using supervised machine learning and big data. Quality Quantity **55**, 19–37 (2020). https://doi.org/10.1007/s11135-020-00992-w
18. Rinne, P., Nygren, A.: From resistance to resilience: media discourses on urban flood governance in Mexico. J. Environ. Policy Plann. **18**(1), 4–26 (2016). https://doi.org/10.1080/1523908X.2015.1021414
19. Sievert, C., Shirley, K.E.: LDAvis: a Method for visualizing and interpreting topics. In: Workshop on Interactive language Learning, Visualisation, and Interfaces, pp. 63–70. ACL, USA (2014)
20. Tannier, X., Vernier, F.: Creation, Visualization and Edition of Timelines for Journalistic Use, IJCAI, pp. 101–106 (2016)
21. Tsegai, D., Augenstein, P., Huang, Z.: Global Drought Snapshot. The need for proactive action. CNULCD, United Nations (2023)
22. Venturini, and al.: An unexpected journey: A few lessons from sciences Po médialab's experience, Big Data in Society (2017)
23. Wilhite, D., Pulwarty, R. S. (eds.): Drought and water crises: integrating science, management, and policy, CRC Press (2018)
24. Yun, B., Lim, J., Yun, M.: An exploratory study of net zero discourse based on South Korean newspapers: a topic modeling and sentiment analysis approach. Soc. Netw. Anal. Min. **13(142)** (2023). https://doi.org/10.1007/s13278-023-01150-4

# Comparison of the Effectiveness of ANN and CNN in Image Classification

Arkadiusz Mirakowski[✉]

Department of Information Systems, Faculty of Management and Quality Science,
Gdynia Maritime University, Morska 83, 81-225 Gdynia, Poland
a.mirakowski@wznj.umg.edu.pl
https://umg.edu.pl

**Abstract.** Considering that both ANN and CNN networks are used in image classification, this paper aimed to evaluate the effectiveness of both types of networks in the binary classification of selected datasets containing color and monochrome samples with variable resolutions. Additionally, it was examined whether there are factors influencing the effectiveness of the applied ANN and CNN networks. Based on the obtained results, it can be unequivocally stated that CNN networks proved to be a decidedly better, almost ideal, tool for image classification compared to ANN networks. Factors that affected the effectiveness of ANN networks, such as sample resolution, the percentage content of samples labeled 0 and 1 in the dataset, and whether the sample was color or monochrome, did not affect the effectiveness of CNN networks. Moreover, it was observed that the choice of research metrics in the form of the number of correctly classified samples and the value of the AUC coefficient proved to be an effective and stable metric of classification quality. The results obtained encourage further exploration for new image classification tools or the development of existing ones.

**Keywords:** machine learning · big data · binary classification · image classification · Artificial Neural Networks (ANN) · Convolutional Neural Networks (CNN)

## 1 Introduction

In recent decades, machine learning has transformed the way we perceive problems associated with solving complex issues in various fields of life. Undoubtedly, the ability of various machine learning techniques to analyze vast datasets, often exceeding the capabilities of human analysis, has opened new horizons in the application of artificial intelligence. Certainly, a reason for this is the increasing digitization and growth of available digital data, which, thanks to the advancing development of civilization, is likely to gain momentum in the next few years [1]. In light of the above, the broad concept of machine learning is gaining increasing importance.

© The Author(s), under exclusive license to Springer Nature Switzerland AG 2024
N.-T. Nguyen et al. (Eds.): ICCCI 2024, CCIS 2166, pp. 210–220, 2024.
https://doi.org/10.1007/978-3-031-70259-4_16

Despite the impressive capabilities of various machine learning techniques, it faces a number of challenges, including issues related to the overtraining of neural networks, hardware resource requirements of computing units, and ethical and legal considerations. Consequently, it can be observed that research in machine learning focuses not only on creating increasingly advanced algorithms and architectures of selected neural networks but also on developing methods for more efficient training of neural networks and interpretability of models.

Machine learning is widely used in image classification, where various techniques and algorithms are applied to recognize patterns and categories in image data. Convolutional Neural Networks (CNNs) and Artificial Neural Networks (ANNs), two machine learning techniques, are extensively used in image classification, although they play different roles in this process. CNNs are specifically designed to process data in the form of multi-dimensional arrays, such as images. They are exceptionally effective in extracting features from images, which is crucial for image classification. ANNs, on the other hand, are a more general model of neural networks, though they can be used for image classification.

In light of the above, considering that both ANN and CNN networks are used in image classification, this study aimed to evaluate the effectiveness of both types of networks in the binary classification of datasets containing an equal number of color and monochrome samples with variable resolutions, differing only in the way content is presented. Furthermore, it was intriguing to determine whether there are factors influencing the effectiveness of the applied techniques.

Through the lens of this article, an analytical approach was undertaken to examine Artificial Neural Networks (ANN) and Convolutional Neural Networks (CNN), with particular attention paid to their potential application in binary image classification. By analyzing the presented research results of various case studies, the capabilities and limitations of both artificial and convolutional neural networks were demonstrated.

## 2    Researching the Implementation of Artificial Neural Networks

Analyzing reports related to the implementation of ANN networks, one can perceive immense possibilities and a wide spectrum of their applications. ANNs are widely used in image recognition systems, such as facial identification, medical image analysis, including diagnosing diseases based on medical images, clinical data, or genetic analyses [2]–[8]. ANNs are also a key tool in natural language processing (NLP), assisting in machine translation, text generation, and understanding of natural speech. In the field of NLP, neural networks are foundational in creating modern speech recognition systems, such as voice assistants and transcription systems. In voice generation, ANNs enable the creation of realistic and naturally sounding speech synthesizers [9]–[11]. ANNs are increasingly used in meteorology, for weather forecasting, predicting phenomena such as precipitation, temperature, atmospheric pressure, or behavior of storm systems, making it easier to prevent natural disasters and plan in agriculture [12]–[15]. Not only weather forecasting has become a domain of ANNs, but also predicting various risks in the financial sector. ANNs are used for stock price prediction, credit risk analysis, and fraud detection [16]–[18]. The last area worth mentioning is robotics, where ANNs are used for robot control, enabling them to learn and adapt to new tasks and environments. In autonomous vehicles, neural networks enable the interpretation of sensor data and real-time decision-making [19]–[23].

On the other hand, analyzing reports dedicated to the implementation of CNNs, the largest group pertains to image recognition and analysis, including locating and identifying multiple objects in a single image, which is key in surveillance systems, satellite image analysis, or autonomous vehicles [24]–[27]. Similar to ANNs, CNNs are engaged in climate-related topics, mainly for monitoring environmental changes, analyzing land cover, or detecting natural disasters [28]–[30]. In medicine, CNNs are used for the analysis of medical imaging, such as radiographies, MRIs, or CT scans, aiding in the detection of diseases like cancers, heart diseases, or vascular changes. In diagnostic imaging, CNNs can contribute to faster and more precise identification of various health conditions [31]–[34]. Robotics is also an important area of application for CNNs, as they are used for vision processing, enabling robots to recognize and manipulate objects, navigate spaces, and avoid obstacles. In autonomous vehicles, CNNs analyze data from cameras and other sensors to predict and respond to road environments [35]–[37].

After analyzing reports concerning ANN networks, in all studies based on ANN, both for color and monochrome samples, the same neural network model was used, consisting of one input layer, one hidden layer with relu activation function and Dropout regularization, another hidden layer with relu activation function, and an output layer utilizing the sigmoid activation function, rmsprop optimizer, and binary crossentropy as the loss function. The number of epochs in the model was set to 10. Similarly, in all studies based on CNN, for both color and monochrome samples, the same neural network model was applied, consisting of one convolutional layer with relu activation function along with pooling and Dropout regularization, one hidden layer with relu activation function and Dropout regularization, and an output layer using the sigmoid activation function, rmsprop optimizer, and binary crossentropy as the loss function. Like the ANN model, the number of epochs was set to 10.

## 3  Computational Experiment Assumptions

In this study, three base datasets containing color samples with a resolution of $227 \times 277$px were used as the research foundation. Each dataset contained a total of 40,000 samples. The first was the Fresh and Stale Images of Fruits and Vegetables [38] from which 56 representative color samples of healthy apples and 56 representative color samples of rotten apples were selected. The reason for this choice was the large diversity of the dataset in terms of resolution and the varied way of presenting fruits and vegetables. Each sample of healthy apples selected from the 56 samples was rotated in the range of $1–359°$ in steps of

1°, resulting in a subset of 20,000 samples, labeled 0. The same procedure was repeated for the 56 samples of rotten apples, assigning them label 1. As a result of combining these created subsets, the first dataset was formed.

The subsequent datasets used were Tomato leaf disease detection [39], from which a dataset consisting of 20,000 color samples of healthy leaves and 20,000 color samples of diseased leaves was created, and Surface Crack Detection [40], from which a dataset was created consisting of 20,000 color samples of concrete without cracks and 20,000 color samples of concrete with cracks. The procedure for creating both datasets was the same as for the Fresh and Stale Images of Fruits and Vegetables dataset. Representative samples of the three datasets are presented in the Fig. 1.

**Fig. 1.** Representative sample examples used in the measurements

In light of the curiosity about whether a change in the resolution of color samples would impact the research results, a transformation of the base resolution of all three datasets - apples, leaves, and concrete samples - was performed, as per Fig. 2.

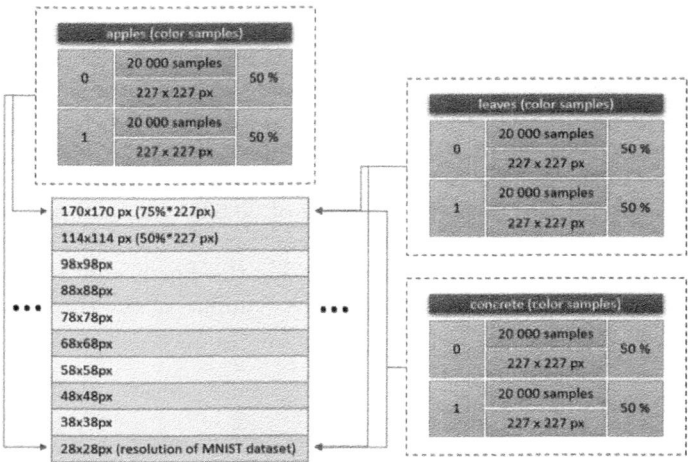

**Fig. 2.** Scheme for transforming the resolution of color samples in base datasets

The next question that necessitated generating new datasets was whether the percentage of samples with label 0 and label 1 would affect the research results. Consequently, additional sets of color samples were created with a changed percentage share: 30% (label 0) and 70% (label 1), and 30% (label 0) and 70% (label 1) for each of the three sets. The last doubt was how a color or monochromatic sample would affect the research outcome. Therefore, sets of monochromatic samples were created, analogous to the color samples. The final summary of the datasets used in the research, as exemplified by the Fresh and Stale Images of Fruits and Vegetables set, is presented in the Table 1.

**Table 1.** Summary of the datasets using the example of the apple collection

| No. | Resolution [px] | Label 0 [%] x Label 1 [%] color samples | Label 0 [%] x Label 1 [%] color samples | Label 0 [%] x Label 1 [%] color samples | Label 0 [%] x Label 1 [%] mono-chrome samples | Label 0 [%] x Label 1 [%] mono-chrome samples | Label 0 [%] x Label 1 [%] mono-chrome samples |
|---|---|---|---|---|---|---|---|
| 1. | $227 \times 227$ | $30 \times 70$ | $50 \times 50$ | $70 \times 30$ | $30 \times 70$ | $50 \times 50$ | $70 \times 30$ |
| 2. | $170 \times 170$ | $30 \times 70$ | $50 \times 50$ | $70 \times 30$ | $30 \times 70$ | $50 \times 50$ | $70 \times 30$ |
| 3. | $114 \times 114$ | $30 \times 70$ | $50 \times 50$ | $70 \times 30$ | $30 \times 70$ | $50 \times 50$ | $70 \times 30$ |
| 4. | $98 \times 98$ | $30 \times 70$ | $50 \times 50$ | $70 \times 30$ | $30 \times 70$ | $50 \times 50$ | $70 \times 30$ |
| 5. | $88 \times 88$ | $30 \times 70$ | $50 \times 50$ | $70 \times 30$ | $30 \times 70$ | $50 \times 50$ | $70 \times 30$ |
| 6. | $78 \times 78$ | $30 \times 70$ | $50 \times 50$ | $70 \times 30$ | $30 \times 70$ | $50 \times 50$ | $70 \times 30$ |
| 7. | $68 \times 68$ | $30 \times 70$ | $50 \times 50$ | $70 \times 30$ | $30 \times 70$ | $50 \times 50$ | $70 \times 30$ |
| 8. | $58 \times 58$ | $30 \times 70$ | $50 \times 50$ | $70 \times 30$ | $30 \times 70$ | $50 \times 50$ | $70 \times 30$ |
| 9. | $48 \times 48$ | $30 \times 70$ | $50 \times 50$ | $70 \times 30$ | $30 \times 70$ | $50 \times 50$ | $70 \times 30$ |
| 10. | $38 \times 38$ | $30 \times 70$ | $50 \times 50$ | $70 \times 30$ | $30 \times 70$ | $50 \times 50$ | $70 \times 30$ |
| 11. | $28 \times 28$ | $30 \times 70$ | $50 \times 50$ | $70 \times 30$ | $30 \times 70$ | $50 \times 50$ | $70 \times 30$ |
| Number of datasets | 11 | 11 | 11 | 11 | 11 | 11 | 11 |

Ultimately, 66 sets of apples, 66 sets of leaves, and 66 sets of concrete samples were developed. Supervised binary classification was conducted on each set using ANN and CNN. The analysis of the research results focused mainly on two metrics: the confusion matrix [[41]–[43]] and the AUC coefficient [41]. The confusion matrix provided the number of correctly and incorrectly classified samples, while the AUC value, obtained through the ROC curve, gave information about the effectiveness level of the applied algorithm. All studies using ANN and CNN neural networks were conducted on a computing cluster with the following hardware parameters: 2 Xeon Platinum 8268 2.9 GHz processors and 192 GB of RAM. Python 3.12.0, Anaconda3 2023.09-0 environment, and Tensorflow 2.13.0 library were used for numerical calculations.

## 4 Results of the Experiment Obtained Using ANN

In the experiment, each of the prepared sets of color and monochromatic samples within a set resolution range was trained using the ANN model (p.2) to determine the number of correctly classified samples with and without defects. Additionally, it was examined whether there was primarily a correlation between the number of correctly classified samples and the value of the AUC coefficient. An exemplary result of classification studies conducted for datasets containing color samples is presented in Fig. 3, which remains correlated with the results obtained for the other datasets.

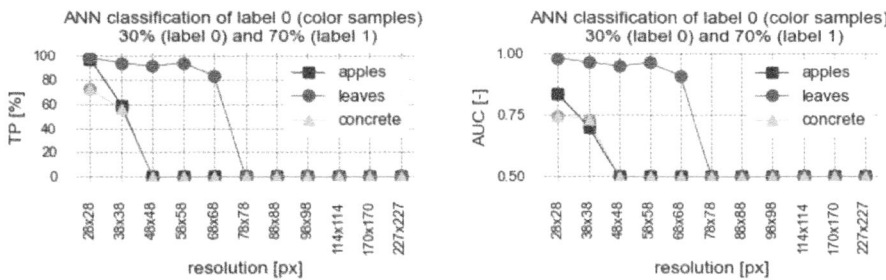

**Fig. 3.** ANN classification of label 0 in color sample sets with a percentage composition of 30% label 0 and 70% label 1

Based on the obtained results, it must be unequivocally stated that on one hand, the ANN neural network is a tool that can be used in image classification, but on the other hand, there are factors that affect its varying effectiveness. The first significant factor appears to be the resolution of the samples. In the case of the used sets, regardless of the classified label, the classification effectiveness is convergent, and the ANN network is effective up to a resolution of 48 × 48px. The same cannot be said for samples from the leaf set, where the ANN network classifies successfully up to a resolution of 78 × 78px. Likely, this is due to differences in the content of selected colors in each group of samples - different in apple samples, different in leaf samples, and yet different in concrete samples. Despite these differences, there is a common denominator for the different configurations of the three basic data sets, which is the correctness related to the fact that as the resolution increases, the effectiveness of the ANN neural network decreases. Additionally, looking at the correlation between the number of correctly classified samples and their corresponding AUC values, a close relationship can be observed, confirming the proper choice from the perspective of selecting measurement metrics. It is also worth looking at the percentage content of samples labeled 0 and 1. In cases where there was a percentage balance between the content of samples with label 0 and 1, the number of correctly classified samples was similar and stable. However, the situation changed when the proportions changed, and then a more effective classification was observed in the case of samples that constituted 70% of the volume of the studied set. An exemplary result of classification studies conducted for datasets containing monochromatic samples is presented in Fig. 4, which remains correlated with the results obtained for the other datasets.

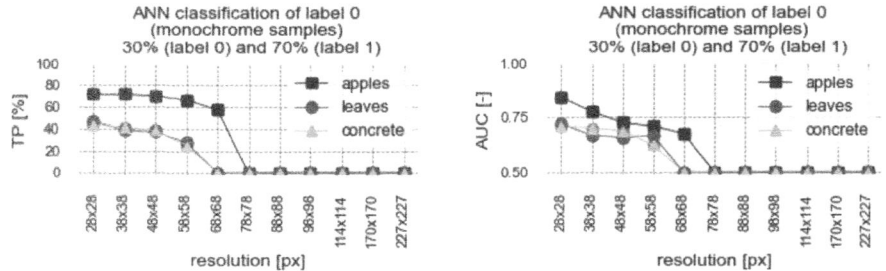

**Fig. 4.** ANN classification of label 0 in monochrome sample sets with a percentage composition of 30% label 0 and 70% label 1

The results obtained for monochromatic samples confirmed that the ANN neural network is a tool that can be used in image classification, and that its effectiveness depends on the same factors as in the case of color samples. Similarly, as with color samples, the effectiveness of the ANN neural network in classifying monochromatic samples decreases with increasing resolution, although its effectiveness improved slightly only in the case of the leaf set and concrete samples. There is also a convergence and thus independence between the effectiveness of the ANN network and the type of label. Looking at the correlation between the number of correctly classified monochromatic samples and their corresponding AUC values, it can be assumed that the choice of measurement metrics was correct. As with the study of color samples, when there was a percentage balance between the content of samples with label 0 and 1, the number of correctly classified samples was similar and stable. Analogous to color samples, the situation changed when the proportions changed, and then a more effective classification was observed in the case of samples that constituted 70% of the volume of the studied dataset.

## 5    Results of the Experiment Obtained Using CNN

In the experiment concerning image classification using CNN, each of the prepared sets of color and monochromatic samples within a set resolution range was trained using the CNN model (p.2) to obtain the numbers of correctly classified samples with and without defects. Furthermore, it was primarily examined whether there is a correlation between the number of correctly classified samples and the value of the AUC coefficient. An exemplary result of classification studies conducted for datasets containing color samples is presented in Fig. 5, which remains correlated with the results obtained for the other datasets containing both color and monochromatic samples.

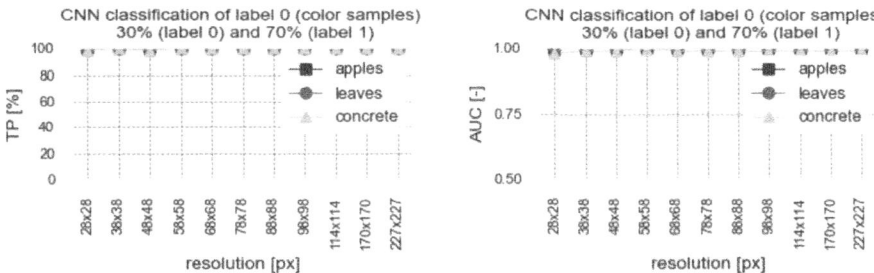

**Fig. 5.** CNN classification of label 0 in color sample sets with a percentage composition of 30% label 0 and 70% label 1

Based on the aforementioned results obtained for CNN classification, it must be stated that convolutional neural networks proved to be a very good tool for classifying color samples, as the level of correctly classified samples was close to 100%, which was confirmed by the AUC coefficient value close to or equal to 1. Factors that affected the effectiveness of ANN neural networks, in the case of CNN, turned out not to impact its effectiveness, which remained stable at a high level for all sets in the studied resolution range. A similar impression can be obtained by looking at the exemplary results for the CNN classification of monochromatic samples.

## 6    Summary

The study attempted to compare the effectiveness of artificial neural networks and convolutional neural networks in classifying color and monochromatic image samples of variable resolutions. Based on the obtained results, it can be unequivocally stated that CNNs proved to be a much better, almost ideal, tool for image classification compared to ANNs. Factors that affected the effectiveness of ANNs, such as sample resolution, the percentage content of samples labeled 0 and 1 in the dataset, and whether the sample was color or monochromatic, did not impact the effectiveness of CNNs. Moreover, it was noted that the choice of research metrics in the form of the number of correctly classified samples and the value of the AUC coefficient proved to be an effective and stable metric of success. With the research goal set to test the effectiveness of both ANN and CNN types in binary classification of three selected sets containing the same amount of color and monochromatic samples of the same resolutions, the goal was achieved. The analysis clearly demonstrated that CNNs are a far better tool in classifying both color and monochromatic samples than ANNs. Furthermore, the analysis proved that the factors affecting the effectiveness of ANNs, mentioned above, do not influence the effectiveness of CNNs. The obtained research results encourage and prompt further studies related to image classification using ANNs, but in the sense of multi-class classification possibly using ensemble learning.

# References

1. ScienceDaily Homepage. https://www.sciencedaily.com/releases/2013/05/130522085217.htm. Accessed 20 Jan 2024
2. Elo, P., Saarinen, J., Värri, A., Nieminen, H., Kaski, K., Aleksander, I., Taylor, J.: Classification of epileptic EEG by using self-organizing maps. Artif. Neural Netw. **2**(1 and 2), 1147–1150 (1992)
3. Bankman, I. N., Sigillito, V. G., Wise, R. A., Smith, P. L.: Feature-based detection of the K-complex wave in the human electroencephalogram using neural networks. IEEE Trans. Biomedical Eng. **39**(12), 1305–1310 (1992). https://doi.org/10.1109/10.184707
4. Andina, D., Álvarez-Vellisco, A., Jevtic, A., Fombellida J.: artificial metaplasticity can improve artificial neural networks learning. Intell. Autom. Soft Comput. **15**(4), 683–696 (2009). https://doi.org/10.1080/10798587.2009.10643057
5. Katritzky, A., et al.: Quantitative correlation of physical and chemical properties with chemical structure: utility for prediction. Chem. Rev. **110**, 5714–5789 (2010)
6. Cho, D.W., Lee, S.J., Chu, C.: The state of machining process monitoring research in Korea. Int. J. Mach. Tools Manuf. **39**(11), 1697–1715 (1999). https://doi.org/10.1016/S0890-6955(99)00026-7
7. Srinivasa, P., Nagabhushana, T.N, Raj, Rao, B.K.N.: Tool condition monitoring using acoustic emission, surface roughness and growing cell structures neural network. Mach. Sci. Technol. **16**(4), 653–676 (2012). https://doi.org/10.1080/10910344.2012.731954
8. Panchalingam, R., Chan, K.C.: A state-of-the-art review on artificial intelligence for Smart Buildings. Intell. Build. Int. **13**(4), 203–226 (2021). https://doi.org/10.1080/17508975.2019.1613219
9. van Hattem, R.: Mastering Python. Write Powerful and Efficient Code Using the Full Range of Python's Capabilities - Second Edition. Packt Publishing (2022)
10. Agbotiname, L.I., Hemanth, J., Do, D.T., Nath Sur, S.: Explainable Artificial Intelligence in Medical Decision Support Systems. Institution of Engineering and Technology (2022)

11. Sabharwal, N., Agrawal, A.: Hands-on Question Answering Systems with BERT. Apress, Berkeley, CA (2021). https://doi.org/10.1007/978-1-4842-6664-9
12. Chen, C.S., Tzeng, Y.M., Cho, M.Y.: The application of artificial neural network to distribution substation load forecasting and temperature sensitivity analysis. J. Chin. Inst. Eng. **19**(2), pp. 171–177 (1996). https://doi.org/10.1080/02533839. 1996.9677777
13. Asbury, C.E.: Weather load model for electric demand and energy forecasting. IEEE Trans. Power Appar. Syst. **94**, 1111–1116 (1975)
14. Hsu, C.T., Tzeng, Y.M., Chen, C.S., Cho, M.Y.: Distribution feeder loss analysis by using an artificial neural network, Electric Power Syst. Res. **34**(2), 85–90 (1995). https://doi.org/10.1016/0378-7796(95)00959-X
15. Lu, C.N., Wu, H.T., Vemuri, S.: Neural network based short term load forecasting. IEEE Trans. Power Syst. **8**(1), 336–342 (1993)
16. Wagdi, O., Salman, E., Albanna, H.: Integration between technical indicators and artificial neural networks for the prediction of the exchange rate: evidence from emerging economies. Cogent Econ. Finance **11**(2) (2023). https://doi.org/10.1080/ 23322039.2023.2255049
17. Coakley, J. R., Brown, C. E.: Artificial neural networks in accounting and finance: modeling issues. Intell. Syst. Account. Finance Manage. **9**(2), 119–144 (2000)
18. Kong, H., et al.: Constructing a personalized recommender system for life insurance products with machine-learning techniques. Intell. Syst. Account. Finance Manage. **29**(4), 242–253 (2022)
19. Schmelter, S., Roehrig, C.: Two staged ANN Based UWB ranging error mitigation for real time self localization on mobile robots. In: ISR Europe 2022; 54th International Symposium on Robotics, pp. 1–7. Munich, Germany (2022)
20. Dahiya, R., Ozioko, O., Cheng, G.: Sensory Systems for Robotic Applications. Institution of Engineering and Technology (2022)
21. Kumarakulasingam, P., Agah, A.: Neural network-based single sensor sound localization using a mobile robot. Intell. Autom. Soft Comput. **14**(1), 89–103 (2008). https://doi.org/10.1080/10798587.2008.10642985
22. Arslan, G., Sakarya, A.: A unified neural-network-based speaker localization technique. IEEE Trans. Neural Networks **11**(4), 997–1002 (2000)
23. Chalapathy, N., Young, E.: Neural network models of sound localization based on directional filtering by the pinna. J. Acoust. Soc. Am. **92**(6), 3140–3156 (1992). https://doi.org/10.1121/1.404210
24. Khalid, O., Khan, S.U., Zomaya, A.Y.: Big Data Recommender Systems, Volume 2 - Application Paradigms. Institution of Engineering and Technology (2019)
25. Yan, Z., Xu, Z., Dai, J.: The big data analysis on the camera-based face image in surveillance cameras. Intell. Autom. Soft Comput. (2017). https://doi.org/10. 1080/10798587.2016.1267251
26. Solanki, A., Kumar, A., Nayyar, A.: Digital Cities Roadmap - IoT-Based Architecture and Sustainable Buildings. John Wiley & Sons (2021)
27. Xu, Z., Hu, Ch., Mei, L.: Video structured description technology based intelligence analysis of surveillance videos for public security applications. Multimedia Tools Appl. **75**, 12155–12172 (2015)
28. Ahmad, S., Murray, R.: World Environmental and Water Resources Congress 2023 - Adaptive Planning and Design in an Age of Risk and Uncertainty, American Society of Civil Engineers (2023)
29. Butler, J.J., Xiong, X., Gu, X.: Earth Observing Systems XXVI, Proceedings of SPIE Volume 11829 (2021)

30. Dubey, A.K., Narang, S.K., Srivastav, A.L., Kumar, A., García-Díaz, V.: Artificial Intelligence for Renewable Energy Systems. Elsevier (2022)
31. Kose, U., Gupta, D., Chen, X.: Explainable Artificial Intelligence for Biomedical Applications. River Publishers (2023)
32. Kalaskar, D.M.: 3D Printing in Medicine (2nd Edition). Elsevier (2023)
33. Kumar, A., Dubey, A.K., Bhatia, S., Kumar, S.A., Le, D.-N.: Evolving Predictive Analytics in Healthcare - New AI Techniques for Real-Time Interventions. Institution of Engineering and Technology (2022)
34. Singh, A.K., Zhou, H.: Medical Information Processing and Security - Techniques and Applications. Institution of Engineering and Technology (2022)
35. Alberola, A.M., Gallego, G.M., Maestre, U.G., Artificial Vision and Language Processing for Robotics. Packt Publishing (2019)
36. Chen, K.C.: Artificial Intelligence in Wireless Robotics. River Publishers (2020)
37. Dahiya, R., Ozioko, O., Cheng, G.: Sensory Systems for Robotic Applications. Institution of Engineering and Technology (2022)
38. Kaggle Homepage. https://www.kaggle.com/datasets/raghavrpotdar/fresh-and-stale-images-of-fruits-and-vegetables. Accessed 20 Jan 2024
39. Kaggle Homepage. https://www.kaggle.com/datasets/kaustubhb999/tomatoleaf. Accessed 20 Jan 2024
40. Kaggle Homepage. https://www.kaggle.com/datasets/arunrk7/surface-crack-detection. Accessed 20 Jan 2024
41. Geron, A.: Uczenie maszynowe z użyciem Scikit-Learn i TensorFlow, Helion (2018)
42. Karayaneva, Y., Hintea, D.: Object recognition in Python and MNIST dataset modification and recognition with five machine learning classifiers. J. Image Graph. **6**(1) (2018)
43. Kohavi, R., Provost, F.: Glossary of terms. Appl. Mach. Learn. Knowl. Discov. Process. **30**, 271–274 (1998)

# Study Neural Model for Recognition of Ancient Turkic Orkhon Runes

Meirbek Mukashev[ID] and Ualsher Tukeyev[✉] [ID]

Al-Farabi Kazakh National University, Almaty, Kazakhstan
ualsher.tukeyev@gmail.com

**Abstract.** More than three hundred Turkic runic inscriptions were discovered in various regions of Central Asia and adjacent regions. Although a significant number of Turkic-runic manuscripts have been read, there are still many undeciphered Turkic manuscripts. There are also many controversial deciphered Turkic manuscripts. Therefore, using automatic text recognition systems for ancient Turkic manuscripts is relevant. An urgent problem is the development of neural models for recognizing texts of ancient Turkic Orkhon runes. This research focuses on developing and implementing a computational approach for recognizing Ancient Turkic Runic Orkhon inscriptions. In response to these challenges, this study presents a framework employing advanced image processing and machine learning techniques to efficiently and reliably recognize Orkhon runic inscriptions.

**Keywords:** Neural Model · Recognition · Ancient Turkic Orkhon Runes

## 1 Introduction

Turkic languages make up a family including more than 35 languages, which are spoken by more than 200 million people across several countries. The Turkic group of languages includes state languages like Azerbaijan, Kazakh, Kyrgyz, Uzbek, Turkish, and Turkmen. The languages of the subjects of the states are Altai, Balkar, Bashkir, Karakalpak, Crimean Tatar, Kumyk, Nogai, Tatar, Tuvan, Uyghur, Khakass, Shor, and Yakut.

All Turkic languages were formed from the ancient Turkic language, which had its written language, which is confirmed since 1889 by the discovery of runic monuments in honor of Bilge Kagan and his younger brother Kultegin in the Orkhon River basin. In 1893, the Danish scientist V. Thomson managed to decipher the Orkhon-Yenisei runic alphabet. These runic monuments of Bilge Kagan and Kultegin were first read and translated into German by V.V. Radlov [1].

Subsequently, more than three hundred Turkic runic inscriptions were discovered in various regions of Central Asia and adjacent regions. Although a significant number of Turkic-runic manuscripts have been read, there are still many undeciphered Turkic manuscripts. There are also many controversial deciphered Turkic manuscripts.

Therefore, using automatic text recognition systems for ancient Turkic manuscripts is relevant. This paper examines the issue of using neural models for automatic character recognition of ancient Turkic Orkhon runic symbols.

© The Author(s), under exclusive license to Springer Nature Switzerland AG 2024
N.-T. Nguyen et al. (Eds.): ICCCI 2024, CCIS 2166, pp. 221–233, 2024.
https://doi.org/10.1007/978-3-031-70259-4_17

## 2   Related Works

Intensive research is being conducted in the field of recognition of ancient, lost writings and runes [1–4]. However, these studies have neglected the ancient Turkic writings. Even though the study of ancient Turkic writings and runes is conducted in the direction of linguistics, they have a humanitarian character. Recently, the issue of creating models for recognizing ancient scripts and runes using deep learning methods and neural models has been considered. In recent years, researchers have received increasing attention from deep-learning-based approaches thanks to their ability to handle complex and difficult image classification tasks. Nowadays, advances in Artificial Intelligence (AI), especially in machine and deep learning, present new opportunities to build tools that support the work of specialists in areas far from the information technology field [5–10]. However, existing models for recognizing and translating ancient Turkic scripts and runes with the proposed method have not yet been found. This problem is a high priority since the topic of research on Turkic writing is extensive.

Nevertheless, there are enough humanitarian works, as mentioned earlier. In these works, research is usually carried out on the peculiarities of the ancient Turkic language, spelling rules, the study of runes, collecting information analyzing primary material, and synthesizing data. Also, an important point of these studies is to find similarities with other ancient regional languages and writing. In general, those works aim at understanding the formation of Old Turkic scripts up to the present day, treating them as a primary research tool based on scientific concepts [11–13].

Based on this information, the problem of high priority is the creation of an accurate model for the recognition of ancient Turkic Orkhon runes. Ancient Turkic Orkhon inscriptions are crucial for understanding the history, culture, and language of the ancient Turkic peoples, but their interpretation relies heavily on experts who can read and decipher these runes. Since ancient Turkic Orkhon inscriptions contain some of the earliest written records of the Turkic languages, deciphering them accurately can provide valuable information about the linguistic evolution and historical context of the Turkic peoples.

## 3   Method

To develop the technology of the recognition of ancient Turkic Orkhon runes based on neural models, the following tasks are set in the work:

– development and research of neural models and algorithms for the recognition of ancient Turkic Orkhon sentences;
– research of the symbols of the ancient Turkic Orkhon alphabet and the features of the symbols of the ancient Turkic Orkhon alphabet;
– formation of a data set of Orkhon runes for training and fitting to the created model;
– creating a script to segment input from an image containing Orkhon runes.

The task of developing and researching neural models and algorithms for the recognition of ancient Turkic Orkhon sentences will be solved by:

- development of a CNN neural model for the recognition of ancient Turkic Orkhon runes;
- designing a CNN architecture optimized for image recognition, comprising convolutional layers followed by pooling layers, ultimately leading to fully connected layers for classification;
- conduct experiments with various architectures, taking into account factors such as depth, kernel sizes, and the number of filters in each layer;
- select an optimizer, such as Adam or SGD, and establish an initial learning rate. Adjusting the learning rate as needed during the training process.

The task of researching the symbols of the ancient Turkic Orkhon alphabet and the features of the symbols of the ancient Turkic Orkhon alphabet will be solved by:

- scrutinizing primary sources, such as the inscriptions on Orkhon monuments, by accessing images, transcriptions, and translations;
- studying the context in which these inscriptions were made, considering historical events, cultural practices, and the purpose of the monuments;
- exploring linguistic aspects of the ancient Turkic Orkhon script. Understand how phonemes are represented, the presence of vowels, and any diacritic marks used;
- examining the script's relationship with other writing systems, exploring potential influences and adaptations (Yenisei, Talas, Altai);
- identifying variations in symbols based on their position within a word or sentence;
- checking the ancient Turkic Orkhon script's status in the Unicode Standard to understand its standardized digital representation for modern technology use.

The task of forming a dataset of Orkhon runes for training and fitting to the created model will be solved by:

- gathering a diverse set of Orkhon-run images found in historical documents, books, or online resources;
- data preprocessing such as resizing an image to a consistent size for uniformity and converting the image to grayscale;
- labeling each image with the corresponding Orkhon rune;
- using data generation by augmenting the dataset in favor small dataset by applying random rotations, flips, and other transformations to increase variability;
- dataset splitting into training and test sets.

The task of creating a script to segment input from an image containing Orkhon runes will be solved by:

- selecting text segmentation tool by image;
- writing a segmentation algorithm for the segmentation of Orkhon runes;
- saving segmented characters.

## 3.1 Development of a Neural Model for Recognition of Ancient Turkic Orkhon Runes

Since this task is an image classification task, the creation of architecture plays an important role in the development of CNN for ancient Turkic Orkhon runes. The proposed neural network model contains a complex structure. Complex models often refer to architectures that are deeper, more intricate, or incorporate advanced features and techniques to achieve superior performance in various tasks such as image classification and object detection.

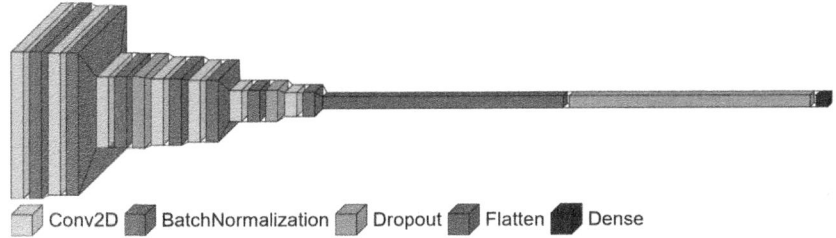

Conv2D    BatchNormalization    Dropout    Flatten    Dense

**Fig. 1.** The architecture of the proposed neural model.

Figure 1 shows a visualization of the architecture of the proposed neural model. This model consists of 4 groups of layers, the input layer, the first convolutional block, the second convolutional block and the output layer:

- The input layer of the model is the first layer that receives the input data, which in this case is images.
- The first convolutional block is designed to extract features from the input image while gradually reducing its spatial dimensions and introducing non-linearity and regularization through activation functions and dropout, respectively.
- The second convolutional block is designed to extract higher-level features from the input while gradually reducing its spatial dimensions and introducing non-linearity and regularization through activation functions and dropout.
- The third block of the model is the final layer that produces the model's predictions. The third contains a flattened and dense layer with a SoftMax activation function. The third block produces a probability distribution over the different classes, making it suitable for multi-class classification tasks. The class with the highest probability is chosen as the model's prediction.

Conv2D contained in the input layer uses 32 filters. The first and second convolutional blocks in the provided model share similarities in their structure but differ mainly in the number of filters used in their convolutional layers and in the complexity of the features they extract. The first convolutional block uses 32 filters for each Conv2D layer contained, and the second convolutional block uses 64 filters for each Conv2D layer contained. While both blocks serve the purpose of feature extraction and hierarchical representation learning, the second convolutional block, with its increased number of filters, can potentially capture more intricate details and complex patterns in the input data compared to the first block. This increased complexity may lead to the model being able to learn more discriminative features, especially in tasks where the input data is more complex or varied. Both convolutional blocks contain Dropout. These dropout layers are placed after the convolutional layers inside each block, which allows for regularization at different stages of feature extraction. By using dropout after convolutional layers, the model can prevent overfitting when learning hierarchical representations based on input data. Conv2D, which is in the third block, uses 128 filters. The filters are essentially the parameters that the network learns during training. Each filter is a small matrix applied across the input image through convolution to produce feature maps. These feature maps represent different patterns or features detected by the network.

Figure 2 presents the parameters of the architecture layers of the proposed neural model.

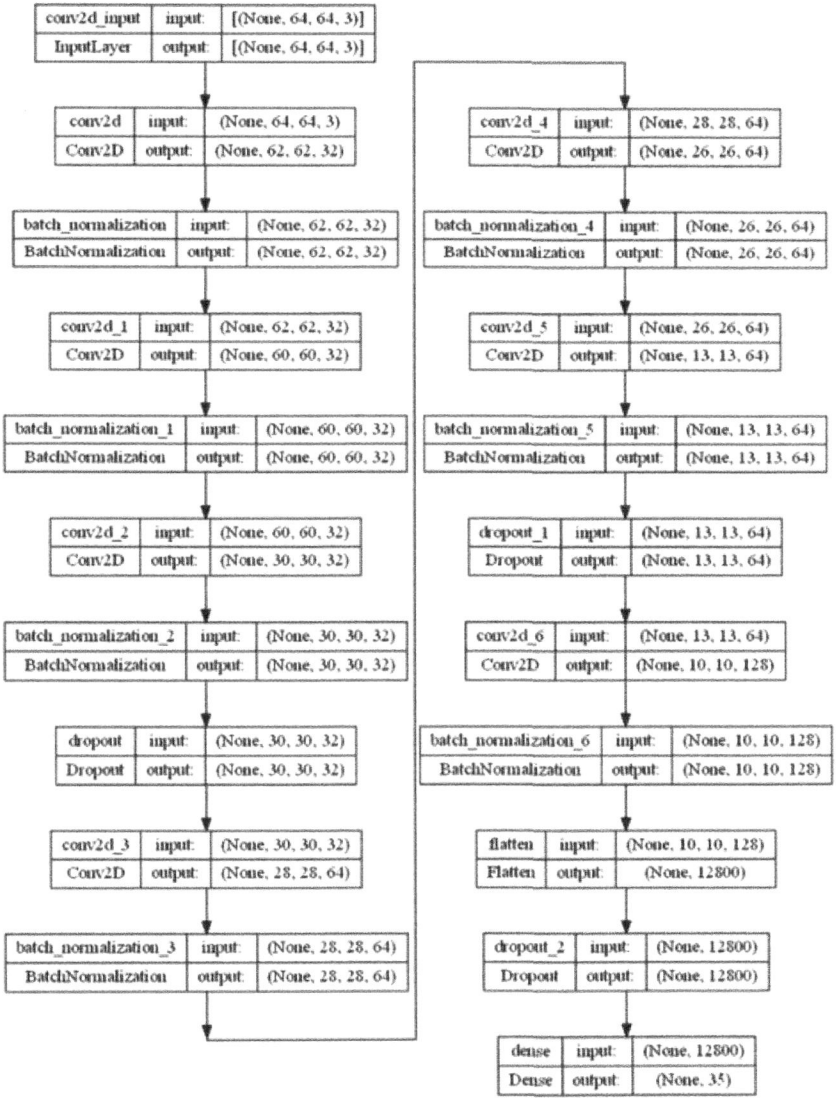

**Fig. 2.** Parameters of the architecture layers of the proposed neural model

### 3.2    Researching the Signs of the Ancient Turkic Orkhon Alphabet and Forming a Dataset

The script of the ancient Turkic Orkhon has 38 signs, which included both consonants and vowels. The Orkhon script is an alphabetic writing system, meaning each symbol represents a sound (phoneme) rather than a whole word. Also, the Orkhon script is a runic script, written from right to left. The script includes special markers to indicate the end of words or phrases. There are eight vowels in the ancient Turkic Orkhon alphabet. These eight vowels were compactly labeled with four characters. These were used to distinguish a word or word endings. Most consonantal signs of the ancient Turkic Orkhon alphabet have two pairs, thick and thin consonants. These pairs define a strict and timid reading of the word. The signs of the Orkhon script are also very similar to the signs of the Yenisei, Talas, Turpan, and Altai with minor differences.

To train a convolutional neural network (CNN) to recognize Orkhon symbols, a large dataset of images containing these symbols is needed. The process of compiling a dataset is based on data collected from various sources, ranging from rock paintings and handwritten documents to books and digital records. When collecting data, all instances are preprocessed, that is, their size is changed to a consistent size to ensure uniformity, as well as convert the image to grayscale to remove unnecessary contamination and increase the clarity of the data. To collect symbols of the same types, classification is performed. Classification involves the process of categorizing data into different classes or groups based on certain criteria or features. When it comes to symbols, classification can be used to group symbols of the same type together based on their characteristics, such as shape, color, or meaning.

For the neural model to show the highest accuracy, it is necessary to avoid repetitions in the dataset. It is quite simple to avoid repeating the data by changing the size, and position and adding a little turbidity and contamination to the images. However, generating data based on available data for position and orientation changes is very efficient.

This implies that data augmentation techniques can be effectively employed to create diverse datasets without the need to collect entirely new data. These parameters are sufficient to avoid duplication of data in the dataset, and intentionally not severely distorting the data will help the model recognize even severely damaged scripts. Symbols of different sizes, collected from different sources and ready for classification, are shown in Fig. 3.

As usual, the collected data is divided into training and testing of the model in a ratio of 70% to 30%. The collected data is divided into training and testing sets in a ratio of 70% to 30%, which is a common practice in machine learning to evaluate model performance.

### 3.3    Creating a Script to Segment Input from an Image Containing Orkhon Runes

Segmentation of words and sentences in an image, in this case Orkhon runes segmentation involves extracting these characters from the rest of the image content. The following are the general steps to perform segmentation:

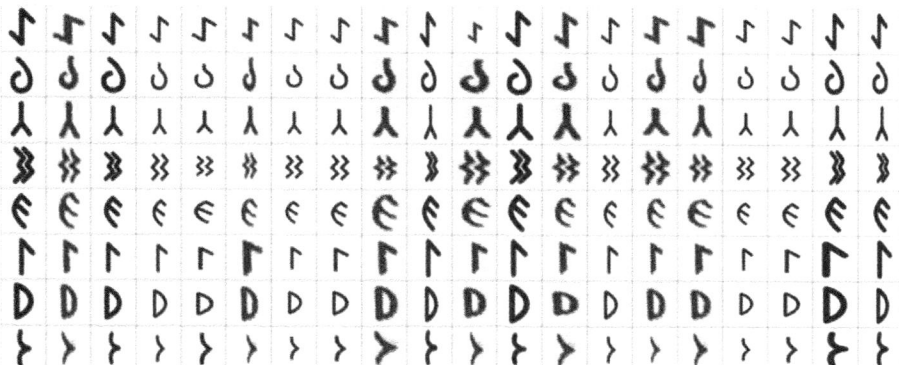

**Fig. 3.** Signs collected from different sources.

– Preprocessing step: In this step, the image is converted to grayscale to simplify processing. All necessary filters are applied (e.g. Gaussian blur) to reduce noise. If necessary, the contrast is increased.
– Contour detection step: The contours are highlighted inside the image. Contours based on area, aspect ratio, and other characteristics are filtered to remove noise and non-rune elements.
– Segmentation step: It is necessary to trim the areas inside the bounding boxes to extract individual Orkhon runes. It is also necessary to resize the cropped images to the standard size for further processing. After completing these steps, saving the segmented characters is needed.

The segmentation of sentences containing Orkhon runes occurs in stages for each line, which is quite similar to building a hierarchy. A segmentation system has been proposed for the segmentation of Orkhon proposals:

– Line segmentation: Line segmentation involves dividing the text into lines. This is particularly important when dealing with paragraphs or blocks of text, where individual lines need to be identified and processed separately. In OCR (Optical Character Recognation) systems, line segmentation helps to isolate each line of text from the image or document, making it easier to recognize individual words and letters accurately. Techniques for line segmentation often involve detecting white spaces or gaps between lines, as well as analyzing the alignment and orientation of text.
– Word segmentation: Word segmentation is the process of breaking down a line of text into individual words. This is crucial for tasks such as text recognition and language analysis. In OCR systems, word segmentation involves identifying spaces between words or analyzing the shapes and patterns of characters to determine word boundaries.
– Letter segmentation: Letter segmentation refers to the division of words into individual letters or characters. In OCR systems, letter segmentation is necessary to recognize each character accurately. This step can be challenging, especially in handwritten text or when characters are closely connected (as in cursive writing). Techniques for letter segmentation often involve analyzing the spatial relationships between strokes and

identifying features such as ascenders, descenders, and serifs to distinguish between letters.

Figure 4 shows a visualization of the segmentation process, the Orkhon Rune segmentation process system consists of 4 stages, sample input, line contouring, word contouring, letter contouring, and extracting.

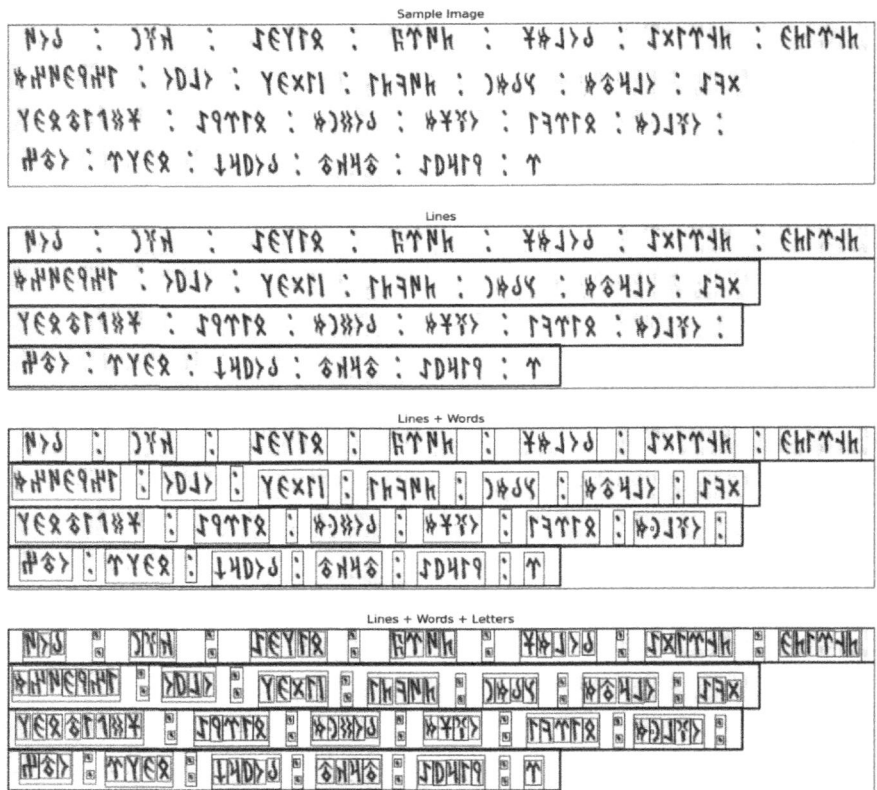

**Fig. 4.** Visualization of the segmentation process.

As shown in Fig. 5, the first two stages of the process, "Sample Processing" and "Line Segmentation" are presented. The first stage for segmentation is to process the sample, remove noise, and convert it to grayscale to enhance clarity, as shown in "Sample Processing". A sample containing the Orkhon script is taken, and preprocessing takes place. The sample is taken from the book [15]. The second stage is the line segmentation process, during which the lines are outlined, after which they will be divided into parts, as shown in "Line Segmentation". The task of this stage is to contour the lines and cropping.

As a result of the "Line Segmentation" process, the selected line will be processed to separate spaces and words. After cropping the line, the word segmentation process

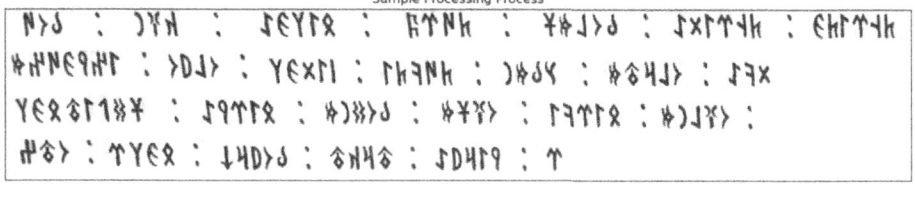

**Fig. 5.** Sample Processing and Line Segmentation Processes.

starts, words and separating spaces between the words are marked and separated. To separate words, the script includes special markers to mark the end of words or phrases, usually a colon as a separator between words in Orkhon inscriptions. The cropped line and "Word Segmentation" process are shown in Fig. 6. In the "Word Segmentation" process, words and colons of the line are contoured and cropped for the next process.

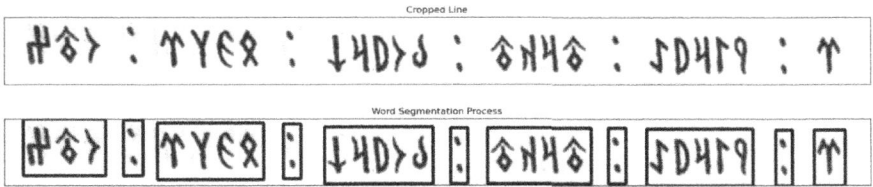

**Fig. 6.** Cropped Line and Word Segmentation Process.

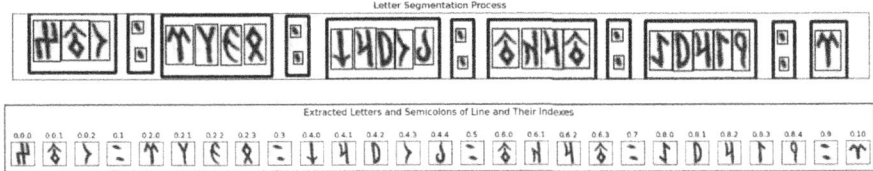

**Fig. 7.** Letter Segmentation Process.

The last stage is letter segmentation. At this stage, all letters and separators will be contoured and extracted. The "Letter Segmentation" process in Fig. 7 shows the contoured words, colons, and letters of each word. After these stages, all letters and separators contained in the image will be sorted and indexed, as shown in the "Extracted Letters and Semicolons of Line and Their Indexes" of Fig. 7.

Overall, line-word-letter segmentation plays a vital role in converting textual information from images or documents into a format that can be processed and analyzed by computers, enabling a wide range of applications, including document digitization and text extraction.

## 4    Experiments and Results

The dataset on which the model is trained is presented in Fig. 9. The dataset consists of 35 types of symbols of the Orkhon inscription, 20 samples each. The dataset is assembled manually from various sources. In total, there are 700 samples of manual collection in the dataset. The dataset was divided for training and testing in a ratio of 70% to 30%.

During the process of fitting the model to the data, the ImageDataGenerator function was used. This function is commonly used in the context of Convolutional Neural Networks (CNNs) for image data augmentation and preprocessing during the training of CNN models. The ImageDataGenerator is configured to perform various data augmentation techniques such as rescaling, rotation, shifting, shearing, zooming, and horizontal flipping. The quantity of generated data varies based on the batch size, the number of epochs, and the steps per epoch. The experiment specified batch size $= 32$ and epochs $= 1000$, steps_per_epoch were selected by the model automatically. Table 1 presents the sample images, the recognized sentences, Unicode, and their evaluation.

Categorical cross-entropy loss function and accuracy metric were used for evaluating the recognized sentences. Categorical cross-entropy computes the cross-entropy loss between the labels and predictions and it is used when there are two or more classes of labels. The wrong-recognized symbols are marked in bold. The first column of Table 1 contains the sample images containing the Orkhon inscriptions that must be recognized and the result of sentence recognition. The third column of the table presents the Unicode of the recognized sentence. The last column represents the model recognition accuracy and loss. The recognition model works fine, but there are small recognition errors in a larger volume of text.

## 5    Discussion

The accuracy of the model for recognizing the Orkhon script was evaluated. The total number of characters including separators on the image samples is 622. The total number of erroneous characters in recognition is 3. The percentage of recognition errors will be $3/622 = 0.4\%$. The errors are because some types of symbols were not added to the dataset. Average accuracy for all 5 samples is 99.5%, average loss for all 5 samples is 0,46%. In general, judging by the accuracy and the minimality of loss, the quality of recognition of Orkhon runes using the proposed model seems to be quite good. The main limitation of the proposed system at the moment is errors in recognizing long sentences, in many ways this is a problem in the process of character segmentation, but nevertheless the error can occur precisely from a small dataset. The runes of the Orkhon symbols are quite similar to each other, therefore the model may be slightly confused, but such cases have not been observed. In theory, this model, with some changes in the dataset, can be used to recognize the writing of other ancient Turkic runes, such as Altai, Talas, etc.

The further goal is to improve and enhance the proposed system: to solve the dataset problem and enhance the architecture of the model.

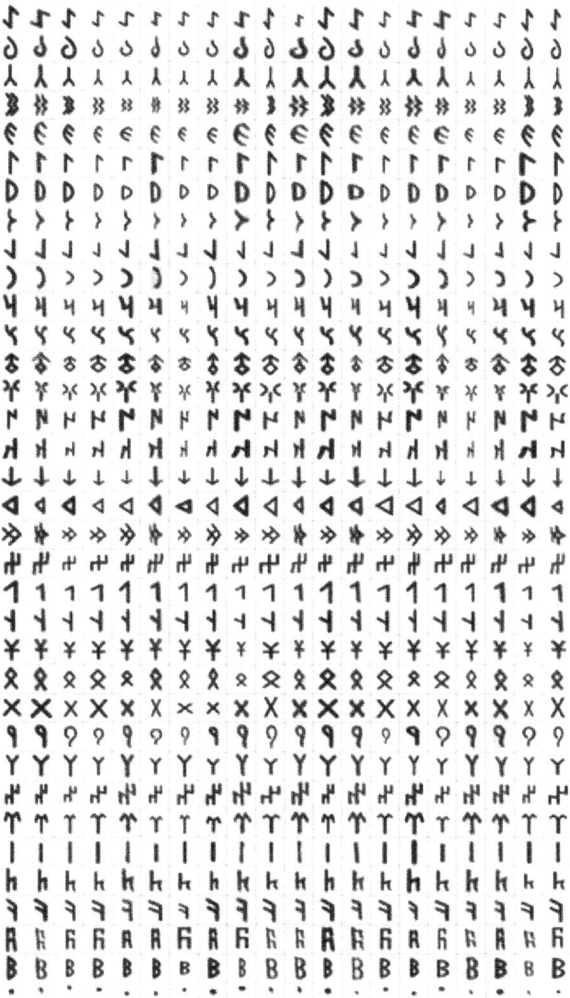

**Fig. 9.** The dataset.

**Table 1.** Sample images, recognized sentences, Unicode, and their evaluation.

| Sample image and recognized sentence | Unicode | Evaluation |
|---|---|---|
| Sample image: *(Old Turkic runic script)*<br><br>Recognized sentence: *(Old Turkic runic script)* | *(Old Turkic runic Unicode text)* | Accuracy: 99,9%<br>Loss: 0.11% |
| Sample image: *(Old Turkic runic script)*<br><br>Recognized sentence: *(Old Turkic runic script)* | *(Old Turkic runic Unicode text)* | Accuracy: 99,8%<br>Loss: 0.19% |
| Sample image: *(Old Turkic runic script)*<br><br>Recognized sentence: *(Old Turkic runic script)* | *(Old Turkic runic Unicode text)* | Accuracy: 99,6%<br>Loss: 0.43% |
| Sample image: *(Old Turkic runic script)*<br><br>Recognized sentence: *(Old Turkic runic script)* | *(Old Turkic runic Unicode text)* | Accuracy: 99,2%<br>Loss: 0.80% |
| Sample image: *(Old Turkic runic script)*<br><br>Recognized sentence: *(Old Turkic runic script)* | *(Old Turkic runic Unicode text)* | Accuracy: 99,2%<br>Loss: 0.80% |

## 6   Conclusion

In conclusion, considering the problems of the Turkic civilization, it is necessary to note the importance of considering the Turkic writing and their religion not only as the leading history of the past but also in the direction directly serving the formation of historical consciousness that meets the spiritual demands of modern life. Currently, there are very few specialists around the world who can read ancient Turkic inscriptions from the original text, and properly functioning information systems/programs/applications for text recognition and translation of ancient Turkic writing do not exist at all until now. This research bridges the gap between technology and cultural heritage, showcasing the potential of computational methods in unraveling the linguistic and historical complexities of Ancient Turkic Runic Orkhon inscriptions. The outcomes of this study have implications for the broader field of digital humanities, offering a valuable tool for scholars and researchers engaged in the preservation and interpretation of ancient scripts.

# References

1. Amanzholov, A.S.: History and theory of ancient Turkic writing, 2nd edn., edited and added. Almaty: Mektep (2010). 368 pp.
2. Narang, S.R., Jindal, M.K., Kumar, M.: Ancient text recognition: a review. Artif. Intell. Rev. **53**, 5517–5558 (2020). https://doi.org/10.1007/s10462-020-09827-4
3. Krithiga, R., Varsini, S., Joshua, R.G., Om Kumar, C.U.: Ancient character recognition: a comprehensive review. IEEE Access **4**, 1–11 (2023)
4. Gordin, S.: Optical character recognition for ancient non-alphabetic scripts. Open Access Government **36**(1), 280–281 (2022)
5. Wang, Z.: Ancient character recognition with deep learning techniques. Appl. Comput. Eng. **9**(1), 194–203 (2023)
6. Cilia, N.D., De Stefano, C., Fontanella, F., Marrocco, C., Molinara, M., Scotto Di Freca, A.: An end-to-end deep learning system for medieval writer identification. Pattern Recogn. Letters **129**, 137–143 (2020)
7. Barucci, A., Cucci, C., Franci, M., Loschiavo, M., Argenti, F.: A deep learning approach to ancient egyptian hieroglyphs classification. IEEE Access **9**, 123438–123447 (2021)
8. Elsawy, A., Loey, M., El-Bakry, H.: Arabic handwritten characters recognition using convolutional neural network. WSEAS Trans. Comput. Res. **5**(1), 11–19 (2017)
9. Rizky, A.F., Yudistira, N., Santoso, E.: Text recognition on images using pre-trained CNN. arXiv e-prints, pp 1–11 (2023)
10. Kumar, G., Bhatia, P.K.: Neural network based approach for recognition of text images. Int. J. Comput. Appl. **62**(14), 0975–8887 (2013)
11. Sadibekov, AK, Baiymbetova, R.K., Myrzakhanova, A.K, Yergubekova, Z., Abilova, G., Sadykova, A.D.: Teaching genesis of the old Turkic alphabet and its connection with Turkic tribe Tamgas. Jurnal Ilmiah Pendidikan **42**(3), 666–682 (2023)
12. Mizhit, L.: The written monuments of the ancient Turks and the Tuvinian poetry. Arts Hum. Open Access J. **2**(6), 397–400 (2018)
13. Rogozhinsky, A.E., Cheremisin, D.V.: The tamga signs of the Turkic nomads in the Altai and Semirechye: comparisons and identifications. Archaeol. Ethnol. Anthropol. Eurasia **47**(2), 48–59 (2019)
14. Proposal for encoding the Old Turkic script in the SMP of the UCS (2008)
15. Orkhon Inscriptions. Kyul-Tegin. Bilge Kagan. Tonyukuk. Abai International Club. Semey (2001)

# Topic Modeling with Variable Neighborhood Search

Alymzhan Toleu[1,2(✉)], Gulmira Tolegen[1,2], and Rustam Mussabayev[1,2]

[1] Satbayev University, Almaty, Kazakhstan
**alymzhan.toleu@gmail.com**
[2] Institute of Information and Computational Technologies, Almaty, Kazakhstan

**Abstract.** In this study, we introduce a topic modeling approach using variable neighborhood search (VNS), it is denoted as VNStopic. VNS is global optimization meta-heuristics, the idea behind it is to find neighborhood solutions and gradually change these solutions to escape from valleys that contain local minima. VNStopic is proposed to alleviate the problems of traditional topic modeling approaches that often rely on local search techniques for parameter estimation. The proposed approach uses VNS to systematically explore areas that are further away from the previous best solutions/topics initially identified. Experimental results show that the proposed approach outperforms baselines in topic coherence.

**Keywords:** Topic modeling · Variable neighborhood search · Global optimization

## 1  Introduction

Topic modeling is a task that involves analyzing a collection of documents to uncover latent thematic structures within them, often referred to as topics. These topics are collections of words that describe specific thematic categories. A developed topic model can be a versatile tool in text analysis that finds application in various fields, including but not limited to information retrieval, document classification [2] and media analysis [13].

Existing approaches for topic modeling are based on matrix factorization, probabilistic topic modeling, clustering-based, or based on neural networks. Most of those models' parameters are estimated with local search techniques, e.g. Gibbs sampling [3], an expectation-maximization algorithm utilized in pLSA [12], LDA [2], ARTM [23], and topic models based on neural networks [8,17] employ gradient-based optimization [15], and the clustering-based approaches utilize local search clustering algorithms like k-means [14].

In this paper, we introduce an approach for topic modeling, employing variable neighborhood search (VNS) [9] as its core global optimization mechanism. VNS is a meta-heuristic for global optimization, the idea behind it is to find neighborhood solutions and gradually change them to escape from valleys that contain local minima and to find better solutions. Different from the traditional

N.-T. Nguyen et al. (Eds.): ICCCI 2024, CCIS 2166, pp. 234–246, 2024.
https://doi.org/10.1007/978-3-031-70259-4_18

local search methodologies which gradually explore the solution space in a constrained manner, VNS doesn't adhere to a specific trajectory but instead explores farther neighborhoods of the current best solution, jumping to a new one solely when an enhancement is observed. The proposed approach gradually changes the neighborhood structure of topics with step size and takes topic coherence score as the objective. In the proposed approach, different step sizes are defined to change the neighborhood structure of topics gradually, it allows for a fine-grained exploration of topics. Topic coherence score is utilized as the primary objective function that serves as a quantifiable indicator of the quality of topics, ensuring the optimization process is firmly adhered to in enhancing the interpretability and consistency of the obtained topics. Experimental results showed that the proposed approach outperforms baselines in topic coherence and by enlarging the number of topics, it also can achieve fine-grain topics with their importance to the corpus.

The rest of the paper is organized as follows: Sect. 1 presents an introduction and Sect. 2 describes the related work for topic modeling from four categories. The approach is described in Sect. 3; Sect. 4 reports the experimental results and Sect. 5 concludes the work.

## 2    Related Work

Many existing approaches to topic modeling have been based on i) matrix factorization techniques and ii) expectation-maximization (EM) algorithm [5], the latter of which is frequently associated with probabilistic topic modeling. Some approaches are based on iii) word network clustering algorithms, whereas others utilize iv) topic modeling techniques with neural networks, taking advantage of the recent advancements in large language models [6]. In what follows, we will focus on these four categories to outline the current methods.

**Matrix Factorization.** Latent semantic analysis (LSA) [4] is a matrix factorization method. It employs singular value decomposition (SVD) [16] to derive the best low-rank approximation of a token-document occurrence matrix. The basic idea is to decompose the token-document matrix into two matrices: document-topic and topic-token. Each row of the token-document matrix represents a document and each column represents a word. Each value of this matrix can simply be the co-occurrence of a token in the corpus, However, it is not significant in providing information about topics. For example, the word "finance" informs us about the topic than the word "people", even the latter has a higher frequency than the former. To alleviate this issue, there has been a revival of interest in these techniques, enriched by the integration of probabilistic models, leading to the development of Probabilistic latent semantic analysis (pLSA) [12].

**Probabilistic Topic Modeling.** Unlike LSA, which relies on SVD, pLSA is based on constructing a probabilistic model to approximate the occurrence matrix of documents, focusing on uncovering the hidden thematic structure within them. pLSA tries to estimate the two latent parameters of a matrix with a probabilistic method. It considers two probabilistic assumptions:

- given a document $d$, the topic $t$ is presented with a probability of $p(t|d)$;
- given a topic $t$, tokens are drawn from the probability of $p(w|t)$.

More formally, pLSA holds the following joint probability:

$$p(D, W) = p(D)p(t|d)p(w|t) \tag{1}$$

where $D$ is a collection of documents, and the equation tells us how likely to see the documents based upon the distribution of topic-document and topic-token. $P(D)$, $P(t|d)$, $P(w|t)$ are the parameters of pLSA, which can be estimated with the EM algorithm.

Latent Dirichlet allocation (LDA) [2] is a Bayesian version of pLSA, it considers a prior initialization for topic-document and topic-token distribution. It is initialized with Dirichlet distribution [20], which is a probability distribution sampling over a probability simplex, which makes a role of sampling with a distribution that fits to practice.

$$P(x|\alpha) = \frac{\Gamma\left(\sum_{i=1}^{K} a_i\right)}{\prod_{i=1}^{K} \Gamma(a_i)} \prod_{i=1}^{K} \theta_i^{a_i} \tag{2}$$

Eq. 2 defines $k - 1$ simplex and, $a_1...a_k$ are the parameters of the topic model, when $a < 1$ the majority of the probability mass is in the"corners" of the simplex, generating mostly documents that have a small number of topics. For example, for the mixture of three topics, in most cases, a document is going to be heavily related to one of these topics, and have light or no relation to other topics. But does not have a uniform relation with all of the three topics. This is essentially what a Dirichlet distribution provides in LDA models.

In pLSA, first of all, a document is sampled, then a topic is sampled based on that document, and finally, a word is sampled based on the chosen topic. The probability of a document is considered a fixed point in the dataset. This makes it difficult for pLSA to generalize to unseen documents. LDA solves the problem of pLSA, and the dataset serves as a training dataset for Dirichlet distribution of topic-document and if it encodes an unseen document, it can sample from Dirichlet distribution easily.

ARTM [23] stands for additive regularization of topic model, it is a multi-objective topic modeling approach. It takes into account the features of topics in the objective function, namely topic coherence, decorrelation, parsing regularization, etc. As described above, LDA generates topics from prior Dirichlet distribution and, the conjugacy of Dirichlet and multinomial distribution reduces complexity and facilitates Bayesian inference. But if LDA is combined with multiple objectives, it may cause a case of non-conjugate prior, and this overcomplicates Bayesian inference. Another issue with LDA is that the Dirichlet prior conflicts with the natural assumption of sparsity: a document related to a small number of topics and topics consists of a small number of domain-specific words, therefore there are zero probabilities may obtained for most of the words and topics. Dirichlet prior can not produce a vector with zero value, it contradicts zero probabilities. ARTM is an approach for solving these problems, and it chooses pLSA as its base model.

**Clustering-Based Approaches.** Word clustering approach for topic modeling is another attempt [21]. It analyzes the word networks which are constructed with co-occurrence from a collection of documents. Each topic is considered a community and utilizes a community detection algorithm adapted for these networks to detect topics. The main idea of this method is to split the general search space of topics $|D| \times |T| \times |W|$, into two smaller spaces: $|T| \times |W|$ and $|D| \times |T|$. In the first space, topics are detected, whereas in the second space, topics are assigned. To construct the word networks, different word representations were utilized: TFIDF, co-occurrence, and word embedding. The experiments from topic detection showed that co-occurrence and TFIDF representations give a higher topic coherence score compared to others. For topic assignments, the Jaccard metric and cosine similarity-based distance approaches were applied. The former is discrete features and the latter is dense features, as shown in the results for topic assignments, the latter method outperforms the former in most cases.

**Neural Topic Models.** Neural topic models (NTM) have shown improvements in overall coherence evaluation [1,7,8,17–19]. Most of the NTMs are based on variational autoencoder (VAE), which takes bag-of-words (BOW) representation of documents as input, learning hidden distribution with layers of neural network and at the decoder part reconstructing the bag-of-words representation. NVDM-GSM [18] solves topic modeling with VAE, the authors proposed neural networks conditioned on a draw from a multivariate Gaussian distribution. Evaluation results show that their model achieved good performance on a range of standard document corpora. SAE-NTM [17] is a sentence-aware NTM, it incorporates the fine-grained sentence embeddings as external knowledge using large language models. SAE-NTM outperforms other neural topic models in topic coherence on three benchmark datasets. BertTopic [8] is one of the popular topic models that incorporate the recent development of Bert. First of all, it generates document embeddings with pre-trained large language models, then does a dimension reduction process and clustering of the documents. Considering each cluster of documents as a large document, and run the typical TFIDF over these new collections of documents to infer the topics.

## 3   Approach

### 3.1   Definition

A collection of documents is denoted as $D$, and $W$ for unique tokens from this dataset, each token can be n-grams. Each document $d \in D$ consists of a sequence of $n_d$ words $w_1, ..., w_{n_d}$. Topic modeling aims to find the possible $T$ topics contained within the dataset the $D$ and $W$. Topics $T$ are hidden variables, and $D$ represents the observable variables.

Given the observable variables $d \in D$ and $d = w_1, ..., w_{(n_d)}$, assume that each token $w_{(i_d)} \in d$ corresponds to a latent topic from the finite set of $T$ topics. Each topic $t_j \in T$ is represented as a bag of tokens, with the top-M relevant tokens in each topic being as cohesive and semantically similar as possible.

## 3.2   Variable Neighborhood Search

Variable Neighborhood Search (VNS) [9] is a meta-heuristic framework that is used to find global minima of optimization problems. The concept of VNS involves identifying neighborhood solutions of the problem and gradually changing these neighborhood solutions to escape from valleys that contain local minima. Algorithm 1 shows a basic VNS process and it can be described as follows: As it is shown, it starts with an initial solution $x$. Using the current solution as input, neighborhood solution $x'$ will be generated by shaking and locally optimized to $x''$. If some improvements are found within the neighborhood solution, then the set neighborhood solution is the current solution. if no improvements are observed, then enlarge the neighborhood structure for shaking. This process will be interactively processed until the final termination condition is met.

---

**Algorithm 1.** Basic Variable Neighborhood Search (VNS)

---

```
 1: Function VNS(x, k_max, t_max)
 2: repeat
 3:     k ← 1
 4:     repeat
 5:         x' ← Shake(x, k)                          // Shaking
 6:         x'' ← BestImprovement(x')                 // Local search
 7:         x ← NeighbourhoodChange(x, x'', k)        // Change neighbourhood
 8:         k ← k + 1
 9:     until k > k_max
10:     t ← CpuTime()
11: until t > t_max
```

---

## 3.3   Topic Modeling with VNS

As we know, VNS [10] is a global optimization meta-heuristic. The purpose of using VNS in topic modeling is to address the limitations of the existing approaches. For instance, LDA, which employs the Gibbs sampling method for parameter estimation, and various neural networks that utilize the gradient descent method for optimization, most of them are only guaranteed to find local minima [3,15].

To apply VNS in topic modeling, we change the task into a clustering problem and the proposed VNS-based topic model can be described with the following several modules:

- vocabulary extraction process, that creates a vocabulary and filters out unimportant words from documents, like stopwords, special characters, punctuations, etc.
- representation learning, which involves representing the words in the vocabulary and documents with numerical vectors.
- the objective function, which evaluates the current solution.
- neighborhood structure generation, which is for changing the solutions with different step size.

**Vocabulary Extraction.** Extracting vocabulary from a collection of documents is a vital step in the proposed approach since the final topics will be obtained over this vocabulary through a clustering approach. We choose to use the term frequency-inverse frequency (TF-IDF) that is aimed at identifying the importance of words within a collection of documents. It calculates the term frequency and inverse term frequency and then combines these two.

Term frequency (tf) calculates how a word occurs in a document. Inverse document frequency (idf) assesses the importance of a word across the entire document collection. The idf diminishes the weight of words that appear too frequently across documents, as these are less informative or unique. i.e., if a word is common, it will occur across all documents, or if a word is too specific, it will only appear in a set of related documents not all of them. For each document, we take the words with the top 50 highest tfidf values as candidates. Moreover, external part-of-speech filtering is also applied, specifically by filtering out adverbs and adjectives, including their superlative and comparative forms.

**Representation Learning.** As a representation learning method for the vocabulary and documents, a distributional representation can be utilized, they are co-occurrence and tfidf, and representation from both of them is sparse and high-dimension. Co-occurrence calculates how many times a pair of the tokens have appeared together in a document. Besides simple distributional representations, the pre-trained language models of Bert for generating the representation of words and documents are utilized. Bert proposes a masked language modeling objective; where some tokens of an input sentence are randomly masked, which is aimed to predict these masked positions. Bert utilizes self-attention layers, and multiple of these layers make multi-headed attentions, which is referred to as transformer [22]. Another advantage of Bert is to add next sentence prediction, to the objective, and it aims to improve sentence modeling. To obtain document representation, we average the sentence's representation within the document.

**Objective Function and Neighborhood.** To calculate the initial solution for VNS approach, we use the k-means [11] clustering algorithm as the local search for simplicity. However, within the proposed framework, other more advanced fuzzy algorithms can also be employed. A local search algorithm will generate clusters of words using vectors generated with the representation learning step. We assume this initial solution is a local minima, which may not fully represent topics of the corpus, it may only be semantically similar words. To escape from this local minima and obtain better topics, we need to apply VNS meta-heuristic, this requires answering the following questions: i) How do we evaluate the current solution? ii) in optimization, in which direction should we proceed? iii) How far should we go?

Question (i) relates to evaluating the current solution, which involves assessing the quality of the topics. In practice, we use topic coherence to assess the quality of obtained topics which evaluates how cohesive the inner words in each topic are. The widely used topic coherence scores for topic quality evaluation metrics are Umass, Uci, NPMI, etc. In the optimization step, we try to optimize topic coherence directly using VNS. For coherence, we choose Umass as its simplicity, it calculates how often two words, $w_i$ and $w_j$ appear together in the corpus:

$$TC_{discrete} = \frac{2}{N \cdot (N-1)} \sum_{i=2}^{N} \sum_{j=1}^{i-1} log \frac{P(w_i, w_j) + \epsilon}{P(w_j)} \tag{3}$$

Umass topic coherence relies on the discrete words, but to adapt this metric for dense vector spaces, we introduce a dense-based topic coherence as follows:

$$TC_{dense} = \frac{1}{N} \sum_{i=1}^{N} \left( \frac{1}{M_i} \sum_{j=1}^{M_i} sim(w_1, w_2)_{ij} \right) \tag{4}$$

where $N$ is the total number of topics. $M_i$ as the number of unique pairs of terms in topic $i$. $sim(w_1, w_2)_{ij}$ calculates the cosine similarity score of $j$-th pair in topic $i$.

---

**Algorithm 2.** Topic Modeling with VNS

---

1: Calculate the initial topics $T$ using the k-means algorithm
2: $S \leftarrow$ Get centroids of topics $T$
3: $f(T, D) \leftarrow$ Compute an initial of coherence scores
4: **for** $iteration \leftarrow 1$ **to** max_iter **do**
5:     **for** each $step\_size$ in $step\_sizes$ **do**
6:         $S' \leftarrow$ Shaking the centers of each topic
7:         $T' \leftarrow$ Local search and obtain the new topics
8:         $f'(T', D) \leftarrow$ Compute the new topic coherence scores
9:         **if** $f'(T', D) > f(T, D)$ **then**
10:             $T \leftarrow T'$ Set new topics
11:             $S \leftarrow S'$ Set new centroids
12:             **break**
13:         **end if**
14:     **end for**
15: **end for**
16: **return** $S$, $T$

---

Then, the objective function can be defined as follows:

$$\max f(T, D) = TC_{discrete|dense} \tag{5}$$

Question (ii) relates to the feasibility of attaining any neighborhood solution or every local minima (valley). One of the simplest ways is to choose a direction at random [9]. Since we choose k-means as a local search and it has centers of topics, in this work, we simply shake the centroids of clusters with a certain value with different step sizes that answer the question (iii). Algorithm 2 shows the topic modeling process with VNS. After identifying topics, the document-topic assignment can be obtained by calculating the similarity between topic and document representations.

## 4  Experiments

Three sets of experiments are reported in this section: i) comparing the discrete and dense vector-based objective function with different tokens' representation including co-occurrence, tfidf and Bert. ii) evaluating the topic coherence of the proposed approach and comparing it with the baselines. iii) experiments related to document classification tasks with obtained topic models.

The proposed approach's performance is evaluated in the following two angles: i) topic quality using coherence score; ii) topic relevance, document classification with the topic model that evaluates how the detected topics are relevant to the corpus. Since the dataset has categories that allow us to compute following metrics for the document clustering: i) adjusted rand score; ii) normalized mutual info score. It should be noted that the proposed model, $VNS_{topic}$ is denoted as VNS for simplicity with different representation settings in the following descriptions.

The proposed approach is evaluated on the 20newsgroups corpus, which is used as a standard corpus for topic modeling. Table 1 shows the statistics of the dataset.

**Table 1.** Dataset statistics.

| Name | 20newsgroups |
|---|---|
| #topic | 20 |
| #documents of training set | 11314 |
| #tokens | 59143 |
| #words filtered with tfidf | 10184 |

### 4.1  Results

To show how the VNS-based approach improves across its interactions, Fig. 1 presents the improvements over each interaction with different topic coherence

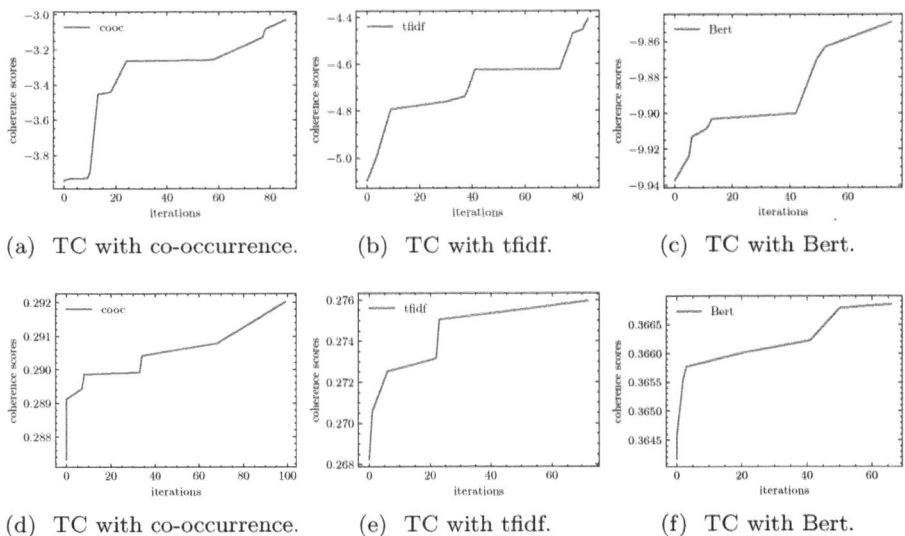

(a)  TC with co-occurrence.    (b)  TC with tfidf.    (c)  TC with Bert.

(d)  TC with co-occurrence.    (e)  TC with tfidf.    (f)  TC with Bert.

**Fig. 1.** Improvement of topic coherence scores across iterations with VNS. Note: the first row of the figure presents the results for $TC_{\text{discrete}}$, and the second row presents the results for $TC_{\text{dense}}$.

and representations. In this experiment, we run 100 iterations to all different settings of the approach, and since they have different *step_size* and neighborhood changes, it obtains improvement in different iterations. We choose specific *step_size* = $[0.001, 0.01, 0.1]$ for the neighborhood change when it utilizes co-occurrence representation. For tfidf, the step size is set to *step_size* = $[0.001, 0.01]$. Considering that Bert's representation is dense and low-dimension compared with others, it requires more fine-grained *step_size*. In the approach, *step_size* is the hyper-parameters that should be fine-tuned. Another parameter is that *topM*, during the optimization process, we only calculate topic coherence for the top *topM*=20 words. Enlarging this parameter could lead to a good result globally, but runs slower.

Comparing topic coherence in the first row of Fig. 1, it can be seen that compared with Bert and tfidf-based model, the co-occurrence-based model gives a high topic coherence. However, Bert-based model gives a lower TC among the three models in the first-row settings since it is related to discrete topic coherence. Both co-occurrence and tfidf representation contain the information about word co-occurrence. In contrast, Bert's representation contains more dense and semantic information. From the second row (for dense coherence), it can be seen that the model based on Bert's representations gives a higher topic coherence.

Table 2 shows the topic coherence results for the proposed approach and the baselines, which include LSA [4], pLSA [12], LDA [2], ARTM [23], word network clustering (WNC) [21] and BertTopic [8]. Top *TopM* = $[10, 20, 30]$ words of

**Table 2.** Results of topic coherence.

| Models\TopM | 10 | 20 | 30 | 10 | 20 | 30 |
|---|---|---|---|---|---|---|
| | Discrete TC | | | Dense TC | | |
| LSA | -2.79 | -3.89 | -3.99 | 0.297 | 0.292 | 0.298 |
| pLSA | -2.14 | -2.59 | -2.85 | 0.335 | 0.319 | 0.311 |
| LDA | **-1.98** | **-2.52** | -2.67 | 0.321 | 0.312 | 0.307 |
| ARTM | -2.24 | -2.75 | -3.05 | 0.344 | 0.323 | 0.311 |
| WNC | -3.17 | -3.69 | -5.74 | 0.262 | 0.250 | 0.245 |
| BertTopic | -2.97 | -3.46 | -3.62 | 0.334 | 0.309 | 0.294 |
| VNS_cooc | -2.44 | -2.53 | **-2.65** | 0.292 | 0.282 | 0.273 |
| VNS_tfidf | -2.51 | -2.86 | -3.12 | 0.275 | 0.265 | 0.261 |
| VNS_bert | -3.04 | -3.44 | -3.87 | **0.366** | **0.351** | **0.344** |

**Table 3.** Results of document classification with obtained topics.

| Models | ARI | NMI |
|---|---|---|
| LSA | 0.171 | 0.294 |
| pLSA | 0.18 | 0.31 |
| LDA | **0.198** | **0.332** |
| ARTM | 0.168 | 0.305 |
| WNC | 0.123 | 0.225 |
| BertTopic | 0.148 | 0.272 |
| VNS_cooc | 0.162 | 0.305 |
| VNS_tfidf | 0.179 | 0.307 |
| VNS_bert | 0.169 | **0.317** |

each topic are extracted and calculated for their topic coherence. It can be seen from the table, for discrete TC, LDA still outperforms others in most of the trials. $VNS_{cooc}$ and $VNS_{tfidf}$ outperforms $VNS_{bert}$ and achieves comparable results with the previous best. In the case of dense TC, as seen that $VNS_{bert}$ outperforms others and the baselines.

Table 3 shows the results of document classification with the obtained topics. In this experiment, we choose $topM = 200$ words for each topic to represent them, and use these words to calculate the topic vector with Bert and for the document, then compute the cosine similarity between documents and topics, classify the documents to the shortest distances with topics. It can be seen from the table that $VNS_{bert}$ achieves very competitive performance with LDA in terms of NMI score and outperforms other baselines.

One of the findings through these experiments is that the VNS-based approach is likely able to cluster all the semantic similar words: e.g. it obtains a large topic can be $[car, motor, motorcycle...]$ as a "transportation" topic when

**Table 4.** Examples of the first 20 important topics of 50 with a fine-grained degree extracted by VNS_bert.

| No. | First five words of the topics | | | | |
|-----|------|------|------|------|------|
| 0 | christianity | religion | christian | theology | Worship |
| 1 | hitter | outfielder | fielder | baseman | baseball |
| 2 | computer | hardware | dell | gigabyte | chipset |
| 3 | israel | palestine | syria | jordanian | arab |
| 4 | crypt | crypto | cryptography | cryptology | encryption |
| 5 | church | pastor | congregation | evangelist | preacher |
| 6 | automobile | car | vehicle | chevrolet | porsche |
| 7 | pistol | gun | rifle | firearm | handgun |
| 8 | prosecution | crime | justice | court | conviction |
| 9 | motorcycle | bike | rider | ridden | bicycle |
| 10 | buyer | purchasing | sale | buying | business |
| 11 | samuelsson | mckenzie | morris | goaltender | bowman |
| 12 | paper | printing | print | document | printer |
| 13 | music | audio | bass | sounder | instrument |
| 14 | program | software | unix | gnu | code |
| 15 | xtsetarg | xdm | xtwindow | xwindow | xdefaults |
| 16 | harassment | accusation | retaliation | provocation | displeasure |
| 17 | drug | medication | medicine | heroin | poison |
| 18 | newspaper | journalist | journalism | news | reporter |
| 19 | stanley | pavel | andreychuk | bob | pete |
| 20 | transmit | telecommunication | transmits | telecom | reception |

setting a small fix topic number. When enlarging the fixed topic number, the approach decomposes the large topic into smallar different topics for example, the previous topic can be divided into "automobile" and "motorcycle" topics. It can be observed in Table 4, that topic_6 and topic_9 are two fine-grained topics, but it could be merged as a single topic if we set a small topic number. Setting a large number of topics, the approach is also able to identify the top-N important topics across the corpus, and further enlarging the number of topics also can achieve fine-grain topics with their importance to the corpus as shown in Table 4.

## 5   Conclusion

This paper presents a topic modeling approach with variable neighborhood search, which is a global optimization meta-heuristics. It explores distant neighborhoods of the incumbent solution, not just simply following local search trajectory. The proposed approach gradually changes the neighborhood structure of topics with different step sizes and takes the topic coherence score as the objective function. Different representations are employed to explore the effectiveness of the approaches. To assess the approach, we introduce a dense topic coherence score that captures more semantic information compared to a discrete score. Experimental results showed that the proposed approach outperforms baselines in topic coherence.

The future work can be divided into two directions: i) To find better neighborhood solutions, more advanced neighborhood structure methods will be utilized. This includes considering topic-specific objectives such as combining sparsity, smoothing, and decorrelation factors, and selecting domain-specific topics. ii) Instead of using a random shaking strategy for topic centroids, more informative methods will be utilized, such as those based on averaging the historical trajectory of topic centroids and the task-specific shaking strategies.

**Acknowledgement.** The work was funded by the Committee of Science of Ministry of Science and Higher Education of the Republic of Kazakhstan under the grant BR21882268.

# References

1. Bianchi, F., Terragni, S., Hovy, D.: Pre-training is a hot topic: contextualized document embeddings improve topic coherence. In: Zong, C., Xia, F., Li, W., Navigli, R. (eds.) Proceedings of the 59th Annual Meeting of the Association for Computational Linguistics and the 11th International Joint Conference on Natural Language Processing (Volume 2: Short Papers), pp. 759–766. Association for Computational Linguistics, Online (2021)
2. Blei, D.M., Ng, A.Y., Jordan, M.I.: Latent Dirichlet allocation. J. Mach. Learn. Res. **3**(null), 993-1022 (2003)
3. Cheng, X., Wang, B., Zhang, J., Zhu, Y.: Fast conditional mixing of MCMC algorithms for non-log-concave distributions (2024)
4. Deerwester, S., Dumais, S.T., Furnas, G.W., Landauer, T.K., Harshman, R.: Indexing by latent semantic analysis. J. Am. Soc. Info. Sci. **41**(6), 391–407 (1990)
5. Dempster, A.P., Laird, N.M., Rubin, D.B.: Maximum likelihood from incomplete data via the EM algorithm. J. Roy. Stat. Soc. B **39**, 1–38 (1977)
6. Devlin, J., Chang, M., Lee, K., Toutanova, K.: BERT: pre-training of deep bidirectional transformers for language understanding. CoRR **abs/1810.04805** (2018)
7. Dieng, A.B., Ruiz, F.J.R., Blei, D.M.: Topic modeling in embedding spaces. Trans. Assoc. Comput. Linguist. **8**, 439–453 (2020)
8. Grootendorst, M.: BERTopic: neural topic modeling with a class-based TF-IDF procedure (2022)
9. Hansen, P., Mladenović, N.: Variable Neighborhood Search, pp. 211–238. Springer US, Boston, MA (2005). https://doi.org/10.1007/978-3-319-07153-4_19-1
10. Hansen, P., Mladenović, N.: Variable neighborhood search: principles and applications. Eur. J. Oper. Res. **130**(3), 449–467 (2001)
11. Hartigan, J.A., Wong, M.A.: A k-means clustering algorithm. JSTOR: Appl. Stat. **28**(1), 100–108 (1979)

12. Hofmann, T.: Probabilistic latent semantic indexing. In: Gey, F., Hearst, M., Tong, R. (eds.) Proceedings of the 22nd Annual International ACM SIGIR Conference on Research and Development in Information Retrieval (SIGIR '99), August 15-19, 1999, Berkeley, CA, USA, pp. 50–57. ACM Press, New York, NY, USA (1999)

13. Hong, L., Davison, B.D.: Empirical study of topic modeling in twitter. In: Proceedings of the First Workshop on Social Media Analytics, pp. 80–88. SOMA '10, Association for Computing Machinery, New York, NY, USA (2010)

14. Jain, A.K., Murty, M.N., Flynn, P.J.: Data clustering: a review. ACM Comput. Surv. **31**(3), 264–323 (1999). https://doi.org/10.1145/331499.331504

15. Jiao, Z., Keller-Ressel, M.: Emergence of heavy tails in homogenized stochastic gradient descent (2024)

16. Klema, V., Laub, A.: The singular value decomposition: its computation and some applications. IEEE Trans. Autom. Control **25**(2), 164–176 (1980). https://doi.org/10.1109/TAC.1980.1102314

17. Liu, H., Gao, J., Xiang, S., Liu, T., Fu, Y.: SAE-NTM: sentence-aware encoder for neural topic modeling. In: Strube, M., Braud, C., Hardmeier, C., Li, J.J., Loaiciga, S., Zeldes, A. (eds.) Proceedings of the 4th Workshop on Computational Approaches to Discourse (CODI 2023), pp. 106–111. Association for Computational Linguistics, Toronto, Canada (2023). https://doi.org/10.18653/v1/2023.codi-1.14

18. Miao, Y., Grefenstette, E., Blunsom, P.: Discovering discrete latent topics with neural variational inference. In: Proceedings of the 34th International Conference on Machine Learning - Volume 70, pp. 2410–2419. ICML'17, JMLR.org (2017)

19. Nan, F., Ding, R., Nallapati, R., Xiang, B.: Topic modeling with Wasserstein autoencoders. In: Korhonen, A., Traum, D., Màrquez, L. (eds.) Proceedings of the 57th Annual Meeting of the Association for Computational Linguistics, pp. 6345–6381. Association for Computational Linguistics, Florence, Italy (2019). https://doi.org/10.18653/v1/P19-1640

20. Ongaro, A., Migliorati, S.: A generalization of the Dirichlet distribution. J. Multivar. Anal. **114**, 412–426 (2013)

21. Tolegen, G., Toleu, A., Mussabayev, R., Krassovitskiy, A.: A clustering-based approach for topic modeling via word network analysis. In: 2022 7th International Conference on Computer Science and Engineering (UBMK), pp. 192–197 (2022)

22. Vaswani, A., et al.: Attention is all you need. In: Guyon, I., Luxburg, U.V., Bengio, S., Wallach, H., Fergus, R., Vishwanathan, S., Garnett, R. (eds.) Advances in Neural Information Processing Systems, vol. 30. Curran Associates, Inc. (2017)

23. Vorontsov, K.V.: Additive regularization for topic models of text collections. Dokl. Math. **89**(3), 301–304 (2014). https://doi.org/10.1134/S1064562414020185

# A Neuro-Symbolic Classification Algorithm Using Neural Cell Assemblies

Florin Leon[✉]

Department of Computers, Faculty of Automatic Control and Computer Engineering, "Gheorghe Asachi" Technical University of Iaşi, Iaşi, Romania
florin.leon@academic.tuiasi.ro

**Abstract.** This paper introduces a neuro-symbolic classification algorithm inspired by the SUSTAIN model in cognitive psychology. The proposed algorithm is implemented using neural cell assemblies, which can be seen as an intermediate level between individual neurons and higher-level cognitive functions. They offer the advantages of neural representations but can also be interpreted symbolically. Two methods are introduced to transform inputs into cell assembly representations. The algorithm creates prototypes from training instances and automatically estimates the weights of the problem dimensions. Although prototypes are typically used to classify new instances based on similarity, the paper also suggests a method to extract explicit, simple rules from the prototypes, encoded using the same cell assembly representation.

**Keywords:** neuro-symbolic models · prototypes · rules · cognitive psychology · SUSTAIN categorization algorithm · cell assemblies

## 1 Introduction

Artificial intelligence (AI) has witnessed remarkable advancements with the introduction of deep learning models, especially in tasks related to image, sound and language processing. While these models excel at data-driven tasks, their limitations in abstract reasoning and causal understanding are likely to hinder the progress towards artificial general intelligence (AGI). Moreover, deep learning models often require vast amounts of labeled data, which makes them impractical for scenarios with limited data.

In contrast, neuro-symbolic systems aim to use symbolic reasoning to facilitate learning with smaller datasets. They achieve this by incorporating abstract reasoning, understanding causality, and employing compact rules for effective knowledge generalization. This approach combines the strengths of neural networks with symbolic reasoning.

Insights from neuroscience and cognitive psychology have the potential to advance neuro-symbolic models, bridging the gap between symbolic reasoning and neural computation. By integrating cognitive principles with neural network architectures, these models could lead to the development of more robust, interpretable AI systems, and enhance our understanding of both human cognition and machine intelligence.

N.-T. Nguyen et al. (Eds.): ICCCI 2024, CCIS 2166, pp. 247–259, 2024.
https://doi.org/10.1007/978-3-031-70259-4_19

In neuroscience, cell assemblies (CA) – also known as neuronal ensembles – are groups of neurons whose activations represent specific concepts or perform specific functions. This can be seen as an intermediate level of representation between individual neurons and higher-level cognitive functions. CAs are composed of neurons, which offer the typical benefits of neural networks in AI, such as the capability to learn from data, to automatically extract relevant features, robustness to noise, and scalability. In addition, they can also be interpreted symbolically, which enables explicit reasoning, consistency and low data requirements.

This paper proposes a new neuro-symbolic classification algorithm inspired by a well-known categorization algorithm in cognitive psychology called SUSTAIN. The proposed algorithm uses an internal representation based on cell assemblies. Two methods are presented for the transformation of inputs into CA representations. One is biologically-inspired but has higher variability, and the other one is more systematic and its mathematical properties enable more consistent algorithm results.

The internal processing of training instances is based on prototypes, which can be applied for both supervised classification and unsupervised clustering, but the focus here is on classification. Prototypes incorporate similar instances and have both a mean value given by the average of instance values, and a variance measure which is shown to be related to feature weights.

A typical use of such a method is the classification of new instances based on their similarity to existing prototypes. However, we also introduce a rule extraction method from the prototypes. These rules are explicit and simple, following the typical characteristics of rules described in cognitive psychology, but represented within the same neural framework using cell assemblies.

## 2  Related Work

While the potential of neuro-symbolic systems is vast, the best path for their development is still unclear. Due to space limitations, this section only highlights a few impactful attempts that could pave the way for future AI, and possibly AGI.

In position paper [1] an architecture and training methods are proposed for constructing autonomous intelligent agents, inspired by biological learning. It presents a differentiable cognitive architecture for predictive world models, a non-contrastive self-supervised learning paradigm, and a hierarchical predictive architecture for hierarchical planning under uncertainty. These contributions address challenges such as learning from observation, combining reasoning with gradient-based learning, and building hierarchical representations of perceptions and actions.

Another position paper [2] discusses the hypothesis that biological intelligence can be explained by a few principles, similar to the principles in physics, and their understanding could aid in building intelligent machines. It focuses on inductive biases exploited by humans and animals, and proposes their integration into AI systems to improve out-of-distribution and systematic generalization, especially in "system 2" tasks that require explicit reasoning. Notable biases include the connection between high-level variables and natural language, modular decomposition of knowledge, causal interpretation of actions, and sparse dependencies between high-level variables.

DeepCTRL [3] is a training method that integrates rules into deep learning models with controllable rule strengths during inference. It employs separate encoders for data and rules, which allow the gradual adjustment of rule strengths without retraining. DeepCTRL increases rule verification ratio while maintaining or improving accuracy, and thus enhances model trust and reliability. It enables hypothesis testing on data samples and unsupervised adaptation based on shared rules between datasets. This technique is agnostic to data type and model architecture, and demonstrates its effectiveness in various domains such as physics, retail and healthcare.

The Differentiable Inductive Logic Programming ($\partial$ILP) framework [4] combines the strengths of connectionist systems and ILP. It is an end-to-end differentiable architecture that enables learning of explicit human-readable symbolic rules while maintaining robustness to noise and ambiguity in data. It reinterprets ILP as a binary classification problem for which it uses cross-entropy loss as an error measure. $\partial$ILP can solve moderately complex tasks that require recursion and predicate invention, such as the "fizz-buzz" problem, and generalizes well to unseen data. Unlike symbolic ILP systems, $\partial$ILP exhibits robustness to mislabeled instances and can handle ambiguous or fuzzy data when connected to convolutional neural networks.

The Neuro-Symbolic Concept Learner (NS-CL) [5] jointly learns visual concepts, words and semantic parsing from images and question-answer pairs, without explicit supervision. It includes perception, semantic parsing and symbolic execution modules. NS-CL learns via curriculum learning, i.e., it gradually acquires object-based and relational concepts. It achieves state-of-the-art performance on the CLEVR dataset, and exhibits robust combinatorial generalization across various tasks and datasets.

Relation Networks (RN) [6] are proposed as plug-and-play modules to solve tasks reliant on relational reasoning in neural networks. RN-augmented networks attain state-of-the-art performance in visual question answering (CLEVR), text-based question answering (bAbI), and reasoning about dynamic physical systems. RNs enable neural networks to implicitly discover and reason about entities and their relations.

We can also mention several reviews: [7] covers neuro-symbolic learning systems, while [8] and [9] focus on research in neuroscience and cognitive psychology that could be applied to AI or AGI.

## 3   The Proposed Neuro-Symbolic System

### 3.1   A Brain-Inspired Representation

In this section and the following one, we will introduce two original methods to create cell assembly representations for the inputs of the algorithm.

The inputs can be either numerical or symbolic. These types are encoded in different ways, but the final structure of the encoding is the same.

For numerical inputs, the model considers neurons as detectors, each tuned to respond maximally to a particular number. Similar to how the brain encodes numbers, these detectors have preferred values and receptive fields that define their sensitivity range. In the human brain, these fields have approximately Gaussian shape and the mapping can be logarithmic, but neurons as detectors in general can have arbitrary tuning curves.

In our model we consider a simplified case, where the tuning curve of a neuron has the shape of a symmetric triangular fuzzy number, which roughly approximates a Gaussian form, and thus the activation $a$ of a neuron $i$ for a numerical (real value) $x$ is:

$$a_i(x) = \max\left(0,\, 1 - \frac{|x - m_i|}{s_i}\right) \quad (1)$$

where $m_i$ is the preferred value of the neuron and $s_i$ is its receptive field width.

Different detectors cover various numerical ranges. When a numerical input is presented, the activation of each detector reflects how close the input is to its preferred value. The combined activation pattern of all detectors encodes the numerical input.

In case of symbolic (categorical) inputs, the model assumes that distinct subsets of neurons are dedicated to specific values. When a certain input value is presented, the corresponding subset fires with all its neurons maximally activated ($a = 1$), while the other subsets remain inactive ($a = 0$). This creates distinct activation patterns for different values.

After this preliminary input encoding, the input layer is projected to another high-dimensional layer (or CA in the same neural space) through a random weight matrix, followed by a process of lateral inhibition, also inspired by biology. For example, this mechanism reflects how the fruit fly (Drosophila melanogaster) identifies different types of odorants, i.e., smell molecules. Modeling the inhibition process, however, is not straightforward. While methods have been proposed to achieve a balance between excitation and inhibition, careful consideration is necessary in modeling the connections and proportions of the two types of neurons.

Using rate-based neurons (neurons with real-valued activations, instead of spiking neurons) simplifies the process, and techniques such as k-Winners Take All (kWTA) have been adopted. However, kWTA selects the same number of active neurons regardless of their pre-inhibition level of activation. In general, it is possible for fewer neurons than this pre-specified number to be active above a sensible threshold. In addition, the temporal aspects characteristic of spiking neurons are no longer considered.

Here, we propose a representation that accounts for this temporal aspect while operating with rate-based neurons. In terms of biological plausibility, as we work at the CA level, the activation of rate-based neurons can be considered as the average activation of a larger population of spiking neurons. Following the initial activations through *random projection*, an *iterative lateral inhibition* process occurs. First, the neuron with the highest activation inhibits the neurons in its neighborhood. The shape of this high-dimensional CA can be a 1D or a 2D lattice. The size of the inhibition neighborhood is a user-defined parameter. There are two options for this process: either to deactivate all the neurons in the neighborhood completely, or to decrease their activations by a specific factor, possibly depending on the distance to the inhibiting neuron. Afterwards, the next highest activated neuron initiates the lateral inhibition process in its neighborhood. This process repeats for the remaining neurons until all activation levels stabilize. The result is a sparse activation pattern that represents the encoding of the input.

For numerical input values, normalized between 0 and 1, these representations *preserve topology*, i.e., the more similar the inputs, the more similar their representations. Moreover, by finding suitable parameter values, the similarity between encodings measured by the Jaccard index, decreases somewhat exponentially as the difference between the inputs increases linearly. This resembles a finding from cognitive psychology experiments [10], where similarity is an exponential negative function of the distance between stimuli.

However, the resulting quasi-exponentially decreasing similarity function lacks smoothness and precision, and it can vary significantly from one trial to another due to the random operations involved. Therefore, an equivalent representation that exhibits a more regular mathematical behavior and clearer interpretation is introduced next.

### 3.2 A (Somewhat) Equivalent Explicit Representation

The alternative representation for numerical inputs is based on a group of neurons, or a *cell assembly module* (CAM), such as the one illustrated in Fig. 1. It assumes that the input range is discretized, e.g., into 10 intervals. Each value is encoded by a contiguous subset of neurons. Between two adjacent values, most of the active neurons are shared, but the number of common active neurons decreases as the difference between the values increases, until the two extreme values no longer have any common neurons. For a precision of 0.1, 20 neurons are used, with 10 neurons active for a specific value. Such a CAM can be seen as a small part of a high-dimensional neuron array with sparse activations. The representation can be easily expanded to enhance precision. To achieve a precision of $10^{-p}$, $2 \cdot 10^{p}$ neurons, among which $10^{p}$ are active, are necessary. In general, there is no requirement for the active neurons to be contiguous for a particular value, but this feature is particularly helpful for visualizing the encoding.

For binary values, the representation assumes disjunctive subsets. For symbolic (categorical) values, one-hot encoding can be employed to transform them into multiple dimensions, or a smaller number of disjunctive active neurons can be used within a single CAM.

Given two *binary* encodings $\mathbf{x}$ and $\mathbf{y}$, their similarity is computed using the Jaccard similarity measure:

$$S_b(\mathbf{x}, \mathbf{y}) = \frac{\sum_i \mathbb{I}(x_i = 1 \, \text{and} \, y_i = 1)}{\sum_i \mathbb{I}(x_i \neq 0 \, \text{or} \, y_i \neq 0)} = \frac{\sum_i \mathbb{I}(x_i \cdot y_i = 1)}{\sum_i \mathbb{I}(x_i + y_i \neq 0)} \tag{2}$$

which is defined as the ratio between the number of common active neurons (the intersection) and the number of positions in the CAM where at least one neuron in $\mathbf{x}$ or $\mathbf{y}$ is active (the union). In Eq. (2), $\mathbb{I}(\cdot)$ denotes the indicator function which is equal to 1 if the condition is true and 0 otherwise.

The algorithm described in Sect. 3.3 operates with *real-valued* neurons whose activations belong to the [0, 1] interval. In this case, the similarity measure follows the same idea but has a slightly different expression:

$$S(\mathbf{x}, \mathbf{y}) = \frac{\sum_i 1 - |x_i - y_i|}{\sum_i \mathbb{I}(x_i + y_i \neq 0)} \tag{3}$$

As Fig. 2 shows, the use of Eq. (2) yields a similarity function that closely resembles an exponential negative function. Since the algorithm is inspired from the field of cognitive psychology, it is notable that such a similarity function emerges from simple comparisons of neural activations within a CAM.

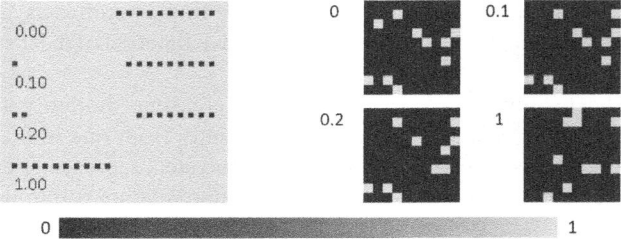

**Fig. 1.** The encoding of normalized numerical values in a cell assembly module and a possible equivalent sparse encoding resulting from iterative lateral inhibition (using the "viridis" color palette).

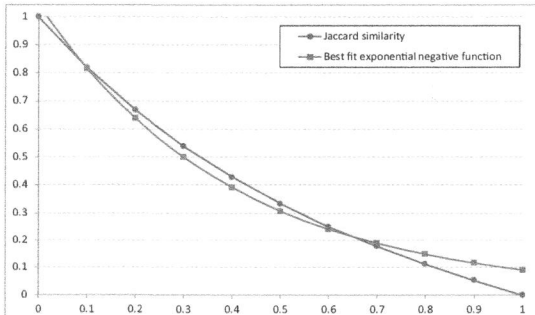

**Fig. 2.** Jaccard similarity as an approximate measure of cognitive similarity.

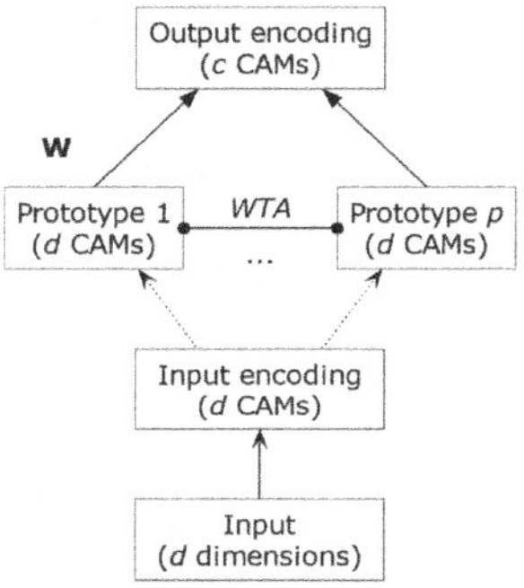

**Fig. 3.** The general architecture of the proposed model.

### 3.3 Learning Prototypes

A significant body of research in cognitive psychology has focused on developing computational models that capture the diverse types and aspects of human learning, particularly categorization, which corresponds to classification in machine learning (ML). In general, models have been proposed based on exemplars, prototypes, and rules (or combinations thereof). Accordingly, people classify an instance (or stimulus) based on its similarity to previously learned instances, its similarity to typical instances characteristic of certain groups, and decision thresholds, usually on a small subset of problem dimensions.

Prototypes are particularly interesting because they can be seen as general cases that include exemplars, but can also be interpreted as approximate rules by imposing certain thresholds on similarity values. Moreover, such methods are applicable to both supervised and unsupervised learning.

The algorithm proposed here is primarily inspired by SUSTAIN [11] and, to a lesser extent, by ART [12]. In SUSTAIN, prototypes have receptive fields for each symbolic value of a dimension, and their expressions are exponential negative functions whose parameters are adjusted through gradient descent. In our approach, similarity is implicit, arising from the representation itself, and thus its mathematical formulation is simpler. The basic idea is that similar instances aggregate into prototypes, while dissimilar individual exemplars may also exist. Once formed, prototypes compete to classify an input.

Unlike SUSTAIN and ART, the proposed algorithm relies on populations of neurons to represent concepts, rather than individual units. Consequently, it can accommodate noise and variations similar to conventional neural networks. The general similarity algorithm works by associating input instances with prototypes, updating weights between prototypes and output CAMs during training, and making predictions based on the learned associations. Unlike other ML classification algorithms, the proposed technique is capable of online learning and does not need retraining when additional instances are encountered.

The algorithm works as follows. The weight matrices between prototype CAMs and output CAMs are initialized with small random values.

When a training instance is presented, if no prototypes exist, the instance initializes the values of the first prototype. Otherwise, the algorithm identifies the winning (i.e., the most similar) prototype and its activation for the current instance. If the activation of the winner is below a user-defined *similarity threshold*, a new prototype corresponding to the current instance is created. In our experiments, this threshold was set to 0, to reduce the number of prototypes. If the class of the presented instance is different from the class of the winning prototype, a new prototype is added.

During the training phase, the winning prototype activates the output CAMs. Their activations are compared to the target activations, corresponding to the output classes, and the "humble teacher" method, also used in [11], is applied for learning. If the activations are lower than 1 and need to be increased, or if they are greater than 0 and need to be decreased, the *delta rule* with a user-defined *learning rate* is employed to adjust the weights. Otherwise, the weights remain unchanged. Although the resulting activations may fall outside the [0, 1] interval, they are subsequently constrained to this interval.

The prototype values are then adjusted using a *merge* operation that moves them closer to the current instance values. When a new instance $\mathbf{x}$ is merged into an existing prototype $\mathbf{p}$, the prototype becomes:

$$\mathbf{p} \leftarrow (1 - \mu) \cdot \mathbf{p} + \mu \cdot \mathbf{x} \tag{4}$$

where $\mu$ is a user-defined *merge rate*. In the experiments presented in Sect. 4, the learning rate and the merge rate are set to 0.1.

A new prototype can also be formed when the similarity of the current instance to the winning prototype is below the similarity threshold, even if the instance and the prototype belong to the same class. The algorithm can work for unsupervised problems as well, based on this condition.

Similar to feature weights, the total activation of a CAM acts as an indicator of the importance of a problem dimension. This aspect is closely tied to the confidence in the relevance of a specific CAM or dimension for the classification. The weight of a dimension $i$ is computed as:

$$w_i(\mathbf{x}) = \frac{\sum_j \max(a_{ij} - \theta, 0)}{\sum_d \sum_j \max(a_{dj} - \theta, 0)} \tag{5}$$

where $d$ indexes the dimensions and $j$ indexes the neurons in the CAM. $\theta$ is a user-defined *activation threshold*, which in our case was set to 0.5.

Given these feature weights, the similarity function between an instance $\mathbf{x}$ and a prototype $\mathbf{p}$ (and, in general, between any two instances) becomes:

$$S_w(\mathbf{x}, \mathbf{p}) = \sum_i w_i \cdot S(\mathbf{x}_i, \mathbf{p}_i) \tag{6}$$

In Eq. (6), $\mathbf{x}_i$ and $\mathbf{p}_i$ are the representations of a specific problem dimension $i$, which are still vectors of neuron activations.

### 3.4  Decoding the Representations

The prototype formation involves an averaging process, which is especially evident for numerical dimensions. Similar to clustering, the prototype value vector eventually represents an average of the instance values that contributed to its generation.

If prototype formation results in an encoding, a corresponding decoding procedure is also necessary. We suggest the following expression to extract a representative value from the encoding:

$$D(\mathbf{x}) = \frac{\sum_n n \cdot H(S(\mathbf{x}, \mathbf{a}(n)))}{\sum_n H(S(\mathbf{x}, \mathbf{a}(n)))} \tag{7}$$

$$H(m; p) = m^p \tag{8}$$

In Eq. (7), $D(\mathbf{x})$ represents the decoding of the vector representation $\mathbf{x}$. The variable $n$ iterates through the distinct number values, for instance, comprising 11 values (0.0, 0.1,..., 1.0) when the input range is represented with 10 intervals. $\mathbf{a}(n)$ represents the encoding of $n$, $S$ is the similarity function, and $H$ is a function used to amplify the influence of more similar representations. $p$ is a hyperparameter that must be greater than 1 and can typically be set to values such as 5 or even 20. A higher value for $p$ yields more precise averages when instances are distinctly clustered together but may overlook the smaller activations of certain instances, which could still have significance.

While the decoded value is similar to the mean of a normal distribution, the dimension weights, computed by Eq. (5), have an effect similar to the variance. Let us examine two 1D cases: one prototype formed by presenting values from 0.1 to 0.5, and another prototype formed solely by presenting the 0.3 value (Fig. 4). In the first scenario, the activations are proportional to the relative frequency of 1s in the representations. Both prototypes have a decoding of 0.3, yet the first one has a lower weight due to the increased variance of data resulting from averaging more numbers. Hence, the weights can also serve as confidence estimations.

Unlike in SUSTAIN, prototypes have an additional parameter reminding of the latency of recall (in a way also related to recency) used in the ACT-R cognitive architecture. In the proposed algorithm, this parameter decreases at the start of each learning epoch for all prototypes and increases only when a prototype is adjusted, i.e., includes a training instance. Over a longer time span, such as every 5 or 10 epochs, prototypes with negative recency values are removed. This ensures that accidental prototypes formed, e.g., due to an unfavorable order of instance presentation, are eliminated, and this leads to an optimal or near-optimal number of prototypes.

**Fig. 4.** The difference in encoding between a prototype formed from several instances with a mean of 0.3 and a prototype corresponding to a single instance with the value 0.3.

### 3.5  Learning Rules

Beside similarity-based classification, rules can be derived from the information encapsulated by the prototypes. Similar to human cognition, rules are derived through a search process or hypothesis testing. Studies in cognitive psychology suggest that in general, rules are verbalizable, i.e., expressible in words, typically involving only a few dimensions, e.g., one or two.

The process of extracting explicit rules applies to both categorical and numerical data using similarity thresholds. When prototypes from the same class exhibit a wide range of values on a dimension, that dimension is considered irrelevant. The remaining dimensions with common values are then compared against prototypes from different classes, following a contrastive approach. If prototypes from different classes share identical values on a dimension, that dimension is ignored. Conversely, dimensions with common values across all prototypes within a class, but distinct from other classes are selected for rule extraction.

First, this process iteratively examines individual dimensions, and then advances to pairs of dimensions. This method results in prototypes typically featuring one or two active dimensions. Activating such a prototype can be viewed as satisfying the precondition of a rule. In our case, the postcondition is the activation of the class in an output CAM. In general, such a CAM could represent anything, e.g., it could trigger an action if linked to a motor buffer, in the sense used in the ACT-R cognitive architecture.

## 4  Case Studies

In this section, we consider two simple problems to illustrate both prototype formation and rule extraction. We use the representation described in Sect. 3.2 for an easier interpretation of the results. However, if biological plausibility is the primary concern, the representation outlined in Sect. 3.1 can be employed with similar results, although involving greater variability.

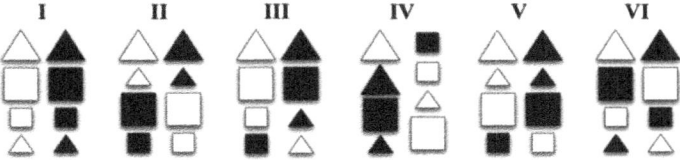

**Fig. 5.** The six Shepard problems [10].

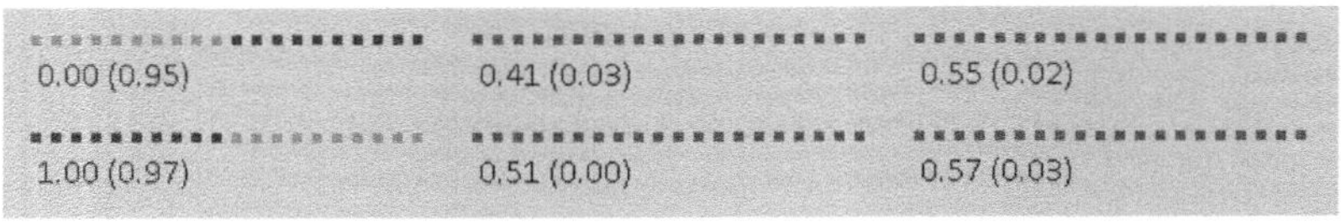

| 0.00 (0.95) | 0.41 (0.03) | 0.55 (0.02) |
| 1.00 (0.97) | 0.51 (0.00) | 0.57 (0.03) |

**Fig. 6.** The encoding of two prototypes for Problem 1. The two prototypes belong to different classes. Only the first dimension is relevant for the classification, and has a definite value (0 and 1, respectively). The other two dimensions are averages of opposing values, which yield dimension weights close to 0 (i.e., these two dimensions are ignored).

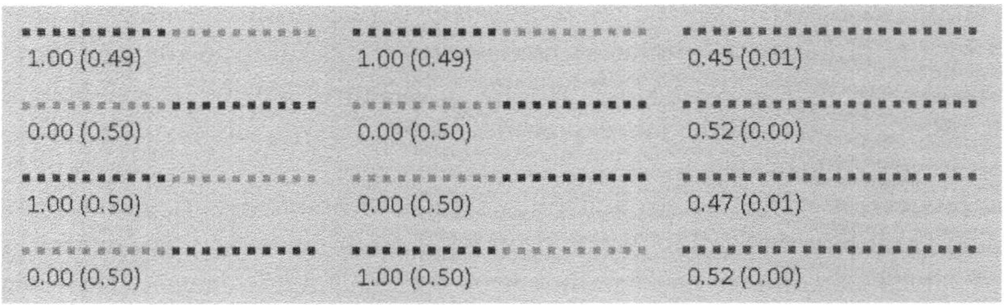

| 1.00 (0.49) | 1.00 (0.49) | 0.45 (0.01) |
| 0.00 (0.50) | 0.00 (0.50) | 0.52 (0.00) |
| 1.00 (0.50) | 0.00 (0.50) | 0.47 (0.01) |
| 0.00 (0.50) | 1.00 (0.50) | 0.52 (0.00) |

**Fig. 7.** The encoding of four prototypes for Problem 2. Since this is a 2D XOR problem, the first two dimensions are equally relevant and the third dimension is ignored.

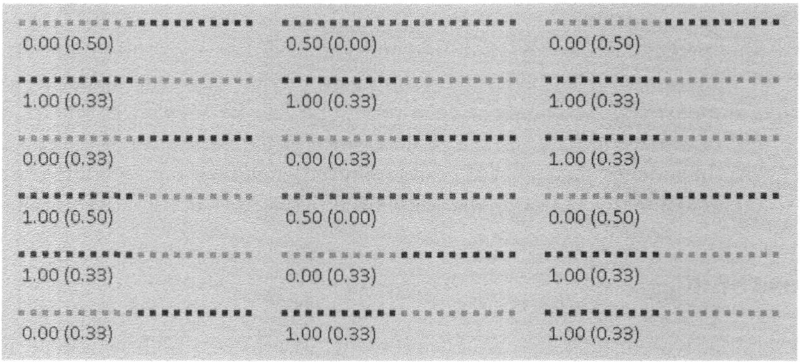

| 0.00 (0.50) | 0.50 (0.00) | 0.00 (0.50) |
| 1.00 (0.33) | 1.00 (0.33) | 1.00 (0.33) |
| 0.00 (0.33) | 0.00 (0.33) | 1.00 (0.33) |
| 1.00 (0.50) | 0.50 (0.00) | 0.00 (0.50) |
| 1.00 (0.33) | 0.00 (0.33) | 1.00 (0.33) |
| 0.00 (0.33) | 1.00 (0.33) | 1.00 (0.33) |

**Fig. 8.** The encoding of six prototypes for Problem 5. The resulting prototypes are similar to the ones provided by the standard SUSTAIN algorithm. Depending on the order of instance presentation, this problem can yield up to eight prototypes, but due to the pruning method based on recency values, only six prototypes remain after repeated training.

The first case study is a classic set of benchmark problems in cognitive psychology known as the six Shepard problems [10] (Fig. 5). The objective is to classify two groups of 3D instances distinguished by color, shape and size. These problems vary in difficulty. For instance, for Problem 1 the classification can be achieved by considering only one dimension, such as color. Problem 2 resembles a XOR operation, taking into account both shape and color while ignoring size. Problems 3–5 represent the rule-plus-exception type. Problem 6 lacks any underlying structure and thus all the instances must be stored.

The results of the algorithm for problems 1, 2 and 5 are presented in Figs. 6, 7 and 8. Problems 3 and 4 are similar to problem 5. For Problem 6 the eight resulting prototypes are equal to the training instances.

For the second case study, we use the dataset presented in Table 1, proposed in [13] for a symbolic rule-extraction algorithm.

**Table 1.** The Season classification dataset [13].

| Weather | Trees | Temperature | Season (class) |
|---------|-------|-------------|----------------|
| rainy (0.5) | yellow (0.5) | average (0.5) | autumn (C2) |
| rainy (0.5) | leafless (0) | low (0) | winter (C3) |
| snowy (0) | leafless (0) | low (0) | winter (C3) |
| sunny (1) | leafless (0) | low (0) | winter (C3) |
| rainy (0.5) | leafless (0) | average (0.5) | autumn (C2) |
| rainy (0.5) | green (1) | high (1) | summer (C1) |
| rainy (0.5) | green (1) | average (0.5) | spring (C0) |
| sunny (1) | green (1) | average (0.5) | spring (C0) |
| sunny (1) | green (1) | high (1) | summer (C1) |
| sunny (1) | yellow (0.5) | average (0.5) | autumn (C2) |
| snowy (0) | green (1) | low (0) | winter (C3) |

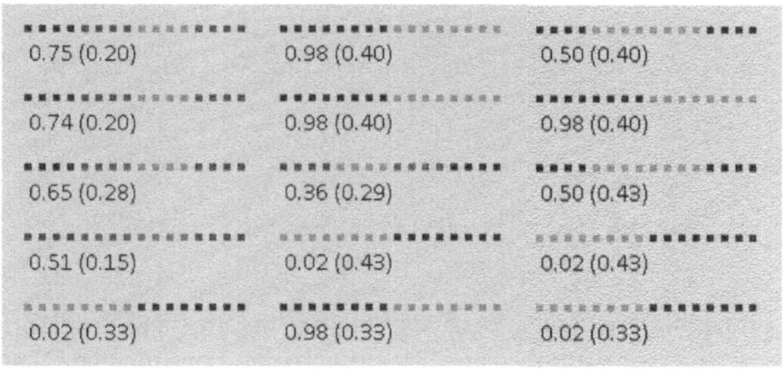

| | | |
|---|---|---|
| 0.75 (0.20) | 0.98 (0.40) | 0.50 (0.40) |
| 0.74 (0.20) | 0.98 (0.40) | 0.98 (0.40) |
| 0.65 (0.28) | 0.36 (0.29) | 0.50 (0.43) |
| 0.51 (0.15) | 0.02 (0.43) | 0.02 (0.43) |
| 0.02 (0.33) | 0.98 (0.33) | 0.02 (0.33) |

**Fig. 9.** The five prototypes representing the Seasons problem. The first four prototypes are highly connected to the neuron groups that reflect the classes 0, 1, 2 and 3, respectively. The last one is a single-instance prototype representing class 3.

These problems can be represented using one-hot encoding, but the higher number of dimensions would render the representation more complex to interpret. This is why we opt for an ordinal encoding with three values (low = 0, medium = 0.5 and high = 1), as depicted in the brackets in the table. However, similar results can be obtained with one-hot encoding.

As a first step, prototypes are formed from these instances (Fig. 9).

Then, the rule extraction algorithm attempts to find specific dimension values that are common to the prototypes pointing to the same class, but are different from those pointing to different classes. For brevity, let us denote dimension $i$ as $D_i$ and use $x$ to represent a dimension that is ignored.

Assuming the prototypes are ordered as depicted in Fig. 9, the process goes as follows. For the first prototype, since $D_1$ has a small weight, it tests $D_2$ to see if it can define class $C_0$. However, $P(x, 1, x)$ in $C_0$ conflicts with $P(0.75, 1, 1)$ in $C_1$. $D_2$ is inhibited and the process moves to $D_3$. Now $P(x, x, 0.5)$ in $C_0$ conflicts with $P(0.67, 0.33, 0.5)$ in $C_2$. Next, a pair of dimensions is considered. $P(x, 1, 0.5)$ has no conflict, and is selected as a rule for $C_0$.

While the inhibition of dimensions is implemented computationally rather than neurally here, recent studies on ART have identified such mechanisms, rendering them biologically plausible. Moreover, simple value comparisons using neural processes are possible.

In a similar way, we consider $C_1$. $P(x, 1, x)$ conflicts with $P(0.75, 1, 0.5)$ in $C_0$, but $P(x, x, 1)$ is acceptable for $C_1$. Therefore, testing $P(x, 1, 1)$ is no longer needed.

For $C_2$, $P(x, x, 0.5)$ conflicts with $P(0.75, 1, 0.5)$ in $C_0$. The other dimensions do not have individual values, but averages, thus $C_2$ remains defined by a prototype where classification is performed only by similarity assessment.

For $C_3$, we have two prototypes: $P(0.5, 0, 0)$ and $P(0, 1, 0)$. The common value is $P(x, x, 0)$, which is not conflicting, therefore it represents a rule for $C_3$.

The three rules are presented in Fig. 10, where the neurons of irrelevant dimensions are displayed in gray.

**Fig. 10.** The rules found for classes $C_0$, $C_1$ and $C_3$.

A similar procedure could have been applied to Problem 1 and Problem 2 mentioned in the first case study. However, the rule extraction method outlined here cannot identify a rule-plus-exception situation. Nonetheless, other more sophisticated search methods could be devised for this purpose.

It is important to emphasize that the problems chosen as case studies are intentionally simple in order to better understand the behavior of the algorithm, but it can be applied to any dataset. Our focus at this stage was not on error metrics, execution time, or other standard experimental evaluations of ML models. The algorithm is designed to be involved in secondary processing, and thus the features of the data are considered to be already extracted by primary processing techniques, such as the current deep learning methods.

## 5   Conclusions

The classification algorithm described in this paper integrates fundamental ideas from SUSTAIN and ART, with a specific emphasis on using neural cell assemblies as the building block of representation. CAs have the advantage of their flexible, noise tolerant neural side together with the possibility of representing symbolic concepts. In the proposed algorithm, the same representation is used for prototypes and explicit rules. Since the representation is neural, it would be natural to connect, e.g., the encoding of a convolutional neural network autoencoder for MNIST images to the internal representation of our approach. An interesting prospect for future exploration is the addition of a form of top-down control that could change the perception itself based on disambiguation cues from the reasoning component.

## References

1. LeCun, Y.: A path towards autonomous machine intelligence (2022). https://openreview.net/pdf?id=BZ5a1r-kVsf. Accessed 15 Feb 2024
2. Goyal, A., Bengio, Y.: Inductive biases for deep learning of higher-level cognition. Proceedings. Math. Phys. Eng. Sci. **478**(2266) (2022). https://doi.org/10.1098/rspa.2021.0068
3. Seo, S., Arik, S.O., Yoon, J., Zhang, X., Sohn, K., Pfister, T.: Controlling neural networks with rule representations. arXiv [cs.LG] (2021). http://arxiv.org/abs/2106.07804
4. Evans, R., Grefenstette, E.: Learning explanatory rules from noisy data. J. Artif. Intell. Res. **61**, 1–64 (2018). https://doi.org/10.1613/jair.5714
5. Mao, J., Gan, C., Kohli, P., Tenenbaum, J. B., Wu, J.: The neuro-symbolic concept learner: interpreting scenes, words, and sentences from natural supervision. arXiv [cs.CV] (2019). http://arxiv.org/abs/1904.12584
6. Santoro, A., et al.: A simple neural network module for relational reasoning. arXiv [cs.CL] (2017). http://arxiv.org/abs/1706.01427
7. Yu, D., Yang, B., Liu, D., Wang, H., Pan, S.: A survey on neural-symbolic learning systems. Neural Networks: Off. J. Int. Neural Network Soc. **166**, 105–126 (2023). https://doi.org/10.1016/j.neunet.2023.06.028
8. Hassabis, D., Kumaran, D., Summerfield, C., Botvinick, M.: Neuroscience-inspired artificial intelligence. Neuron **95**(2), 245–258 (2017). https://doi.org/10.1016/j.neuron.2017.06.011
9. Leon, F.: A review of findings from neuroscience and cognitive psychology as possible inspiration for the path to artificial general intelligence. arXiv [cs.AI] (2024). http://arxiv.org/abs/2401.10904
10. Shepard, R.N., Hovland, C.I., Jenkins, H.M.: Learning and memorization of classifications. Psychol. Monographs **75**(13), 1–42 (1961). https://doi.org/10.1037/h0093825
11. Love, B. C., Medin, D. L., Gureckis, T. M.: SUSTAIN: a network model of category learning. Psychol. Rev. **111**(2), 309–332 (2004). https://doi.org/10.1037/0033-295x.111.2.309
12. Grossberg, S.: Adaptive Resonance Theory: How a brain learns to consciously attend, learn, and recognize a changing world. Neural Networks Off. J. Int. Neural Network Soc. **37**, 1–47. https://doi.org/10.1016/j.neunet.2012.09.017 (2013)
13. Pham, D.T., Aksoy, M.S.: RULES: a simple rule extraction system. Expert Syst. Appl. **8**(1), 59–65 (1995). https://doi.org/10.1016/s0957-4174(99)80008-6

# Transforming Challenges: Siamese-Based Vision Transformers for Robust Occluded Face Recognition

Laila Ouannes[1][(⊠)] [iD], Anouar Ben Khalifa[1,2],
and Najoua Essoukri Ben Amara[1]

[1] Université de Sousse, Ecole Nationale d'Ingénieurs de Sousse, LATIS- Laboratory of Advanced Technology and Intelligent Systems, 4023 Sousse, Tunisie
{laila.ouannes,anouar.benkhalifa}@eniso.u-sousse.tn,
Najoua.BenAmara@eniso.rnu.tn
[2] Université de Jendouba, Institut National des Technologies et des Sciences du Kef, 7100 Le Kef, Tunisie

**Abstract.** Face recognition systems are essential in various applications. Still, dealing with deteriorated situations such as fluctuations in head posture, lighting, facial expressions, and partial occlusion, presents great difficulty for them. In this work, we provide a novel method for reliable face identification under challenging circumstances by leveraging the Siamese network-based vision transformer architecture. The Siamese network is known for its ability to learn powerful representations from pairs of input data, making it suitable for handling complex variations in face images. We introduce a Transformer-based architecture that integrates Siamese networks to capture long-range dependencies and spatial relationships in facial features effectively. Our method focuses on learning discriminative features from degraded face images, enabling accurate recognition even in challenging conditions. According to experimental findings, our suggested approach works better than current practices in recognizing faces under various degradation factors. The proposed Siamese network-based transformer shows promising results on the two publicly available datasets the EKFD and the IST-EURECOM LFFD offering a reliable solution for face recognition in real-world scenarios with degraded conditions.

**Keywords:** Vision Transformers · Siamese networks · Face recognition · degraded conditions

## 1 Introduction

Face recognition has become a critical component in various applications, ranging from security and surveillance to biometric authentication and human-computer interaction [15]. However, the performance of conventional face recognition systems can be significantly compromised under degraded conditions, where factors

N.-T. Nguyen et al. (Eds.): ICCCI 2024, CCIS 2166, pp. 260–272, 2024.
https://doi.org/10.1007/978-3-031-70259-4_20

such as partial face occlusions, head pose variations, illumination variations, and facial expression variations pose significant challenges [14]. Indeed, partial face occlusions occur when certain regions of the face are obscured by objects, hands, or other people. These occlusions can hinder the accurate detection and recognition of facial features, leading to degraded performance in face recognition systems. Head pose variations, on the other hand, refer to changes in the orientation of the face to the camera. This includes rotations, tilts, and scale changes, which can alter the geometric configuration of the face and make it challenging to match against stored templates. Illumination variations present another major challenge in face recognition, where changes in lighting conditions can result in uneven illumination across the face. This can affect the appearance of facial features, leading to errors in feature extraction and matching [21]. Additionally, facial expression variations, such as smiles, frowns, or squints, introduce further complexity by altering the appearance of the face and making it difficult to establish accurate matches.

In recent years, significant strides have been made in the field of face recognition, particularly in addressing the challenges posed by degraded conditions [24]. Researchers have increasingly turned to deep learning architectures to tackle these challenges, leveraging their ability to learn complex patterns and representations from data [16]. Among the various deep learning architectures employed, Recurrent Neural Networks (RNNs) [7, 25] have emerged as a popular choice due to their capability to capture temporal dependencies in sequential data, making them well-suited for modeling the temporal dynamics of facial features. Following the success of RNNs, Long Short-Term Memory (LSTM) networks [3, 12], have garnered attention for their ability to capture long-term dependencies in sequential data while mitigating the vanishing gradient problem associated with traditional RNNs. LSTMs have demonstrated promising performance in face recognition tasks, particularly in scenarios involving temporal variations such as facial expression changes and head pose variations. In recent years, Transformers [10, 11] have emerged as a powerful class of deep learning architectures, initially popularized in natural language processing tasks. Indeed, Transformers excel at capturing long-range dependencies and capturing complex relationships between input tokens, making them well-suited for modeling spatial dependencies in images. When applied to face recognition tasks, Transformers have shown remarkable performance in capturing intricate patterns and dependencies between face patches by using its variant Vision Transformers (ViT) [5], thereby enhancing recognition accuracy, particularly in the presence of partial occlusions and illumination variations.

Despite the success of deep learning architectures, one common challenge remains the requirement for large-scale labeled datasets for training. The reliance on extensive data can be prohibitive, particularly in scenarios where annotated data is scarce or expensive to obtain. To address this challenge, researchers have turned to Siamese networks [6], a class of neural networks designed for learning similarity between pairs of data points also designed for dimensionality reduction. Siamese networks offer the advantage of requiring fewer labeled samples for training, as they learn to discriminate between similar and dissimilar pairs of images directly from the data [17].

In this paper, our proposed approach, a Siamese-based Vision Transformers network for robust occluded face recognition, leverages the synergies between Siamese architecture [6] and ViT networks [5] to address the challenges posed by occlusions in face recognition tasks. By combining the strengths of these two architectures, our method aims to enhance the accuracy and robustness of face recognition systems in scenarios where facial features are obscured or partially occluded. The Siamese architecture facilitates the learning of discriminative features from paired images and enhances the problem of big dataset required when using deep learning models, while the ViT network efficiently captures long-range dependencies and patterns in the input data.

The remainder of this paper is organized as follows: Sect. 2 discusses related work, while Sect. 3 introduces our proposed approach. In Sect. 4, we present the evaluation and discussion of various results obtained from both the EURECOM Kinect Face Dataset (EKFD) [13] and the IST-EURECOM Light Field Face Dataset (IST-EURECOM LFFD) [20]. Finally, Sect. 5 concludes our work.

## 2 Related Work

With the significant advancement in deep learning architectures and techniques, researchers have increasingly turned to them to tackle various challenges in computer vision. Face recognition in particular faces challenges including head posture fluctuations and partial occlusion, which can be caused by wearing a surgical mask or sunglasses, that are tough for standard approaches to handle well. One promising solution is using neural networks capable of learning dependencies between face patches. Initially, RNNs were employed for this purpose. In a study by Han et al. [7], an auto-associative memory model was proposed for face recognition, employing recurrent neural networks. The model operated under the condition that the face database was replaced by its model parameters. Another innovative approach was introduced by Yan et al. [25], presenting a novel multi-feature fusion and decomposition framework for age-invariant face recognition. The framework aimed to learn more discriminative and robust features while reducing intra-class variants. Originally, a face time sequence was sampled from several face images of the same identity but at different ages. Then, from the facial feature series that the backbone network had extracted, contextual information was extracted using multi-head attention. Finally, to make sure that the generated age-independent features accurately captured the identification information of the face and showed more resilience against the aging process, feature decomposition and fusion based on the face time sequence were integrated.

Despite the success of RNNs in modeling temporal dependencies, these networks faced limitations in capturing long-range dependencies due to vanishing gradient problems and difficulty in retaining context information. LSTM networks were introduced as an improvement over RNNs, offering better capability in handling sequential data. In [3] an integrated deep learning model, termed convolutional neural network-long short-term memory (CNN-LSTM) was introduced for precise face image recognition by leveraging optimal features. Additionally, Maafiri et al. [12] proposed a robust face recognition model, DeepWTPCA-L1, which utilized WTPCA-L1 features in conjunction with a CNN-LSTM architecture.

However, LSTM networks still suffer from computational inefficiency and struggle to capture complex dependencies in high-dimensional data. In recent years, Transformer-based architectures [2,9] have gained attention for their success in natural language processing tasks [28]. The challenges of heterogeneous face recognition, including domain discrepancies and small training datasets leading to overfitting, were addressed by proposing a novel approach in [11] that formulated image translation as a "one-to-many" generation problem. In this method, reference images were used to guide the process, integrating a memory module to capture prototypical style patterns and a style transformer module for subtle content fusion. In [22] Su et al. introduced a novel Hybrid token Transformer module in their work to identify key facial semantics for recognition by combining atomic and holistic tokens. Atomic tokens were generated from small fixed-size regions, enabling the learning of fine-grained core representation.

Transformers models excel in capturing long-range dependencies through self-attention mechanisms but may require large datasets and computational resources, making them less practical for certain applications [18]. To address this challenge, researchers have explored alternative approaches, one of which involves the use of Siamese networks [4] [23]. Unlike traditional deep learning architectures, Siamese networks are designed to work with smaller datasets by learning from pairs of input samples rather than relying solely on large labeled datasets [17]. This approach enables the network to learn the similarity between pairs of samples, making it more efficient in handling limited data availability [1]. In [27] an alternative enhancement solution was proposed by incorporating contrasted attention into the training process for both negative and positive face pairs. Additional attention was generated through clustering-based algorithms for creating face pairs. In [8], face recognition utilizing a few pictures was shown using a metric-based methodology that included several iterations of model training. In the first step, a generic facial recognition dataset such as DigiFace-1 m was used to train the basic model. The characteristics that the model had learned were further optimized and refined in the second step by applying a metric learning loss, such as triplet loss. With fewer data samples, this two-stage method improved the model's efficacy for real-time face recognition. Thus, Siamese networks offer a practical solution for face recognition tasks in scenarios where acquiring large labeled datasets is challenging or impractical [29].

Building upon the advancements in deep learning architectures and inspired by the effectiveness of Transformers in capturing dependencies between facial patches, coupled with the capability of Siamese networks to mitigate the requirement for extensive datasets, we introduce a novel Siamese-based vision Transformers for improving face recognition performance in the presence of its inherent challenges. By combining the advantages of Transformers and Siamese networks, this novel strategy aims to improve the accuracy and resilience of face recognition systems in degraded conditions. The next section provides a more thorough explanation of the Siamese and ViT networks that were employed.

## 3   Proposed Approach

Our approach introduces an innovative framework for degraded face recognition by implementing a Siamese network [6] based on Vision Transformer (ViT) [5]. The main flowchart is depicted in Fig. 1.

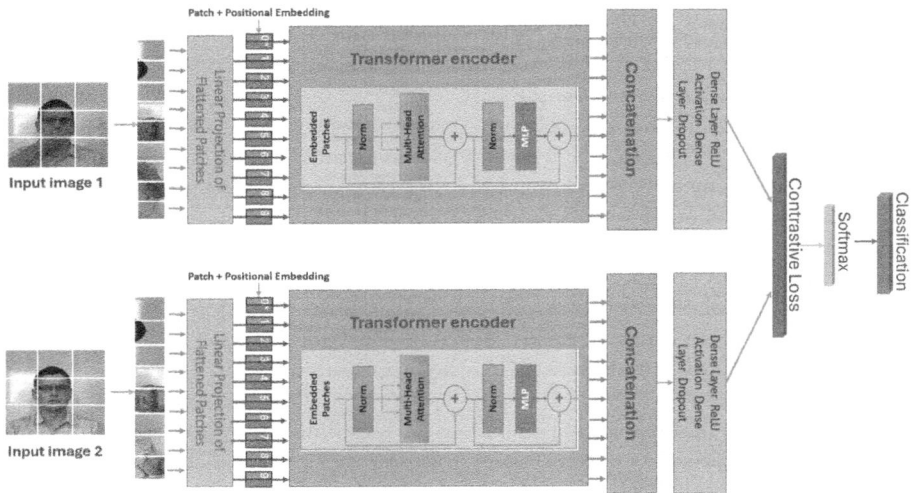

**Fig. 1.** Proposed framework for degraded face recognition.

Our suggested framework incorporates a process to create patch embeddings along with a Transformers encoder to represent the input face image. Initially, the input image is divided into patches of fixed size, and each patch is projected into vector chips. These chips are then fed into the transformer in a way that doesn't depend on their order. To encode location information, a learnable vector for location encoding is added to the input tokens. Additionally, a learnable token is included to provide a consistent global representation of the image. The Transformer consists of multiple layers, each containing a multi-headed self-attention layer and a feed-forward network. Both layers are normalized and connected in a way to preserve information flow [5]. In the next step, the distances between the outputs of the Transformer branches are calculated using the Euclidean distance to minimize the contrastive loss. Finally, for the classification step, we use a Softmax layer.

## 3.1   Siamese-Based ViT for Degraded Face Recognition

A subtype of artificial neural networks known as Siamese networks [6] are usually made up of two or more similar sub-networks that share parameters and weights. To generate similar output vectors, these networks operate by sequentially processing two separate input vectors. All sub-networks have the same updates done to them, and they are all set with comparable parameters and weights. Siamese networks use either triplet loss [19] or contrastive loss [6] functions to train the network when each sub-network gets an image. The contrastive loss function utilizes pairs of inputs that may be similar or dissimilar, but the triplet loss function uses three inputs: the anchor, positive, and negative samples, sampled according to their classes. We employ the contrastive loss function in our investigation since it is consistent with the Siamese network architecture. Large training datasets and retraining on the complete dataset are common requirements for adding or deleting classes in traditional neural networks. On the other hand, Siamese networks acquire similarity functions that allow them to identify new dataset classifications and forecast how similar two photos are to each other without requiring a lot of retraining.

In contrast to other loss functions that assess the model's performance overall training samples in the dataset, the contrastive loss calculates the difference between two input pictures processed by the Siamese network. Equation (1) is utilized to compute the contrastive loss, as described by Hadsell et al. [6].

$$Loss(x, y) = \frac{1}{2}yx^2 + \frac{1}{2}(1 - y)\{max(margin - x, 0)\}^2 \tag{1}$$

The label given to the pair value is represented by the variable $y$ in the equation, where $y = 1$ implies comparable pictures and $y = 0$ denotes dissimilar images. The Euclidean distance between the outputs of the two Siamese network sub-networks is represented by the variable $x$. To further impose a restriction, the parameter $margin$ is added. A loss will result if the distance between two disagreeing photos in a pair is less than the margin value. The contrastive loss function has two terms, often only one of which is non-zero for a given pair of network inputs. When the inputs are comparable, the first term becomes non-zero, and this term is reduced by decreasing the distance between the inputs. In contrast, when the inputs differ, the second term becomes non-zero, and reducing this term entails extending the input distance to a minimum margin distance. Less dissimilarity between the pair before any loss is realized is indicated by smaller margin values.

In this study, we employ the same ViT architecture [5] and its associated parameters across all utilized datasets. The primary aim is to train the Siamese-based ViT model to extract features from face images by leveraging knowledge transferred from prior datasets. Specifically, we construct an embedding space to characterize face images, ensuring resilience against both inter-class similarity and intra-class variance. This standardized training approach is applied uniformly across all datasets under consideration.

## 3.2   ViT Encoder

The degraded face recognition process leverages sophisticated neural network architectures, particularly the ViT framework [5], to effectively determine dependencies between different occluded and non-occluded facial features. At its core, this approach utilizes self-attention mechanisms to prioritize different regions within the input data, denoted as $I$ for the input face images and $E$ for the embedding vector obtained from the Transformer encoder.

The Transformers encoder operates through a series of steps aimed at extracting discriminative features from face images. Employing self-attention mechanisms, the encoder assesses the importance of various regions within facial data, allowing it to focus on relevant features despite occlusions. As depicted in Fig. 1, this process unfolds across multiple layers, gradually extracting high-level representations. By considering dependencies between different input image patches, the encoder effectively discerns complex patterns contributing to facial structure. The resulting embedding vector encapsulates essential information about discriminative features within the occluded facial data.

This embedding vector serves as a condensed representation of the input occluded face, laying the groundwork for face recognition. With self-attention mechanisms guiding the process, the Transformers encoder excels in capturing relevant facial information, facilitating accurate facial features. The operation of the encoder is summarized by equation (2), highlighting its ability to focus on crucial facial features essential for successful degraded face recognition.

$$E = \text{Encoder}(I) \tag{2}$$

## 3.3   Recognition Stage

The face recognition step is realized by using a Softmax layer.

Mathematically, the classification layer comprises a linear transformation followed by a Softmax activation function. Denoting the output of the final encoding layer as $\mathbf{X}$, which represents the reconstructed and refined features of the face image, the classification layer can be mathematically expressed in equation (3) :

$$\mathbf{U} = \text{softmax}(\mathbf{W} \cdot \mathbf{X} + \mathbf{b}) \tag{3}$$

In this context, $\mathbf{W}$ denotes the weight matrix, $\mathbf{b}$ represents the bias vector, and $\mathbf{U}$ signifies the output probabilities across various classes. Through the Softmax function, the output is transformed into a probability distribution, where each element denotes the likelihood of the input belonging to a specific class. Training the classification layer aims to minimize the categorical cross-entropy loss, which quantifies the dissimilarity between the predicted probabilities and the actual labels.

By incorporating the classification layer, the face recognition model evolves into a multi-task network, concurrently tackling degraded face classification objectives. This integration enables the model to harness hierarchical features for precise reconstruction while gaining insights into the recognized facial attributes or traits.

### 3.4   Optimization Process

In the context of optimizing the training of a model tasked with face recognition, the Mean Squared Error (MSE) loss serves as a foundational mathematical criterion for quantifying the identification. The MSE loss is calculated as the average of the squared differences between individual pixel values, encapsulating the spatial divergence across the entire image dataset. Mathematically, the MSE loss is articulated in equation (4):

$$\text{MSE Loss} = \frac{1}{N} \sum_{i=1}^{N} (y_i - \hat{y}_i)^2 \tag{4}$$

Here, N signifies the total number of pixels within the images under consideration, $y_i$ denotes the pixel intensity in the ground truth image, and $\hat{y}_i$ represents the corresponding pixel intensity in the predicted image.

The MSE loss operates as a pivotal optimization objective during training, compelling the model to minimize the quadratic disparities between the predicted and ground truth pixel values. By minimizing this loss, the model endeavors to converge toward face images that closely align with the ground truth at the pixel level.

## 4   Experiments AndDiscussion

### 4.1   Datasets

We assess the proposed approach by conducting experiments on two distinct datasets: the EKFD [13] and the IST-EURECOM LFFD [20]. 52 people-14 girls and 38 males-were photographed for 468 photographs throughout two sessions to make up the EKFD. Information from 100 volunteers-66 men and 34 women-obtained using a Lytro ILLUM camera makes up the IST-EURECOM LFFD dataset. Each participant exhibits 20 variations in facial expressions, head poses, illumination conditions, and partial occlusions. Figure 2 presents some examples of face images from both datasets.

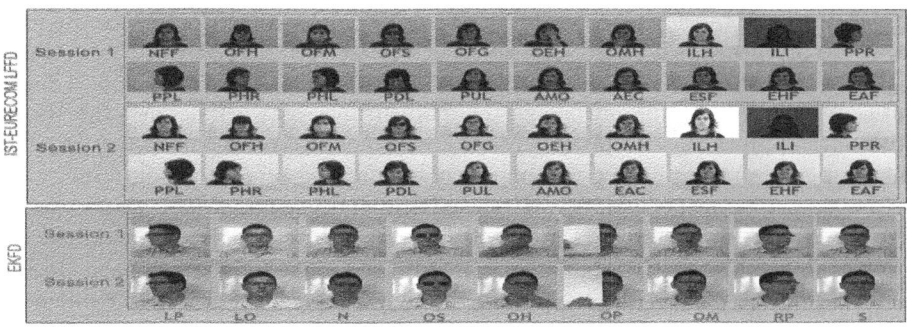

**Fig. 2.** Samples of facial variations from the IST-EURECOM LFFD and the EKFD datasets.

## 4.2   Results and Discussion

We use every 2D image from the first training session to evaluate our system, while reserving those from the subsequent session for testing. The evaluation metric used for identification is the facial recognition rate given by the expression of the equation (5) as follows:

$$RR\ (\%) = \frac{Number of identified classes}{Total number of classes} \times 100 \tag{5}$$

Our approach involves implementing a Siamese network based on the ViT network for degraded face recognition. The attained outcomes are assessed by comparison with the techniques found in the literature. Table 1 displays the EKFD recognition rate numbers.

**Table 1.** Evaluation of recognition rate (RR%) for primary approaches on EKFD

| Approaches | RR (%) |
|---|---|
| 2D-QSPCA [24] | 69.87 |
| ALexnet [14] | 50.64 |
| VGG-16 [14] | 38.46 |
| VGG-19 [14] | 37.18 |
| Inception-v3 [14] | 72.01 |
| NMR [26] | 75.00 |
| Siamese network [17] | 78.10 |
| CycleGANs [16] | 68.59 |
| Transformer network [18] | 95.10 |
| **Proposed Siamese-based ViT network** | **96.66** |

The results depicted in Table 1 illustrate the superior efficacy of our proposed network in contrast to existing methods in the literature.

To validate our methodology with the IST-EURECOM LFFD dataset, we juxtaposed our transformer-based face recognition network with other techniques from the literature. The outcomes are briefly presented in TABLE 2.

We find that our network consistently beats these outcomes across both datasets after comparing our suggested Transformer network against the state-of-the-art approaches and procedures. The efficacy of Transformers networks in enhancing face recognition performance is highlighted by this validation.

To provide a deeper analysis of our experimental findings, we categorize the face images into five distinct clusters: variations in illumination (I), partial occlusion (PO), head pose (HP), neutral (N), and facial expression (FE). Each cluster serves a specific purpose in evaluating the model's performance under different conditions. The Neutral cluster comprises the baseline representation of each individual, while the HP cluster assesses performance under varied head pose orientations. The PO cluster evaluates robustness in handling occluded facial

**Table 2.** Evaluation of recognition rate (RR%) for primary methods on IST-EURECOM LFFD

| Methods | RR (%) |
|---|---|
| ALexnet [17] | 43.95 |
| VGG-16 [17] | 61.90 |
| VGG-19 [17] | 46.90 |
| Inception-v3 [17] | 62.05 |
| PCA [21] | 51.50 |
| LBP [21] | 55.60 |
| Siamese network [17] | 70.38 |
| CycleGANs [16] | 53.60 |
| Transformer network [18] | 85.38 |
| **Proposed Siamese-based ViT network** | **89.34** |

regions, while the I cluster examines performance under diverse lighting conditions. Finally, the FE cluster provides insights into the model's ability to recognize faces across various emotional states. By categorizing the experimental data into these distinct clusters, we aim to comprehensively evaluate our proposed methodology's performance under various challenging conditions. Figure 3 presents the detailed results for the above-mentioned challenges.

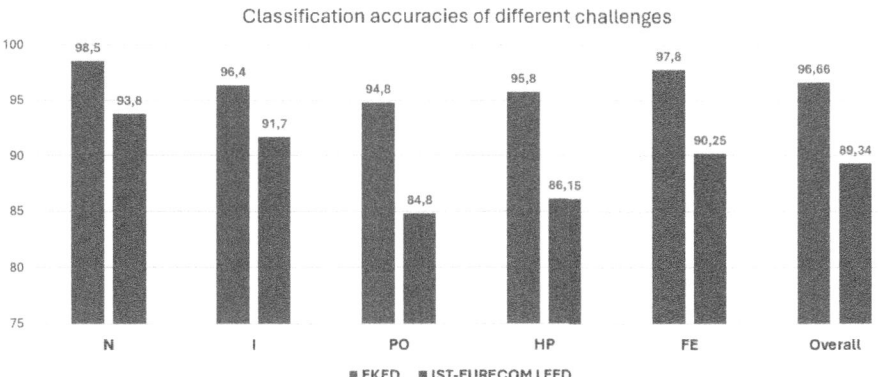

**Fig. 3.** Classification accuracies of different challenges for both used datasets

Through a comprehensive evaluation of our approach's performance across various challenges, we observe that head pose variations and partial occlusions pose significant obstacles, leading to compromised accuracy due to the absence of crucial facial information.

# 5   Conclusion

In conclusion, our study presents a comprehensive framework for face recognition in degraded conditions, addressing challenges such as partial occlusion, head pose variations, illumination changes, and facial expression variations. Leveraging the power of deep learning architectures, including Siamese networks and Vision Transformers, we have demonstrated promising results in enhancing face recognition performance, which can guarantee the effectiveness of facial recognition applications. Through extensive experimentation on diverse datasets, we have validated the efficacy of our methodology in overcoming the limitations of traditional face recognition systems. Nevertheless, the proposed method presented a small weakness by identifying a few images that present partial occlusion or a critical head pose variation. Moving forward, our research opens avenues for further exploration and refinement of deep learning techniques for robust face recognition systems by exploring occluded face reconstruction.

# References

1. Albayati, A.M., Chtourou, W., Zarai, F.: Leveraging a two-level attention mechanism for deep face recognition with siamese one-shot learning. J. Robot. Control (JRC) **5**(1), 92–102 (2024)
2. Chaudhari, A., Bhatt, C., Krishna, A., Mazzeo, P.L.: ViTFER: facial emotion recognition with vision transformers. Appl. Syst. Innov. **5**(4), 80 (2022)
3. Chengathir Selvi, M., Maruthupandi, J., Bhuvaneswari, T., Manonmani, T.: An adaptive gravitational search optimization (AGSO)-based convolutional neural network-long short-term memory (CNN-LSTM) approach for face recognition and classification. Concurrency Comput. Pract. Experience **36**(3), e7916 (2024)
4. Chicco, D.: Siamese neural networks: an overview. In: Cartwright, H. (ed.) Artificial Neural Networks. MMB, vol. 2190, pp. 73–94. Springer, New York (2021). https://doi.org/10.1007/978-1-0716-0826-5_3
5. Dosovitskiy, A., et al.: An image is worth $16 \times 16$ words: transformers for image recognition at scale. arXiv preprint arXiv:2010.11929 (2020)
6. Hadsell, R., Chopra, S., LeCun, Y.: Dimensionality reduction by learning an invariant mapping. In: 2006 IEEE Computer Society Conference on Computer Vision and Pattern Recognition (CVPR 2006), vol. 2, pp. 1735–1742. IEEE (2006)
7. Han, Q., et al.: Research on face recognition method by autoassociative memory based on RNNs. Complexity **2018**, 1–12 (2018)
8. Jain, S., Pundir, A., Singh, S., Saxena, G.J.: Navigating the face recognition: unleashing the power of few-shot learning through metric-based insights. Multimedia Tools Appl., 1–23 (2024). https://doi.org/10.1007/s11042-024-18671-5
9. Khan, S., Naseer, M., Hayat, M., Zamir, S.W., Khan, F.S., Shah, M.: Transformers in vision: a survey. ACM Comput. Surv. (CSUR) **54**(10s), 1–41 (2022)
10. Liu, H., Lu, J., Feng, J., Zhou, J.: Two-stream transformer networks for video-based face alignment. IEEE Trans. Pattern Anal. Mach. Intell. **40**(11), 2546–2554 (2017)
11. Luo, M., Wu, H., Huang, H., He, W., He, R.: Memory-modulated transformer network for heterogeneous face recognition. IEEE Trans. Inf. Forensics Secur. **17**, 2095–2109 (2022)

12. Maafiri, A., Elharrouss, O., Rfifi, S., Al-Maadeed, S.A., Chougdali, K.: DeepWTPCA-l1: a new deep face recognition model based on WTPCA-l1 norm features. IEEE Access **9**, 65091–65100 (2021)
13. Min, R., Kose, N., Dugelay, J.L.: KinectfaceDB: a Kinect database for face recognition. IEEE Trans. Syst. Man Cybern. Syst. **44**(11), 1534–1548 (2014)
14. Ouannes, L., Ben Khalifa, A., Essoukri Ben Amara, N.: Deep learning vs handcrafted features for face recognition under uncontrolled conditions. In: 2019 International Conference on Signal, Control and Communication (SCC), pp. 185–190 (2019). https://doi.org/10.1109/SCC47175.2019.9116159
15. Ouannes, L., Ben Khalifa, A., Essoukri Ben Amara, N.: Facial recognition in degraded conditions using local interest points. In: 17th IEEE International Multi-Conference on Systems, Signals & Devices 2020 (SSD 2020), pp. 404–409 (2020). https://doi.org/10.1109/SSD49366.2020.9364124
16. Ouannes, L., Ben Khalifa, A., Essoukri Ben Amara, N.: Comparative study based on de-occlusion and reconstruction of face images in degraded conditions. Traitement du Signal **38**(3), 573–585 (2021). https://doi.org/10.18280/ts.380305
17. Ouannes, L., Ben Khalifa, A., Essoukri Ben Amara, N.: Siamese network for face recognition in degraded conditions. In: 6th International Conference on Advanced Technologies for Signal and Image Processing (ATSIP), pp. 1–6 (2022). https://doi.org/10.1109/ATSIP55956.2022.9805878
18. Ouannes, L., Ben Khalifa, A., Essoukri Ben Amara, N.: Enhancing face recognition in degraded conditions via vision transformer. In: 10th International Conference on Control, Decision and Information Technologies (CoDIT 2024) (2024, in press)
19. Schroff, F., Kalenichenko, D., Philbin, J.: FaceNet: a unified embedding for face recognition and clustering. In: 2015 IEEE Conference on Computer Vision and Pattern Recognition (CVPR), pp. 815–823 (2015). https://doi.org/10.1109/CVPR.2015.7298682
20. Sepas-Moghaddam, A., Chiesa, V., Correia, P.L., Pereira, F., Dugelay, J.L.: The IST-EURECOM light field face database. In: Biometrics and Forensics (IWBF), 2017 5th International Workshop on, pp. 1–6. IEEE (2017)
21. Sepas-Moghaddam, A., Correia, P.L., Nasrollahi, K., Moeslund, T.B., Pereira, F.: Light field based face recognition via a fused deep representation. In: 2018 IEEE 28th International Workshop on Machine Learning for Signal Processing (MLSP), pp. 1–6 (2018). https://doi.org/10.1109/MLSP.2018.8516966
22. Su, W., Wang, Y., Li, K., Gao, P., Qiao, Y.: Hybrid token transformer for deep face recognition. Pattern Recogn. **139**, 109443 (2023). https://doi.org/10.1016/j.patcog.2023.109443, https://www.sciencedirect.com/science/article/pii/S0031320323001437
23. Valero-Mas, J.J., Gallego, A.J., Rico-Juan, J.R.: An overview of ensemble and feature learning in few-shot image classification using siamese networks. Multimedia Tools Appl. **83**(7), 19929–19952 (2024)
24. Xiao, X., Zhou, Y.: Two-dimensional quaternion PCA and sparse PCA. IEEE Trans. Neural Netw. Learn. Syst. **30**(7), 2028–2042 (2019). https://doi.org/10.1109/TNNLS.2018.2872541
25. Yan, C., et al.: Age-invariant face recognition by multi-feature fusion and decomposition with self-attention. ACM Trans. Multimedia Comput. Commun. Appl. (TOMM) **18**(1s), 1–18 (2022)
26. Yang, J., Luo, L., Qian, J., Tai, Y., Zhang, F., Xu, Y.: Nuclear norm based matrix regression with applications to face recognition with occlusion and illumination changes. IEEE Trans. Pattern Anal. Mach. Intell. **39**(1), 156–171 (2017). https://doi.org/10.1109/TPAMI.2016.2535218

27. Yu, H.Q.: Attention enhanced siamese neural network for face validation. In: Artificial Intelligence and Applications, vol. 2, pp. 21–27 (2024)

28. Zhong, Y., Deng, W.: Face transformer for recognition. arXiv preprint arXiv:2103.14803 (2021)

29. Zhu, J.: One-shot deformed face recognition via siamese neural network. In: International Conference on Algorithm, Imaging Processing, and Machine Vision (AIPMV 2023), vol. 12969, pp. 605–612. SPIE (2024)

# Recent Methods and Algorithms in Speech Segmentation Tasks

Dina Oralbekova[1,2,3,5](✉) (ID), Orken Mamyrbayev[1,4,5] (ID), Turdybek Kurmetkan[1,4] (ID),
and Nurdaulet Zhumazhan[5] (ID)

[1] Institute of Information and Computational Technologies, Almaty, Kazakhstan
[2] Institute of Information Security Problems, Almaty, Kazakhstan
[3] Almaty University of Power Engineering and Telecommunications, Almaty, Kazakhstan
d.oralbekova@aues.kz
[4] Al-Farabi Kazakh National University, Almaty, Kazakhstan
[5] U.Joldasbekov Institute of Mechanics and Engineering, Almaty, Kazakhstan

**Abstract.** The article addresses challenges in human-computer interaction through natural language, particularly in the context of collaborative conversations. The issue of overlapping audio data affects the accuracy of speech recognition and synthesis systems, especially in scenarios like meetings and negotiations. Emphasis is placed on the need for segregating and clustering speech from multiple speakers, highlighting challenges arising from diverse sound conditions in everyday life. The article then delves into the task of diarization, underscoring the importance of segmenting and processing speech data for effective voice control of devices. Subsequently, it explores the combination of GMM and i-vectors, as well as the evolution of approaches using deep learning, including convolutional and recurrent neural networks. Considering recent trends, the authors analyze the application of Transformers for handling long-term dependencies in data. The concluding section of the article provides a comprehensive overview and analysis of contemporary diarization methods, encompassing algorithms, error evaluation metrics, and descriptions of popular tools, with a focus on more modern approaches. This work constitutes a significant contribution to the field of speech diarization research, covering more current methods and trends compared to previous reviews in this domain.

**Keywords:** Diarization · Speech Recognition · Speaker Separation · Segmentation · Speaker Clustering

## 1 Introduction

Interaction with computers through natural language relies on deep learning methods embedded in various speech recognition and synthesis systems. The research findings of enhanced models, along with the utilization of ready-made frameworks and APIs, are progressively improving the accuracy of computer understanding in diverse conditions. However, the overlay of audio data poses a significant challenge to speech recognition

N.-T. Nguyen et al. (Eds.): ICCCI 2024, CCIS 2166, pp. 273–283, 2024.
https://doi.org/10.1007/978-3-031-70259-4_21

results, especially in the context of multi-party conversations during meetings and nego-tiations, critically impeding the functionality of recognition systems. Systems failing to capture and distinguish between different speakers become ineffective in such tasks. Despite the rarity of isolation from external sounds in real-world scenarios, parallel loud conversations from surrounding people can impact the clarity of speech.

Moreover, the need arises for systems to detect and cluster individual speakers. In most conversational situations, people interact with others, making this scenario com-mon due to the influence of extraneous sounds on ongoing conversations, distorting the original essence when converting audio data into text. Recognition challenges arise when multiple voices overlap during speech tasks, even with optimal filters to clean various noises and interferences. The unresolved issue of separating overlaid voice data impedes efficient and high-quality speech recognition, commonly referred to as diariza-tion. Correctly segmenting and processing speech data is essential for managing device voice control in systems implementing such technology. Extracting meaningful speech segments from data streams is a pertinent task in speaker identification, verification, biometric search, and other domains requiring human-machine interaction.

The article explores the combination of Gaussian Mixture Models (GMM) and i-vectors, commonly used in early diarization methods. Over time, more sophisticated and efficient methods, including deep neural networks such as convolutional and recurrent networks, have emerged. Deep learning methods offer flexibility in modeling complex dependencies in data, and the popularity of end-to-end approaches has risen. In the con-text of diarization, end-to-end methods directly Model the mapping from input audio data to output speaker activity labels without explicit feature extraction or clustering stages, aiming for systems trainable end-to-end without complex intermediate data processing.

With the advent of Transformers and their variants, the ability to process long sequences of data has facilitated their application in analyzing long-term dependencies in speech. These methods may incorporate contextual attention to account for prolonged interactions.

The article provides a review and analysis of contemporary speech diarization meth-ods, including algorithms, diarization error evaluation metrics, and descriptions of com-monly used tools and databases. While numerous works have addressed diarization methods [1–5], this research distinguishes itself by covering more recent advancements in speech diarization methods.

## 2   Standard Speech Diarization System

The diarization process serves as an intermediary stage in voice identification. Speaker separation is achieved through segmentation and clustering methods. A standard speech diarization system consists of four fundamental steps, including:

– Feature vector extraction from the input speech (commonly using mel-frequency cepstral coefficients)
– Speech activity detection, where speech is classified into speech and non-speech data, determining the speaker;
– Segmentation method, detecting speaker changes;
– Clustering method, creating clusters of speakers (Fig. 1).

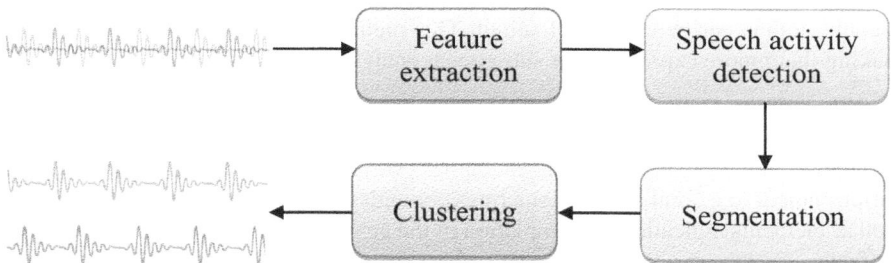

**Fig. 1.** Standard speech diarization system

### General Diarization Implementation Algorithm

1. Data Preparation – the initial step involves preparing audio files in the WAV format, representing speech data;
2. Conversion to single-channel: In the case of having dual-channel or multi-channel audio data, it is converted to a single-channel format for consistent processing.
3. Speech activity detection – the audio data is passed through a function that identifies segments with active speech. These segments are extracted and saved in WAV format in the specified directory.
4. Diarization with speaker recognition – if diarization with speaker recognition is required, all audio recordings are checked for affiliation with specific speakers. Recordings attributed to a particular speaker are moved to the directory assigned to that speaker.
5. MAP (Maximum a Posteriori) Model Adaptation [6] – MAP adaptation of the model is performed for all audio recordings using shift vectors. This process refines the model by considering the specific characteristics of individual audio recordings. Subsequently, model adaptation is applied to all audio recordings, yielding shift vectors.
6. Clustering of shift vectors – the obtained shift vectors are passed to a function where clustering takes place. This step groups shift vectors based on their similarity, facilitating subsequent analysis and identification of different speakers in the speech data.

**Feature Extraction from Audio Files.** For effective speech diarization, a crucial step involves extracting features from the audio signal, which constitutes a significantly important component capable of substantially influencing the accuracy of the diarization process. These features must be resilient to noise, exhibit notable differences between individual speakers, and enable efficient and straightforward calculations based on the audio signal [7].

To prepare the features, the audio signal is segmented into short intervals ranging from 20 to 40 ms. Subsequently, transformations based on various signal parameters, including spectral, temporal, and cepstral parameters, are applied. One of the main methods widely used in this context is the Mel-Frequency Cepstral Coefficients (MFCC) method [8]. MFCCs provide a compact representation of the audio signal, taking into account human ear perception of sound. This method highlights key speech features, making them more expressive for subsequent analysis and diarization.

**Speech Activity Detection.** Speech activity detection is a crucial stage in the diarization process, aimed at categorizing audio recordings into "speech" and "non-speech" segments. "Non-speech" segments encompass various audio elements such as silence, various noises (e.g., dial tones, phone rings), sounds characterized by impact or click, as well as elements indicating contemplation of the next statement, music, and other sound artifacts.

One common approach involves using Gaussian Mixture Models (GMM) with maximum likelihood estimation for classifying "speech" and "non-speech" segments by training the model on a training dataset. However, this method has the drawback of dependence on training data, affecting adaptation to changes in acoustic recording conditions [9]. To overcome this issue, hybrid schemes have been proposed. An initial training dataset is created using signal energy-based classification, and this dataset is then used to train the maximum likelihood method with GMM. This approach provides more flexible adaptation to various acoustic conditions, enhancing system robustness to changes in the recording environment [10]. Subsequently, a method was proposed for constructing diarization systems with i-vectors [11]. In this method, GMMs are used to model acoustic features and extract i-vectors representing speakers.

**Segmentation.** Segmentation for highlighting segments where speaker changes occur involves various algorithms employing approaches such as threshold values based on feature characteristics, machine learning methods, or even deep neural networks. Speaker changes typically manifest as alterations in the characteristics of the audio signal. Moments of speaker changes can be identified, for example, as points where signal characteristics undergo abrupt changes. This may include variations in speech timbre, intonation, or other acoustic features. To enhance segmentation accuracy, post-processing methods such as smoothing, outlier filtering, or the utilization of contextual information may be applied to refine segment boundaries [12]. The output of the segmentation stage yields temporal intervals where speaker changes occur. These intervals can be used in subsequent stages for creating speaker clusters..

**Clustering.** The final phase in a speech diarization system is clustering, which groups audio segments assigned to the same speaker [13]. There are numerous clustering methods, and the specific choice depends on task characteristics and data properties. Common methods include hierarchical clustering [14], k-means clustering [15], and others. For each pair of feature vectors, a similarity measure is calculated, which may involve Euclidean distance, cosine similarity, or other distance metrics. These measures are used to determine the degree of similarity between audio segments. The chosen clustering algorithm is applied to the feature vectors, grouping them into clusters based on their similarity. Clusters represent groups of segments that, according to the algorithm, belong to the same speaker. Some clustering methods require predefining the number of clusters (e.g., k-means), while others like DBSCAN can automatically determine the number of clusters. Optimal cluster selection can be based on various metrics such as the elbow method. The obtained clusters represent groups of segments presumably belonging to different speakers. Subsequently, post-processing may be conducted, including refining cluster boundaries or making decisions based on additional criteria. Ultimately, the clusters correspond to different speakers. The output of the clustering stage provides identified groups of audio segments, presumably belonging to different speakers. These results can be used for generating a final report or further analyses. This stage concludes the speech diarization process, offering information about speakers and their segments in the audio recording.

## 3   Methodology

**I-vector.** An i-vector represents a compact description of a speaker in the feature space [16]. The computation of an i-vector involves utilizing a Gaussian Mixture Model (GMM) to extract parameters such as the mean vector ($\mu$), covariance matrix, and mixture component weights. Subsequently, statistical estimation is applied to calculate a vector (y) for each segment of the audio signal. It can be expressed as a weighted average combination of Gaussian mixtures, and it can be formulated as follows (1):

$$M = \mu + T \cdot y \tag{1}$$

where $M$ is the $i$-vector, $\mu$ is the mean vector across all speakers, $T$ is the transformation matrix, and $y$ is the vector of Gaussian mixture components.

An i-vector serves as a compact representation for comparing speakers. In the diarization process, extracted i-vectors for various segments of an audio file can be clustered using methods such as agglomerative clustering or k-means. This facilitates the identification and differentiation of various speakers in the audio recording. However, it requires extensive training data, which can be problematic for languages with limited data.

**X-vector.**  An algorithm based on x-vectors involves extracting spectrograms from the input audio stream and forming a frame-level representation using deep learning models. Frequency filtering may be applied during preprocessing to eliminate irrelevant frequencies unrelated to human speech [17]. X-vectors benefit from specialized neural network training for extracting features specific to a particular speaker, overcoming the high variability with which speech can be pronounced.

A system based on x-vectors [18, 19] utilizes deep neural networks (DNN) for speaker recognition. The system comprises two main modules. The first module extracts frame-level features from the speech signal sequence using a deep neural network with temporal dynamics to account for sequential information between frames [20, 21]. This module extracts features at the level of individual frames, considering the structural information of the temporal sequence in the speech signal.

The second module extracts feature at the segment level from the obtained voice characteristics at the frame level. A statistical layer is applied to the features, calculating the mean and standard deviation of these characteristics. This mechanism transforms frame-level functions into segment-level functions and normalizes features of different frame lengths, creating features of equal sentence-level length.

In subsequent stages, two fully connected layers are combined, and a softmax output layer is applied. The extracted feature vectors from the first fully connected layer represent x-vectors of voice representation [22]. Systems using x-vectors as features have demonstrated higher performance compared to those using i-vectors, owing to their more representative characteristics, deeper speech modeling, increased model complexity, and enhanced computational hardware capabilities.

Drawbacks. 1. Lack of direct optimization for diarization errors. These systems cannot be directly optimized to minimize diarization errors since the clustering procedure falls into the category of unsupervised learning methods.

2. Inefficiency with overlapping segments. Such systems struggle with situations involving overlapping segments containing replicas of multiple speakers in one fragment, often neglecting such cases. These issues are addressed by more recent speech diarization methods presented in the following subsections..

**End-to-End Models.** End-to-end (E2E) models represent a relatively modern technology in the field of machine learning that proposes a complex system of training a single model, presenting a complete target system while bypassing the intermediate levels typically present in traditional systems [23]. The E2E approach involves using a unified optimization criterion to enhance the system. E2E diarization algorithms differ in that they do not utilize intermediate sound representations. Instead, these algorithms take a spectrogram as input and return speaker labels.

An E2E model can be conceptualized as a function F:X → Y, where X is the space of input audio signals, and Y is the space of corresponding speaker labels. Training such a model often involves minimizing a loss function L, which measures the difference between predicted labels and actual labels (2):

$$Min_F \sum_{(x,y)\in D} L(F(x), y) \tag{2}$$

where $D$ is the training dataset. This training process typically involves using gradient descent or other optimization methods. One example of an E2E approach in speech diarization is the application of Recurrent neural networks (RNN) or Convolutional neural networks (CNN) to process input audio signals and output speaker labels.

**Transformer.** The Transformer model was introduced in (24) and initially designed for machine translation tasks. However, due to its efficiency in processing sequences, it can also be successfully applied to speech diarization tasks. The main component of Transformer is the self-attention mechanism, which allows the model to focus on different parts of the input sequence when generating the output. The general formula for the attention mechanism in the Transformer model looks as follows (3) [25]:

$$Attention\left(U^Q, U^K, U^V\right) = softmax\left(\frac{U^Q U^{KT}}{\sqrt{d_{attention}}}\right) U^V \tag{3}$$

where Q, K, and V are the query, key, and value matrices, respectively, and $d_{attention}$ is the dimensionality of the keys.

The algorithm for applying the Transformer model in speech diarization is as follows [26]:

1) The original audio signal is divided into small time frames, and features are extracted for each frame, for example, using Mel-Frequency Cepstral Coefficients (MFCC) or a spectrogram.

2) The input features are fed into the Transformer model. The model may consist of multiple Transformer layers, each applying the attention mechanism.
3) The attention mechanism allows the model to highlight important segments of the input sequence related to specific speakers. This is particularly useful in diarization tasks where the goal is to differentiate voices of different speakers.
4) The results from the Transformer model are used to generate speaker labels based on detected patterns in the input features.

**Conformer [27].** The Conformer model is a neural network architecture designed for efficient sequence processing, such as audio signals. It includes attention layers, convolutional layers, and sequence processing mechanisms, making it a powerful tool for handling audio data in tasks like speech diarization.

The Conformer model encompasses key features that contribute to its effectiveness in processing sequential data, such as audio signals. It leverages a Multi-Head attention mechanism, enabling the model to focus on diverse aspects of the input sequence through multiple parallel attention subspaces. Additionally, the incorporation of convolutional blocks enhances the model's ability to extract local features effectively. Complementary to these blocks, normalization and feedforward network layers contribute to the overall performance. Self-attention mechanisms empower the model to selectively attend to various segments of the input sequence, a critical capability for addressing distinct speakers in diarization tasks. Moreover, Conformer employs normalization and regularization mechanisms, utilizing layer normalization to stabilize training processes and normalize data, further contributing to its robust performance in sequence-based applications.

Applying Conformer in speech diarization tasks involves feeding the audio signal into the model and training it on the output speaker labels.

## 4   Metrics

In diarization, the Diarization Error Rate (DER) is the key metric for evaluating errors and defined as [1–5] (4):

$$DER = E_{SPKR} + E_{FA} + E_{MISS} + E_{OVL} \tag{4}$$

Here $E_{SPKR}$ represents the error of misclassifying a speaker, indicating the percentage of time incorrectly assigned to the wrong speaker, $E_{FA}$ is the false alarm error, signifying the percentage of time allocated to speakers that should have been categorized as "non-speech", $E_{MISS}$ corresponds to the missed speech error, representing the percentage of time when speech was identified as "non-speech". Lastly, $E_{OVL}$ stands for the overlapped speaker error, revealing the percentage of time when some speakers in a segment were not assigned to any specific speaker.

## 5   Speaker Diarization Frameworks

There are several speaker diarization frameworks and libraries available that facilitate the implementation of speaker diarization systems. These frameworks often provide tools for feature extraction, clustering, and other essential tasks in the diarization pipeline. Here are some notable speaker diarization frameworks:

1. *pyAudioAnalysis* – a Python library for audio feature extraction, classification, segmentation, and various audio analysis tasks. It includes functionality for speaker diarization.
2. *Pyannote* is widely acknowledged as one of the most utilized libraries for speaker diarization. Its pre-trained models are crafted from the VoxCeleb datasets, capturing recordings of celebrities sourced from YouTube. The audio quality in these recordings is notably high, making the models well-suited for tasks with clear and crisp audio. However, for applications dealing with diverse data, such as recorded phone calls, retraining may be necessary. Pyannote's inference pipeline excels in identifying multiple speakers concurrently, enabling multi-label diarization. Notably, the library does not allow for the predefined specification of the number of speakers before applying the clustering algorithm. This characteristic may result in either an overestimation or underestimation of the number of speakers, particularly if such information is available beforehand.
3. *LIUM SpkDiarization* – Developed by the Laboratoire d'Informatique de l'Université du Maine (LIUM), this speaker diarization toolkit is implemented in Java. It provides tools for speaker diarization and has been widely used in research.
4. *Kaldi* – Kaldi is a powerful toolkit for speech recognition that also includes tools for speaker diarization. It is widely used in both research and industry for various speech processing tasks.
5. *ALIZE* is a speaker recognition toolkit that includes components for speaker diarization. It provides a set of C++ libraries and tools for speaker-related tasks.
6. *pyAudioDiarization* – another Python library for speaker diarization tasks. It leverages machine learning techniques and provides an easy-to-use interface for diarization tasks
7. The *NVIDIA NeMo* toolkit offers distinct collections tailored for Automatic Speech Recognition, Natural Language Processing, and Text-to-Speech models. These pre-trained networks have been trained on datasets such as VoxCeleb, Fisher, and Switch-Board, comprising English telephone conversations. This makes NeMo particularly well-suited for initiating fine-tuning procedures in call-center scenarios when compared to the pre-trained models utilized in pyannote. For more details on the pre-trained models, refer to the provided link. NeMo facilitates seamless integration of diarization results with ASR outputs, enabling the generation of speaker-aware transcripts. Moreover, it allows the user to specify the anticipated number of speakers in advance, enhancing the precision of diarization outcomes

## 6  Main Challenges Requiring Improvement in Speaker Diarization

Despite the fact that continuous work is underway to improve models and the emergence of new effective models, there are tasks and issues that have not yet been resolved or require improvement.

*1) Overlapping Speech Detection*

The presence of overlapping speech segments in audio recordings, where multiple speakers talk simultaneously, often poses a challenge for enhancing diarization accuracy. There is a growing focus on research in this area, especially in scenarios like meetings and debates. Many modern systems limit the assignment of each segment to only one speaker, potentially leading to missed errors as certain segments may not be definitively attributed to a specific speaker. Overlapping segments can complicate the identification of various noises, echoes, and other external factors. Algorithms developed to address this issue need to ensure high processing speed to be applicable in online diarization applications.

*2) Handling Noise and Heterogeneous Conditions*

Complex acoustic conditions such as noise, echoes, and variations in sound quality can significantly impact the diarization process. Developing algorithms resilient to such conditions is crucial for improving the overall performance of diarization systems.

*3) Dealing with Non-standard Dialects and Languages*

Diarization systems should effectively operate with different dialects and languages, an essential aspect in the context of multilingual and culturally diverse environments.

*4) Accelerating Diarization*

The temporal complexity of a diarization system is a key factor in determining its efficiency. This aspect is particularly significant for systems performing real-time diarization. In light of this trend, contemporary research actively aims to reduce the processing time of speech recognition systems to enhance their performance.

Improving these aspects in speaker diarization contributes to the development of more efficient and versatile systems, holding significant value across various domains, including telecommunications, data analysis, and artificial intelligence systems.

# 7   Conclusion

In this review article, contemporary methods of speaker diarization were examined, representing an integral part of the development of artificial intelligence systems and natural human-computer interaction. Natural language interaction is becoming increasingly realistic thanks to the use of deep learning in speech recognition and synthesis systems. However, the challenge of audio data overlap in scenarios involving multiple speakers remains relevant, hindering effective speech recognition. The significance of speaker diarization in scenarios such as meetings, debates, and negotiations underscores the necessity for segregating and identifying voices of individual speakers. Various factors affecting diarization quality were analyzed, including speech and non-speech noises, background speech, accents, intonations, and more. Methods ranging from traditional GMM and i-vectors to more modern approaches using deep neural networks, including Transformers, were presented.

The review encompassed existing research and approaches in speaker diarization, highlighting current trends and achievements in this field. Progress toward end-to-end methods and the utilization of Transformers provide opportunities for more effective handling of long-term dependencies in speech. Thus, the development of speaker diarization methods plays a key role in enhancing the accuracy and efficiency of recognition and interaction systems, opening new perspectives in the fields of speaker identification, verification, and other applications.

**Acknowledgement.** This research has been funded by the Science Committee of the Ministry of Science and Higher Education of the Republic of Kazakhstan (Grant No. AP19174298).

**Disclosure of Interests.** The authors have no competing interests to declare that are relevant to the content of this article.

# References

1. Tranter, S.E., Reynolds, D.A.: An overview of automatic speaker diarization systems. IEEE Trans. Audio Speech Lang. Process. **14**(5), 1557–1565 (2006). https://doi.org/10.1109/TASL. 2006.878256
2. Anguera, X., Bozonnet, S., Evans, N., Fredouille, C., Friedland, G., Vinyals, O.: Speaker diarization: a review of recent research. IEEE Trans. Audio Speech Language Process. **20**, 356–370 (2012). https://doi.org/10.1109/TASL.2011.2125954
3. Moattar, M., Homayounpour, M.: A review on speaker diarization systems and approaches. Speech Commun. **54**(10), 1065–1103 (2012)
4. Basu, J., et al.: An overview of speaker diarization: approaches, resources and challenges. In: 2016 Conference of the Oriental Chapter of International Committee for Coordination and Standardization of Speech Databases and Assessment Techniques (O-COCOSDA), Bali, Indonesia, pp. 166–171 (2016). https://doi.org/10.1109/ICSDA.2016.7919005
5. Park, T.J., Kanda, N., Dimitriadis, D., Han, K.J., Watanabe, S., Narayanan, S.S.: A review of speaker diarization: recent advances with deep learning. arXiv:abs/2101.09624 (2021)
6. Pereyra, M.: Revisiting maximum-a-posteriori estimation in log-concave models. SIAM J. Imaging Sci. **12**, 650–670 (2016)
7. Nogales, R.E., Benalcázar, M.E.: Analysis and evaluation of feature selection and feature extraction methods. Int. J. Comput. Intell. Syst. **16**, 153 (2023). https://doi.org/10.1007/s44 196-023-00319-1
8. Prabakaran, D., Sriuppili, S.: Speech processing: MFCC based feature extraction techniques-an investigation. J. Phys. Conf. Ser. 1717 (2021)
9. Weng, Z., Li, L., Guo, D.: Speaker recognition using weighted dynamic MFCC based on GMM. In: 2010 International Conference on Anti-Counterfeiting, Security and Identification, Chengdu, China, pp. 285–288 (2010). https://doi.org/10.1109/ICASID.2010.5551341
10. Rahulamathavan, S., Yao, X., Yogachandran, R., Cumanan, K., Rajarajan, M.: Redesign of Gaussian mixture model for efficient and privacy-preserving speaker recognition. In: 2018 International Conference on Cyber Situational Awareness, Data Analytics and Assessment (Cyber SA), pp. 1–8 (2018). https://doi.org/10.1109/CyberSA.2018.8551477
11. Ibrahim, N.S., Ramli, D.A.: I-vector extraction for speaker recognition based on dimensionality reduction. Procedia Comput. Sci. **126**, 1534–1540 (2018). https://doi.org/10.1016/j.procs. 2018.08.126
12. Teimoori, F., Razzazi, F.: Incomplete-data-driven speaker segmentation for diarization application; a help-training approach. Circuits Syst. Signal Process, **38**, 2489–2522 (2019). https:// doi.org/10.1007/s00034-018-0974-6
13. Gupta, A., Purwar, A.: Analysis of clustering algorithms for Speaker Diarization using LSTM. In: 2022 1st International Conference on Informatics (ICI), Noida, India, pp. 19–24 (2022). https://doi.org/10.1109/ICI53355.2022.9786928
14. Singh, P., Ganapathy, S.: Deep self-supervised hierarchical clustering for speaker diarization (2020). arXiv:2008.03960v1, https://doi.org/10.48550/arXiv.2008.03960
15. Ikotun, A.M., Ezugwu, A.E., Abualigah, L., Abuhaija, B., Heming, J.: K-means clustering algorithms: a comprehensive review, variants analysis, and advances in the era of big data. Inf. Sci. **622**, 178–210 (2023)
16. Mtibaa, A., Petrovska-Delacrétaz, D., Boudy, J., Hamida, A.: Privacy-preserving speaker verification system based on binary I-vectors. IET Biometrics (2021). https://doi.org/10.1049/ bme2.12013

17. Snyder, D., Garcia-Romero, D., Sell, G., Povey, D., Khudanpur, S.: X-Vectors: robust DNN embeddings for speaker recognition. In: 2018 IEEE International Conference on Acoustics, Speech and Signal Processing (ICASSP), Calgary, AB, Canada, pp. 5329–5333 (2018). https://doi.org/10.1109/ICASSP.2018.8461375

18. Neururer, D., Dellwo, V., Stadelmann, T.: Deep neural networks for automatic speaker recognition do not learn supra-segmental temporal features. arXiv:abs/2311.00489 (2023)

19. Chakroun, R., Frikha, M.: A deep learning approach for text-independent speaker recognition with short utterances. Multimed Tools Appl. **82**, 33111–33133 (2023). https://doi.org/10.1007/s11042-023-14942-9

20. Gao, M., Zhang, X.: Improved convolutional neural network–time-delay neural network structure with repeated feature fusions for speaker verification. Appl. Sci. **14**, 3471 (2024). https://doi.org/10.3390/app14083471

21. Farsiani S., Izadkhah H., Lotfi S.: An optimum end-to-end text-independent speaker identification system using convolutional neural network. Comput. Electr. Eng. 100, ISSN 0045-7906, https://doi.org/10.1016/j.compeleceng.2022.107882 (2022)

22. Mamyrbayev, O., Kydyrbekova, A., Alimhan, K., Oralbekova, D., Zhumazhanov, B., Nuranbayeva, B.: Development of security systems using DNN and i & x-vector classifiers. Eastern-Europ. J. Enterp. Technol. **4**(9 (112)), pp. 32–45 (2021)

23. Oralbekova, D., Mamyrbayev, O., Othman, M., Kassymova, D., Mukhsina, K.: Contemporary approaches in evolving language models. Appl. Sci. **13**(23), 12901 (2023). https://doi.org/10.3390/app132312901

24. Vaswani, A., et al.: Attention is all you need. In: Proceedings of the 31st International Conference on Neural Information Processing Systems (NIPS'17), Curran Associates Inc., Red Hook, NY, USA, pp. 6000–6010 (2017)

25. Mamyrbayev, O., Alimhan, K., Oralbekova, D., Bekarystankyzy, A., Zhumazhanov, B.: Identifying the influence of transfer learning method in developing an end-to-end automatic speech recognition system with a low data level. Eastern-Eur. J. Enterp. Technol. **19**(115), 84–92 (2022). https://doi.org/10.15587/1729-4061.2022.252801

26. Lai, Y., Tang, X., Fu, Y., Fang, R.: End-to-end speaker diarization with transformer. arXiv: 2112.07463 (2021). https://doi.org/10.48550/arXiv.2112.07463

27. Xia, W., Lu, H., Wang, Q., Tripath, A., López-Moreno, I., Sak, H.: Turn-to-diarize: online speaker diarization constrained by transformer transducer speaker turn detection. In: ICASSP 2022 - 2022 IEEE International Conference on Acoustics, Speech and Signal Processing (ICASSP), pp.8077–8081 (2022)

# Knowledge Engineering and Application for Industry 4.0

# Testing Usability of Different Implementations for VR Interaction Methods

Marek Kopel(✉)📧, Maciej Walczyński📧, and Piotr Wyrostek

Faculty of Information and Communication Technology, Wroclaw University of Science and Technology, Wybrzeże Wyspiańskiego 27, 50-370 Wroclaw, Poland
marek.kopel@pwr.edu.pl

**Abstract.** This paper presents a study analyzing the differences between various ways in which people can interact with objects in virtual reality (VR). The paper mainly focuses on how useful these methods are and what experiences they offer to the user. The main goal is to find solutions to problems related to these methods of interaction and to evaluate the effectiveness of currently available solutions. The study analyzes twelve key interaction methods in VR, such as movement, rotation, resizing, throwing, pushing, pulling, pressing, typing, pointing, selecting, archery, and hitting one object with another. The evaluation criteria for these methods include their utility, perceived reality they offer, and overall user experience. The study employs a methodical approach in which participants must try out different versions of these interaction methods in the VR environment and then complete a detailed questionnaire about their experiences. This study focuses specifically on the experiences offered by VR headsets, such as the HTC Vive with haptic controllers, while ignoring non-standard controllers or other non-standard devices. In this way, it provides a comprehensive overview of the current state of VR technology and its potential applications for the average user, focusing on key aspects that could contribute to the development of even more immersive and engaging virtual reality environments.

**Keywords:** virtual reality · interaction · usability · UX

## 1 Introduction

*Note*: the terms "UX" (user experience) and "usability" are treated in this paper more or less equally, even though the UX is a newer and broader term. Usually usability addresses aspects of UX.

Virtual Reality (VR) represents a groundbreaking technology that has the potential to change the way we interact with digital content. By leveraging the advancements in computer graphics, hardware, and user interfaces, VR allows users to immerse themselves in a simulated environment and interact with it in ways that were previously impossible. This is achieved by stimulating the users'

N.-T. Nguyen et al. (Eds.): ICCCI 2024, CCIS 2166, pp. 287–300, 2024.
https://doi.org/10.1007/978-3-031-70259-4_22

senses such as vision and hearing, and increasingly even touch and smell. The VR industry has rapidly expanded, with major companies such as HTC, Oculus, and Valve leading the way in providing advanced VR systems to consumers.

As VR continues to evolve, it is increasingly being adopted across various fields such as gaming, education, healthcare, training, and entertainment. Each of these fields presents unique challenges and requirements for interaction with objects within the virtual environment. Understanding the mechanics of these interactions and how they impact user experience is crucial for the development of applications that are not only functional but also intuitive and immersive.

The objective of this study is to identify different implementations of methods of interaction with objects in virtual reality through an extensive literature overview and compare them, focusing on their usability and user experience. The goal of this research is to understand the challenges associated with different methods of interaction and evaluate the currently available solutions for these challenges.

## 2    Related Works

In the comprehensive analysis of relevant literature pertaining to interaction in virtual reality, it becomes apparent that there is a considerable deficiency in the body of scientific work that elucidates the diverse approaches for implementing virtual reality interaction methodologies, particularly with a focus on usability. This scarcity of research in the field denotes the paramount importance of this study and serves as an impetus for a more extensive and profound understanding of the implications of various augmented reality interaction techniques on usability. Moreover, it is worth mentioning that the existing body of literature is primarily concentrated on individual or a limited set of interaction methods, often explored within the confines of a single research study. This fragmentation of knowledge further underscores the necessity to bridge the gaps in the current understanding of virtual reality interaction methodologies and their ramifications on user experience. [12] describes position manipulation as one of the most important forms of interaction in virtual reality. It allows to specify an object's position in the virtual world and is also referred to as translation or positioning. For interaction with objects that are not within reach in the virtual environment [3] introduces two categories of techniques for interacting with remote objects: arm-extension techniques and ray casting techniques. Arm-extension techniques involve extending the user's virtual arm to reach distant objects. The techniques is called the "go-go" technique [14]. Ray casting techniques involve using a virtual light ray for grabbing objects.

Rotation is an another form of fundamental interaction with objects of virtual reality [10]. Its goal is to change the orientation of an object in the virtual world [9]. Scaling is the last of interaction methods grouped along positioning and rotating as "basic operations". One approach to scaling is the "Hand-in-Middle" (HIM) technique described by [16]. In this method, one hand is used for translation and rotation, whereas engaging an hand activates the scaling mode. The scaling is manipulated based on the distance between two hands - increasing the distance makes the object uniformly bigger and equivalently decreasing the distance makes the object uniformly smaller. A very similar approach can be found in [7] named as One-Handed with Two-Hand Scaling (OTS) and Air TRS. All of these methods follow the similar principle of grabbing an object with two hands and moving them closer or further away from each other to change the scale of an object. A different approach to scaling was proposed in [4]. It's authors propose a way to allow 3°C of freedom scaling, separately for each axis using 3D widgets.

[17] focuses on the problem of throwing in virtual reality. The authors point out that there hasn't been much research focused on solving the problems of throwing in VR. While some studies address throwing aspects, there isn't much work addressing fundamental issues. The study highlights the challenges of transferring a natural action such as throwing into a virtual reality space.

Pushing, as a method of interaction with objects in virtual reality, involves the application of force to move objects away from the user. The problem is especially prominent when using the controllers provided by the most popular virtual reality head mount displays. As stated in [8], the interaction is not as intuitive and appealing as using the bare hands for interaction.

Pressing is one of the fundamental interaction techniques in both physical and virtual environments. In the context of virtual reality, pressing generally involves the user performing an action that simulates pressing a button or key in the virtual space. This simulation can be achieved through various means such as hand-held controllers, mid-air hand gestures, or even gaze-based interactions [15]. Authors of [2] in their study explored different aspects of pressing in virtual environments focusing on aspects that demand attention during button design. [6] provides a deep analysis and classification of existing methods for text entry, while [15] describes an evaluation of four distinct text-entry techniques in virtual reality.

Pointing, as an interaction technique, facilitates the user's ability to select and manipulate objects by pointing towards them, thus simulating the natural human behavior of indicating an object of interest. This mechanism is highly integral to VR systems as it forms the backbone of object manipulation, selection, and navigation. Numerous pointing techniques have been developed and integrated into virtual reality environments, each with its set of advantages and limitations. In [5] author introduces deep analysis of other studies and a classification of 3D pointing techniques. He reviews over 12 different pointing techniques and split them into main three categories - virtual hand-based, ray-based and spotlight-based.

## 3    Environment Setup

The experiment environment is designed to compare and evaluate different methods of interaction in VR concerning their usability. The inspiration for creating this environment came from [1]. The VR interactions usually are with 3D models or UI elements. The models and elements are intentionally kept simple, emphasizing the interaction aspect rather than detailed modeling. This ensures that the collected data is no biased by graphics, resolution and other visual aspects. The study aims at simulating practical applications a HTC Vive was chosen as one of the SOTA Head-Mounted Display (HMD) products. The system runs on a PC with an RTX 2070 graphics card, i7-4771 processor, and 16 GB of RAM, ensuring adequate stability and performance optimizations for a frame rate of 90FPS to match the HMD's 90 Hz refresh rate. On the software side, Unity is used as the primary development environment due to its versatility and wide support for VR development. The XR Interaction Toolkit SDK facilitates various interactions in VR. OpenXR is used to provide a standard interface for VR devices, ensuring compatibility with HTC Vive. As presented in Fig. 1 separate scenes are constructed for each interaction method. Each scene contains various rooms, each dedicated to different implementations within the interaction method. Each room contains an identical task coupled with information regarding the current implementation.

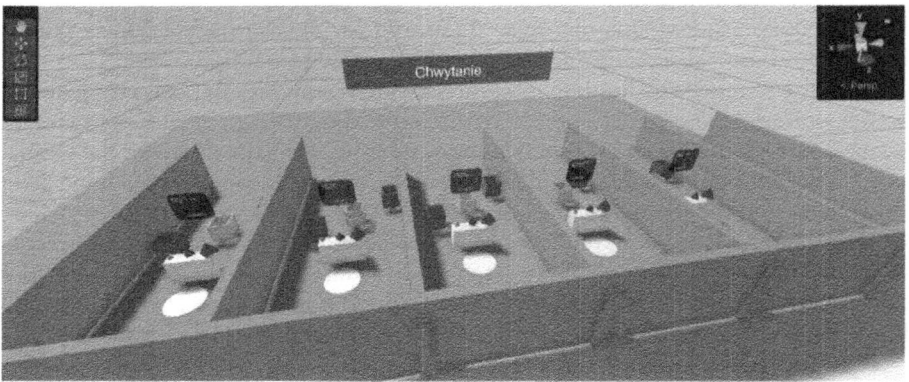

**Fig. 1.** A screenshot of the experiment VR environment for a method implementations' evaluation. The method presented is position manipulation (or positioning) and there are 5 implementations in the test (see the first row of Tab. 1). The task here is to put objects from the white block into colour matching containers. White circle marks the starting position of the VR user for each method implementation and the black rectangle is a screen with the instructions.

## 4    Experiment

The aim of this study is to thoroughly analyze the challenges associated with interaction methods in the virtual world and to find the most suitable approaches for specific applications in terms of usability. The experiment participants are students, mostly from Wrocław University of Science and Technology. Within 10 participants, there are 6 males and 4 females, mostly experienced with video games, but with little to no experience with VR gaming. The age group of the participants reflects the typical profile of a gamer. Participants took a questionnaire before and after completing each task being a specific method implementation. The gathered information are used to evaluate the strengths and weaknesses of each method implementation. The questionnaire for each method is containing questions related to the System Usability Scale (SUS), as well as additional questions regarding precision, user experience (satisfaction), and realism.

For each of the studied method of interaction a questionnaire with the System Usability Scale (SUS) questions (items 1–10) was administered, along with additional questions focusing on precision, satisfaction (enjoyment), and realism (items 11–13). Questionnaire items are as follows:

1. "I think that I would like to use this method frequently." - This question is crucial in determining the appeal and utility of the interaction method. A rating of 1 would indicate strong disagreement, suggesting the participant does not foresee using the method often, whereas a rating of 5 would indicate strong agreement, suggesting the participant finds the method highly useful and would likely use it frequently.

2. "I find the method unnecessarily complicated." - Complexity can be a barrier to usability, and this question helps in understanding if the interaction method is perceived as too complex. A score of 1 indicates strong disagreement, implying that the participant finds the method straightforward, whereas a 5 would suggest that they find it excessively complicated.

3. "I find the method easy to use." - This question aims to assess the usability of the interaction method. A rating of 1 represents strong disagreement, suggesting difficulty in usage, whereas a rating of 5 represents strong agreement, indicating ease of use. It's critical to understand this aspect as ease of use often correlates with a better user experience.

4. "I think that I would need technical support to be able to use this method." - Evaluating if users feel the need for external help is essential in understanding the self-sufficiency of the interaction method. A rating of 1 would indicate strong disagreement, suggesting that the participant feels capable of using it independently, while a rating of 5 would suggest a high need for support.

5. "I think that the different aspects of this method of interaction are well integrated." - A well-integrated system is essential for a seamless user experience. A rating of 1 indicates strong disagreement, implying poor integration, whereas a 5 indicates strong agreement, suggesting that the participant found the method to be well-composed.

6. "I think there are many inconsistencies in this method." - It's essential to gauge whether users find the interaction method to be inconsistent in its behavior. Inconsistencies can greatly hamper the user experience. A rating of 1 would indicate strong disagreement, suggesting that the participant finds the method consistent, while a rating of 5 would indicate that they find it to be full of inconsistencies.

7. "I think that most people will be able to master this method very quickly." - This question addresses the learnability of the interaction method. A rating of 1 would indicate strong disagreement, suggesting that the participant

believes it would take time for most people to learn the method, while a 5 indicates that they think it can be quickly and easily mastered by most users.

8. "I find the method uncomfortable to use." - Comfort is a critical aspect of user experience, especially in VR where physical interactions are simulated. A rating of 1 represents strong disagreement, implying that the participant finds the method comfortable, whereas a 5 represents strong agreement, indicating discomfort.

9. "I feel confident using this method." - Confidence in using an interaction method can indicate its intuitiveness and ease of use. A score of 1 indicates a lack of confidence, whereas a score of 5 indicates high confidence. This helps in understanding if the interaction method empowers the user.

10. "I had to learn a lot of things before I started working with the method." - This question gauges the learning curve associated with the interaction method. A rating of 1 indicates that the user did not have to learn much to get started, whereas a 5 indicates a steep learning curve.

11. "I believe that this method allowed me to reach my goal precisely." - Precision is key in interaction methods, especially in VR. A rating of 1 indicates that the participant found the method imprecise, whereas a 5 indicates that they found it very precise in achieving their goals.

12. "I find the experience of using this method enjoyable." - Enjoyment is an essential component of user experience. A rating of 1 indicates that the participant did not enjoy using the method, whereas a 5 indicates a highly enjoyable experience.

13. "I believe that the experience of using this method was realistic." - This question measures the realism of the interaction method, which is vital for immersion in VR. A rating of 1 indicates that the participant found the method unrealistic, whereas a 5 indicates a very realistic experience.

Participants rated the questions using a Likert scale from 1 to 5, where 1 indicated strong disagreement and 5 indicated strong agreement. For each method participants were asked to perform a series of tasks in VR environment, with each task designed to test different method implementation, specifically:

- **Position manipulation** - Participants were asked to pick up various objects of different shapes and sizes from a table and place them into a box.
- **Rotation** - Users had to rotate objects to a specific orientation.
- **Scaling** - In this task, users had to scale objects to a specific size.
- **Throwing** - This task required participants to hit targets at varying distances by throwing objects.
- **Pushing** - It tests whether pushing feels natural and whether the virtual objects react to the force in a realistic manner.
- **Pulling** - Similar to pushing, pulling is another fundamental human action. In this task, the user was asked to pull blocks into certain positions to solve puzzles.
- **Pressing** - In this task, participants had to press virtual buttons in a specific sequence.

- **Typing** - Users were presented with a virtual keyboard and asked to type a given sentence.
- **Pointing** - This task involved users pointing at specific objects in the virtual environment in a certain order.
- **Selection** - Participants were tasked with selecting specific objects in the virtual environment in a specified order.
- **Archery** - In this task, participants took part in a target shooting game where they had to hit targets with a virtual bow and arrow at various distances and sizes.
- **Hitting an object with another object** - Users were tasked with hitting a series of targets using a virtual, oblong object resembling a baseball bat.

These tasks were structured and repeatable, ensuring an objective and accurate assessment for each participant. Each method is represented by several implementations. Each implementation may use different types of metaphors and interaction mechanics popular in VR environment. The most popular are: virtual hand and virtual pointer. The virtual hand metaphor allows users to reach and grab objects, mimicking real-world hand movements. Instead of controllers held by user, he sees 3D models of hands (see Fig. 2ACEGH) which can bend fingers and grasp objects by mapping these actions to controller buttons of sensors. The Go-Go interaction technique is a flexible extension of the virtual hand technique that allows for distance selection of objects that are out of natural reach.

Virtual pointer metaphor involves users interacting with objects by pointing at them. This metaphor is especially useful for interacting with objects that are out of reach and might be implemented through a variety of devices. It does not need as sophisticated controller as for virtual hands. It only needs to tell the direction user is pointing. The pointed direction and object is usually visualised and implemented with a raycast - a line coming out of the controller or virtual hand on one end and colliding with the pointed object on the other end (see Fig. 2BD). The pointed object is selected can be manipulated which is called anchor control. If the selection is held for a moment the selected object be manipulated automatically, e.g. gradually rotating or scaling "over time". But the raycast does not need to be operated with controller in hand. Gaze ray casting, which is similar to eye-tracking, uses the eye orientation as the root of the ray, but it requires special equipment and extra eye calibration. To avoid this, head ray casting is proposed in [13], using the orientation of the head (HMD) as the root of the ray.

To address the precision limitations of the simple virtual hand method, a more advanced technique can be employed, known as the Air TRS method, described in [11]. It adapts the Translation-Rotation-Scale (TRS) operations to three dimensions for object manipulation in mid-air. With one hand, users can directly grab and move an object, while a second hand allows rotation around the first hand. It's particularly useful for close-range interactions and provides more precise, and a wider range of rotation manipulation than the simple virtual hand method.

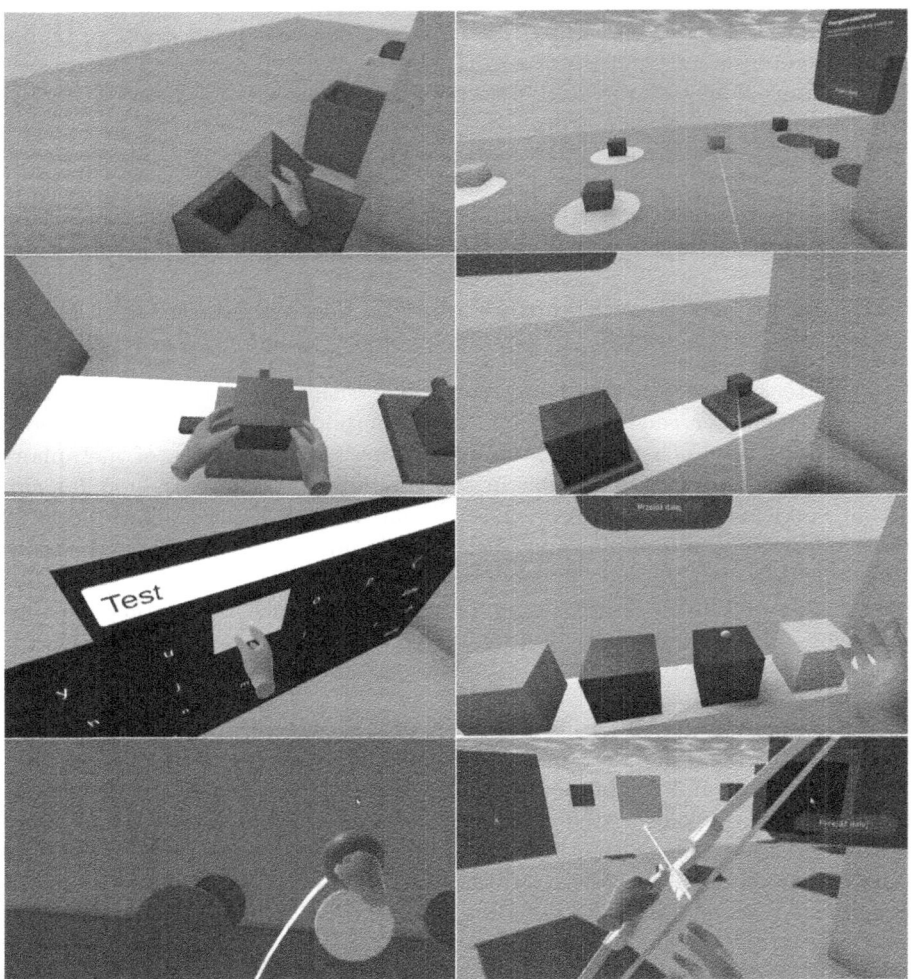

**Fig. 2.** Screenshots of some of the tested VR interaction method implementations: A - virtual hand positioning, B - raycast pulling, C - 2 virtual hands rotation, D - raycast scale over time, E - virtual hand typing - poking 2D keyboard, F - Head ray casting with white 3D bubble cursor, G - throwing at circle targets and showing white trajectory, H - archery.

## 5   Results

The data gathered via Google Forms has been divided into 12 parts, corresponding to a different method of interaction - each with several implementations - and aggregated with average metric. The result is presented in Tab. 1.

**Table 1.** The experiment result for 12 VR interactions with alternative implementations: an aggregated percentage score of System Usability Scale (SUS) questionnaire, and 3 Likert scale questions on precision, satisfaction and realism.

| Method Implementation | SUS | Precision | Satisfaction | Realism |
|---|---|---|---|---|
| **Positioning**: Virtual hand | 90 | 4.0 | 3.9 | 4.5 |
| Raycast - force pull | 82 | 4.1 | 4.0 | 3.6 |
| GoGo technique | 80 | 4.4 | 4.2 | 3.1 |
| Raycast - traditional | 71 | 3.7 | 3.4 | 2.9 |
| Raycast - anchor control | 62 | 3.6 | 3.3 | 2.1 |
| **Rotation**: Raycast - rotate over time | 86 | 3.8 | 4.2 | 4.6 |
| 1 virtual hand | 77 | 4.5 | 3.2 | 3.5 |
| Raycast - match direction | 72 | 3.4 | 3.7 | 4.2 |
| 2 virtual hands (Air TRS) | 71 | 4.1 | 3.6 | 4.0 |
| 2 raycasts (Air TRS) | 39 | 2.7 | 2.1 | 2.7 |
| **Scaling**: 2 virtual hands (Air TRS) | 85 | 4.6 | 4.0 | 4.1 |
| Raycast - scale over time | 68 | 3.7 | 3.8 | 4.4 |
| 2 raycasts (Air TRS) | 52 | 2.7 | 2.8 | 2.7 |
| **Throwing**: Default | 80 | 3.2 | 3.6 | 4.0 |
| Mass based - scale | 75 | 4.5 | 3.6 | 4.2 |
| Mass based - smoothing | 75 | 4.3 | 3.5 | 3.9 |
| Mass based - smoothing head assistance | 71 | 3.1 | 3.3 | 4.0 |
| Mass based - smoothing displaying trajectory | 52 | 3.4 | 3.1 | 3.8 |
| **Pushing**: Virtual hand - grabbing | 85 | 4.7 | 3.5 | 4.3 |
| Raycast - traditional | 80 | 3.6 | 4.5 | 4.5 |
| Raycast - anchor control | 70 | 2.3 | 3.7 | 4.3 |
| Virtual hand - physical hand | 56 | 3.7 | 2.3 | 2.6 |
| **Pulling**: Raycast - traditional | 86 | 3.2 | 3.9 | 4.1 |
| Virtual hand - grabbing | 85 | 4.2 | 3.8 | 4.0 |
| Raycast - anchor control | 75 | 2.6 | 4.1 | 4.0 |
| Virtual hand - physical hand | 49 | 3.2 | 1.9 | 2.1 |
| **Pressing**: 3D button - raycast | 95 | 3.5 | 4.5 | 4.8 |
| 2D button - raycast | 91 | 3.2 | 4.4 | 4.7 |
| 3D button - virtual hand - grabbing | 89 | 3.9 | 4.1 | 4.1 |
| 2D button - virtual hand - grabbing | 84 | 3.7 | 3.8 | 4.3 |
| 3D button - virtual hand - poking | 54 | 3.9 | 2.7 | 2.9 |

(*continued*)

**Table 1.** (*continued*)

| Method Implementation | SUS | Precision | Satisfaction | Realism |
|---|---|---|---|---|
| **Typing**: 2D keyboard - raycast - press | 86 | 3.9 | 4.1 | 4.3 |
| 2D keyboard - head ray casting | 80 | 3.3 | 4.7 | 4.0 |
| 3D keyboard - virtual hand - poke | 67 | 3.7 | 3.7 | 3.0 |
| Physical keyboard | 64 | 4.8 | 3.2 | 3.4 |
| 2D keyboard on hand - virtual hand - poke | 54 | 3.8 | 2.5 | 3.4 |
| 2D keyboard - virtual hand - poke | 38 | 3.2 | 1.8 | 2.6 |
| **Pointing**: Raycast - 3D bubble | 90 | 3.9 | 3.6 | 4.5 |
| Raycast - traditional | 90 | 3.5 | 3.9 | 4.3 |
| GoGo technique | 78 | 3.2 | 3.7 | 4.1 |
| Virtual hand | 72 | 4.5 | 3.1 | 4.2 |
| Head ray casting | 71 | 3.5 | 3.5 | 3.8 |
| **Selection**: Raycast - press to select | 90 | 3.8 | 4.2 | 4.5 |
| Virtual hand | 89 | 4.7 | 3.2 | 4.3 |
| GoGo technique | 82 | 3.8 | 3.7 | 4.2 |
| Raycast - hover to select | 81 | 3.3 | 3.7 | 3.9 |
| Head ray casting - hover to select | 76 | 4.1 | 4 | 4.1 |
| **Archery**: Automatically loaded arrows | 97 | 3.6 | 4.4 | 4.9 |
| Arrows loaded by pressing | 88 | 3.2 | 4.2 | 4.8 |
| Separate quiver | 87 | 4.2 | 3.8 | 4.2 |
| Quiver behind the back | 80 | 4.9 | 3.9 | 4.7 |
| **Hitting with an object**: Physics based | 100 | 4.8 | 4.6 | 5 |
| No physics | 94 | 3.6 | 3.9 | 4.8 |
| No collision | 93 | 3.1 | 3.5 | 4.6 |
| Object getting stuck | 92 | 4.6 | 4.2 | 4.7 |

The results can now be analyzed specifically for each of the 12 methods and their implementations (techniques):

– **Position manipulation** - Four different techniques were assessed, including Virtual Hand, Raycast - Force Pull, GoGo Technique, and Raycast - Anchor Control. The highest SUS score was achieved by the Virtual Hand (89.50), followed by Raycast - Force Pull (82.25), GoGo Technique (80.25), and Raycast - Anchor Control (62.25). The perceived precision, satisfaction, and realism were also generally higher in the Virtual Hand and Raycast - Force Pull, with the GoGo Technique performing acceptably and Raycast - Anchor Control being the least effective.

– **Rotation** - Five techniques were studied - Raycast - Rotate Over Time, 1 Virtual Hand, Raycast - Match Direction, 2 Virtual Hands (Air TRS), and 2 Raycasts (Air TRS). Among these techniques, Raycast - Rotate Over Time had the highest SUS score (85.75) and was closely followed by 1 Virtual Hand (77.00). The remaining techniques Raycast - Match Direction (72.25), 2 Virtual Hands (Air TRS) (70.75), and 2 Raycasts (Air TRS) (38.50) had lower SUS scores and varied performance in perceived precision, satisfaction, and realism.

– **Scaling** - In scaling manipulation, the 2 Virtual Hands (Air TRS) technique achieved the highest SUS score (85.25), significantly higher than the Raycast - Scale Over Time (68.25) and the 2 Raycasts (Air TRS) (51.50) techniques. The perceived precision, satisfaction, and realism were also better in the 2 Virtual Hands (Air TRS) technique compared to the other two techniques.

– **Throwing** - Five different techniques were evaluated - Default, Mass Based - Scale, Mass Based - Smoothing, Mass Based - Smoothing - Head Assistance, and Mass Based - Smoothing - Displaying Trajectory. The Default technique performed the best with the highest SUS score (79.75), closely followed by the Mass Based - Scale (75.25) and Mass Based - Smoothing (74.75) techniques. The remaining techniques achieved lower SUS scores and varied performance levels in perceived precision, satisfaction, and realism.

– **Pushing** - Four techniques were assessed in pushing manipulation - Virtual Hand - Grabbing, Raycast - Traditional, Raycast - Anchor Control, and Virtual Hand - Physical Hand. Virtual Hand - Grabbing had the highest SUS score (84.75), significantly higher than the other three techniques, which were Raycast - Traditional (79.75), Raycast - Anchor Control (70.25), and Virtual Hand - Physical Hand (55.75). The perceived precision, satisfaction, and realism were also generally higher in the Virtual Hand - Grabbing technique compared to the other techniques.

– **Pulling** - Four techniques were evaluated - Raycast - Traditional, Virtual Hand - Grabbing, Raycast - Anchor Control, and Virtual Hand - Physical Hand. Among these techniques, Raycast - Traditional (86.25) had the highest SUS score, closely followed by the Virtual Hand - Grabbing (84.50). The remaining techniques, Raycast - Anchor Control (74.75), and the Virtual Hand - Physical Hand (48.75) had lower SUS scores and performed poorly in perceived precision, satisfaction, and realism.

– **Pressing** - Five techniques were studied - 3D Button - Raycast, 2D Button - Raycast, 3D Button - Virtual Hand - Grabbing, 2D Button - Virtual Hand - Grabbing, and 3D Button - Virtual Hand - Poking. Among these techniques, the 3D Button - Raycast had the highest SUS score (94.75), followed by the 2D Button - Raycast (91.25), 3D Button - Virtual Hand - Grabbing (89.25), and the 2D Button - Virtual Hand - Grabbing (84.25). The 3D Button - Virtual Hand - Poking technique achieved the lowest SUS score (53.75) and performed poorly in perceived precision, satisfaction, and realism.

– **Typing** - Six techniques were evaluated - 2D Keyboard - Raycast - Press, 2D Keyboard - Head Raycasting, 3D Keyboard - Virtual Hand - Poke, Physical Keyboard, 2D Keyboard Placed on Hand - Virtual Hand - Poke, and 2D Keyboard - Virtual Hand - Poke. The 2D Keyboard - Raycast - Press technique had the highest SUS score (85.75), followed by the 2D Keyboard - Head Raycasting (80.25) and the 3D Keyboard - Virtual Hand - Poke (66.50). The remaining three techniques - Physical Keyboard (64.00), 2D Keyboard Placed on Hand - Virtual Hand - Poke (54.00), and 2D Keyboard - Virtual Hand - Poke (38.25) had lower SUS scores and varied performance in perceived precision, satisfaction, and realism.

- **Pointing** - Five techniques were compared - Raycast - 3D Bubble, Raycast - Traditional, GoGo Technique, Virtual Hand, and Head Raycasting. The Raycast - 3D Bubble achieved the highest SUS score (90.00), closely followed by the Raycast - Traditional (89.50). The remaining techniques, GoGo Technique (77.75), Virtual Hand (71.75), and Head Raycasting (71.25), had lower SUS scores and varied performance levels in perceived precision, satisfaction, and realism.

- **Selection** - Five techniques were assessed - Raycast - Press to Select, Virtual Hand, GoGo Technique, Raycast - Hover to Select, and Head Raycasting - Hover to Select. The Raycast - Press to Select technique achieved the highest SUS score (90.00), closely followed by Virtual Hand (89.25). The remaining techniques, GoGo Technique (82.00), Raycast - Hover to Select (81.25), and Head Raycasting - Hover to Select (76.00), had lower SUS scores and varied performance in perceived precision, satisfaction, and realism.

- **Archery** - Four techniques were studied - Automatically Loaded Arrows, Arrows Loaded by Pressing, Separate Quiver, and Quiver Behind the Back. Among these techniques, Automatically Loaded Arrows achieved the highest SUS score (96.75), followed by Arrows Loaded by Pressing (88.00), Separate Quiver (87.25), and Quiver Behind the Back (80.00). The techniques also showed varying performance levels in perceived precision, satisfaction, and realism.

- **Hitting with an object**- Four techniques were compared - Physics Based, No Physics, No Collision, and Object Getting Stuck. The Physics Based technique had the highest SUS score (99.75), significantly higher than all other techniques. The No Physics (93.75), No Collision (92.50), and Object Getting Stuck (92.25) techniques had lower SUS scores and varied performance in perceived precision, satisfaction, and realism.

## 6    Conclusions

This paper goal is to identify and compare different methods of interaction with objects in virtual reality, focusing on their usability and user experience. The study analyzed twelve interaction methods in VR applications to assess usability, precision, satisfaction, and realism. The results demonstrated varying levels of efficacy for different interaction methods. Virtual Hand techniques were identified as particularly effective for position manipulation tasks, while raycast was found to be optimal for object selection and rotation tasks. Physics-based methods proved to be more realistic for interactions like hitting objects.

These findings have significant implications for the design and development of VR applications, particularly in gaming, where smooth interaction is crucial for immersion. Through a comprehensive evaluation involving user studies, expert reviews, and quantitative analyses, this research has highlighted the strengths and weaknesses of different interaction methods. It resulted in providing an overview of the most problematic methods and identifying the most suitable methods for specific use cases.

The results analysis that not all interaction methods are equally effective. Depending on the nature of the VR application, different interaction methods may be more suitable. For example, in VR games where precision and realism are vital, the Virtual Hand technique for position manipulation and the physics-based method for hitting objects seem to be preferred. On the other hand, for educational or professional applications where usability and efficiency are more critical, methods such as raycast may be more appropriate. Additionally, there was a positive correlation between perceived realism and usability scores, indicating that interaction methods that felt more realistic were generally regarded as more usable.

By providing empirical data on the performance of various interaction methods, it helps to bridge the gap between theoretical concepts and practical applications. The study's findings can inform designers and developers about which interaction methods are currently the most effective and realistic, enabling them to make informed decisions during the development of VR applications.

Understanding the nuances of interaction methods may also lead to the development of specialized controllers and input devices for VR. These devices could be optimized for specific interaction methods, thereby improving the overall user experience in the particular method of interaction. The findings of this study have significant implications for VR-based training and simulation programs. By selecting interaction methods that maximize precision and realism, training programs can achieve higher levels of efficacy. This is especially relevant in fields such as healthcare, aviation, and the military, where simulation-based training is critical.

# References

1. B-Reel: simulating weight in VR (2016). https://medium.com/@Breel.co/simulating-weight-in-vr-d161e87990b. Accessed 10 March 2024
2. Bermejo, C., Lee, L.H., Chojecki, P., Przewozny, D., Hui, P.: Exploring button designs for mid-air interaction in virtual reality: a hexa-metric evaluation of key representations and multi-modal cues. Proc. ACM Hum.-Comput. Interact. **5**(EICS), 1–26 (2021)
3. Bowman, D.A., Hodges, L.F.: An evaluation of techniques for grabbing and manipulating remote objects in immersive virtual environments. In: Proceedings of the 1997 Symposium on Interactive 3D Graphics, pp. 35–ff (1997)
4. Conner, B.D., Snibbe, S.S., Herndon, K.P., Robbins, D.C., Zeleznik, R.C., Van Dam, A.: Three-dimensional widgets. In: Proceedings of the 1992 Symposium on Interactive 3D Graphics, pp. 183–188 (1992)
5. Dang, N.T.: A survey and classification of 3D pointing techniques. In: 2007 IEEE International Conference on Research, Innovation and Vision for the Future, pp. 71–80. IEEE (2007)

6. Dube, T.J., Arif, A.S.: Text entry in virtual reality: a comprehensive review of the literature. In: Kurosu, M. (ed.) HCII 2019. LNCS, vol. 11567, pp. 419–437. Springer, Cham (2019). https://doi.org/10.1007/978-3-030-22643-5_33

7. Feng, J., Cho, I., Wartell, Z.: Comparison of device-based, one and two-handed 7dof manipulation techniques. In: Proceedings of the 3rd ACM Symposium on Spatial User Interaction, pp. 2–9 (2015)

8. Höll, M., Oberweger, M., Arth, C., Lepetit, V.: Efficient physics-based implementation for realistic hand-object interaction in virtual reality. In: 2018 IEEE Conference on Virtual Reality and 3D User Interfaces (VR), pp. 175–182. IEEE (2018)

9. Kruijff, E., LaViola, J.J., Poupyrev, I.: 3D user interfaces: theory and practice. IEEE Trans. Visual Comput. Graphics **18**(4), 565–572 (2012)

10. Mendes, D., Caputo, F.M., Giachetti, A., Ferreira, A., Jorge, J.: A survey on 3D virtual object manipulation: from the desktop to immersive virtual environments. In: Computer Graphics Forum, vol. 38, pp. 21–45. Wiley Online Library (2019)

11. Mendes, D., Fonseca, F., Araujo, B., Ferreira, A., Jorge, J.: Mid-air interactions above stereoscopic interactive tables. In: 2014 IEEE Symposium on 3D User Interfaces (3DUI), pp. 3–10. IEEE (2014)

12. Mine, M.R.: Virtual environment interaction techniques. UNC Chapel Hill CS Dept (1995)

13. Nickel, K., Stiefelhagen, R.: Pointing gesture recognition based on 3d-tracking of face, hands and head orientation. In: Proceedings of the 5th International Conference on Multimodal Interfaces, pp. 140–146 (2003)

14. Poupyrev, I., Billinghurst, M., Weghorst, S., Ichikawa, T.: The go-go interaction technique: non-linear mapping for direct manipulation in VR. In: Proceedings of the 9th Annual ACM Symposium on User Interface Software and Technology, pp. 79–80 (1996)

15. Speicher, M., Feit, A.M., Ziegler, P., Krüger, A.: Selection-based text entry in virtual reality. In: Proceedings of the 2018 CHI Conference on Human Factors in Computing Systems, pp. 1–13 (2018)

16. Ulinski, A., Zanbaka, C., Wartell, Z., Goolkasian, P., Hodges, L.F.: Two handed selection techniques for volumetric data. In: 2007 IEEE Symposium on 3D User Interfaces. IEEE (2007)

17. Zindulka, T., Bachynskyi, M., Müller, J.: Performance and experience of throwing in virtual reality. In: Proceedings of the 2020 CHI Conference on Human Factors in Computing Systems, pp. 1–8 (2020)

# Retrofitting a Legacy Cutlery Washing Machine Using Computer Vision

Hua Leong Fwa[(✉)] [iD]

Singapore Management University, 81 Victoria Street, 188065 Singapore, Singapore
hlfwa@smu.edu.sg

**Abstract.** Industry 4.0, the digitalization of manufacturing promises to lead to lowered cost, efficient processes and even discovery of new business models. However, many of the enterprises have huge investments in legacy machines which are not 'smart'. In this study, we thus designed a cost-efficient solution to retrofit a legacy conveyor belt-based cutlery washing machine with a commodity web camera. We then applied computer vision (using both traditional image processing and deep learning techniques) to infer the speed and utilization of the machine. We detailed the algorithms that we designed for computing both speed and utilization. With the existing operational constraints of our client, frequent re-training of the deep learning model for object detection is not feasible. Thus, we compared the generalizability of the two techniques across 'unseen' cutleries and found traditional image processing to be generalizable across 'unseen' images. Our proposed final solution uses traditional image processing for computation of utilization but a hybrid of traditional image processing and deep learning model for speed computation as it is more reliable. Our client has implemented our proposed solution for one conveyor belt-based cutlery washing machine and will be planning to scale this to multiple conveyor belt-based cutlery washing machines.

**Keywords:** Industry 4.0 · Computer Vision · Deep Learning · Image Processing

## 1 Introduction

Industry 4.0 is characterized by digitalization of manufacturing premised on the use of technologies such as Internet of Things (IOT), Cloud Computing, Data Analytics and Artificial Intelligence (AI) [11,12,17]. With Industry 4.0, the advent of smarter and more autonomous machines promises to make manufacturing more efficient which further leads to optimized processes, lowered cost and even possible discovery of new business models. Notwithstanding the many benefits of Industry 4.0, the move to Industry 4.0 has been challenging for many companies.

Many of these companies have made substantial investments into the legacy machines which are not data ready or 'smart'. However, these legacy machines

N.-T. Nguyen et al. (Eds.): ICCCI 2024, CCIS 2166, pp. 301–313, 2024.
https://doi.org/10.1007/978-3-031-70259-4_23

are still functional and continue to fulfil the operational needs of the companies. The replacement of these legacy machines with smarter ones entails high investment in equipment and technology which may not be feasible especially for small and medium enterprises with tight budgets. Retrofitting or equipping these legacy machines with sensors to enhance their functionality or performance may be a practical solution for these companies to jump on the industry 4.0 bandwagon without huge capital investments [23]. The integration of these new technologies into the companies' traditional processes in the digitalization journey in turn offers new opportunities for these companies to redesign business processes and facilitate data driven decision making.

In our study, we used computer vision techniques for enhancing a traditional conveyor belt-based cutlery washing machine with the capability of tracking the utilization and running speed of the machine. The original conveyor belt-based cutlery washing machine is a legacy machine which cleans cutleries. The various cutleries are first placed on the conveyor belt of the machine and when the machine is switched on, the conveyor belt then moves at three different levels of pre-set speeds (controlled via a knob) to transport the cutleries into a steam cleaning compartment. The actual speed of the conveyor belt varies though with the weight of the cutleries placed on it i.e. the conveyor belt slows to a speed lower than the preset one when heavier cutleries are placed on it. Sensing the actual speed of the conveyor belt however, is secondary as our client wanted instead to monitor prolonged instances of waste and careless usage. These are situations where the cutlery washing machine is running but only some or no cutleries were actually placed on it (waste usage) or where cutleries were placed on the conveyor belt but the machine was intermittently stopped (careless usage). The intermittent starting and stopping of the cutlery washing machine, according to the client, will result in higher incidences of machine failure and thus undesirable. With the use of a commodity web camera, we managed to retrofit the conveyor belt-based cutlery washing machine not only to be data ready but also to accomplish the above stated objectives of our client at a low cost.

Our client wanted to track the utilization as well as the running speed of each of their conveyor belt-based cutlery washing machines to optimize their use and lower the running cost but this cannot be achieved without the installation of additional sensors. Our proposed solution is to use computer vision techniques to process images captured through a web camera installed on top of the conveyor belt of the machine for computing the utilization and speed of the machine. With real-time tracking of utilization and speed of the conveyor belt-based cutlery washing machines, our client would be able to visualize and better optimize both the operations and running cost of the machines.

## 2   Related Studies

Li et al. [13] adapted the You Only Look Once (YOLO) network [19] for detection of six types of surface defects on steel strips. Using the improved network, they were able to achieve 97.55% mAP and 95.86% recall rate for the six types of defects at a detection speed of 83 FPS.

Pham et al. [18] proposed a real-time packaging defect detection system also based on YOLO network for detecting defects on package boxes moved using conveyor belts. In their study, the authors were able to achieve 78.6% mAP when evaluating the YOLO network over a test set of 40 images.

In another study by Li et al. [14], a traditional digital image processing technique was proposed to recognize congestion on conveyor belts. The authors used a static-edge detection method and then applied statistical techniques to analyse the extracted features of packages for determining whether a congestion had occurred. The authors contended that deep learning techniques would not be adequate for industrial parks with thousands of cameras due to the need for massive computing resources to train the models and the requirement for a massive set of training images. Their results also showed that they were able to achieve higher precision and recall as compared to the performance of deep learning models such as Inception [22] and YOLO.

The study by Liu and Qu [16] compared the use of traditional digital image processing technique against a deep learning based convolutional neural network for the defect classification of PCB boards. The authors compared a traditional classification algorithm based on digital image processing against one using Convolutional Neural Network [7] on a set of 1818 PCB defect images. The results indicated that the algorithm based on convolutional neural network achieved a higher classification accuracy of 95.7% which is higher than the traditional method.

The above studies demonstrated the feasibility of both traditional image processing techniques and deep learning algorithms for object detection in a factory environment. In recent years, deep learning algorithms notably are gaining popularity for use in industrial defect detection due to their ability to achieve higher accuracy. However, as highlighted by Li et al. [14], albeit the higher accuracy achieved by deep learning algorithm, deep learning algorithms may not be adequate in all industrial contexts due to their need for re-training, massive image data sets and high computing resource requirements for training of the models.

In our study, we thus compare between the use of traditional image processing technique versus deep learning algorithm in terms of their generalizability. With higher generalizability of the detection algorithm, we postulate that re-training (arising from new and unseen objects) will not need to be frequent. Finally, we also adapted and formulated algorithms to compute both utilization and running speed of the conveyor belt-based cutlery washing machine from the detected objects on the conveyor belt in this study.

# 3   Methodology

## 3.1   Hardware Setup

A commodity web camera is mounted on the ceiling on top of the conveyor belt-based cutlery washing machine. The web camera is configured to record video

**Fig. 1.** Web camera mounted on ceiling to capture top-down video of conveyor belt-based cutlery washing machine

with resolution of 1920 by 1080 pixels covering the entire conveyor belt of the conveyor belt-based cutlery washing machine at a rate of 30 frames per second. The web camera is in turn connected via USB to a mini-PC running Windows 11. The mini-PC is equipped with an Intel i7-1260P processor, 32 GB of RAM and a 500GB SSD drive.

The cutlery washing machine has a steam compartment within it which uses water and steam for cleaning and sterilizing the cutleries. The web camera cannot be placed on or within close proximity to the cutlery washing machine due to moisture and temperature conditions. The high temperature of steam also causes potential fogging of the camera lens due to condensation. Due to these constraints, both the web camera and mini-PC thus have to be placed a safe distance away from the cutlery washing machine and this explains our placement of the web camera and mini-PC. A picture of the mounting of the web camera is shown in Fig. 1. The mini-PC is residing within the ceiling and thus hidden from view.

Our client runs several conveyor belt-based cutlery washing machines and intends to deploy this solution for all the machines. The cutlery washing machines are distributed far apart on the factory floor and connecting the web cameras mounted on top of each cutlery washing machines using usb cables to a single mini-PC is not feasible due to the distance. On top of this, the factory floor is not provisioned with adequate wireless infrastructure which rules out wireless connectivity between the cameras and mini-PCs. We thus intend to retrofit each cutlery washing machine with their own individual set of mini-PC and web camera for the final deployment.

## 3.2   Object Detection

The main objectives for this study are to compute the utilization of the conveyor belt-based cutlery washing machine and the speed of movement of the machine's conveyor belt. We postulate to evaluate both computer vision and deep learning objection detection techniques for the detection of the cutleries. Specifically, with the localization of the objects by bounding boxes, the utilization can be computed from the sum of areas of the predicted bounding boxes that enclose the cutleries while the speed can be derived from tracking the difference in distance in pixels (for the detected objects) across a fixed number of frames of captured video.

In recent years, deep learning has been widely adopted for object detection. Most state-of-the-art object detectors utilize deep learning networks as their backbone and detection network to extract features automatically from images for classification and localization [10,24]. One of the popular deep learning-based object detector is YOLO [1,2,8]. In YOLO, convolutional layers are used to extract image features which are in turn used to predict both the bounding boxes (for locating the object) and learn the class probabilities. YOLO divides every image into a grid of S x S and every grid predicts N bounding boxes and confidence. The confidence denotes the precision of the bounding box and whether the bounding box contains an object or not. YOLO is one of the fastest detector with the capability to process images in real-time at 45 frames per second and inference of up to 300 times faster as compared to other detectors [5]. In addition, YOLO also outperforms other detectors in terms of its ability to generalize from natural images to images in other domains [19]. As the cutleries are placed on a moving conveyor belt, the speed of detecting and recognizing the objects is instrumental to the accurate calculation of utilization and speed. The real-time detection speed of YOLO is thus one of the key consideration for its use in this study.

The Faster-RCNN [20] is another popular architecture that is used for object detection. Faster-RCNN uses a backbone network for learning a convolutional feature map before using a separate network to predict the region proposals. The predicted region proposals are then reshaped using a Region of Interest (RoI) pooling layer used for both classifying the image within the proposed region and predicting the offset values for the bounding boxes.

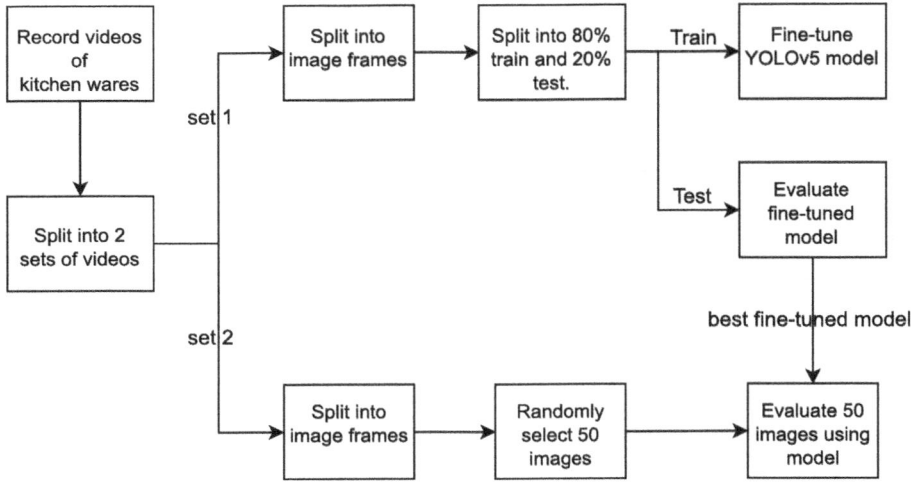

**Fig. 2.** Processing and evaluation workflow of videos using pre-trained YOLOv5 model

We will compare between the use and accuracies of YOLO and Faster-RCNN for the detection of the cutleries in our context.

### 3.3   YOLOv5 and Faster-RCNN

We first captured videos of different cutleries being cleaned on the conveyor belt-based cutlery washing machine. The processing and evaluation workflow for the videos using YOLOv5 is shown in Fig. 2.

We recorded videos of various cutleries placed on the conveyor belt-based cutlery washing conveyor belt for washing. A total of 6 videos were recorded with each recording different sets of cutleries. The videos vary in lengths from 2 to 5 min. The videos were split into 2 sets with the second set reserved for evaluating the generalizability of the YOLOv5 model.

Using the first set of the recorded videos, we split the videos into image frames using Python OpenCV library and then manually labelled the bounding boxes in each image as the ground truth. 80% of the images were randomly selected as training image to be used for fine-tuning the model while 20% of the images (validation images) were used for evaluating the performance of the fine-tuned model. Fine-tuning or transfer learning [3] is a technique where a model pre-trained for one task is tuned or tweaked to perform a second similar task. We used pre-trained weights for YOLOv5s model from Ultralytics [9] and fine-tuned the model using our training images. YOLOv5s were pre-trained using the Microsoft Common Objects in Context (COCO) [15] dataset. We selected YOLOv5 out of the many object detection models as it offers fast inference speed of about 100 ms running on a CPU.

Using the training images, we fine-tuned the pre-trained YOLOv5s model by running the training loop for a total of 100 epochs with Stochastic Gradient

Descent (SGD) as optimizer, a batch size of 64 and evaluated its performance on the validation images with early stopping. After training for 100 epochs, we were able to achieve a mean Average Precision (mAP) of 0.9655 on the validation images at an intersection-over-union (IOU) threshold of 0.5. The mAP is a metric that is commonly used for evaluating object detection models.

Using similar protocol, we fine-tuned Faster-RCNN by using MobileNet v2 [21] as the model backbone (with pre-loaded weights). The Faster-RCNN model which is trained for 100 epochs with SGD as optimizer, achieved a mAP of 0.899 for an IOU threshold of 0.5 with early stopping on the validation images. We thus selected YOLOv5 as the object detection model for our study.

In this study, we further evaluated YOLOv5 against a traditional image processing technique - Colour Image Segmentation [4,6] for the detection of cutleries on the conveyor belt-based cutlery washing machine. The generalizability of the object detection techniques is important as the set of cutleries to be cleaned is not fixed and new collections of cutleries may be introduced from time to time. Due to operational requirements, our client would also prefer minimizing down time and effort to re-train the object detection model.

### 3.4    Color Image Segmentation (CIS)

We converted a sample set of 10 images from Red, Green and Blue (RGB) to Hue, Saturation and Value (HSV) colour space. Using the OpenSegment Streamlit application (https://kxborg-open-segment-hsv-segment-idubxp.streamlit.app), we first establish the HSV lower and upper boundaries color range for masking out the conveyor belt background. Having masked out the pixels within the HSV color range, the masked image is further converted to a binary image and then inverted so that the objects on the conveyor belt will be white while the conveyor belt is black. Selecting only the white pixels will then give us the boundaries or segments of the objects.

This technique works in this context as the conveyor belt is uniformly light yellow and can be differentiated in colour from the objects to be detected i.e. the cutleries. In addition, the conveyor belt-based cutlery washing machine is also deployed by our client in a large industrial kitchen floor with good lighting conditions. We recognize that the detection will deteriorate for cutleries that are similar in colour to the conveyor belt or in poorly lit conditions. Notwithstanding the 2 constraints, an advantage of this over the deep learning technique

**Table 1.** No. of objects missing detection by YOLOv5 and Color Image Segmentation

| Type | No. of images | No. of Objects | YOLOv5 | Color Image Segmentation (CIS) |
| --- | --- | --- | --- | --- |
| | | | No. missed | No. missed |
| 'Unseen' images | 28 | 152 | 70 (46.05%) | 11 (7.24%) |
| 'Seen' images | 22 | 155 | 10 (6.45%) | 14 (9.03%) |

(YOLOv5) is elimination of the need to re-train the deployed model with additional images should there be a need to detect new sets of cutleries.

## 4 Results and Discussion

### 4.1 Generalizability

To evaluate the generalizability of both techniques, we selected 50 images randomly from the second set of videos. We consider that if the object's bounding box (as detected by the algorithm) has an overlap of more than 50% with the ground truth bounding box, the object is considered as detected. It will be considered as missed detection otherwise. These images contain a mix of cutleries that were 'seen' (used in the training of the model) and 'unseen' (not used in the training of the model) by the YOLOv5s model. The generalizability results for the 2 techniques is shown in Table 1

From Table 1, we can see that for the 'unseen' images, YOLOv5 did not manage to detect 46.05% of the 152 objects while CIS only missed out 7.24% of the 152 objects. YOLOv5 performed slightly better for the 'seen' images by missing out 6.45% of the 155 objects as compared to CIS missing out 9.03% of the 155 objects. Thus, we surmise that CIS performs better in terms of generalizability and will be the preferred technique used for computation of utilization of conveyor belt-based cutlery washing machine.

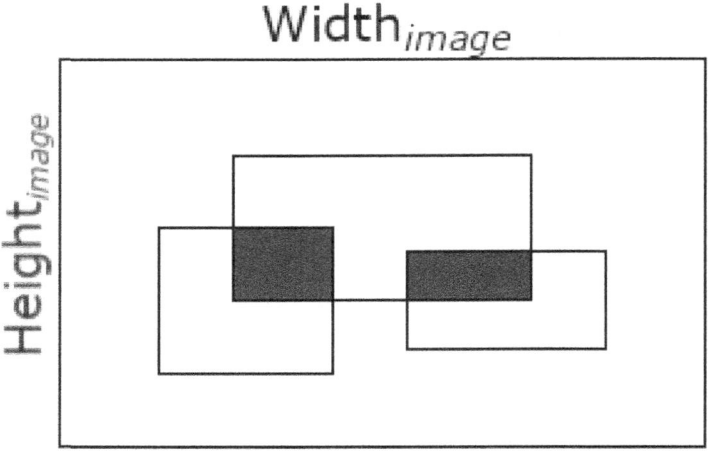

**Fig. 3.** Computation of utilization

## 4.2   Utilization and Speed Computation

The utilization of the conveyor belt-based cutlery washing machine which can be derived from the areas of the boundaries for each detected object using the CIS technique is given by.

$$Utilization = \frac{\sum Areas_{boundaries} - \sum Areas_{overlapped}}{Width_{image}.Height_{image}} \tag{1}$$

The overlapped areas are the gray shaded areas while the boundary areas are the areas of the whole rectangles (both shaded and unshaded) as depicted in Fig. 3.

To detect the running speed of the conveyor belt-based cutlery washing machine, we employ the centroid tracker for tracking the objects.

The centroid tracker algorithm is detailed in Algorithm 1.

---

**Algorithm 1.** Centroid tracker algorithm

---

**Input**: $currImageFrame$
**Output**: $storedObjects$

 1: Detect objects for $currImageFrame$ as $currObjects$
 2: Compute centroids of $currObjects$ as $currCentroids$
 3: **while** $inputCentroid$ in $currCentroids$ **do**
 4:    **if** $trackedObjects$ is empty **then**
 5:        Add 1 to $objectId$
 6:        Insert $objectId$ and $inputCentroid$ as $trackedObj$ into $trackedObjects$
 7:    **else**
 8:        Compute euclidean distance between $inputCentroid$ and centroid of $trackedObjects$ as $distDiff$
 9:
10:        **if** $\min(distDiff)$ <threshold **then**
11:            Update centroid coordinates of object corresponding to $\min(distDiff)$ in $trackedObjects$
12:        **else**
13:            Assign new object id to object
14:            Add new object with centroid coordinate to $trackedObjects$
15:        **end if**
16:    **end if**
17: **end while**

---

We employ the centroid tracking algorithm for object tracking as it runs fast and works well in our context. The accuracy of object tracking is however in turn dependent on accurate detection of the objects.

With the identification and tracking of the objects, we proceed to compute the speed for every 30 frames. The distances moved by objects is calculated as the Euclidean distance between the coordinates of the tracked objects for frame 1 and frame 30 for every 30 frames of images. The speed is then computed as median of the computed distances over the 30 frames. As the webcam is set to record at 30 frames per second, the unit of measurement for computed speed will be the number of pixels moved per second. The algorithm for speed computation is detailed in Algorithm 2.

---

**Algorithm 2.** Speed computation algorithm

---

**Input**: *prevStoredObjects*, *currStoredObjects*
**Output**: *speed*
 1: **while** *currObject* in *currStoredObjects* **do**
 2:     **while** *prevObject* in *prevStoredObjects* **do**
 3:         Compute euclidean distance between *prevObject* and *currObject* as *dist*
 4:         **if** *dist* <*minDist* **then**
 5:             *minDist = dist*
 6:         **end if**
 7:     **end while**
 8:     Insert *minDist* into *distances*
 9: **end while**
10: Compute median(*distances*) as *speed*
11: **return** *speed*

---

As one of the primary objective of our client is to monitor instances of careless usage (intermittent starting and stopping of the machine), we further evaluated that our proposed speed computation algorithm is able to detect all instances where the speed is zero (where the conveyor belt slowed from a running state to a complete stop) and where the conveyor belt starts moving from a stopped state albeit with the constraint that there is at least a cutlery that is 'seen' by the camera. When combined with the utilization measure, the proposed speed algorithm is also able to detect instances of low or waste usage where speed is non-zero but utilization is low. This would be a situation where the cutlery washing machine is running but only few cutleries are placed on the conveyor belt.

### 4.3 Constraints

One shortcoming of the CIS detection technique is that for objects that are placed closed to one another, it tends to detect the multiple objects as a single joint object. An example of this is shown in Fig. 4. The lower row of 3 baskets in Fig. 4 is detected as a single object which spans the entire width of the image as opposed to YOLOv5 which managed to detect the individual baskets. This will affect the speed computation as across the image frames, the coordinates of the centroid of the tracked object will be the same and the computed speed will thus be zero. Thus, for speed computation, we switched to the YOLOv5 detection model for situations when CIS detects an object which spans the entire width of the image (within a tolerance of 5 pixels). However, we recognize that there will

(a) Detection of objects as one by CIS     (b) Detection of same objects by YOLOv5

**Fig. 4.** Multiple objects detected as one by CIS as compared to YOLOv5

still be situations when YOLOv5 cannot detect the objects and CIS detects the objects as spanning the entire width of the image. In such situations, the speed computed by our proposed algorithm will unfortunately still be zero.

Another constraint of the CIS detection technique is for objects that are similar in color to the conveyor belt or objects that are translucent in color, either part or all of the object may not be detected. This will result in inaccurate calculation of utilization.

The final constraint is with the use of computer vision techniques for inferring the speed of the conveyor belt-based cutlery washing machine. When no objects are placed on a moving conveyor belt, both the utilization and speed will be zero as computed using the proposed techniques. The reason for this is that no objects will be detected thus resulting in zero computed distance or speed between frames. A possible solution that we proposed to our client to tackle this issue is to install an energy socket power monitoring sensor to monitor the energy usage of the power socket that the cutlery washing machine is plugged into. For situations when the cutlery washing machine is operating but no objects are placed on the conveyor belt, the energy consumption sensor will detect that the machine is running and yet no objects are on it.

## 5   Conclusion

We have retrofitted the legacy conveyor belt-based cutlery washing machine using commodity web-camera and computer vision techniques to equip it with additional capabilities of tracking both utilization and speed of the machine. We proposed object detection using both traditional image processing and deep-learning techniques for the computation of utilization and speed. With the existing operational constraints of our client, frequent re-training of the deep learning model for object detection is not feasible. Thus, we also evaluated the generalizability of both object detection techniques and found colour image segmentation to be more generalizable and thus adequate for use in utilization computation. For speed computation however, a hybrid of both colour image segmentation and YOLOv5 object detection was adopted.

With the enhanced data capability, our client will be able to visualize the automatically collected real-time utilization and speed data and optimize its operations in a data driven manner. Our client can collate and compare the utilization rate of individual conveyor belt-based cutlery washing machines across different times of the day and across the different locations for optimizing the operations of the conveyor belt-based cutlery washing machines. With the capability of detecting the speed of the machines, our client is also able to monitor instances of waste (where the machine is running but underutilized) or careless usage of the cutlery washing machine (where the machine is intermittently turned on and off) and thus minimize wastage, down-time and reduce the maintenance costs of the machines. A commodity camera was used in this proof of concept application as we were constrained by client's budget. We recognize that a higher end industrial grade camera would likely achieve better performance. Another possible future extension of this work is the investigation of newer and faster deep learning models that is more generalizable across 'unseen' images.

# References

1. Adarsh, P., Rathi, P., Kumar, M.: YOLO v3-Tiny: object detection and recognition using one stage improved model. In: 2020 6th International Conference on Advanced Computing and Communication Systems (ICACCS), pp. 687–694. IEEE (2020)
2. Ahmad, T., Ma, Y., Yahya, M., Ahmad, B., Nazir, S., Haq, A.U.: Object detection through modified yolo neural network. Sci. Program. **2020**, 1–10 (2020)
3. Bengio, Y.: Deep learning of representations for unsupervised and transfer learning. In: Proceedings of ICML Workshop on Unsupervised and Transfer Learning, pp. 17–36. JMLR Workshop and Conference Proceedings (2012)
4. Busin, L., Vandenbroucke, N., Macaire, L.: Color spaces and image segmentation. Adv. Imaging Electron Phys. **151**(1), 1 (2008)
5. Diwan, T., Anirudh, G., Tembhurne, J.V.: Object detection using yolo: challenges, architectural successors, datasets and applications. Multimedia Tools Appl. **82**(6), 1–33 (2022)
6. Hema, D., Kannan, D.S.: Interactive color image segmentation using HSV color space. Sci. Technol. J **7**(1), 37–41 (2019)
7. Hinton, G.E., Krizhevsky, A., Sutskever, I.: ImageNet classification with deep convolutional neural networks. Adv. Neural. Inf. Process. Syst. **25**(1106–1114), 1 (2012)
8. Jia, X., Tong, Y., Qiao, H., Li, M., Tong, J., Liang, B.: Fast and accurate object detector for autonomous driving based on improved YOLOv5. Sci. Rep. **13**(1), 9711 (2023)
9. Jocher, G.: ultralytics/yolov5: v6.0 - YOLOv5n 'Nano' models, Roboflow integration, TensorFlow export, OpenCV DNN support (2021). https://doi.org/10.5281/zenodo.5563715
10. Kaur, R., Singh, S.: A comprehensive review of object detection with deep learning. Digit. Signal Proc. **132**, 103812 (2023)
11. Lee, I., Lee, K.: The Internet of Things (IoT): applications, investments, and challenges for enterprises. Bus. Horiz. **58**(4), 431–440 (2015)

12. Lee, J., Bagheri, B., Kao, H.A.: A cyber-physical systems architecture for industry 4.0-based manufacturing systems. Manufact. Lett. **3**, 18–23 (2015)
13. Li, J., Su, Z., Geng, J., Yin, Y.: Real-time detection of steel strip surface defects based on improved yolo detection network. IFAC-PapersOnLine **51**(21), 76–81 (2018)
14. Li, Y., Niu, Y., Liu, Y., Zheng, L., Wang, Z., Zhe, W.: Computer vision based conveyor belt congestion recognition in logistics industrial parks. In: 2021 26th IEEE International Conference on Emerging Technologies and Factory Automation (ETFA), pp. 1–8. IEEE (2021)
15. Lin, T.-Y., et al.: Microsoft COCO: common objects in context. In: Fleet, D., Pajdla, T., Schiele, B., Tuytelaars, T. (eds.) ECCV 2014. LNCS, vol. 8693, pp. 740–755. Springer, Cham (2014). https://doi.org/10.1007/978-3-319-10602-1_48
16. Liu, Z., Qu, B.: Machine vision based online detection of PCB defect. Microprocess. Microsyst. **82**, 103807 (2021)
17. Molano, J.I.R., Lovelle, J.M.C., Montenegro, C.E., Granados, J.J.R., Crespo, R.G.: Metamodel for integration of Internet of Things, social networks, the cloud and industry 4.0. J. Ambient Intell. Humanized Comput. **9**, 709–723 (2018)
18. Pham, D.L., Chang, T.W., et al.: A yolo-based real-time packaging defect detection system. Procedia Comput. Sci. **217**, 886–894 (2023)
19. Redmon, J., Divvala, S., Girshick, R., Farhadi, A.: You only look once: unified, real-time object detection. In: Proceedings of the IEEE Conference on Computer Vision and Pattern Recognition, pp. 779–788 (2016)
20. Ren, S., He, K., Girshick, R., Sun, J.: Faster R-CNN: towards real-time object detection with region proposal networks. In: Advances in Neural Information Processing Systems, vol. 28 (2015)
21. Sandler, M., Howard, A., Zhu, M., Zhmoginov, A., Chen, L.C.: MobileNetV2: inverted residuals and linear bottlenecks. In: Proceedings of the IEEE Conference on Computer Vision and Pattern Recognition, pp. 4510–4520 (2018)
22. Szegedy, C., Vanhoucke, V., Ioffe, S., Shlens, J., Wojna, Z.: Rethinking the inception architecture for computer vision. In: Proceedings of the IEEE Conference on Computer Vision and Pattern Recognition, pp. 2818–2826 (2016)
23. Tran, T.A., Ruppert, T., Eigner, G., Abonyi, J.: Retrofitting-based development of brownfield industry 4.0 and industry 5.0 solutions. IEEE Access **10**, 64348–64374 (2022)
24. Zaidi, S.S.A., Ansari, M.S., Aslam, A., Kanwal, N., Asghar, M., Lee, B.: A survey of modern deep learning based object detection models. Digit. Signal Proc. **126**, 103514 (2022)

# On Plagiarism and Software Plagiarism

Rares Folea[1,2(✉)] and Emil Slusanschi[1]

[1] Department of Computer Science and Engineering, Faculty for Automatic Control and Computers, National University of Science and Technology Politehnica Bucharest, Bucharest, Romania
`emil.slusanschi@cs.pub.ro`
[2] Doctoral School of Engineering and Applications of Lasers and Accelerators (S.D.I.A.L.A.), Bucharest, Romania
`rares.folea@stud.acs.upb.ro`

**Abstract.** This paper explores the complexities of automatic detection of software similarities, in relation to the unique challenges of digital artifacts, and introduces Project Martial, an open-source software solution for detecting code similarity. This research enumerates some of the existing approaches to counter software plagiarism by examining both the academia and legal landscape, including notable lawsuits and court rulings that have shaped the understanding of software copyright infringements in commercial applications. Furthermore, we categorize the classes of detection challenges based on the available artifacts, and we provide a survey of the previously studied techniques in the literature, including solutions based on fingerprinting, software birthmarks, or code embeddings, and exemplify how a subset of them can be applied in the context of Project Martial.

**Keywords:** code similarity · software plagiarism · software fingerprints · software birthmarks · code embeddings · plagiarism detection

## 1 Introduction

In its most general definition, *plagiarism*[1] is the act of presenting another person's work as original, without proper attribution. While there were many cases of *plagiarism* in history, the first documented case of using the word *plagiarius* [31] was during the life of the roman poet Martial, when exact copies of his poems and epigrams started to appear presented as personal work authored by some other, obscure writers. He is considered to be the first person to claim authorship rights, in an ancient world where intellectual property and copyright laws not only were not enforced, but not even considered.

**Software plagiarism** is just one of the many branches of plagiarism, but due to the nature of software development, which is an almost entirely digital,

---

[1] The word originates from the Latin *"plagiarius"*, meaning torturer, oppressor or kidnapper.

N.-T. Nguyen et al. (Eds.): ICCCI 2024, CCIS 2166, pp. 314–326, 2024.
https://doi.org/10.1007/978-3-031-70259-4_24

fast-paced environment, where duplication activities can be performed really fast, the number of acts related to software plagiarism is raising swiftly. The main motivation behind plagiarizing software is to avoid the effort required to develop a novel, original solution from scratch.

Software plagiarism is a widespread issue affecting both academic world - where students might seek better grades - and commercial space, where intellectual property lawsuits can raise claims for up to several billions of dollars. Despite the vastly different stakes, the core act of plagiarism remains the same: illegitimately claiming authorship of another person's work.

Famous cases of plagiarism attract significant public attention, and software plagiarism is no exception. Remarkably, these debates and allegations might extend beyond the field of software engineering. Some high-profile allegations have been dramatized in films like *"Browser Wars"* and *"The Billion Dollar Code"*. In contrast, cases of alleged plagiarism in introductory computer science courses may topics of discussion among students, sparking conversations about ethics and decisions of the university committees.

In **Sect.** 2 of this paper, we explore the unique aspects of software plagiarism compared to other fields. **Section** 3 examines plagiarism's impact within academia, while **Sect.** 4 addresses legal implications associated with plagiarism. **Section** 5 aims to provide a categorization of possible classes of solutions that may arise to detect plagiarism, based on the availability of source code or binary code, and in **Sect.** 6, we present certain techniques that have been used for automatic detection. Finally, in **Sect.** 7 we discuss promising areas for future research and we introduce Project Martial, an open-source software solution that helps detecting code similarities.

## 2 Comparison with Other Kinds of Plagiarism

How is detecting software plagiarism different than other kinds of plagiarism is a good, yet difficult to answer question. It is hard to believe that a trivial set of rules can cover all the possible situation, because of the nature of software development dependencies. As in the examples mentioned later, where a ruling of the US Supreme Court acknowledged that cases where duplicating the declarative part of the code does without copying the implementation part is not to be considered plagiarism. As a direct consequence, the ruling stated that APIs are not subject to copyrights infringements. This gives a unique characteristic on how plagiarism should be seen in the software industry as opposed to other industries, such as music or literature, where there is no clear correspondence between the declarative and implementing part of the work.

In general, *there is no consensus* [10] *around the difficulty of the problem*, with many multiple arguments pro- and against- the ease of detecting plagiarism in software as opposed to identifying it in other work.

On one hand, there are arguments supporting that similarity between two sources written in natural language text rather than software is much easier to be reasoned for humans. That is because, in the case of human language, references and quotes in the corpus may lead to texts of 10% similarity that are likely not be considered plagiarised, while for software development, analogous working methods may be hard to be labeled, because the distinction between *"same theoretical knowledge"* and *"plain copy"* is more difficult to be established, especially in situations where there are a very few ways of getting things right [29]. A very famous duplication of code without raising the question of plagiarism was presented by Ken Thompson during 1984's Turing Award Lecture [33], when citing his collaboration work with Dennis Ritchie, he mentioned that towards their long collaboration, only one occasion of miscoordination of work has happened. Thompson outlines that he discovered that both have *"written the same 20-line assembly language program"*. He concludes: *"I compared the sources and was astounded to find that they matched character-for-character"*.

On the other hand, arguments that that software plagiarism may be easier identifiable are centered around the language complexity between the two, with [20] reasoning that a text in a human language contains an effectively unlimited number of possible words that can be used, with the both intra- or extra-corporal plagiarism methods, while code plagiarism is constrained to a well defined set of keywords, provided by the minimal vocabulary of a given programming language.

To add some complexity to this problem, one additional constraint arise when there is a scenario with no access to the source code of the suspicious programs, and the only available resource is the executable binary. Because the assembly instructions are far more obfuscated for a human-mind structured analysis, trying to evaluate the originality of an executable is a substantially more challenging problem than the study of source code [42].

Nevertheless, not providing the source code is not the only possible way to obfuscate the analysis. This can also be achieved, by applying iterative changes to the original code, that do not modify the behaviour of the program, but makes it look different. There are numerous obfuscation techniques that have been developed, that can be mainly classified into both control- and data-oriented; starting from naive techniques such as symbols renaming or boolean splitters (refer to the example in Listing), and advancing to more involved changes, such as adding noise instructions, reordering independent instructions, loops unrolling (Listing A1), changes in the control flow conditional branches, branch inversion (Listing A2) or aggregating multiple instruction (Listing A3). The examples, provided in the syntax of the Go programming language, are just some simple academic samples that can alter[2] the original code in a way that the resulting code look substantially different.

To add, studies [24,41] have investigated solutions for adding a set of obfuscations who are mainly aimed at protecting the intellectual property, encapsulated

---

[2] It is worth mentioning that some modifications (such as loop-unrolling or vectorisation) have multiple functionality. These techniques can be used not only as methods for obfuscation, but rather they can serve as optimization techniques, in order to make the code more efficient, rather than trying to make it look different. However, in the scope of the paper only modifications with the goal of producing code obfuscations are being studied.

by the software, by making it extremely hard for a person that only has access to the obfuscated outcome to decode the original logic. Such perplexity techniques can be applied both at source code level, as well as at binary level. Just as the presented techniques presented above, because the outcome of these procedures is to alter the outlook of the software without altering the behaviour, these can also can be used to abuse software plagiarism detection tools into miss detecting similarity [35].

Special considerations about software plagiarism[3] have to be raised in the light of the recent evolution of LLM-based solutions, that proved to be efficient at manipulating large code snippets[4]. Large pretrained language models have already been used and were proved to be efficient in altering the original source code, such that it has remained undetected [2].

This is detailed in [19] which concludes that LLMs *"have a great potential to generate sophisticated text outputs without being well caught by the plagiarism check software"*. Finally, preliminary results using this new technology seems to suggest that pre-LLM tools have only *"scratched the surface of the possibilities compared to what large neural language models can achieve in producing convincing high-quality paraphrases"* [39] and acknowledge that even humans ability to detect machine-based obfuscations *"appear to decrease with increasing model size as they can change sentence structure and word order instead of single word replacements"*.

## 3   Academia

In the academic world, plagiarism, seen as *"an evasion of learning"* [40], because it *"enables students to gain credit for significant portions of assessment without having developed any capacity for understanding of critical evaluation of the material presented for assessment."*, is not a seldom event. As the number of computer science students is still increasing each year, new challenges arise. Data USA states that the total number of degrees awarded in 2020 exceeded fifty-thousand graduates [37], with a 10% growth rate year-to-year, and The British Computer Society noted that [1] the applications to study computer science in United Kingdom for 2022 increased from 140 thousands the previous year to almost 160 thousands this year, with a trend that shows that interest in the computer science related subject kept growing.

Given the scale, there is definitely a need for improving the automatic tools that allow for the identification of possible cases of software plagiarism. As [10]

---

[3] Large language models capabilities perform well in more assignments, such as free form natural language essays, not just on altering software programs, which introduce an additional challenge in other areas of plagiarism too.

[4] Fortunately, there is also a bright side when it comes to new LLM-based tools. There is evidence that these tools help [12] improving the productivity by assisting developers (and not only). OpenAI's GPT-4 [28], Google's Gemini [11] and other tools can nowadays generate hundred of lines of code from a simple prompt.

notes, *"there are too many students and too few staff members [...]. Automating plagiarism detection would help very much"*.

To begin with, various works [16] acknowledges the difficulties for identifying plagiarism in natural language for essays and reports with existing tools, yet software plagiarism has it's own specificity. A survey on the topic of the academic issue associated with software plagiarism [10] presents that only 8% of the staff university did use a automatic software-based approach to try and detect plagiarism, while the vast majority of 80% of the staff still relies on instinct and/or personal experience to tackle this problem. It has been pointed out that the complexity in understanding the nature of plagiarism could result in having solutions *"to reducing plagiarism that may rest more on prevention rather than detection"* [40].

The most used system for automatic software plagiarism detection is Moss[5]. The maintainers present it as *"a way of highlighting the components of programs that are worth a more thorough inspection, that saves instructors and teaching staff a lot of time"* [32]. It's emphasised that *"after specific sections of the programs have been examined, it shouldn't matter whether the questionable code was initially uncovered by Moss or by a person"* and that *"the argument for plagiarism should stand on its own"*. Multiple researchers subscribed to this ethic idea and as [39] points out, while automatic plagiarism detection can *"point out potential plagiarism cases, a team of experts should make a final decision on such cases"*, because, the same work in [39] warns that *"False-positive cases of wrongly accused researchers could ruin their careers forever* and therefore, *"all cases should be carefully evaluated before any final verdict"*.

It is also worth pointing to research such as Mossad [13], that showed that automatic solutions to generate mutants of software versions with the sole purpose of defeating plagiarism detectors are possible.

The core difference between academia and industry is the amount of involvement that the actors are motivated to invest, in the process of plagiarising. While in the commercial context there may be a lot more resources and incentives available, if it is necessary to defend against copyright violation, in an educational context, *"the effort expended by students to hide their plagiarism is likely to be much less"* [18].

## 4   Legal Aspects and Lawsuits

The fast development of the Tech industry led to the development of a relatively new area in Legal, specialised in technology-based litigation. The majority of lawsuits are centered towards determining the eligibility of a given trademark or patent. Well known trademark cases include both lawsuit with hardware-established claims, such as Apple has demanded ownership claim of its iPad

---

[5] Moss stands for Measure Of Software Similarity and is an automatic system for determining code similarity. https://theory.stanford.edu/%7Eaiken/moss/.

trademark in China, but software claims are no exception and, in the corporate world there are many such well known cases and accusations of software plagiarism.

Software copyright- and plagiarism-allegations are seldom straightforward to decide, and most of them requires a trial to reach a verdict and in some special cases, even involvement of the US Supreme Court.

*Oracle America Inc v. Hewlett Packard Enterprise Co* case has been settled between the parts after a *"disputed unauthorized software updates for the Solaris operating system owned by Oracle"*, with alleging accusations that Hewlett Packard Enterprise *"used its copyrighted software without a license and undercut Oracle's support pricing"* [17], with the goal of reducing the cost of new software development. The telecommunication company *Verizon*, according to the lawsuit against it, has allegedly utilized *BusyBox* programming in wireless routers, which were distributed to consumers, but failed to give customers access to the BusyBox source code as required by the GPL [38] licence. This case was also settled, subject to license adherence.

In *Google LLC v. Oracle America Inc.*, the US Supreme Court considered multiple factors, pointing out that Google's use of the Java APIs was transformative and the amount of interfaces copied was limited only to the bare minimum which was necessary to allow developers to work in the new mobile development framework proposed by Android, with the already accumulated knowledge, from the Java programming language. In the preparation of the case, Google argued that *"Open interfaces between programs are the building blocks of many of the services and products we use today, as well as of technologies we haven't yet imagined. An Oracle win would upend the way the technology industry has always approached the important issue of software interfaces. It would, for the first time, grant copyright owners a monopoly power to stymie the creation of new implementations and applications. And it would make it harder and costlier for developers and startups to create more products for people to use"* [3]. On the opposite side, Oracle considered that this trial started *for the simple proposition that stealing-no matter the convenience it may offer to the thief-is not acceptable* [4] and *"attempts to claim that its theft and clear commercial use of an existing technology already in the market is somehow covered by the fair use doctrine"*, with such an argument *"would virtually eliminate copyright altogether because nearly all copying would be <<fair>>"* [4].

The jury considered that the portion of copied code was *"inherently bound together with uncopyrightable ideas"* with only 0.4% of the entire API being copied by Google, an argument in favor of *fair use*. Nevertheless, the Android platform was proved by time not to be a substitute or a rival for Java SE. Within the jury there was also a dissenting opinion on this trial, arguing that the Court should have addressed the question whether Oracle's code is copyrightable and if by copying the code, Google gained multiple advantages and erased a large value of Oracle's partnerships.

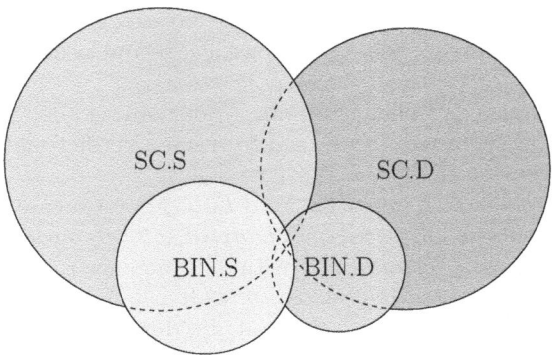

**Fig. 1.** The overlap of techniques, based on the class of problem.

## 5   Classes

As mentioned in the previous sections, the problem of identifying software pla-giarism can require substantially different approaches based on the available resources. We identify two major sets of problems based on that: the class of detection with access to the source code (**SC**) and the class of detection with access to the binary code only (**BIN**). Note that while access to **SC** also implies access to the **BIN** resources, by simply taking the source code and compiling (for languages whose native implementation allows compiling into the machine language) or packing (for interpreted languages), there may be special cases in which the tool used to translate the code to machines is either unavailable, undisclosed or proprietary. This raises a very niche class of problems in which the exact binary cannot be obtained from the source code.

Naturally, the semantics of the code can be much easier analysed when access to the source code is available, as the description of the processes that the pro-gram needs to follow is more expressive in a high-level programming language than in binary machine-specific instructions [6]. A visual representation of the overlap of techniques that can be used to tackle each challenge is captured in Fig. 1.

Therefore, the **SC** class of problems may result in better software plagiarism detectors, due to the additional access to the source code, which can be tackled in more advanced ways. For example, the approaches can perform analysis on the static representation of the code **SC.S**, or dynamically **SC.D**, by monitoring the execution of the program.

Similarly, we can define **BIN.S**, as the class of approaches when the machine code is evaluated directly, using a static approach. Symmetrically, **BIN.D** can be defined, for classes where the machine code is evaluated directly, dynamically, but it almost perfectly overlaps[6]. with **SC.D**.

The problem of automatic software plagiarism detection is far from reaching a generic solution, but among the years of research in literature, some progress towards identifying an automated way of potential cases of plagiarism was made. The traditional approaches are to only proclaim plagiarism when similarities are found[7].

## 6    Techniques in Automatic Plagiarism Detection

The **fingerprint-based** approaches can provide automatic findings of program similarities, but in most of the cases, these methods also lack convincing explainability capabilities in order to determine why certain programs are similar. One well establish algorithm is winnowing [30], a solution for fingerprinting that divides the file into k-grams and has been the core idea for the MOSS software, but there are many other successful usages of fingerprint ideas in other work too, both on **SC** [6,9,27], as well as on **BIN** [6]. In the good faith principle, *it is still up to a person to review the portions of the code that the engine has flagged and make a judgement whether that section is indeed plagiarism* [32,39] or not.

**Table 1.** The main techniques used in plagiarism detection, and known classes of problems where they have been applied to detect plagiarism.

| Techniques | SC.S | SC.D | BIN.S | BIN.D |
|---|---|---|---|---|
| Code analysis (syntactic, semantic) | ✓ | | | |
| API-based | ✓ | | ✓ | |
| Profiling | | ✓ | | ✓ |
| Fingerprint | ✓ | ✓ | ✓ | ✓ |
| Software birthmarks | ✓ | ✓ | ✓ | ✓ |
| Code Embeddings | ✓ | ✓ | ✓ | ✓ |
| LLM-based | ✓ | | | |

---

[6] The only distinction that we have found was, when in the transformation from the source code to the binary, debug traces and symbols are being enabled, which could be missing otherwise, if the binary is the only resource available.

[7] While most of the solutions focus on *identifying similarities* and only declare plagiarism when a similarity is found, a different approach is taken in the work of [42], that starts from the idea that *"as long as we can find one dissimilarity, programs are semantically different"*, yet emphasises that *"if we cannot find any, it is likely a plagiarism case"*. This is a much more ambitious approach, but comes with the drawbacks that it lacks explainability which makes the human analysis proposed in the methodology of [39] significantly more difficult.

**Software birthmark**[8] approaches are meant to overcome some downsides of software fingerprinting. Like the fingerprint-based algorithms, birthmarks have the advantage that they can be applied on both **SC**, source code [6,9,25,26,35] as well as on **BIN**, where only the binary code is provided [6,23,35,35,36]. Birthmark techniques can be enhanced with additional mechanisms in complex programs, such as supervising the construction [34] via Siamese networks to capture perturbations caused by the non-deterministic scheduling of machine instruction in the context of multi-threading applications.

**Code embeddings** are yet another powerful mechanism that can be used in similarity detection in which the program is analysed via its numerical vector representations. They aim to capture the semantic meaning and syntactic structure of code snippets in a condensed format and have potential in placing the code snippets with similar functionality close to each other in the multi-dimensional analysed space. While most approaches aims to tackle **SC** problems [8], they can also be applied for **BIN** [14] classes.

For studying **SC.S** problems, the studies in [7,21] are using a detection method running an **API-labeled dependency** graph analysis.

A encouraging, but yet to be explored territory is the use of Large Language Models as a technique for automatic plagiarism detection in software. Due to the proved good LLMs performance manipulating source code, this is a promising technology in engaging in **SC.S** problems.

Note that final applications may use a mix of these approaches from different solutions for analysing similarity. As long as they assume that the problem is in **SC**, all presented techniques can be applied: **SC.S**, **SC.D**, **BIN.S**, **BIN.D**. If the problem is in **BIN**, the available techniques can belong to **BIN.S** and **BIN.D**. A summary of the main existing approaches has been captured in Table 1.

# 7   Project Martial: Current and Future Work

Part of the authors' research and contributions to the field of similarity detection is focused around finding novel methods that operate on both the static code, but also perform dynamic analysis (for both **SC** and **BIN** classes). Project Martial[9] is a growing, experimental, open-source initiative, that aims to provide automatic assistance in detecting software plagiarism.

Project Martial provides a modular framework that allows easily new detection tools to be added or fine-tune the meta-parameters of the existing models to better fit. At the moment, the stack consists of three unique analysers, with more analysers expected by the end of 2024 and early 2025.

---

[8] Birthmarks are generally computed as a set of fairly invariable features that are extracted from the analysed algorithm, with the overall goal of uniquely identifying the program. Once birthmarks are extracted, the similarity between two programs is simply computed by a function that takes the two resulting birthmarks and compares them.

[9] https://github.com/raresraf/project-martial.

The main analyser fits into the **SC** category and uses natural language processing techniques for computing similarities during the static analysis on the human readable parts of of the comments [15]. The methodology is based on using transformers (RoBERTa [22] and Universal Sentence Encoder [5]) to embed the content of comments, followed by performing detection of potential similarities, based on computing the cosine similarity. As future work in this analyzer, we want to eventually expand these techniques work on code with no human annotations, such as comments or other directives. In this space, novel code analysing techniques based on large language models have proved a good potential to help into expanding machine-native source codes with annotations in human language, essentially acting as a pre-processing phase of the source code before applying the existing model.

The second analyser is based solely on inspection of machine-readable comments(e.g. linter directives), that computes distances in multi-dimensional spaces to compute the estimated code similarity, based on one-hot encoded representations of the source code. [15]. This technique also fits into the **SC** category.

The third approach in project Martial uses dynamic-code complexities [14] for detecting software plagiarism and thus is a contributions to the techniques in **BIN** category. We plan on extending the existing birthmarks (currently based on a limit subset of metrics, such as CPU cycles, branch prediction or cache misses statistics) to capture more and more performance metrics as well as working on optimizing the similarity detection function.

Detecting similarity comparisons based on networking telemetry is an additional area of work. The initial approach focuses on finding commonalities between relational databases and evaluates whether internal database information can be found and compared by examining fingerprints that are taken from client-side network packets. Through the process of recording and examining the distinct features that are contained in these packets-like protocol headers, SQL query patterns, and signatures-it would be feasible to identify linkages between databases relying solely on payload content and the manner of communication.

## Appendix A: Code listings for obfuscation techniques

```
s := []int{1, 2, 3, 4, 5, 6}        s := []int{1, 2, 3, 4, 5, 6}
- for i := 0; i < 6; i += 1 {       + for i := 0; i < 6; i += 3 {
    sum += s[i]                            sum += s[i]
- }                                 +       sum += s[i+1]
                                    +       sum += s[i+2]
                                    + }
```

**Code Listing A1.** Example of a loop unrolling obfuscation.

```
  s, err := f()                       s, err := f()
- if err != nil {                   + if err == nil {
-     return nil, err               +     // main logic
- }                                 + }
- // main logic                     + return nil, err
```

**Code Listing A2.** Example of branch inversion obfuscation.

```
- if '0' <= c && c <= '9' {         + func unhex(c byte) byte {
-     c = byte(c - 'A' + 10)        +     if '0' <= c
- } elif case 'A' <= c && c <= 'F'  +     && c <= '9' {
-     c = byte(c - '0')             +         return c - '0'
- }                                 +     if 'A' <= c && c <= 'F' {
- u := byte('C')                    +         return c - 'A' + 10
                                    +     }
                                    +     return 0
                                    + }
                                    + u := unhex(byte('C'))
```

**Code Listing A3.** Example of obfuscation of the original code by aggregating multiple instructions into a dedicated method

# References

1. BCS, T.C.I.f.I.: Record numbers have applied for UK computer science degrees this year. https://www.bcs.org/articles-opinion-and-research/record-numbers-have-applied-for-uk-computer-science-degrees-this-year/ (2022). Accessed 29 Jan 2023
2. Biderman, S., Raff, E.: Fooling moss detection with pretrained language models. In: Proceedings of the 31st ACM International Conference on Information & Knowledge Management, pp. 2933–2943 (2022)
3. Blog, G.: The case for open innovation. https://blog.google/outreach-initiatives/public-policy/case-for-open-innovation/ (2020). Accessed 29 Jan 2023
4. Blog, O.: Copyright Caveat Emptor. https://www.oracle.com/corporate/blog/oracle-vs-google-021220.html (2020). Accessed 29 Jan 2023
5. Cer, D., et al.: Universal sentence encoder. arXiv preprint arXiv:1803.11175 (2018)
6. Cesare, S., Xiang, Y.: Software similarity and classification. Springer Science & Business Media (2012). https://doi.org/10.1007/978-1-4471-2909-7
7. Chae, D.K., Ha, J., Kim, S.W., Kang, B., Im, E.G.: Software plagiarism detection: a graph-based approach. In: Proceedings of the 22nd ACM international conference on Information & Knowledge Management, pp. 1577–1580 (2013)
8. Chen, Z., Monperrus, M.: A literature study of embeddings on source code. arXiv preprint arXiv:1904.03061 (2019)
9. Chilowicz, M., Duris, E., Roussel, G.: Syntax tree fingerprinting for source code similarity detection. In: 2009 IEEE 17th international conference on program comprehension, pp. 243–247. IEEE (2009)
10. Chuda, D., Navrat, P., Kovacova, B., Humay, P.: The issue of (software) plagiarism: a student view. IEEE Trans. Educ. **55**(1), 22–28 (2011)

11. DeepMind, G.: Gemini - google deepmind. https://deepmind.google/technologies/gemini. Accessed 15 Dec 2023
12. Dell'Acqua, F., et al.: Navigating the jagged technological frontier: field experimental evidence of the effects of AI on knowledge worker productivity and quality. Harvard Bus. School Technol. Oper. Mgt. Unit Working Paper (24-013) (2023)
13. Devore-McDonald, B., Berger, E.D.: Mossad: Defeating software plagiarism detection. Proc. ACM Program. Lang. 4(OOPSLA) (2020). https://doi.org/10.1145/3428206,
14. Folea, R., Iacob, R., Slusanschi, E., Rebedea, T.: Complexity-based code embeddings. In: International Conference on Computational Collective Intelligence, pp. 256–269. Springer (2023). https://doi.org/10.1007/978-3-031-41456-5_20
15. Folea, R., Slusanschi, E.: Code comments: a way of identifying similarities in the source code. Mathematics 12(7) (2024). https://doi.org/10.3390/math12071073
16. Foltýnek, T., et al.: Testing of support tools for plagiarism detection. Int. J. Educ. Technol. High. Educ. 17, 1–31 (2020)
17. HP Enterprise settles Oracle copyright lawsuit after $30 mln verdict. https://www.reuters.com/legal/litigation/hp-enterprise-settles-oracle-copyright-lawsuit-after-30-mln-verdict-2023-01-18/ (2023). Accessed 29 Jan 2023
18. Joy, M., Luck, M.: Plagiarism in programming assignments. IEEE Trans. Educ. 42(2), 129–133 (1999)
19. Khalil, M., Er, E.: Will chatgpt get you caught? rethinking of plagiarism detection. arXiv preprint arXiv:2302.04335 (2023)
20. Lancaster, T., Culwin, F.: Classifications of plagiarism detection engines. Innov. Teach. Learn. Inf. Comput. Sci. 4(2), 1–16 (2005)
21. Liu, C., Chen, C., Han, J., Yu, P.S.: Gplag: detection of software plagiarism by program dependence graph analysis. In: Proceedings of the 12th ACM SIGKDD international conference on Knowledge discovery and data mining, pp. 872–881 (2006)
22. Liu, Y., et al.: Roberta: a robustly optimized BERT pretraining approach. arXiv preprint arXiv:1907.11692 (2019)
23. Lu, B., Liu, F., Ge, X., Liu, B., Luo, X.: A software birthmark based on dynamic opcode n-gram. In: International Conference on Semantic Computing (ICSC 2007), pp. 37–44. IEEE (2007)
24. Madou, M., Anckaert, B., De Bus, B., De Bosschere, K., Cappaert, J., Preneel, B.: On the effectiveness of source code transformations for binary obfuscation. In: Proceedings of the International Conference on Software Engineering Research and Practice (SERP06), pp. 527–533. CSREA Press (2006)
25. Myles, G., Collberg, C.: Detecting software theft via whole program path birthmarks. In: Information Security: 7th International Conference, ISC 2004, Palo Alto, CA, USA, September 27-29, 2004. Proceedings 7, pp. 404–415. Springer (2004)
26. Myles, G., Collberg, C.: K-gram based software birthmarks. In: Proceedings of the 2005 ACM symposium on Applied computing, pp. 314–318 (2005)
27. Narayanan, S., Simi, S.: Source code plagiarism detection and performance analysis using fingerprint based distance measure method. In: 2012 7th International Conference on Computer Science & Education (ICCSE), pp. 1065–1068. IEEE (2012)
28. OpenAI: Gpt-4 technical report (2023)
29. Rosales, F., García, A., Rodríguez, S., Pedraza, J.L., Méndez, R., Nieto, M.M.: Detection of plagiarism in programming assignments. IEEE Trans. Educ. 51(2), 174–183 (2008)

30. Schleimer, S., Wilkerson, D.S., Aiken, A.: Winnowing: local algorithms for document fingerprinting. In: Proceedings of the 2003 ACM SIGMOD international conference on Management of data, pp. 76–85 (2003)

31. Seo, J.M.: Plagiarism and poetic identity in martial. Am. J. Philology, pp. 567–593 (2009)

32. Plagiarism detection. https://theory.stanford.edu/~aiken/moss/. Accessed 23 Sep 2023

33. Thompson, K.: Reflections on trusting trust. Commun. ACM **27**(8), 761–763 (1984)

34. Tian, Z., Wang, Q., Gao, C., Chen, L., Wu, D.: Plagiarism detection of multi-threaded programs via Siamese neural networks. IEEE Access **8**, 160802–160814 (2020)

35. Tian, Z., Zheng, Q., Liu, T., Fan, M., Zhuang, E., Yang, Z.: Software plagiarism detection with birthmarks based on dynamic key instruction sequences. IEEE Trans. Software Eng. **41**(12), 1217–1235 (2015). https://doi.org/10.1109/TSE.2015.2454508

36. Ullah, F., Wang, J., Farhan, M., Habib, M., Khalid, S.: Software plagiarism detection in multiprogramming languages using machine learning approach. Concurrency Comput. Pract. Experience **33**(4), e5000 (2021)

37. USA, D.: Computer Science; STEM Major. https://datausa.io/profile/cip/computer-science-110701. Accessed 29 Jan 2023

38. Open-source legal group strikes again on busybox, suing verizon — computerworld. https://www.computerworld.com/article/2537947/open-source-legal-group-strikes-again-on-busybox--suing-verizon.html. Accessed 23 Sep 2023

39. Wahle, J.P., Ruas, T., Kirstein, F., Gipp, B.: How large language models are transforming machine-paraphrased plagiarism. arXiv preprint arXiv:2210.03568 (2022)

40. Warn, J.: Plagiarism software: no magic bullet! High. Edu. Res. Dev. **25**(2), 195–208 (2006)

41. Wu, Z., Gianvecchio, S., Xie, M., Wang, H.: Mimimorphism: A new approach to binary code obfuscation. In: Proceedings of the 17th ACM conference on Computer and communications security, pp. 536–546 (2010)

42. Zhang, F., Wu, D., Liu, P., Zhu, S.: Program logic based software plagiarism detection. In: 2014 IEEE 25th international symposium on software reliability engineering, pp. 66–77. IEEE (2014)

# Automatic Detection of Ambulance Vehicles in Day and Night Conditions in Surveillance Videos

Kazimierz Choroś[(✉)] [iD]

Faculty of Information and Communication Technology, Department of Applied Informatics, Wrocław University of Science and Technology, Wyb. Wyspiańskiego 27, 50-370 Wrocław, Poland
kazimierz.choros@pwr.edu.pl

**Abstract.** Analysis of surveillance videos is one of the most important task in the computer vision area. Street traffic monitoring is useful for developing the traffic light systems at road intersections, detection of red light violations, traffic intensity study, optimization of traffic organization, and also for ensuring the secure interaction of autonomous vehicles with emergency vehicles. The detection of emergency vehicles mainly such as ambulances, fire trucks, and police cars with flashing lights and sirens, but also some others like tow trucks or snow plows is frequently based on the analysis of vehicle shape or color as well as sound of sirens. The observations and the initial tests have shown that the detection of ambulances in surveillance videos recorded during the day is successful if it is mainly based on the acoustic analysis. Whereas, for surveillance videos recorded at night or late in the evening a more efficient approach is the analysis of rotating and flashing lights installed on ambulances.

**Keywords:** Content-Based Video Analysis · Emergency Vehicles · Ambulance Vehicles · Surveillance Videos · YOLOv3 Model · Color Analysis · Day and Night Conditions

## 1 Introduction

The main goal of content-based analysis of videos from traffic monitoring is to automatically detect the specific actions recorded in the videos such as accidents, danger behaviors and situations, people appearances on the road, vehicles of a given type such as municipal buses, long vehicles, big trucks, or motorbikes, passing in a given place, etc., but also free and secure passage of emergency vehicles. Such a detection task is important for developing the traffic light systems at road intersections, detection of red light violations, traffic intensity study, optimization of traffic organization, and also for ensuring the secure interaction of autonomous vehicles with emergency vehicles.

The detection of emergency vehicles mainly such as ambulances, fire trucks, and police cars with flashing lights and sirens, but also some others like tow trucks or snow plows is frequently based on the analysis of vehicle shapes or colors as well as sounds

N.-T. Nguyen et al. (Eds.): ICCCI 2024, CCIS 2166, pp. 327–337, 2024.
https://doi.org/10.1007/978-3-031-70259-4_25

of sirens. Ambulances and other emergency vehicles use a variety of colored lights depending on national legal regulations. They are typically blue. White lights are usually used in conjunction with another color on emergency vehicles.

The goal of the paper is to examine the opportunity to detect ambulance emergency vehicles only basing on the analysis of visual data in surveillance videos recorded in day and night conditions. The criterion examined in this research is the rotating and flashing emergency blue light installed on the ambulances. This criterion has not been analyzed so far.

The paper is structured as follows. The next section describes related work on the approaches of emergency vehicle detection. The third section discusses specific features of ambulance vehicles. The next section describes the data used in the experiments and presents the results of the tests performed on real videos recorded in a city street and showing what is the efficiency of ambulance detection in day and night conditions. The final conclusions are presented in the fifth section.

## 2  Related Work

Theoretical research as well as experimental studies on surveillance videos have been conducted for many years. Based on 282 references the review is presented in [1] of a semantic understanding of road traffic flow situations and emergencies to detect anomalies and facilitate traffic prediction. The authors claim that the computer vision in traffic mainly focuses on the static detection of vehicles and pedestrians. In this paper all kinds of traffic monitoring analysis methods are classified from the perspective of macro traffic flow including traffic speed, traffic flow, and traffic density, as well as micro road behavior including vehicle behavior, interaction behavior, and temporal reasoning of motion and accident prediction.

Many different approaches, algorithms, and systems have been proposed for automatic detection of emergency vehicles. In the system of an intelligent traffic control described in [2] it was assumed that all individual vehicles are equipped with radio frequency identification (RFID) tags, moreover placed at such a strategic location, that makes it impossible to remove or destroy them. The RFID technology permits to count vehicles passing a given path, to determine a traffic congestion, and to set the green light duration for that path, and even to detect stolen vehicles. Furthermore, the RFID devices can force to turn on the green light when an ambulance is approaching the intersection of streets with traffic lights. Although, the detection system has been proposed by using RFID technology, the authors pointed out that the RFID readers are unreliable due to a very short range of communication. Moreover, this solution makes the system expensive.

An increasingly popular technology of Internet of Things (IoT) creates the new opportunities to send directly and immediately to the ambulance the information on an accident [3]. Such smart systems using IoT technology are also able to warn the drivers of vehicles about an approaching ambulance, and to demand them to make a free path and make easier the quick ambulance ride on the road [4].

In recent years Convolutional Neural Network (CNN) has become one of the very popular methods of vehicle classification and detection of emergency cars on a heavy traffic road [5, 6]. The authors of [7] proposed a system to detect emergency vehicles from CCTV footage using the deep convolutional neural network. They used several pre-trained models: 2-layer CNN, VGG-16, Inception-v3, Xception and received good results. In the experimental research described in [8] the authors applied the same CNN model, however after some modifications of the CNN layer as well as the modification of the very popular object detection and classification algorithm based on the VGG16 architecture [9]. However, the results were varied, some results were false. For example some red passenger cars were identified as fire trucks, some others white as police cars. The authors suggested that maybe it happened because color was a critical feature.

A large part of the methods of emergency vehicle detection is based on the analysis of the acoustic signals. The method proposed for example in [10] detects the ambulance siren in a traffic utilizing neural network. The authors suggested that the drivers should install on their mobile phones a special application that will use the phone's microphone and automatically detect the siren to alert the user. The recorded audio signal is first divided into windows and then the features in both time and frequency domain are extracted. The analysis of the sound in the vehicle should take into account also the contribution of other noise like horn and engine sounds.

In turn, in [11] the authors described another system in which the priority shifts to the lane in which the emergency vehicle was detected. To compute the traffic density digital image processing techniques were used, whereas to detect the appearance of an emergency vehicle acoustic signal processing techniques were applied.

Three deep neural networks models, i.e. dense layer, convolutional neural network, and recurrent neural network with different configurations and parameters have been investigated in [12]. Then, the experimental tests on various configurations with different parameters enabled the authors to design an ensemble model. Such a fully connected model provided a high accuracy of detection. It was also noticed that the advantage of siren analysis is the fact that emergency vehicles start giving warnings from long distances comparing to the analysis of videos recorded by surveillance cameras.

The other approach is a strategy based on the recognition of word 'AMBULANCE" usually written using specific characters in an emergency vehicle front hood [13]. However such a strategy is not efficient because frequently this inscription is not visible, or partially visible, or the light condition makes this recognition impossible.

The detection of emergency vehicles is critical task for autonomous cars to ensure the save automatic driving. Many solutions are proposed for the producers of autonomous cars, some of them are analyzed in a review paper [14]. Furthermore, for example in [15] a system to be incorporated in the vehicles has been proposed. An ambulance detection was achieved by the analysis of the images as well as sounds. The ambulance detection through images was ensured by the camera mounted on top of the vehicle, whereas, microphones placed at the rear end of the car were used to recognize if an ambulance siren was there in the surrounding or not. Then both results were finally combined to determine whether there was an ambulance in close surroundings.

This approach based on both analysis of audio and visual data has been very frequently applied in other studies on emergency vehicle detection [16].

## 3 Ambulance Vehicles

An ambulance is a vehicle used to transport patients to treatment facilities, such as medical clinics, emergency medical services, and hospitals. Such a vehicle is medically-equipped to ensure medical care to the patient during the transport. Ambulances respond to medical emergencies. Therefore, they are considered emergency vehicles and they are equipped with flashing emergency lights and sirens. Most often ambulances are van or pickup truck type vehicles (Fig. 1). Smaller cars like passenger cars are also present in medical services mainly for the transport of blood (Fig. 2) but they are not using flashing emergency lights or sirens.

**Fig. 1.** The most popular type of ambulance. Ambulance is without flashing lights, that means in a non-emergency situation. This photo was taken on the street on a bright sunny day.

The road traffic laws require motorists to move over and slow down when passing a stationary rescue vehicle equipped with emergency lights. Ambulance vehicles using flashing blue lights and sirens have right of way over all other vehicles. Other vehicles should move onto the shoulder of the road. This allows the ambulance to arrive at the emergency site or medical facility more quickly. On the other hand, blue lights designating emergency/priority vehicles may be used only by authorized vehicles.

It might seem obvious that the detection of ambulances can be easily realized basing on the analysis of vehicle shape, color, inscription 'AMBULANCE' on the front hood of the vehicle, or on the detection of the sound of special ambulance sirens. The word 'AMBULANCE' is written in its mirror image in front of the vehicles to allow the drivers of the vehicles in front of the ambulance to read correctly through their rear view mirror. In this way they know that an ambulance is behind them, and therefore, they should give a free path to the ambulance.

What are the main features characterized vehicles and which can be used in the process of automatic detection of emergency vehicles? They are mainly as follows:

- shape of the car typical for the given car brand and model,
- dominant color of the car body (bodywork),

**Fig. 2.** Medical car for blood transport.    **Fig. 3.** Ambulance with only wide blue stripe.

- additional colored stripes on the car body,
- inscription on the car body,
- huge word 'AMBULANCE' written laterally inverted in front of the ambulance,
- rotating and flashing emergency lights.

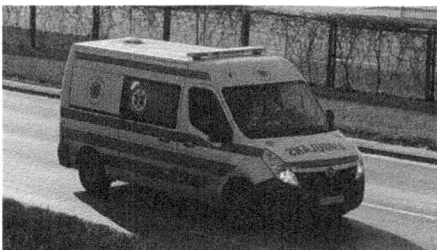

**Fig. 4.** Two examples of similar van type vehicles: ambulance and a vehicle used for other transport purposes.

The detection of ambulances based only on shapes of objects detected in a video is not resultful. There are many efficient algorithms proposed for the recognition of types (shapes) of vehicles recorded in surveillance videos on the road. Unfortunately, the most frequently ambulances are van type vehicles used also in many other types of transport, not only for medical purposes (Fig. 4).

So, detecting a van type vehicle on the road is significant but is not enough to state that the analyzed car is an ambulance.

The analysis of colors of vehicles is also ineffective. Usually the dominant color of an ambulance is white. However, their color recorded in surveillance videos strongly depends not only on their real color but also on the calibration of recording cameras, the type and intensity of street lighting (Fig. 5), weather conditions (rain, snow, fog, etc.), or even on lights of other cars driving in close proximity.

The dominant color of ambulances is white, but red color (Fig. 1) and blue color (Fig. 3) are also present on the car body. Some ambulances have only wide blue stripe on the body.

**Fig. 5.** The ambulances with flashing emergency lights and without emergency lights seem to be different colors (images form the surveillance video used in the tests).

There are two main reasons that the ambulance lights are on but the siren is off. It may be intended to improve the comfort of the patient who may feel stressed and uncomfortable in the ambulance due to noise. Or simply the way is clear and there is no need for the siren especially at night. Therefore, the detection of ambulance vehicles basing on the detection of the sound of sirens may not be successful.

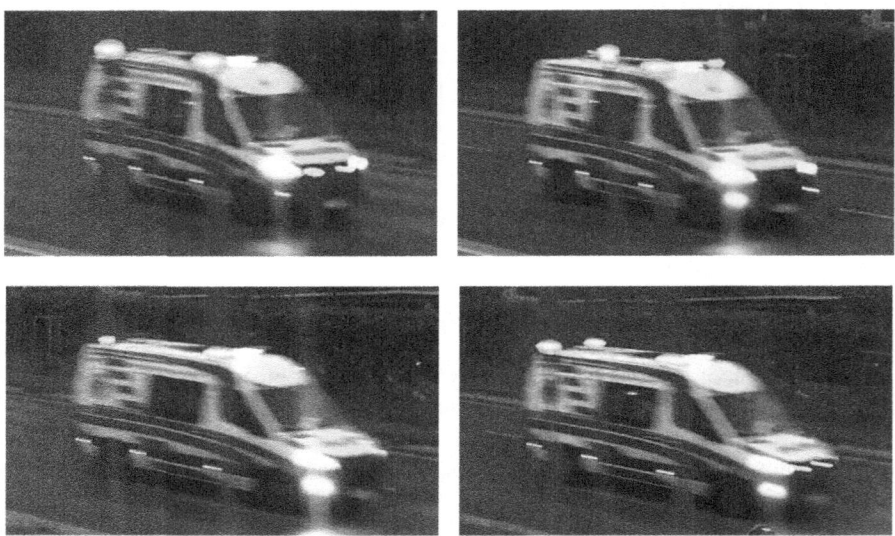

**Fig. 6.** Four frames from the video segment of the ambulance passing the street with flashing emergency lights (images form the surveillance video used in the tests).

All these observations indicate that the most typical, distinguishing, and differentiate features of ambulance vehicles are not shapes or colors of vehicles but rotating and flashing emergency blue lights. Especially since the flashing lights generate very expressive flares in surveillance videos mainly in those recorded at night. The amount of flares varies, depending on the day brightness, influence of other sources of lights, viewing angle, and other factors (Fig. 6).

# 4  Ambulance Emergency Lights Detection

Two surveillance videos were analyzed, both recorded by the same static camera with a resolution of 1920 x 1080 pixels, 30 fps (frames per second), at the same place in Wrocław (avenue with three lanes inside the city), the recording length of about 30 min, the same camera angle and view. However, the first one was recorded during the sunny day, the second one late in the evening (Table 1).

The test videos were first analyzed to detect objects/vehicle. It was achieved using the well-known YOLOv3 model (a unified model for object detection). The training and testing code of YOLOv3 is open source. Moreover, a variety of pre-trained models are also available to download [17, 18]. The YOLOv3 approach is based on three steps: resizing the input image to $448 \times 448$, application of a single convolutional neural network on the image, and finally, assigning a confidence level for the received detections.

**Table 1.** Main characteristics of the test videos.

|  | Video 1 | Video 2 |
|---|---|---|
| Light conditions | sunny day | late evening /night |
| Resolution | 1920 x 1080 | 1920 x 1080 |
| Frame rate | 30 | 30 |
| Length [min:sec] | 32:14 | 30:52 |
| Vehicle number | over 700 | over 350 |

In the tests all vehicles detected were classified into three groups: passenger cars, buses, and trucks including small trucks, van, and big lorries. Buses were excluded from the further analysis because big buses are used for tourist trips as well as for urban or intercity public transports but not as ambulance vehicles.

The YOLOv3 not only detects objects but also determines bounding boxes for them. It is very useful and advantageous because the statistical analysis of the occurrences of blue color pixels, what is very typical for the ambulances with flashing emergency lights, can be performed not for the whole images but only for bounding boxes with detected vehicles.

To optimize this statistical analysis a Savitzky-Golay filter [19] was applied on the detected objects for the purpose of smoothing the data. Such a smoothing process ensures that outstanding data is eliminated.

So, the procedure applied was as follows:

- detection of objects in a video using YOLOv3,
- determining bounding boxes for detected objects,
- classification of vehicles detected,
- changing RGB values to YUV,
- application of a Savitzky-Golay filter,
- calculation of occurrences of blue color,
- occurrence thresholding.

The results of blue color analysis for vehicles detected in videos recorded at night are presented in the form of charts. The difference between the charts drawn for a non-ambulance vehicle (Fig. 7) and an ambulance one (Fig. 8) is very significant. The effect of lights rotating and flashing on an ambulance can be easily noticed. The cyclical changes in intensity of occurrences of blue color are evident, even if these cycles are not regular because the examined ambulance is moving and the lighting conditions on the road are changing.

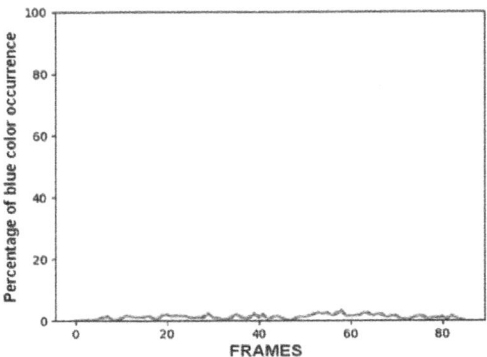

**Fig. 7.** The percentage of blue color occurrence for a non-ambulance vehicle.

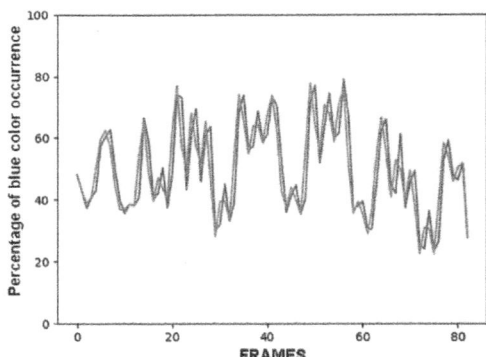

**Fig. 8.** The percentage of blue color occurrence for the ambulance detected in the video

(two drawn lines represent original data and smoothed values using a Savitzky-Golay filter).

In the examined video recorded on a dark evening 423 vehicles were detected and only the real ambulance was classified as ambulance (Table 2).

Unfortunately, the detection of ambulance emergency lights in videos recorded during the day, especially a sunny day, occurred not successful because the blue light is not so intensive to be identified (Fig. 9). It is almost impossible to perceive the emergence

**Table 2.** The results of the detection of ambulances in videos recorded at night.

|                     | Ambulances | Non-ambulances |
| ------------------- | ---------- | -------------- |
| Classified as:      |            |                |
| Ambulances          | 1          | 0              |
| Non-ambulances      | 0          | 422            |

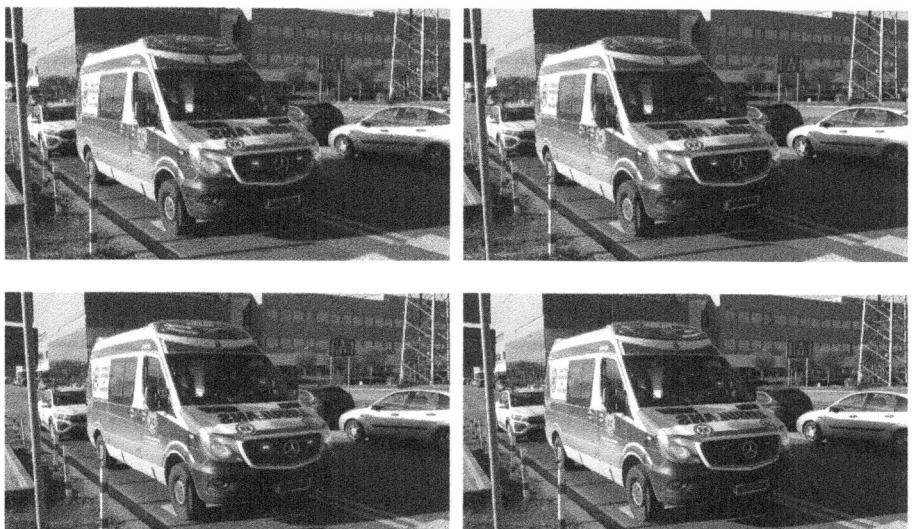

**Fig. 9.** During the sunny day blue flashing lights are very difficult to perceive and even more difficult to automatically detect.

lights because of the brightness and sunlight. The blue flares, so characteristic for night recordings, are not occurring.

The diagrams of percentage of blue color occurrence drawn for a non-ambulance and an ambulance vehicle cannot permit to formulate the right decisions. The great number of detected vehicles other than ambulance was wrongly classified as ambulance. The conclusion is that such a procedure based on the analysis of blue color can be applied but for evening or night videos.

## 5  Conclusions and Further Research

Street traffic monitoring is useful for developing the traffic light systems at road intersections, detection of red light violations, traffic intensity study, optimization of traffic organization, and also for ensuring the secure interaction of autonomous vehicles with emergency vehicles. The detection of emergency vehicles mainly such as ambulances, police vehicles, fire engines, and some others like tow trucks, snow plows, bomb disposal vehicles, or mountain rescue is one of the very important task in the computer vision area.

It could be expected that the detection of emergency vehicles can be easily achieved basing on the analysis of vehicle shapes and colors, or sound of sirens. However, although many tests have been performed to verify these approaches and many sophisticated algorithms have been developed, the results are not fully satisfactory. The problem is still valid.

The observations as well as the experimental studies described in this paper have shown that the detection of ambulances in surveillance videos recorded during the day can be successful if it is mainly based on the acoustic analysis. Whereas, in surveillance videos recorded at night or late in the evening a more efficient approach is the analysis of rotating and flashing lights installed on ambulances. Let us remind that blue rotating or flashing lights may be carried only by emergency vehicles.

The further experimental research will be undertaken to analyze the efficiency of this approach for emergency vehicles other than an ambulance, also using other than blue color lights, i.e. fire trucks and police vehicles.

# References

1. Chen, J., Wang, Q., Cheng, H.H., Peng, W., Xu, W.: A review of vision-based traffic semantic understanding in ITSs. IEEE Trans. Intell. Transp. Syst. **23**(11), 19954–19979 (2022)
2. Sundar, R., Hebbar, S., Golla, V.: Implementing intelligent traffic control system for congestion control, ambulance clearance, and stolen vehicle detection. IEEE Sens. J. **15**(2), 1109–1113 (2015)
3. Shaik, A., Bowen, N., Bole, J., Kunzi, G., Bruce, D., Abdelgawad, A., Yelamarthi, K.: Smart car: An IoT based accident detection system. In: Proc. of the Global Conference on Internet of Things (GCIoT), pp. 1–5. IEEE (2018)
4. Kotronis, C., et al.: Managing criticalities of e-Health IoT systems. In: Proc. of 17th International Conference on Ubiquitous Wireless Broadband (ICUWB), pp. 1–5. IEEE (2017)
5. Deepajothi, S., Rajan, D.P., Karthikeyan, P., Velliangiri, S.: Intelligent traffic management for emergency vehicles using convolutional neural network. In: Proc. of 7th International Conference on Advanced Computing and Communication Systems (ICACCS), vol. 1, pp. 853–857. IEEE (2021)
6. Vishnu, U., Sarma, S., Yashumithaa, M., Kr, S., Gv, V.: CNN based intelligent traffic control system to support emergency vehicles. In: Proc. of the International Conference on Electronics, Computing and Communication Technologies (CONECCT), IEEE, pp. 1–5 (2022)
7. Roy, S., Rahman, M.S.: Emergency vehicle detection on heavy traffic road from CCTV footage using deep convolutional neural network. In: Proc. of the International Conference on Electrical, Computer and Communication Engineering (ECCE), IEEE, pp. 1–6 (2019)
8. bin Che Mansor, M.A.H., Kamal, N.A.M., bin Baharom, M.H., bin Zainol, M.A.: Emergency vehicle type classification using convolutional neural network (2021). In: Proc. of the International Conference on Automatic Control & Intelligent Systems (I2CACIS), pp. 126–129. IEEE
9. Simonyan, K., Zisserman, A.: Very deep convolutional networks for large-scale image recognition (2014). arXiv preprint arXiv:1409.1556
10. Rane, D., Shirodkar, P., Panigrahi, T., Mini, S.: Detection of ambulance siren in traffic. In: Proc. of the International Conference on Wireless Communications Signal Processing and Networking (WiSPNET), pp. 401–405. IEEE (2019)

11. Moka, S.S.P., Pilla, S.M., Radhika, S.: Real time density based traffic surveillance system integrated with acoustic based emergency vehicle detection. In: Proc. of the 4th International Conference on Computer, Communication and Signal Processing (ICCCSP), IEEE, pp. 1–7 (2020)

12. Mittal, U., Chawla, P.: Acoustic based emergency vehicle detection using ensemble of deep learning models. Procedia Comput. Sci. **218**, 227–234 (2023)

13. Gowtham, P., Eswari, P., Arunachalam, V.P.: An investigation approach used for pattern classification and recognition of an emergency vehicle. In: Proc. of the International Conference on Soft-computing and Network Security (ICSNS), pp. 1–7. IEEE (2018)

14. Yu, X., Marinov, M.: A study on recent developments and issues with obstacle detection systems for automated vehicles. Sustainability **12**(8), 3281 (2020)

15. Garg, A., Gupta, A.K., Shrivastava, D., Didwania, Y., Bora, P.J.: Emergency vehicle detection by autonomous vehicle. Int. J. Eng. Res. Technol. **8**(5), 190–194 (2019)

16. Tran, V.T., Tsai, W.H.: Audio-vision emergency vehicle detection. IEEE Sens. J. **21**(24), 27905–27917 (2021)

17. Redmon, J., Divvala, S., Girshick, R., Farhadi, A.: You Only Look Once: Unified, real-time object detection. In: Proc. of the IEEE Conference on Computer Vision and Pattern Recognition, pp. 779–788. IEEE (2016)

18. Redmon, J., Farhadi, A.: YOLOv3: An incremental improvement. arXiv preprint arXiv:1804.02767 (2018)

19. Yang, H., Cheng, Y., Li, G.: A denoising method for ship radiated noise based on Spearman variational mode decomposition, spatial-dependence recurrence sample entropy, improved wavelet threshold denoising, and Savitzky-Golay filter. Alex. Eng. J. **60**(3), 3379–3400 (2021)

# Tools for Identifying and Preventing Loneliness in Older Adults

Christos Mettouris[1](✉), Evangelia Vanezi[1], Leonie Cammerlander[2], Paul Schober[2], Andria Hadjicosta[3], Sotiria Moza[3], Matthias Rohringer[4], Irena Zemaitaityte[5], Jan Kellerer[4], Eva Schulc[4], Raminta Bardauskiene[5], Paolo Zaramella[6], Alberto Maistrello[6], and George Angelos Papadopoulos[1]

[1] Department of Computer Science, University of Cyprus, Nicosia, Cyprus
{mettouris.g.christos,vanezi.evangelia,
papadopoulos.george}@ucy.ac.cy
[2] Hafelekar Unternehmensberatung Schober GmbH, Neulengbach, Austria
{leonie.cammerlander,paul.schober}@hafelekar.at
[3] Materia Group, Nicosia, Cyprus
{andria,sotiria}@materia.com.cy
[4] Department Pflegewissenschaft und Gerontologie, UMIT Tirol, Tirol, Austria
{Matthias.Rohringer,Jan.Kellerer,Eva.Schulc}@umit-tirol.at
[5] Institute of Educational Science and Social Work, Mykolas Romeris University,
Vilnius, Lithuania
irene@mruni.eu2, raminta@mruni.eu
[6] Consulenza Direzionale di Paolo Zaramella, Milan, Italy
paolo.zaramella@studiocentroveneto.com

**Abstract.** Loneliness among the older adults is a pressing problem worldwide. The deaths of loved ones, deteriorating health, the anticipation of one's own death, and the absence of a person one could rely on result in people experiencing loneliness in older age. The goal of the European project "Digi-Ageing" is to combat age-related loneliness through digital tools and training for healthcare professionals. These tools include a digital loneliness identification tool (screening tool) to identify loneliness and a digital loneliness intervention tool (reminiscence tool), supported by a dedicated curriculum for user training. This paper's primary objective is to discuss the research methodology followed to address the issue of loneliness in the context of the Digi-Ageing project, and introduce the Digi-Ageing platform and its two novel tools designed to assess the degree of loneliness among elderly individuals, as well as steps to address them. The paper explores the tools' innovative features and potential applications across various countries. Furthermore, it aims to initiate a discourse on the tools' future prospects, particularly regarding their implementation with diverse vulnerable groups.

**Keywords:** Loneliness Identification · Loneliness Prevention · Screening Tool · Reminiscence Tool · Digital Tools

N.-T. Nguyen et al. (Eds.): ICCCI 2024, CCIS 2166, pp. 338–350, 2024.
https://doi.org/10.1007/978-3-031-70259-4_26

# 1  Introduction

In every EU member state, the share of the population aged 65 and over is already noticeably increasing and accounts for over 20%. As per Eurostat forecast, the proportion of EU residents aged 65–79 is expected to rise to 17% by 2100, compared to the 15% recorded in 2022. The anticipated change is quite significant, as the share of seniors aged 80 years and above is projected to more than double, surging from 6% to 15%. Loneliness among the older adults is a pressing problem in many countries around the world. The deaths of loved ones, deteriorating health, the anticipation of one's own death, and the absence of a person one could rely on provide an essential context for people to experience loneliness in older age. Loneliness can be described as a feeling that arises when there is a perceived difference between the desired and attained levels of social relationships [1]. The quantity (available and reliable for helping social ties) and quality (degree of intimacy and understanding) of social relations are two dimensions of loneliness that is subjectively experienced by an individual [1, 2]. Experiencing loneliness has a great impact on a person's physical and mental health.

Digi-Ageing is a European project funded by the EU's Erasmus+ program and its goal is to combat age-related loneliness through digital tools and training for healthcare professionals. These tools include a digital screening tool to identify loneliness and a digital intervention tool, supported by a dedicated curriculum for user training [3].

Collaborating across five partner countries, each with distinct expertise in gerontology or nursing science, the consortium strives to create a comprehensive solution to a growing issue. The importance of the project lies, not only in its immediate aim to enhance the quality of life for the elderly, but also in raising awareness and establishing an engaged, solution-oriented network.

This paper's primary objective is to discuss the research methodology followed to address the issue of loneliness in the context of the Digi-Ageing project, as well as introduce its two novel tools designed to assess the degree of loneliness among elderly individuals, as well as steps to address loneliness. The paper explores the tools' innovative features and potential applications across various countries. Furthermore, it aims to initiate a discourse on the tools' future prospects, particularly regarding their implementation with diverse vulnerable groups.

The paper is organized as follows: Sect. 2 discusses the research methodology and Sect. 3 describes the related work from the aspect of similar available digital tools. Section 4 presents the Digi-Ageing platform and digital tools from a technical perspective and discusses how caregivers and patients can use them. Section 5 presents the evaluation of the two tools conducted in the context of the Digi-Ageing project, and the paper closes with conclusions and future work.

# 2  Research Methodology

Concepts for detecting loneliness are categorized into uni- and multi-dimensional scales. Uni-dimensional approaches focus on a single dimension of loneliness, such as a lack of social support or social contacts. These treat loneliness as a distinct phenomenon reflected in this one dimension. Conversely, multi-dimensional concepts view loneliness as a complex and multifaceted phenomenon, consisting of various dimensions like emotional loneliness, social isolation, and a lack of belonging. This approach acknowledges the complexity of the loneliness experience and allows for a more detailed and differentiated analysis. However, there is limited evidence for multi-dimensional scales as they have been used sparingly in studies [4, 5].

In a clinical setting, the best-established instruments are the Berkman-Syme Social Network Index (SNI) for measuring social isolation for ages 18-64 and the short version of the UCLA-LS for measuring loneliness [6].

The design of the Digi-Ageing Loneliness screening tool was the end result of the 3rd work package. Further details can be found in the deliverable IO3-A1: [Deliverable name], that provides the specifications of the Digi-Ageing platform and tools.

The research methodology for the creation of the loneliness identification tool (screening tool) was based on a combination of desktop and field research, aimed at comprehensively understanding all stakeholders and synthesizing data from recent literature. Key factors such as age, gender, personal life characteristics and events, and social parameters were considered to develop a comprehensive profile of the average end-user of the application. To achieve the screening objectives, the instrument utilizes a mix of standardized and unstandardized tools, step-by-step tailoring data collection depending on the user's loneliness risk. By integrating multiple data sources and employing diverse research techniques, the methodology aims to create a holistic understanding of the stakeholders' needs and facilitate the development of an application that effectively addresses loneliness among its target audience.

The Screening Tool was based on the:

1) Identification of the demographic, personal, health, social, financial and other risk factors contributing to the prevalence of loneliness among older adults based on recent research.
2) Exploration of technological instruments and gateways which could be appropriate for older adults with a wide range of characteristics and also their caregivers and healthcare professionals.
3) Consideration of the limitations of the ageing user i.e., decreased attention span, visual and hearing acuity, need for high contrast screens, etc.

The Digi-Ageing screening tool was designed to be used by healthcare professionals and caregivers while assessing the end-user beneficiary (i.e. older adult). The tool includes two identification levels: (1) Risk Profile – Quick Loneliness Check, (2) Standardised Loneliness assessment – UCLA Revised Scale. In case the results of these two first steps indicate that the older adult is at moderate or high risk of loneliness then the professional is encouraged to proceed to step (3), namely, the ECOMAP and the creation of a joint (4) Action plan to assist older adults in implementing changes in their everyday life and engage their network to address the signs of loneliness.

The Risk Profile – Quick Loneliness check was created as a quick and handy checklist containing the major loneliness risk factors as identified by literature, in order for the professional to evaluate the overall risk profile of the individual. These factors can be categorised into four main domains, namely, demographic risk factors (old age, female gender, low income, family status (e.g., living alone), recent retirement), environmental risk factors (living status, low transportation accessibility), physical and mental health risk factors (visual, hearing impairments, mobility impairments, mental diseases, recent personal losses) and social risk factors (limited access to internet, member of a minority, access to services and products, global/national crisis) [7–10].

Following the initial screening, the tool provides an option to cross-confirm the results using a standardized questionnaire. Upon evaluating the available options (e.g., de Jong Gierveld Scale, UCLA, and more), the consortium opted to incorporate the UCLA revised questionnaire. This decision was influenced by the fact that standardized translated versions were accessible in all consortium countries, and the questionnaire demonstrated ease of use and sound psychometric properties [11, 12]. By integrating the UCLA revised questionnaire, the tool aims to enhance accuracy and consistency in assessing the relevant factors across diverse populations in the consortium countries.

At the third level, the ECOMAP was selected as a tool to initiate discussions between the older adult and the healthcare professional, offering a visual representation and comprehension of the older adult's existing social network and relationships. This internal subjective evaluation allows the older adult to express their feelings of closeness with each person in their life. All three levels of assessment complement each other, culminating in an aggregated average risk level.

The solution aims to provide a risk assessment, indicating the likelihood of the person under the care of the healthcare professional or caregiver developing loneliness. However, it does not confirm whether the person is currently suffering from loneliness. Therefore, the responses gathered should be tailored to this risk-oriented perspective, focusing on understanding and addressing potential risks rather than confirming the presence of loneliness.

## 3   Related Work

Several digital tools have been developed in recent years, to address the issue of loneliness among older adults from different aspects. Paper [13] discusses existing mobile applications that have the potential to be used daily by older adults, aiming to combat the effects of isolation in their lives. The work, published during the COVID-19 pandemic era, presents 15 apps, divided in 6 categories: (i) Social Networking including popular networking applications like FaceTime and Skype; (ii) Medical app specialised in telemedicine; (iii) Medical apps specialised in prescription management; (iv) Health & Fitness; (v) Food & Drink, and (vi) Visual & Hearing impairment.

Social Networks, communication apps such as Skype or Pinterest and other digital technologies are also discussed as a means of preventing and/or overcoming loneliness [14–16], by preventing social isolation for older adults through increasing communication and connectivity between friends, family, and caregivers. Authors in [17] found in their study group that by using the mobile game app "Gobang", participants were able to network with other players, maintain social contacts and increase their social well-being by 10%. The mobile game app "Gobang" [18] explores in specifics how Whatsapp's basic features can be used by the healthcare ecosystem, aiding remote care and making it more efficient in terms of money and time in addition to improving the quality of life of patients. Whatsapp is perceived as a particularly user-friendly messaging app with a clean and simple to use design – this means that groups that are ordinarily excluded from accessing it, such as older people, are able to use it.

Even though the above studies obtained positive results, it was also shown that problematic use of social media via mobile devices has a negative impact on social isolation among older people with a significant result [19].

While the abovementioned works discuss existing technologies as solutions to loneliness, [20] examines the feasibility of a novel communication technology to enhance social connectedness among older adults in residential care, by developing and evaluating an accessible iPad-based communication app that supports older adults' asynchronous communication with family and friends. Results showed that the technology required an adaptation period but was a feasible communication tool. However, increased social connectedness (meaningful social interaction) was only reported by participants with geographically distant relatives, while the authors did not come to a significant result (p => 0.5) in the quantitative study on reducing the experience of loneliness and social isolation.

World Health Organisation's (WHO) ongoing project "+Simple, digital inclusion for older people" offers a platform that groups content (news, procedures, social networks and pages of interest), while older adults were trained to use the tool. The overall aim was to promote elderly adults' social inclusion through a digital literacy process. At the moment, 106,550 tablets with the "+ Simple" platform were delivered to people over 60 years of age. Subsequently the "+ Simple Community app was integrated, seeking to approach older adults through the creation and participation of events, and the possibility to connect with other older adults.

Connect2affect is a new initiative that aims to create an online network of resources and support that meets the needs of anyone who is isolated or lonely, including older adults. It provides an isolation self-assessment for understanding the visitors risk level and providing the most helpful information and recommendations based on their results.

We observe that most works discuss the usage of existing technologies and applications in combating loneliness in older adults, while a few novel tools were suggested; however, all aim to establish communication or build communities with other older adults as the means for preventing or fighting loneliness and isolation. Our approach aims to offer a new direction for caregivers to work with each individual, analysing in depth their situation through established scientific tools, allowing them to recognise their level of loneliness and to work on fighting loneliness with specialised sessions assisted by the technology, tools, and resources offered by our platform.

## 4    The Digi-Ageing Digital Tools

In the context of the Digi-Ageing project, two tools have been designed and developed: a loneliness identification tool (screening tool) and a loneliness prevention tool (reminiscence tool). In this section we will describe them, focusing on technical and User Interface (UI) related aspects.

Both tools were developed using state-of-the-art web technologies and were integrated on a web-platform that uses WordPress as the basis for development. The Digi-Ageing database (DB) was designed and implemented in MySQL. The two tools and their modules were developed as plugins on the WordPress (WP) web-platform, utilizing the capabilities and power of this platform in terms of user management, content management, privacy, usability and UI design (usage of themes). More to the point, an empty installation of a WP web-platform was set up where it was used by the Digi-Ageing development team to design the Digi-Ageing platform's UI, as well as develop the two tools and their related modules as WP plugins.

Figure 1 shows the software architecture of the Digi-Ageing platform. The Dashboard allows the caregiver to select between using one of the two tools, or access the training. The training provides access to the Digi-Ageing blended learning training course, where the caregiver may receive useful information about the training, as well as navigate through the complete curriculum of the blended learning training course.

For using the screening tool or the reminiscence tool, the caregiver needs first to select one of his/her registered patients. Patients registration is being managed via the patient management module.

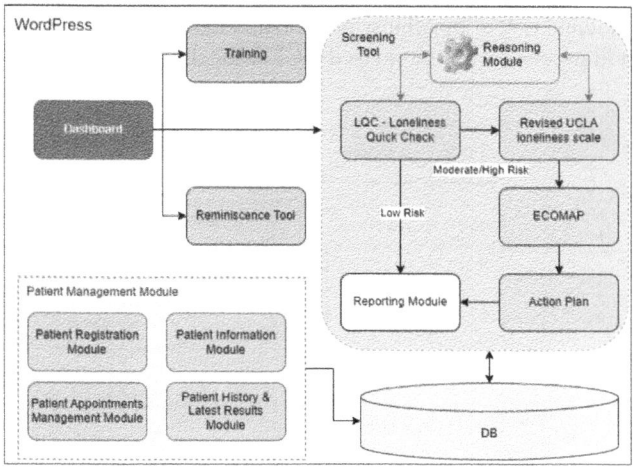

**Fig. 1.** The Digi-Ageing Platform Architecture.

The patient management module includes 4 modules: the patient registration module is responsible for the patient registration that is conducted by the caregivers. To register, the patient selects a login name that could be their real name or not, while their birthday and sex is also requested. Other than this, no patient private data is requested for registering as a patient. A patient being registered is automatically assigned under the organization of the caregiver. The patient information module stores and retrieves the information of the patient, while the patient history and latest results module stores the scores of the patient for LQC, UCLA, ECOMAP, Action Plan, and any other data generated that concern the patient. Displaying and printing the latest results for the patient is also a responsibility of the patient history and latest results module. The patient appointments management module manages the appointments (see reminiscence tool below) of the caregiver with his/her patients via a dashboard. The caregiver may visit his/her past appointments and read any notes or resources saved.

## 4.1  Screening tool

After patient registration, the caregiver may proceed to use the Digi-Ageing tools together with the patient. Firstly, the screening tool will be used. Figure 2 shows the workflow for the tool. The screening tool's first step is to conduct the Loneliness Quick Check (LQC), where the 15 questions are projected to the user in the form shown in Fig. 3. The patient should answer explicitly with a "yes" or "no" to avoid any ambiguity. The caregiver takes care of the interaction with the tool. The tool does not allow for any question to remain unanswered. Depending on the patient's LQC score, the caregiver will be prompted with an informative message with the patient's level of loneliness (low, moderate, or high), as well as the proposed next steps for the patient. In the case of low risk of loneliness, the patient should proceed to access the patient prevention report provided by the reporting module (see Fig. 2), which provides a list of recommendations that may help the patient feel more active, socialized, and tackle the feelings of emotional or social loneliness. If otherwise the patients' risk of loneliness is estimated to be moderate or high (see Fig. 4), the patient should proceed with the revised UCLA loneliness scale.

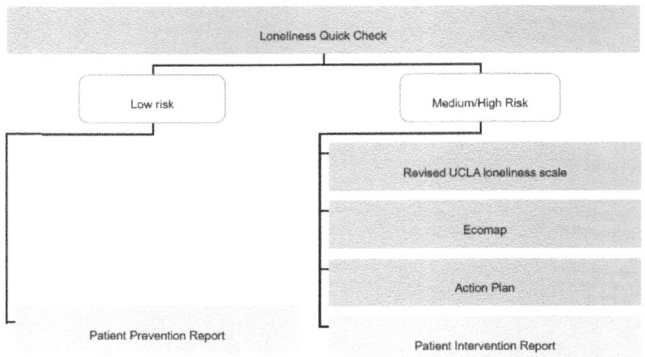

**Fig. 2.** Screening tool workflow.

**Instructions:** Please tick when the statement applies to the person you are assessing.

| | LQC - LONELINESS QUICK CHECK | YES | NO |
|---|---|---|---|
| 1 | The person has recently retired or became unemployed (max two years) | ○ | ○ |
| 2 | The person does not have close family or friends/social network | ○ | ○ |
| 3 | The person does not have an intimate relationship | ○ | ○ |

**Fig. 3.** LQC - The questionnaire.

With the revised UCLA (20 questions) loneliness scale (Fig. 5) the patient indicates how often a set of the statements are descriptive of themselves. The patient may reply using the statements: Never, Rarely, Sometimes or Often. The caregiver fills-in the questionnaire and submits. Figure 6 shows an example of the results when a moderately-high degree of loneliness is detected: the tool suggests proceeding to the next tools, but

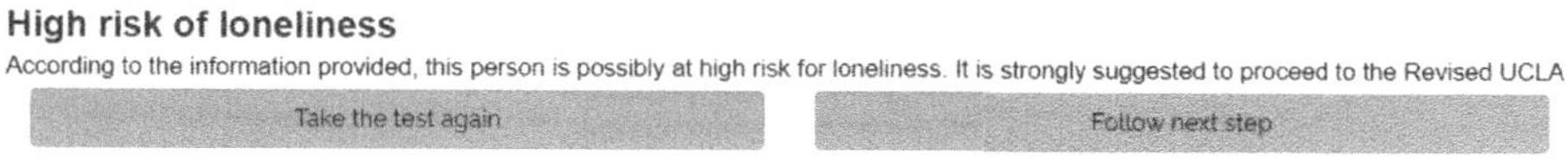

**High risk of loneliness**

According to the information provided, this person is possibly at high risk for loneliness. It is strongly suggested to proceed to the Revised UCLA

| Take the test again | Follow next step |

**Fig. 4.** LQC - The results.

it is important to note that the final decision is left on the caregiver's clinical judgment. The reasoning module is a reasoning engine that computes the patients' scores for the LQC and UCLA that are next being stored in the DB.

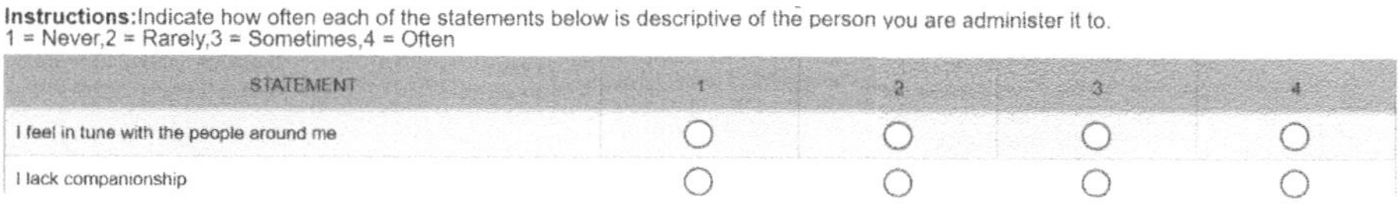

**Instructions:** Indicate how often each of the statements below is descriptive of the person you are administer it to.
1 = Never, 2 = Rarely, 3 = Sometimes, 4 = Often

| STATEMENT | 1 | 2 | 3 | 4 |
|-----------|---|---|---|---|
| I feel in tune with the people around me | ○ | ○ | ○ | ○ |
| I lack companionship | ○ | ○ | ○ | ○ |

**Fig. 5.** UCLA - The questionnaire.

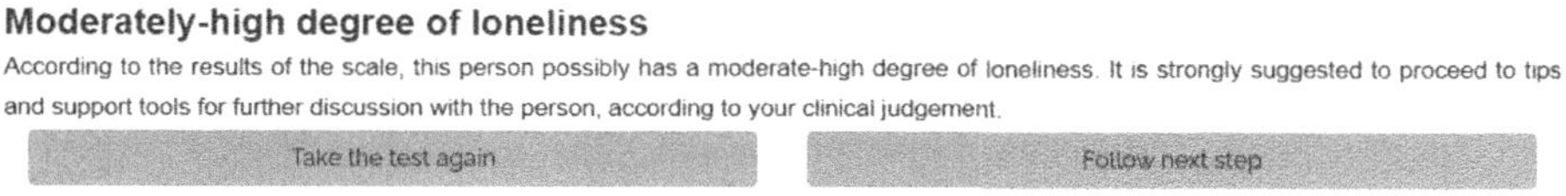

**Moderately-high degree of loneliness**

According to the results of the scale, this person possibly has a moderate-high degree of loneliness. It is strongly suggested to proceed to tips and support tools for further discussion with the person, according to your clinical judgement.

| Take the test again | Follow next step |

**Fig. 6.** UCLA - The results.

Following, the caregiver guides the patient through the ECOMAP tool. ECOMAP allows the patient to position his/her friends, family members, caregivers, social contacts and others as colourful figures upon a canvas that includes three concentric circles, such that the distance of each of these figures with the patient (grey figure in the middle) denotes how important their relationship is: the shorter the distance (i.e., placed in the inner circles) the closer the relationship, see Fig. 7. The positioning of the figures on the canvas is being done by the caregiver by using "drag and drop". The ECOMAP was developed using open-source libraries that allow for dragging and dropping items on a canvas.

Next in the screening tool palette of tools is the action plan. The action plan is an appropriately designed web form where the caregiver can report his/her ideas about the steps the patient should take in order to arrive at the desired results, i.e. reducing the patient's levels of loneliness. The action plan is stored in the DB and can be updated with the progress of the patient.

It is important to note that all Digi-Ageing modules mentioned store the patient's data in the DB and thus, when revisited, the carer and patient may continue from the point they left their session the previous time. This provides the caregiver the flexibility required to stop a session with a patient at any time and be able to continue at a later point in time without any data to be lost.

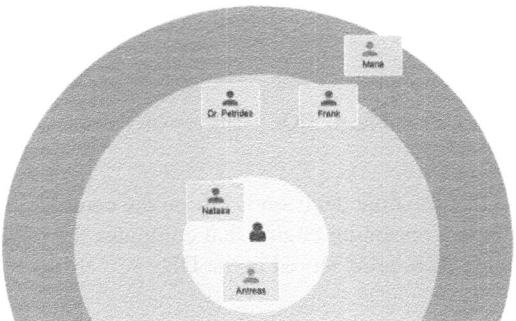

**Fig. 7.** ECOMAP.

## 4.2  Reminiscence tool

Reminiscence is the process of recalling to mind personal memories, involving not only positive, but also negative memories, as well as what a memory means to a person. The loneliness prevention tool or reminiscence tool uses modern technology to allow the caregiver and patient to initiate a discussion between them on a subject relevant to a patient's memory. Reminiscence guidance is intended to empower older adults to overcome loneliness and to find hope, value, and meaning in their lives. Caregivers benefit from improved knowledge of their patients, foster bond between professionals and patients and from a reduction in stress.

The reminiscence tool defines a guided videoconferencing session between the caregiver and the patient, during which the reminiscence takes place. The videoconferencing session is being conducted via ZOOM technology. The reminiscence session can be done face-to-face or even remotely, as the Digi-Ageing platform defines an easy way for the patient to sign-in to the ZOOM session: the caregiver contacts the patient and provides him/her with a link to the platform (e.g., via an SMS). At the time of the reminiscence session, the patient clicks on the link to access a webpage with a text box and a button. The patient enters their login name (alias) and clicks on the button. The following page provides the ZOOM link which he/she can click to access the reminiscence session. The above process assumes that the patient has a computer/mobile device with internet access and the ZOOM software installed (should be done by the caregiver or a family member). Then, the caregiver shares his/her screen and the reminiscence session starts.

The reminiscence process supported by the reminiscence tool is consisted of 4 stages that must be performed sequentially, but can be paused and be continued at any point in time as all data are being stored in the DB. The 4 stages are: stage 1 onboarding and planning stage: the caregiver is being instructed to inform the patient about the aims of the reminiscence session, and that he/she should set a warm, friendly tone, listen actively without rushing or forcing, creating thus positive emotions to the patient. The caregiver should identify themes of interest with the patient, creating thus positive emotions using pictures, short videos and music from suggested topics such as: favourite places/music/food, first day at school, first toy, and first job. Stage 2, elicit memories by using PROPS planning stage: in this stage the caregiver shares his/her screen to help the patient revisit his/her memories by using digital resources like favourite places/music/food. The reminiscence tool provides a plethora of such digital resources created by the project consortium available to be used during reminiscence sessions (Fig. 8).

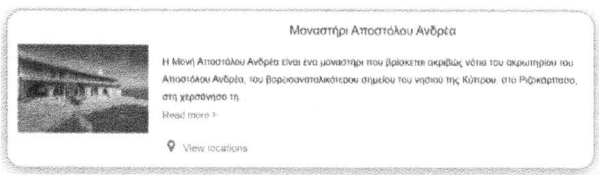

**Fig. 8.** Reminiscence tool: resources for favourite places.

Stage 3 includes the expression and sharing of memories by the patient. The caregiver is instructed by the tool to support the patient by actively listening, establishing trust and rapport, giving the choice to choose memories to share and not to share. Where the patient decides to share a difficult memory, the caregiver has the responsibility to ensure that hurtful feelings are properly listened, as well as to give support and genuine empathy to the patient. The fourth and final stage is the closing stage, where the patient should complete his/her story satisfactory. The caregiver is instructed by the tool to reflect on the positive outcomes and feelings that have been achieved, and to motivate the patient to move on to new projects (sharing memories) and activities. Moreover, the tool enables the caregiver to make notes and evaluate the session by writing text or uploading notes to the appointment's page.

## 5 Evaluation

The Digi-Ageing project's evaluation design, concentrating on the empirical pilot-testing of a newly introduced screening tool, employed qualitative evaluation research principles. These principles aimed at scientifically substantiating shifts in practice while revealing their subsequent impact, thereby following the guidelines outlined by Mayring [21]. The adherence to Prigge et al. [22] comprehensive evaluation guidelines fortified the design's foundation, assuring meticulous operationalization of the project's goals and objectives. This approach was robust, guaranteeing conformity to pivotal criteria such as utility, feasibility, fairness, and accuracy.

Hypothesis generation, integral to the evaluation process, was constructed on four foundational target levels: technical preconditions, applicability, and desired outcomes. These hypotheses underwent empirical examination using questionnaires. Subsequent data analysis in conjunction with iterative feedback loops yielded actionable recommendations. The continuous interaction with project partners across the project's lifespan ensured a shared understanding and consistent refinement of the project's processes and outcomes. The conclusions of the evaluation, formulated as actionable recommendations, were presented at the final conference in Lithuania in May 2023, providing a comprehensive, nuanced view of the project's achievements.

In parallel, the EU project Digi-Ageing conducted a data analysis for the validation of the Loneliness Quick Check (LQC) developed within the project. The LQC aimed to identify loneliness risk factors as predictors of scores on the UCLA Loneliness Scale, used to gauge the severity of loneliness. The pilot phase data were compared against anonymized test subject data from the LQC and the UCLA. With both test results available for 135 individuals, a descriptive analysis showed the LQC predicting a moderate to high presence of loneliness risk factors in over 85% of subjects. In contrast, only 51.8% of subjects ranked their perceived loneliness as moderate or high according to UCLA. Initial analysis indicates the LQC tends towards overestimation of loneliness risk, which might be beneficial in a multi-stage screening process to avoid overlooking potentially affected individuals. However, further statistical analyses employing a binomial logistic regression model didn't show significant predictive power of LQC for UCLA outcomes. Therefore, based on the data set, it cannot be conclusively shown that overall LQC scores or their categorical interpretation could predict actual loneliness as diagnosed by UCLA. However, the statistical calculations of the predictive validity of the LQC are based on an international convenience sample. This sample was neither internationally nor nationally representative for the participating countries. Therefore, it is a further goal to conduct studies on the validation and cultural adaptation of the LQC based on valid national and international data.

## 6 Discussion and Future Work

Researchers [23, 24] point out that using a digital platform to identify and predict loneliness in older people could help to identify a trend, recognise loneliness as a risk factor and/or correlate, make a meaningful contribution to the allocation of resources to where they are needed most, and optimize interventions. According to [25], it is essential that people are willing to record their information, that the information is used in a meaningful way, and that users are able to interpret any results. If users are unwilling to accept the system, the system cannot be effective. Therefore, it is necessary to ensure that the tools/platform are sufficiently inclusive and effective. For platforms to be appealing over the long term, it is important to involve, not only the older adult, but also his/her support network such as family, friends, care workers and others, including the needs/requirements of different stakeholders [26, 27]. The integration of digital health tools to identify and mitigate age-associated loneliness signals a promising trajectory, inviting additional intervention avenues. This study unveils the potential applicability of these tools within domains such as community care, mobile nursing, and integrated care, where the interplay between age-associated loneliness and digital health approaches could offer enriching perspectives. A comprehensive training mobilization targeting healthcare professionals in effectively leveraging these tools and addressing loneliness is deemed crucial in responding to impending challenges. Incorporating nursing students and incumbent personnel may stimulate an intergenerational knowledge exchange, fostering pragmatic implementation. Initiatives such as Digi-Ageing play a crucial role in heightening societal cognizance towards loneliness repercussions, fostering collaborative problem-solving endeavours. Loneliness is an intricate phenomenon, indiscriminately affecting individuals across age brackets and social tiers, a relevance amplified amidst the COVID-19 pandemic onset. Despite this study's concentration on age-associated loneliness, digital health tools possess potential applicability across varying demographics, fortifying the fight against loneliness. Future work should accentuate the transposability of these findings across other application domains and the cultivation of integrative digital resolutions.

Future investigative pursuits are encouraged to implement ex-ante and ex-post evaluations, facilitating the generation of valid conclusions, and ensuring the inclusion of an expanded participant cohort. Moreover, the forthcoming Erasmus+ project, [Forthcoming project name] manifests significant potential in addressing the intricate challenges presented by an aging demographic.

**Acknowledgments.** This work is supported by the European Commission as part of the Digi-Ageing EU project funded by the Erasmus+ Programme under grant agreement 2020-1-AT01-KA202-078084.

# References

1. Peplau, L.A., Perlman, D. (eds.) Loneliness: A sourcebook of current theory, research and therapy. John Wiley & Sons (1986)
2. Weiss, R.S.: Loneliness: The experience of emotional and social isolation. MIT Press (1973)
3. Digi-Ageing platform. http://digi-ageing.eu
4. National Academies of Sciences, Engineering, and Medicine, "Social isolation and loneliness in older adults: Opportunities for the health care system," Washington, D.C: National Academies Press, pp. 107–109 (2020). https://doi.org/10.17226/25663.
5. Schwab, R.: Einsamkeit: Grundlagen für die klinisch-psychologische Diagnostik und Intervention, Bern: Huber, pp. 77–83 (1997)
6. National Academies of Sciences, Engineering, and Medicine, "Social isolation and loneliness in older adults: Opportunities for the health care system," Washington, D.C: National Academies Press, p. 142 (2020). https://doi.org/10.17226/25663.
7. Prieto-Flores, M.E., Forjaz, M.J., Fernandez-Mayoralas, G., Rojo-Perez, F., Martinez-Martin, P.: Factors associated with loneliness of noninstitutionalized and institutionalized older adults. J Aging Health. **23**(1), 177–94 (2011). https://doi.org/10.1177/0898264310382658. Epub 2010 Sep 29 PMID: 20881107
8. Aartsen, M., Jylhä, M.: Onset of loneliness in older adults: results of a 28 year prospective study. Eur. J. Ageing **8**, 31–38 (2011). https://doi.org/10.1007/s10433-011-0175-7
9. Rico-Uribe, L.A., Caballero, F.F., Olaya, B., Tobiasz-Adamczyk, B., Koskinen, S., Leonardi, M., et al.: Loneliness, social networks, and health: a cross-sectional study in three countries. PLoS One (2016)
10. Niedzwiedz, C.L., Richardson, E.A., Tunstall, H., Shortt, N.K., Mitchell, R.J., Pearce, J.R.: The relationship between wealth and loneliness among older people across Europe: is social participation protective? Prev Med. **91**, 24–31 (2016). https://doi.org/10.1016/j.ypmed.2016.07.016. PubMed PMID: 27471027
11. De Jong Gierveld, J., Van Tilburg, T.G.: A six-item scale for overall, emotional and social loneliness: confirmative tests on new survey data. Res Aging. **28,** 582–598 (2006). https://doi.org/10.1177/0164027506289723.
12. Russell, D., Peplau, L.A., Cutrona, C.E.: The revised UCLA Loneliness Scale: Concurrent and discriminant validity evidence. J. Personality Soc. Psychol. **39**, 472–480 (1980)
13. Swechya, B., Healy, M., Goldberg, E.M.: 15 smartphone apps for older adults to use while in isolation during the COVID-19 pandemic. Western J. Emergency Med. **21**(3), 514 (2020)
14. Karina, A.: Social Isolation and Technology: How Technology can be Used to Reduce Social Isolation Among Older Adults in British Columbia (2017)

15. Delello, J.A., McWhorter, R.R.: Reducing the digital divide: connecting older adults to iPad technology. J. Appl. Gerontol. **36**(1), 3–28 (2017)
16. Mary Ann, J., Chipps, J., Padmanabhanunni, A.: "This phone saved my life": Older persons' experiences and appraisals of an mHealth intervention aimed at addressing loneliness. J. psychol. Africa **29**(2), 159–166 (2019)
17. Li, N., Chen, W.: A mobile game for the social and cognitive well-being of elderly People in China. Stud. Health Technol. Inform. **242**, 614–621 (2017)
18. Duque, M.: Learning from WhatsApp Best Practices for Health: Communication Protocols for Hospitals and Medical Clinics. Marília Duque. London: ASSA, 151 p. (2020). ISBN: 978-1-8380069-0-7 (eBook)
19. Meshi, D., Cotten, S.R., Bender, A.R.: Problematic social media use and perceived social isolation in older adults: a cross-sectional study. Gerontology **66**(2), 160–168 (2019). https://doi.org/10.1159/000502577
20. Barbosa Neves, B., Franz, R., Judges, R., Beermann, C., Baecker, R.: Can digital technology enhance social connectedness among older adults? A feasibility study. J. Appl. Gerontology **38**(1), 49–72 (2019)
21. Mayring, P.: Einführung in die qualitative Sozialforschung: eine Anleitung zu qualitativem Denken, (6. Auflage). Basel; Weinheim: Beltz (2016)
22. Prigge, J., Niestroj, M., Schott, D., Schoppmann, U., Halves, E.: Die Revision der Empfehlungen für Aus- und Weiterbildung in der Evaluation (EAUWE) – ein Werkstattbericht aus der Ad-hoc-Gruppe der DeGEval. Zeitschrift Für Evaluation **1**, 158–162 (2022). https://doi.org/10.31244/zfe.2022.01.17
23. Barreto, M., Victor, C., Hammond, C., Eccles, A., Richins, M.T., Qualter, P.: Loneliness around the world: Age, gender, and cultural differences in loneliness. Personality and Individual Differences **169** (2021). https://doi.org/10.1016/j.paid.2020.110066.
24. Doryab, A., Villalba, D.K., Chikersal, P., Dutcher, J.M., Tumminia, M., Liu, X.: Identifying behavioral phenotypes of loneliness and social isolation with passive sensing: statistical analysis, data mining and machine learning of smartphone and fitbit data. JMIR MHealth UHealth **7**(7) (2019). Article e13209, https://doi.org/10.2196/13209.
25. Stuart, A., et al.: Loneliness in older people and COVID-19: applying the social identity approach to digital intervention design. Comput. Hum. Behav. Rep. **6**, 100179 (2022). https://doi.org/10.1016/j.chbr.2022.100179
26. Coelho, J., Duarte, C.: A literature survey on older adults' use of social network services and social applications. Comput. Hum. Behav. **58**(2016), 187–205 (2016). https://doi.org/10.1016/j.chb.2015.12.053
27. Mitton, C., Smith, N., Peacock, S., Evoy, B., Abelson, J.: Public participation in health care priority setting: a scoping review. Health Policy **91**(3), 219–228 (2009). https://doi.org/10.1016/j.healthpol.2009.01.005

# A Clustering Approach for Personalized Coaching Applications

Annika Van Buren[1], Audrey Kwan[1], Harald. H. Rietdijk[2(✉)],
Talko B. Dijkhuis[2], Patricia Conde-Cespedes[3], Hilbrand Oldenhuis[2],
and Maria Trocan[3]

[1] Stanford University, Stanford, CA 94305, USA
[2] Hanze University of Applied Sciences, 9747AS Groningen, Netherlands
`h.h.rietdijk@pl.hanze.nl`
[3] ISEP, 92130 Issy-les-Moulineaux, France

**Abstract.** Insufficient physical activity presents a significant hazard to overall health, with sedentary lifestyles linked to a variety of health issues. Monitoring physical activity levels allows the recognition of patterns of sedentary behavior and the provision of coaching to meet the recommended physical activity standards. In this paper, we aim to address the problem of reducing the time consuming process of fitting classifiers when generating personalized models for a coaching application. The proposed approach consists of evaluating the effects of clustering participants based on their walking patterns and then recommending a unique model for each group. Each model consists of a random forest classifier with a different number of estimators each. The resulting approach reduces the fitting time considerably while keeping nearly the same classification performance as personalized models.

**Keywords:** Personalized coaching · sedentary lifestyle · fitting time optimization · clustering · Random Forests · estimators · variability

## 1 Introduction

Insufficient physical activity poses a significant risk to health and well-being. Sedentary lifestyles are associated with a variety of health problems, such as heart disease, cancer, stroke, and diabetes [1–3]. Through the use of wearable devices, such as smartwatches, it has become easier to gain insight into actual physical activity levels. Monitoring physical activity levels allows individuals to identify patterns of sedentary behavior and assess whether they meet the recommended activity guidelines. Furthermore, by tracking activity levels, individuals would be able to take proactive steps to mitigate potential health risks and make the necessary adjustments to incorporate more movement into their daily routines [4,5]. In addition to its physical benefits, monitoring physical activity also supports mental health and cognitive function. Regular exercise has been shown to alleviate symptoms of depression and anxiety, improve mood, and improve cognitive abilities, such as memory and concentration [6–8].

© The Author(s), under exclusive license to Springer Nature Switzerland AG 2024
N.-T. Nguyen et al. (Eds.): ICCCI 2024, CCIS 2166, pp. 351–363, 2024.
https://doi.org/10.1007/978-3-031-70259-4_27

In recent years, there has been a large number of innovations in the field of personalization and individualization in healthcare and coaching applications. In [9] an overview of physical activity coaching applications can be found, and in [10] a review of behavior-changing therapy and rehabilitation applications is presented. A more specific example of an application that uses wearable devices to monitor physical activity can be found in the works of Dijkhuis *et al.*[11,12]. The authors propose the use of a machine learning-based procedure to train a digital activity coach that can provide information on daily fitness of a person. Specifically, the digital coach will monitor the probabilities throughout the day of the likelihood that the user will reach their goal of physical activity.

The paper [11] evaluates the performance of eight different machine learning algorithms and finds that tree algorithms and tree-based ensemble algorithms are the most promising for training a digital activity coach with personalized models. The results presented in the paper show a significant improvement in performance when one model is fitted per participant instead of using generalized models for the entire group (the approach also suggested in [13]).

Further improvements are expected to be obtained by generating more models that are fitted to specific sections of the data. However, as noted by the authors, *"it is problematic to provide a generalized recommendation for specific algorithms, parameters, or parameter settings"*. The problem could be solved by *"investigating the underlying mechanisms to be able to choose the best algorithm beforehand"*. However, fitting one model per person is too resource consuming. That is why in this paper we approach the problem differently. First, we perform clustering on the participants based on the hypothesis that people inside a cluster share similar patterns, and we fit only one model per cluster. When these insights are obtained, algorithm selection and model fitting can be optimized. The remainder of this paper is organized as follows; first, in Sect. 2, we present an overview of the relevant results and data used in [11], and explain the result we want to achieve. In Sect. 3 we present an overview of our approach which consists of two main parts, clustering and classification. In Sect. 4 we detail the clustering procedure. In Sect. 5 the classification results are presented. Finally, in Sect. 6 a discussion of the results is presented.

## 2    Preliminary Results

In the original study by Dijkhuis *et al.* [11] the step data of 48 participants were collected over a period of 33 weeks, using Fitbit activity trackers. These participants were involved in a health program conducted at the Hanze University of Applied Sciences Groningen. Data from 43 of these participants could be used for the study. These data were used to evaluate the performance achieved when predicting the probability that the participant will reach the step goal at 6 p.m. each day. To compensate for divergent behavior outside working hours, only step data collected between 7 a.m. and 6 p.m. on workdays was considered. Furthermore, to be able to use the data as input for regular machine learning methods, it was necessary to add the cumulative step count. In this way, instead of having to treat each data point as part of a time series, it was possible to use the data as input for a classification process. The resulting features that were used to train the different models were weekday, hour, step count per hour, and cumulative step count up to the given hour.

The open source Scikit-Learn library was used for the study. This package offers the choice of 41 classifiers. Using the flow chart provided by Scikit-Learn [14] and a cheat sheet found on the Microsoft Azure Machine Learning platform [15], a selection of eight classifiers was made; AdaBoost (ADA), Decision Trees (DT), KNeighborsClassifier (KNN), Logistic Regression (LR), Neural Networking(NN), Stochastic Gradient Descent (SGD), Random Forest (RF) and Support Vector Classification (SVC). These classifiers were first used to evaluate the performance of a generalized model across the entire data set, and subsequently a model was generated for each individual participant. In this last step, a selected set of hyperparameters for each classifier were 'fitted' to obtain the best performing model per participant. The comparison of results showed that the performance levels were significantly higher when personalized models were used. The maximum accuracy for the generalized model ranged from 71% to 78%, but reached 94% when using personalized models (considered data split is 70% for training and 30% for validation).

As stated in the discussion of the original study, and can also be seen, for example, in the work of Sarker [16], making general recommendations for classifier selection and hyperparameter settings is problematic. Using the Scikit-Learn grid search functionality provides a tool to determine optimal parameter settings per classifier. However, when this approach is used for each personalized model, the fitting times can become very large, depending on the number of parameters and their range of possible values. Therefore, the objective of this study is to investigate whether it is possible to improve the fitting times while maintaining the performance levels of the personalized approach in the original study. We use the source code of the original research as a baseline for this analysis, as well as the data set that contains the steps per hour of each participant. In the proposed solution, we seek to streamline and improve the training process by applying clustering of the participants based on the results obtained by analyzing the walking patterns of the participants and examining the hyperparameter behavior of each machine learning algorithm used. This analysis should lead to a recommendation of selected parameter settings per cluster, which will reduce the fitting time.

## 3   Description of the Proposed Solution

In Fig. 1 a schematic overview is given of the steps that were taken in the process to elaborate the proposed solution. Our approach is separated in two parts : clustering (first four steps in diagram) and classification (last step).

The clustering procedure was decomposed in the following steps :

1. The first step was to perform a thorough analysis of the preliminary results. The goal of this analysis was to limit the scope of the experimental setup to focus on one single classifier.

2. The second step was to formulate a working hypothesis and define the relevant metrics needed to perform the clustering process.
3. In step three, the available data was analyzed using the metric defined in step two. The results of this analysis form the input for step four, in which the clusters were defined.
4. Step four consisted of applying the clustering to the available data.

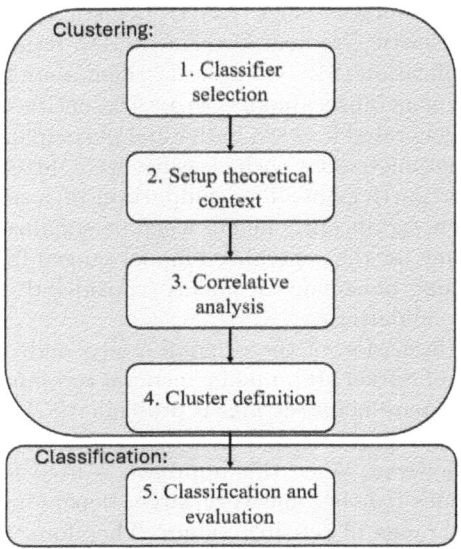

**Fig. 1.** Clustering steps.

Finally, the last step consists of performing a specific classification method for each cluster obtained from the previous steps. All the details are described in Sect. 5.

## 4    Clustering Procedure

In this section we describe all the steps that allowed us to partition the data into clusters, that is, steps 1 to 4 of the diagram in Fig. 1.

### 4.1    Experimental Setup

Our initial step in this research project was to evaluate the performance of various machine learning algorithms to identify the candidate most suitable for optimization. We considered the algorithm performance metrics F1-score, accuracy, and the operational performance metric given by the runtime of the fitting

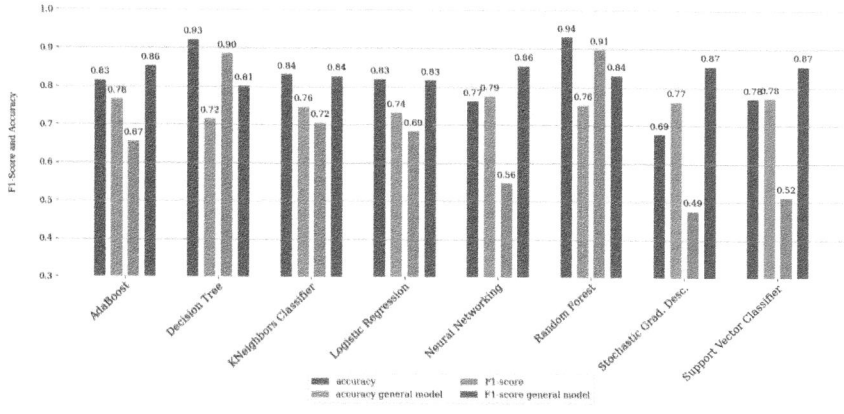

**Fig. 2.** Average F1-scores and accuracy per classifier.

procedure, to fully evaluate each algorithm. The results of the algorithm performance metrics are shown in Fig. 2. The operational metrics are shown in Fig. 3.

The results of the performance metrics show that the Random Forest and Decision Tree algorithms have the highest accuracy and F1-score for the models generated per participant. The proposed solution to cluster participants for model generation should give a better result than the generalized approach. Both classifiers also show a significant improvement in performance when the results of personalized modeling are compared with general modeling, making them suitable candidates for the experiment.

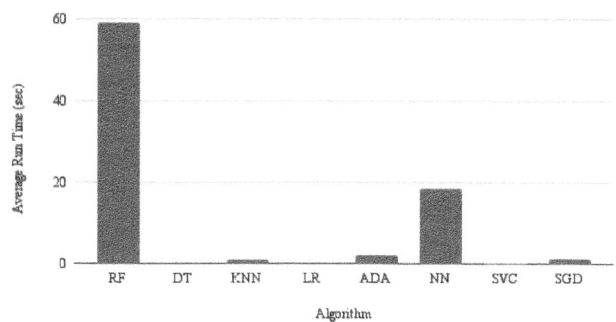

**Fig. 3.** Average run time per model of the fitting procedure for each classifier.

For further selection, we next considered the operational performance of the algorithms. Although the RF algorithm had the highest accuracy and F1-score for most of the participants, it also demonstrated significant computational demands, reflected in its extended runtime. On the other hand, the fitting times for the DT algorithm were already minimal. Since our objective was to enhance the model generation efficiency without compromising the algorithm's predictive capability, taking these results on performance and computational efficiency into consideration led us to select Random Forest for further optimization.

Since the fitting time depends on the number of values supplied for each hyperparameter, the generation efficiency can be further improved by presetting one or several of these values. Taking this into account, the clustering should focus on finding criteria that group participants with identical values for the hyperparameter setting. For the Random Forest algorithm, the parameters used in the original fitting process are the number of trees in the forest (n_estimators), the number of features to consider when looking for the best split (max_features), and the function to measure the quality of the split (criterion). The RF classifier has more parameters [17], but these three were considered to be the most likely to have a significant impact on the performance of the resulting model in the context of the original study.

### 4.2   Theoretical Context Setup

The preliminary clustering approach came from the hypothesis that the greater variability in the distribution of the features of the data set may correlate with higher optimal hyperparameter values for the Random Forest Algorithm, especially n_estimators. The logic underpinning this hypothesis is that a higher day-to-day step count variability might necessitate more estimators to accurately capture fluctuations, thereby reducing model variance and enhancing robustness. Tailoring an n_estimators value for each cluster would allow for a model that adapts to the characteristics of each group, with the aim of reducing over-fitting in low-variability clusters and enhancing robustness in high-variability clusters. To be able to focus on the evaluation of the impact of the characteristics of the data set on the algorithm performance, this approach was taken instead of using more traditional clustering approaches.

The features of the original data set consist of weekday, hour, step count per hour, and cumulative step count. Since weekday and hour should not show any relevant differences per participant, we have to look at the other two features. To investigate the relationship between participant step count variability and optimal n_estimators value, variability was quantified using the following methodology.

- *Data aggregation by interval:* For each participant, steps were aggregated to calculate the average sum steps per hour. The average cumulative sum steps up to each hour was also calculated. Data for the step count for each participant were taken from the training set used in the fitting process.
- *Standard Deviation (SD):* The standard deviation of the sum steps per hour was calculated. This metric quantifies the degree of variation from the average step count within each hour, providing insight into the consistency of the level of activity of the participants. The standard deviation of the cumulative steps for each hour was also calculated.

- *Coefficient of Variation (CV):* To facilitate a normalized comparison of variability that accounts for differences in average activity levels among participants, the coefficient of variation is calculated. This value is calculated by dividing the standard deviation of the aggregated sum by the average step count for the respective interval. The CV provides a relative measure of variability, allowing for an understanding of fluctuation in activity levels in proportion to the mean steps.
- *Hyperparameter values:* The values of n_estimators of the personalized models generated in the original study were used. These models were fitted with possible values 10, 50, 100, or 500 for this parameter.

In Fig. 4 the averages calculated over the aggregated values, cumulative step counts and the coefficient of variation are plotted against the corresponding n_estimators value. The results of the variability measures and averages for each participant were compared to the optimal n_estimators values designated for their respective RF models. The analysis revealed a consistent trend between the variability indicators, highlighting a correlation in which the increase in the variability in step counts was correlated with a preference for higher optimal n_estimators values. The increase in the average step count was also correlated with a preference for higher optimal n_estimators values. Moreover, those with both higher average step counts and notable variability tend to require a higher number of trees to effectively model their data.

Optimal n_estimators values correlated most closely with the average step count per hour. Average cumulative sum steps per hour showed a weaker positive correlation with n_estimators. The correlation coefficient of the average cumulative sum steps per hour showed a weaker negative correlation with n_estimators. When compared to the correlation between average step count per hour and optimal n_estimators values, participants with an unexpectedly high n_estimators were often associated with a low correlation of average cumulative sum steps per hour or high average cumulative sum steps per hour.

### 4.3   Clustering Definition

On the basis of these observations, the following approach for the clustering process was defined. The first step in constructing a structured framework for selecting the value of the n_estimators parameter in Random Forest models is determined by the participant step count data. Therefore, the initial step in the selection process is partitioning based on **Average Sum Steps**. Then adjustments are made according to **Average Cumulative Sum Steps per Hour** and **CV Average Cumulative Sum Steps per Hour**.

In the analysis of the relationship between similar ranges in participant average steps per hour and the value of n_estimators three clusters were determined, each with a specified value of n_estimators of 50, 100, or 500. This classification is shown in Table 1.

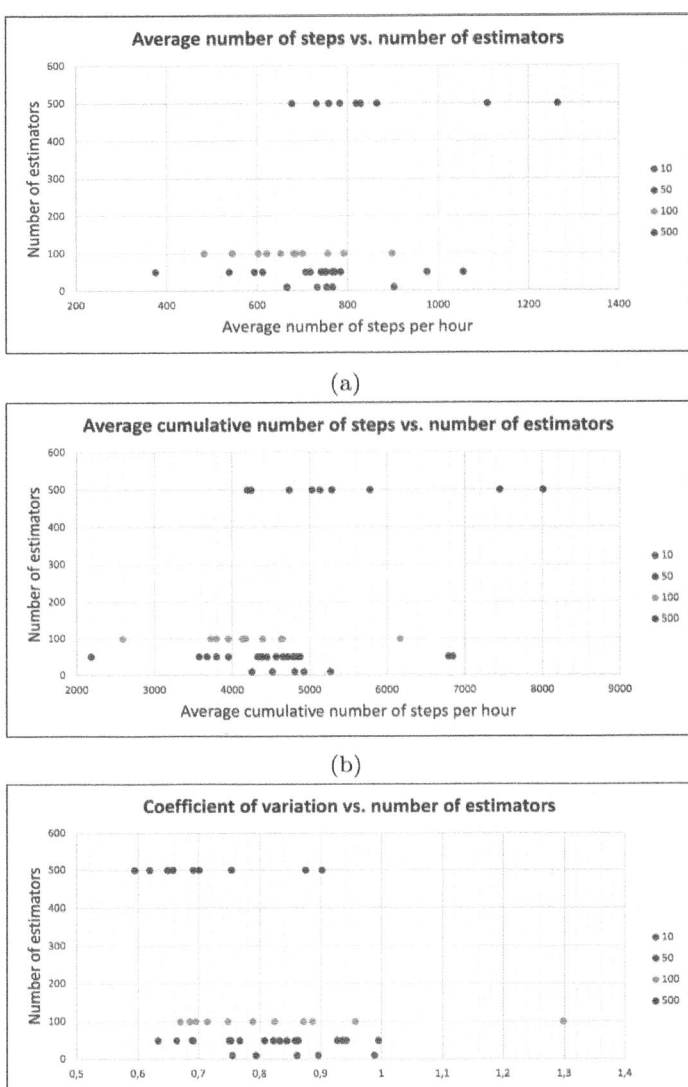

**Fig. 4.** Variability Metrics by **n_estimators**. (**a**) Average sum steps.(**b**) Average cumulative sum steps. (**c**) Hour coefficient of variation.

Further refinement of the setting of **n_estimators** is carried out based on the **Average Cumulative Sum Steps per Hour** and the **CV Average Cumulative Sum Steps per Hour**. Adjustments are only applied to the first two groups and considering the upper limit for the increment of **n_estimators**.

**Table 1.** Initial partitioning based on average sum steps.

| Average Sum Steps value | n_estimators value |
|---|---|
| Up to 657.048 | 50 |
| Between 657.048 and 808.093 | 100 |
| Above 808.093 | 500 |

For the first cluster of participants with steps up to 657.048 that were initially assigned a value of 50 to n_estimators, an increase to 100 is applied if both of the following criteria are satisfied:

- Average Cumulative Sum Steps per Hour $\geq$ 4569.44
  and
- CV Average Cumulative Sum Steps per Hour $\leq$ 0.788.

This adjustment is based on the assumption that increasing n_estimators is justified by the increase in activity levels, provided the variability remains within a reasonable range.

For the second cluster, with steps ranging from 657.048 to 808.093, the initial recommendation of 100 n_estimators is maintained unless activity and variability metrics meet the following criteria:

- Average Cumulative Sum Steps per Hour $\geq$ 5000
  and
- CV Average Cumulative Sum Steps per Hour $\leq$ 0.829.

In this case an elevation to a value of 500 for n_estimators is applied. This is aimed at accommodating significantly elevated activity levels while ensuring that variability is sufficiently managed to optimize the model's performance.

### 4.4    Resulting Clusters

With the clustering framework in place, a data frame was created for each of the three clusters. Each data frame was composed by collecting training data from the original personalized fitting run of the participants in each specific cluster. Figure 5 shows the number of participants in each cluster. From left to right, there are 9 participants in the first cluster (n_estimators = 50), 23 in the second cluster (n_estimators = 100), and 11 in the third cluster (n_estimators = 500). Therefore, each group has at least 20% of the total number of participants in the experiment.

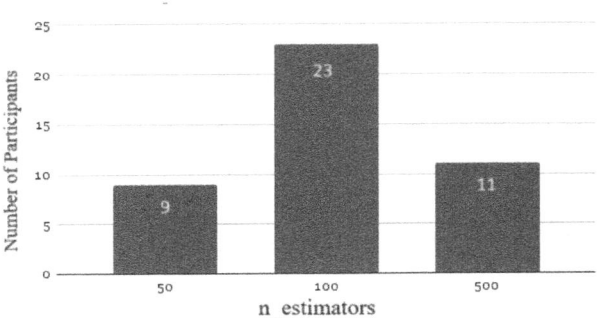

**Fig. 5.** The number of participants in each cluster.

# 5 Classification Results

The RF classifier fitting process was executed with a unique **n_estimators** value for each data frame, and several performance metrics (i.e., accuracy, F1-score, fitting time) were evaluated, comparing them to the original results of Dijkhuis *et al.* [11]. As in the original research, five-fold cross-validation was applied.

## 5.1 Performance Evaluation : Accuracy and F1 Score

Figure 6a and Fig. 6b show the total average accuracy and F1 score results of the original personalized run and the run using our clustering framework. Specifically, "with clustering" indicates the results obtained when one model was generated per clustered group of participants, using the specific **n_estimators** value for that cluster. "Without clustering" represents the original results obtained using a personalized model per participant. The results show that Total average accuracy was extremely similar between the clustering and no clustering trials. However, the clustering trial performed better than the no clustering trial in regards to total average F1-score. This indicates that on a global scale, the clustering method does not adversely affect the performance of the Algorithm.

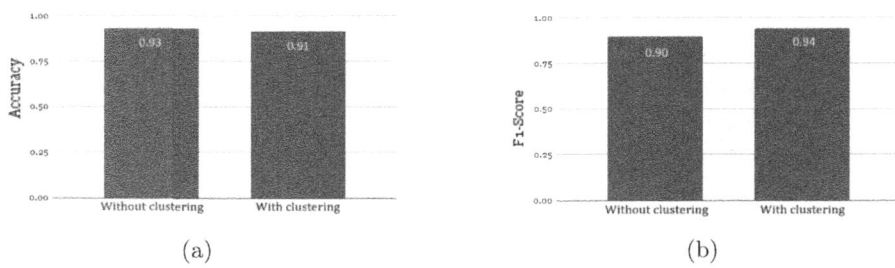

**Fig. 6.** Total average performance for all participants. (a) Accuracy. (b)F1-score.

Figure 7a and Fig. 7b illustrate the accuracy and F1-performance for each cluster. This view provides a more in-depth analysys of the Random Forest algorithm's performance within each cluster. Accuracy is around the same for all of the clusters, with the lowest accuracy being 0.912683 (n_estimators = 50) and the highest accuracy being 0.915141 (n_estimators = 500). Like accuracy, F1-scores were about the same for all of the clusters, with the lowest F1-score being 0.938219 (n_estimators = 50) and the highest F1-score being 0.941708 (n_estimators = 500).

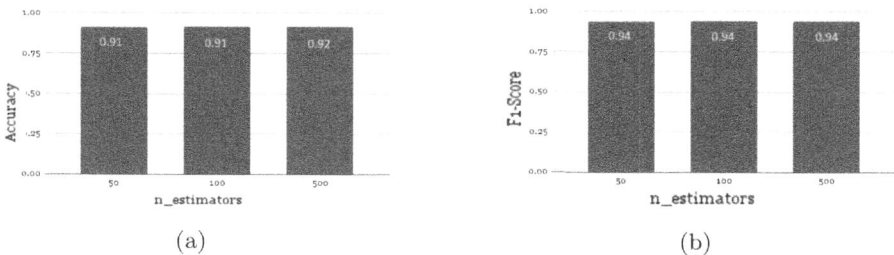

(a)                                              (b)

**Fig. 7.** Per cluster average performance. (**a**) Accuracy. (**b**)F1-score.

## 5.2    Fitting Time

As expected clustering led to a substantial decrease in total fitting time, as can be seen in Fig. 8a. Without clustering, total fitting time was 42.37 min, while with clustering, total fitting time was 5.70 min. Thus, clustering led to an 86.55% decrease in total fitting time, which is a highly significant optimization that will greatly reduce overall computational time.

Looking at fitting time per cluster, presented in Fig. 8b, there is a positive correlation between the number of n_estimators and the amount of the fitting time. This is to be expected since the resulting model is more complex. Therefore, there is a trade-off between performance and computational cost that must be taken into account when determining the n_estimators value for each cluster.

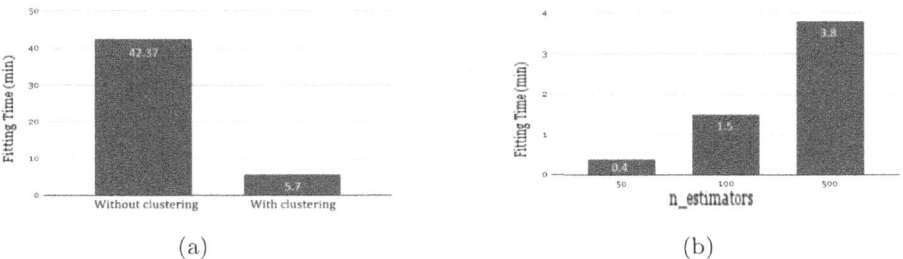

(a)                                              (b)

**Fig. 8.** Fitting times. (**a**) Totals. (**b**)Time per cluster.

# 6     Conslusions and Discussion

As shown in the results, the total accuracy and total F1-scores were about the same or better with than without clustering. This proves that clustering the data by walking variability does not lead to detriments in the performance of the RF algorithm, and instead leads to equally reliable predictions. Total fitting time, though, was significantly reduced by clustering. This is a particularly significant result, since RF was the most computationally expensive algorithm out of all the machine learning algorithms initially considered in [11]. Thus, decreasing fitting time while maintaining RF's high accuracy and F1-scores, through using preselected hyperparameters based on the characteristics of the dataset, would significantly improve the practicality of using machine learning algorithms in a digital physical activity coach.

Several considerations were taken into account to accommodate the size of the available data. For example, at all times the same random seed was used to ensure that the used selections were identical. Furthermore, the clustering train set was constructed using the train sets generated in the personalized run, instead of generating a new train set using all the data per cluster.

# References

1. World Health Organization: Physical activity (2024). https://www.who.int/health-topics/physical-activity
2. Losina, E., Yang, H.Y., Deshpande, B.R., Katz, J.N., Collins, J.E.: Physical activity and unplanned illness-related work absenteeism: data from an employee wellness program. PloS one **12**(5), e0176872 (2017). https://doi.org/10.1371/JOURNAL.PONE.0176872
3. Lee, I.M., Shiroma, E.J., Lobelo, F., Puska, P., Blair, S.N., Katzmarzyk, P.T.: Effect of physical inactivity on major non-communicable diseases worldwide: an analysis of burden of disease and life expectancy. Lancet **380**(9838), 219–229 (2012). https://doi.org/10.1016/S0140-6736(12)61031-9
4. Carter, D.D., Robinson, K., Forbes, J., Hayes, S.: Experiences of mobile health in promoting physical activity: a qualitative systematic review and meta-ethnography. PLoS ONE **13**(12), e0208759 (2018). https://doi.org/10.1371/JOURNAL.PONE.0208759
5. Chatterjee, A., Prinz, A., Gerdes, M., Martinez, S.: Digital interventions on healthy lifestyle management: systematic review. J Med Internet Res **23**(11), e26931 (2021). https://doi.org/10.2196/26931. http://www.ncbi.nlm.nih.gov/pubmed/34787575
6. Ekelund, U., et al.: Does physical activity attenuate, or even eliminate, the detrimental association of sitting time with mortality? A harmonised meta-analysis of data from more than 1 million men and women. Lancet **388**(10051), 1302–1310 (2016). https://doi.org/10.1016/S0140-6736(16)30370-1
7. Sharma, A., Madaan, V.: Exercise for mental health. Primary Care Companion J. Clin. Psychiatry **8**(2), 106–107 (2006). https://doi.org/10.4088/pcc.v08n0208a
8. Kamali, M.E., et al.: Virtual coaches for older adults' wellbeing: a systematic review. IEEE Access **8**, 101884–101902 (2020). https://doi.org/10.1109/ACCESS.2020.2996404

9. Gámez Díaz, R., Yu, Q., Ding, Y., Laamarti, F., El Saddik, A.: Digital twin coaching for physical activities: a survey. Sensors **20**(20), 5936 (2020). https://doi.org/10.3390/s20205936, https://www.mdpi.com/1424-8220/20/20/5936

10. Lauer-Schmaltz, M.W., Cash, P., Hansen, J.P., Maier, A.: Designing human digital twins for behaviour-changing therapy and rehabilitation: a systematic review. In: Proceedings of the Design Society, vol. 2, pp. 1303–1312. Cambridge University Press (2022). https://doi.org/10.1017/pds.2022.132

11. Dijkhuis, T.B., Blaauw, F.J., van Ittersum, M.W., Velthuijsen, H., Aiello, M.: Personalized physical activity coaching: a machine learning approach. Sensors (Switzerland) **18**(2), 623 (2018). https://doi.org/10.3390/s18020623

12. Blok, J., Dol, A., Dijkhuis, T.: Toward a generic personalized virtual coach for self-management: a proposal for an architecture. In: 9th International Conference on eHealth, Telemedicine, and Social Medicine 2017. Hanze University of Applied Sciences, Nice (2017)

13. Chatterjee, A., Pahari, N., Prinz, A., Riegler, M.: Machine learning and ontology in eCoaching for personalized activity level monitoring and recommendation generation. Scientific Reports **12**(1) (2022). https://doi.org/10.1038/s41598-022-24118-4

14. Scikit-Learn: Choosing the right estimator - scikit-learn 1.4.1 documentation. https://scikit-learn.org/stable/tutorial/machine_learning_map/index.html

15. Microsoft: Machine Learning Algorithm Cheat Sheet - designer - Azure Machine Learning — Microsoft Learn. https://learn.microsoft.com/en-us/azure/machine-learning/algorithm-cheat-sheet?view=azureml-api-1

16. Sarker, I.H.: Machine learning: algorithms, real-world applications and research directions. SN Comput. Sci. **2**(3) (2021). https://doi.org/10.1007/s42979-021-00592-x

17. Scikit-Learn: Scikit-Learn documentation on RandomForestClassifier. https://scikit-learn.org/stable/modules/generated/sklearn.ensemble.RandomForestClassifier.html

# Addressing Initialization and Data Ordering Issues in Latent Factor-Based Recommendation Systems

Gia Hong Tiet[1,2] 🆔, Thi Hoang Vy Ho[1,2] 🆔, Thi Thanh Ha Do[1,2] 🆔,
Thi My Hang Vu[1,2] 🆔, Le Thi Kim Nhung Ho[1,2] 🆔, Cuong Pham-Nguyen[1,2] 🆔,
and Nguyen Hoai Nam Le[1,2(✉)] 🆔

[1] Faculty of Information Technology, University of Science, Ho Chi Minh City, Vietnam
{tghong,hthvy,dttha,vtmhang,hltknhung,pncuong,
lnhnam}@fit.hcmus.edu.vn
[2] Vietnam National University, Ho Chi Minh City, Vietnam

**Abstract.** Recommendation systems play a crucial role in helping users navigate information overload, particularly in today's digital era. Their primary objective is to predict users' preferences for items. Latent factor-based recommendation systems achieve this by aligning users and items under latent factors. Previous studies mainly focused on devising effective objective functions for learning these latent factors. However, the accuracy of latent factors also depends on their initialization and the order of the collected data fed into the training. Therefore, in this paper, we propose methods to address these two issues. The experiments were conducted on two standard datasets, Movielens 1M and Yahoo Webscope R4, using the RMSE metric. The experimental results indicated that our proposed methods improve the accuracy of latent factor-based recommendation systems.

**Keywords:** Latent factor · Collaborative filtering · Recommendation system

## 1 Introduction

The recommendation system is an advanced technique designed to find the most suitable items for users. By analyzing users' past preferences, this system can predict items they may be interested in, providing accurate recommendations in the future [1]. The applications of recommendation systems are diverse and have become an indispensable part of Internet platforms. For instance, on Facebook, recommendation systems suggest friends for users to connect with. Amazon has achieved remarkable success in assisting users in making more accurate decisions through recommendation systems [2].

The latent factor model is one of the effective models for implementing recommendation systems. It is based on the principle of identifying a set of latent factors to represent users and items [3]. This representation facilitates predicting a user's preference for an item. Consequently, items with high preferences are recommended to the user.

The process of learning to represent users and items under latent factors relies on item preferences collected from users in the past. These preferences are typically expressed as numerical ratings, ranging from very dissatisfied to very satisfied with the items. Thus, an objective function related to the distance between observed ratings and model-predicted ratings is optimized [4]. Many studies focus on utilizing additional information beyond ratings to construct even better objective functions [3–5].

The optimization process of objective functions in latent factor models primarily relies on Stochastic Gradient Descent (SGD). Its effectiveness depends not only on how the objective function is constructed but also on how valuables are initialized and the order in which observed ratings are fed into SGD. In this paper, we utilize observed ratings and item categories to address the two aforementioned issues. Specifically, the contributions of this paper are as follows:

- We provide an interpretation of latent factors in recommendation systems.
- We introduce a method for selection of significant users. The above interpretation provides a foundation for initializing latent factors based on these significant users.
- We propose a technique that leverages the insights gained from selected significant users to guide the prioritization of observed ratings during the system training phase.

## 2   Related Work

### 2.1   Problem Statement

The inputs to recommendation systems consist of preferences collected in the form of ratings, typically ranging from 1 to 5. These ratings are gathered through surveys following user experiences with items. The system's objective is to predict the unknown ratings of a target user [6]. The items predicted to be most liked are then recommended to the target user. All known and unknown ratings are organized into entries of an item-user rating matrix, denoted as $R_{(n \times m)}$ where $n$ is the number of items and $m$ is the number of users. Figure 1 illustrates an item-user rating matrix and its symbols.

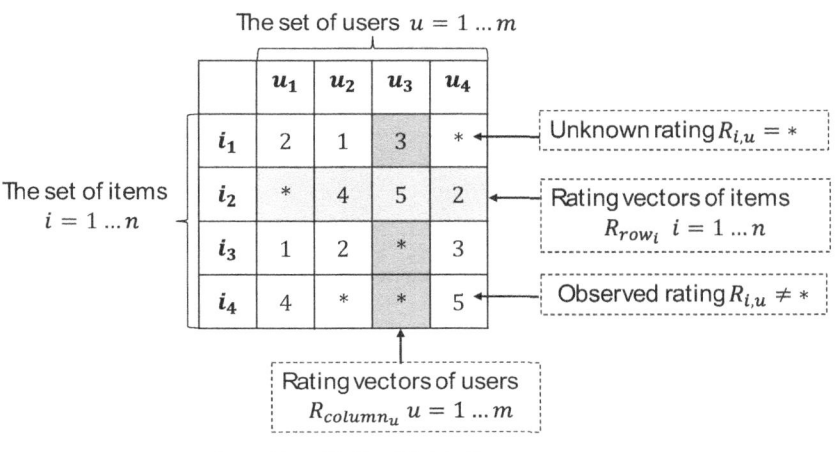

**Fig. 1.**  An item-user rating matrix.

## 2.2  Latent Factor Model

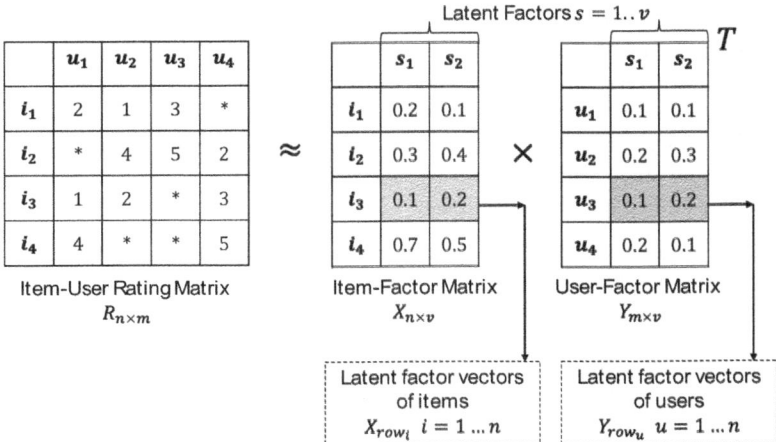

**Fig. 2.** Latent factor model.

Latent factor models aim to learn $n$ vectors representing $n$ items and $m$ vectors representing $m$ users across $v$ latent factors [3, 4, 7]. As presented in Fig. 2, these vectors are organized into the rows of two latent factor matrices: an item-factor matrix, denoted as $X_{(n\times v)}$, and a user-factor matrix, denoted as $Y_{(m\times v)}$. Given an item-user rating matrix $R_{(n\times m)}$, $X_{(n\times v)}$ and $Y_{(m\times v)}$ are the results of the approximation on $R_{(n\times m)}$ as follows:

$$R_{(n\times m)} \approx X_{(n\times v)}.Y^T_{(m\times v)} \tag{1}$$

Because $X$ and $Y$ are fully specified, predicting an unknown rating (i.e., a missing entry in $R$) can be accomplished as follows:

$$R_{i,u} \approx X_{row_i}.Y^T_{row_u} \approx \sum_{z=1}^{v} X_{i,z}.Y_{u,z} \tag{2}$$

## 2.3  Objective Function for Learning Latent Factor Matrices

As presented in Subsect. 2.2, $R$ is a missing matrix with specified entries containing observed ratings. Therefore, its approximation in Eq. (1) is equivalent to optimizing an objective function as follows [7]:

$$\min_{X,Y} \sum_{R_{i,u}\neq *} \left(R_{i,u} - X_{row_i}.Y^T_{row_u}\right)^2 \tag{3}$$

Most research on latent factor models focuses on improving the aforementioned objective function. Specifically, [4] and [8] incorporated data on the items that each user has interacted with ($\mathbb{I}_u u = 1 \dots m$), such as clicks or purchases, into the objective function alongside the rating data. In [8], the aggregation of latent factor vectors of the items that a user has interacted with ($\sum_{j \in \mathbb{I}_u} F_{row_j}$) is used to supplement the latent factor vector of that user in the objective function, as follows:

$$\min_{X,Y,F} \sum_{R_{i,u} \neq *} \left( R_{i,u} - X_{row_i} \cdot \left( Y_{row_u} + |\mathbb{I}_u|^{-\frac{1}{2}} \cdot \sum_{j \in \mathbb{I}_u} F_{row_j} \right)^T \right)^2 \tag{4}$$

The study [4] learns an item-factor matrix ($X'_{(n \times v)}$) and a user-factor matrix ($Y'_{(m \times v)}$) based on an item-user interaction matrix ($R'_{(n \times m)}$). These two matrices are utilized to model the final user-factor and item-factor matrices in the objective function, as follows:

$$R'_{(n \times m)} \approx X'_{(n \times v)} \cdot Y'^T_{(m \times v)}$$

$$\min_{X,Y,Q} \sum_{R_{i,u} \neq *} \left( R_{i,u} - (X_{row_i} + X'_{row_i} \cdot Q) \cdot (Y_{row_u} + Y'_{row_u} \cdot Q)^T \right)^2 \tag{5}$$

The authors in [3] convert the reviews written by users for items into numerical sentiments ($S_{i,u}$). These numerical values are then incorporated as a second rating in the objective function, as follows:

$$\min_{X,Y} \sum_{R_{i,u} \neq *} \left( R_{i,u} - X_{row_i} \cdot Y^T_{row_u} \right)^2 + \left( S_{i,u} - X_{row_i} \cdot Y^T_{row_u} \right)^2 \tag{6}$$

[9] relies on item descriptions to identify items that are similar to a given item ($\mathbb{N}_i i = 1 \dots n$). The latent factor vector of an item is enriched by incorporating the latent factor vectors of the items similar to it ($\sum_{j \in \mathbb{N}_i} F_{row_j}$) within the objective function, as follows:

$$\min_{X,Y,F} \sum_{R_{i,u} \neq *} \left( R_{i,u} - \left( X_{row_i} + |\mathbb{N}_i|^{-\frac{1}{2}} \cdot \sum_{j \in \mathbb{N}_i} F_{row_j} \right) \cdot Y^T_{row_u} \right)^2 \tag{7}$$

In [5], the authors consider text reviews (Bert vector: $W_{i,u}$) as contextual information. These text reviews directly influence user ratings within the objective function, as follows:

$$\min_{X,Y} \sum_{R_{i,u} \neq *} \left( R_{i,u} - X_{row_i} \cdot Y^T_{row_u} - X_{row_i} \cdot W^T_{i,u} - Y_{row_u} \cdot W^T_{i,u} \right)^2 \tag{8}$$

## 2.4  Initialization of Latent Factor Matrices

Algorithm 1. An SGD process to optimize the objective function of the latent factor model.

Input: item-user rating matrix: $R_{n \times m}$; the number of latent factors: $v$
Output: item-factor matrix: $X_{n \times v}$; user-factor matrix: $Y_{m \times v}$
Initialize $X$ and $Y$
Get observed ratings: $\mathbb{E} = \{ (i, u) \mid i = 1 \ldots n \wedge u = 1 \ldots m \wedge R_{i,u} \neq * \}$
While (Not satisfying the convergence criterion):
   Randomly shuffle the elements in $\mathbb{E}$
   For each pair $(i, u)$ in $\mathbb{E}$:
      Update the latent factor vector of $i$: $X_{row_i} \leftarrow X_{row_i} - \alpha . \nabla_{X_{row_i}} J(i, u)$
      Update the latent factor vector of $u$: $Y_{row_u} \leftarrow Y_{row_u} - \alpha . \nabla_{Y_{row_u}} J(i, u)$

Most studies optimize the objective function of the latent factor model using Stochastic Gradient Descent (SGD). One advantage of SGD is its utilization of only one observed rating at a time for learning. This approach is particularly suitable when deploying recommendation systems in contexts with a high number of users and items, especially for generating real-time recommendations [10, 12]. Specifically, as shown in Algorithm 1, for each data point $(i, u)$ corresponding to an observed rating $(R_{i,u} \neq *)$, SGD operates solely on the objective function at this point, denoted as $J(i, u)$, and iteratively updates each variable in $J(i, u)$ using the partial derivative of $J(i, u)$ concerning that variable ($\nabla_{X_{row_i}} J(i, u)$ and $\nabla_{Y_{row_u}} J(i, u)$) with a learning rate $\alpha$.

As outlined in Algorithm 1, the SGD requires initialization. Effective initialization is crucial for SGD to produce user-factor and item-factor matrices that accurately reflect users' preferences and item characteristics. Many studies on latent factor models often initialize these two matrices randomly, thus leading to non-deterministic results for rating predictions. Therefore, some studies leverage external information for initialization. For example, [12] learned the embedding matrix of items from their textual descriptions, which was then used as the initial item-factor matrix. Additionally, [4] utilizes the item-factor and user-factor matrices learned from user behaviors (as referred to in Eq. (5)) for initialization in the objective function. Moreover, for text review data, [5] aggregates all reviews of a user and for an item, vectorizes them using Bert, and then utilizes them to initialize the latent factor vectors of corresponding users and items.

## 3 Motivation

To mitigate random initialization in SGD during the training process of the latent factor model, previous studies have leveraged various additional sources of information, such as purchasing behavior, item descriptions, textual reviews, etc. However, such supplementary data may not always be available, limiting the flexibility of recommendation systems across different domains. Therefore, in this paper, we introduce a method for initializing latent factor matrices in training the latent factor model. Our proposed method relies solely on the fundamental information typically collected by recommendation systems: ratings and item categories.

In this study, we treat the item-user rating matrix as a dataset of items, where the categories of the items serve as data labels. Therefore, a supervised feature selection method can be employed to identify significant users within the system. Then, we will reinterpret the meaning of the user-factor and item-factor matrices in the latent factor model to leverage the significant users in initializing them during SGD-based system training. These details will be presented in Subsects. 4.1 and 4.2. Furthermore, in Subsect. 4.3, we also use the significant users to determine the order of observed ratings fed into SGD rather than being completely random.

# 4   Proposed Method

## 4.1   Interpretation of Latent Factor Matrices

The core of the latent factor model is to approximate a missing item-user rating matrix $R_{(n \times m)}$ into two latent factor matrices: an item-factor matrix $X_{(n \times v)}$ and a user-factor matrix $Y_{(m \times v)}$, as presented in Eq. (1). We rewrite Eq. (1) in a column-wise direction as follows:

$$R_{(n \times m)} \approx X_{(n \times v)}.Y_{(m \times v)}^{T} \Leftrightarrow R_{column_u} \approx \sum_{j=1}^{v}(Y_{u,j}.X_{column_j})  \ u = 1 \ldots m \quad (9)$$

It can be interpreted from Eq. (9) that the $v$ columns of matrix $X$, i.e., $X_{column_j}$ $j = 1...v$, serve as base users. They are axes for reconstructing the $m$ original users, i.e., $R_{column_u} u = 1 \ldots m$. Each entry $Y_{u,j} j = 1 \ldots v$ represents the coordinate of the original user $u$ concerning the base user $X_{column_j} j = 1 \ldots v$. Figure 3 illustrates the roles of base users and user coordinates. Now, the item-factor matrix $X$ and user-factor matrix $Y$ can be referred to as the base user matrix and user coordinate matrix, respectively.

## 4.2   Initialization of Base User Matrix and User Coordinate Matrix (UCInit)

In this subsection, we propose a method to **Init**ialize the base **U**ser matrix $X_{(n \times v)}$ and user **C**oordinate matrix $Y_{(m \times v)}$, named **UCInit**. Specifically, we identify significant users to serve as initial base users, i.e., the initialization of the columns of matrix $X$, as shown in Fig. 4. With a large number of items across various categories, the users' preferences are very diverse. Therefore, the preferences of base users must also be diverse enough to span all preferences of original users in the initial iterations of SGD. Based on these observations, first, we extract the top $p$ original users who have provided the highest number of ratings in the system. These $p$ users are then considered as features representing the items. This presentation and item categories become the labeled data for the item classification. With this setup, a supervised feature selection method is performed to identify the most significant features (users) across various item categories. Eventually, the top $v$ significant users are chosen as the $v$ initial base users.

As depicted in Subsect. 4.1, each entry in the user coordinate matrix $Y$ represents the coordinate of an original user concerning a base user. Thus, as shown in Fig. 5, matrix $Y$ can be initialized by the correlation between pairs of users, with one being an original user $u$ and the other being the significant user $u'$ selected as the initial base user. In this paper, we use cosine to compute their correlation, as follows:

$$cosine_{u,u'} = \frac{\sum_{i \in \mathbb{I}_u \cap \mathbb{I}_{u'}} R_{i,u} \times R_{i,u'}}{\sqrt{\sum_{i \in \mathbb{I}_u} R_{i,u}^2} \times \sqrt{\sum_{i \in \mathbb{I}_{u'}} R_{i,u'}^2}} \quad (10)$$

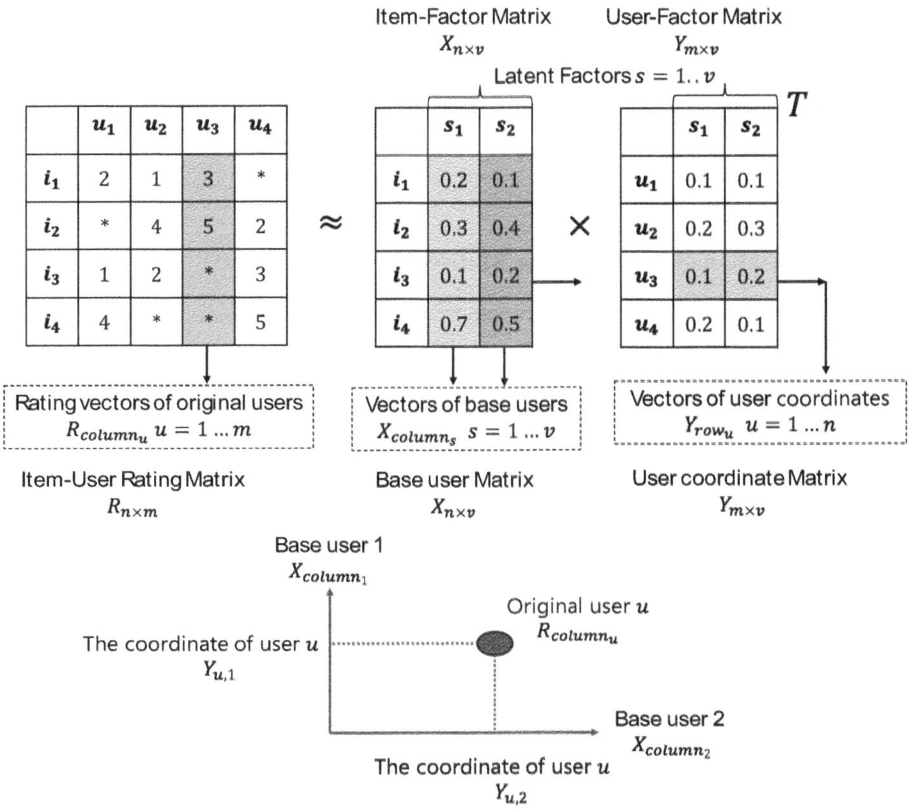

Item-Factor Matrix
$X_{n\times v}$

User-Factor Matrix
$Y_{m\times v}$

Latent Factors $s = 1..v$

|     | $u_1$ | $u_2$ | $u_3$ | $u_4$ |
|-----|-------|-------|-------|-------|
| $i_1$ | 2 | 1 | 3 | * |
| $i_2$ | * | 4 | 5 | 2 |
| $i_3$ | 1 | 2 | * | 3 |
| $i_4$ | 4 | * | * | 5 |

$\approx$

|     | $s_1$ | $s_2$ |
|-----|-------|-------|
| $i_1$ | 0.2 | 0.1 |
| $i_2$ | 0.3 | 0.4 |
| $i_3$ | 0.1 | 0.2 |
| $i_4$ | 0.7 | 0.5 |

$\times$

|     | $s_1$ | $s_2$ |
|-----|-------|-------|
| $u_1$ | 0.1 | 0.1 |
| $u_2$ | 0.2 | 0.3 |
| $u_3$ | 0.1 | 0.2 |
| $u_4$ | 0.2 | 0.1 |

$T$

Rating vectors of original users
$R_{column_u}$  $u = 1 \ldots m$

Vectors of base users
$X_{column_s}$  $s = 1 \ldots v$

Vectors of user coordinates
$Y_{row_u}$  $u = 1 \ldots n$

Item-User Rating Matrix
$R_{n\times m}$

Base user Matrix
$X_{n\times v}$

User coordinate Matrix
$Y_{m\times v}$

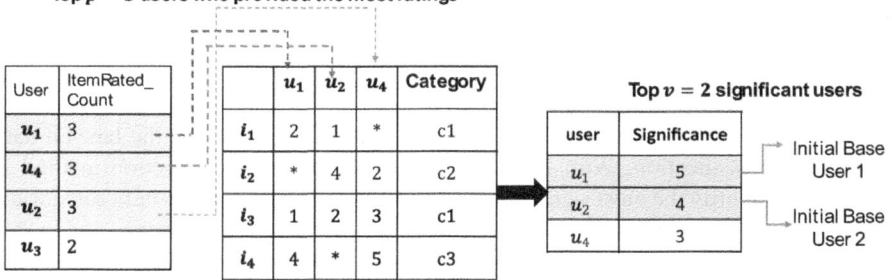

Base user 1
$X_{column_1}$

Original user $u$
$R_{column_u}$

The coordinate of user $u$
$Y_{u,1}$

Base user 2
$X_{column_2}$

The coordinate of user $u$
$Y_{u,2}$

**Fig. 3.** Base user matrix and user coordinate matrix.

Top $p = 3$ users who provided the most ratings

| User | ItemRated_Count |
|------|-----------------|
| $u_1$ | 3 |
| $u_4$ | 3 |
| $u_2$ | 3 |
| $u_3$ | 2 |

|     | $u_1$ | $u_2$ | $u_4$ | Category |
|-----|-------|-------|-------|----------|
| $i_1$ | 2 | 1 | * | c1 |
| $i_2$ | * | 4 | 2 | c2 |
| $i_3$ | 1 | 2 | 3 | c1 |
| $i_4$ | 4 | * | 5 | c3 |

Top $v = 2$ significant users

| user | Significance |
|------|--------------|
| $u_1$ | 5 |
| $u_2$ | 4 |
| $u_4$ | 3 |

Initial Base User 1

Initial Base User 2

**Fig. 4.** The initialization of the base user matrix

where $\mathbb{I}_u$ and $\mathbb{I}_{u'}$ are the item sets rated by $u$ and $u'$, respectively. However, computing the cosine between each original user and each initial base user can be computationally

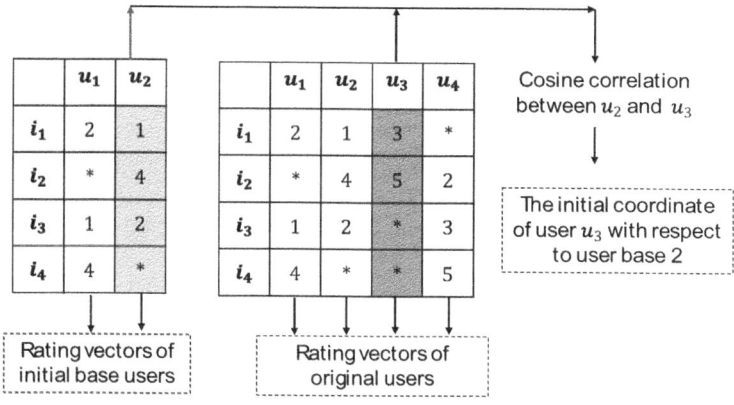

**Fig. 5.** The initialization of the user coordinate matrix

expensive, especially as the number of users increases. Therefore, we have designed distributed processing on Hadoop for implementing the initialization of the user coordinate matrix, as shown in Fig. 6.

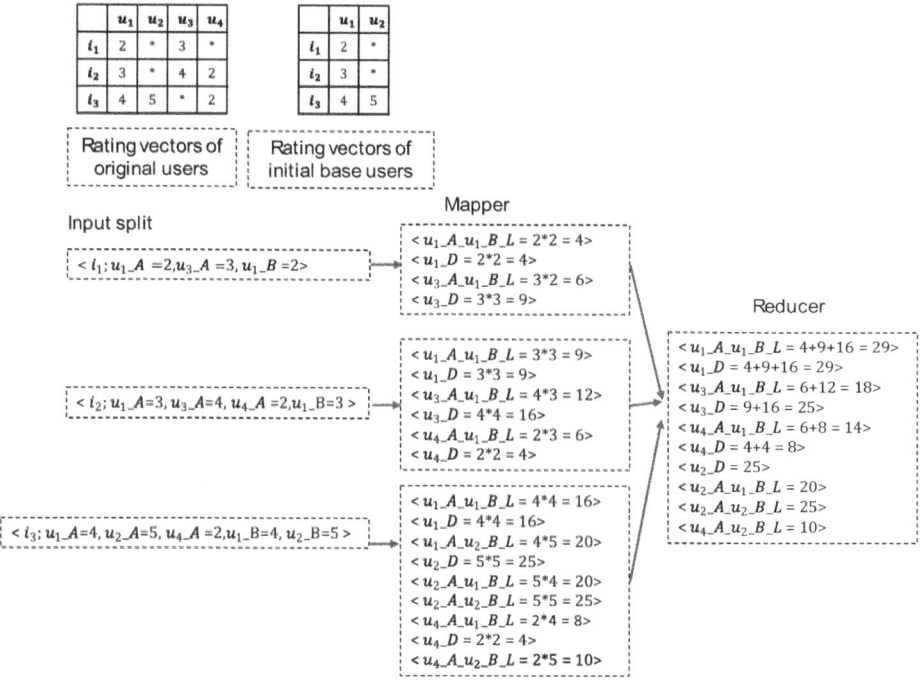

**Fig. 6.** Implementing the initialization of the user coordinate matrix on Hadoop.

Specifically, firstly, the rating data is organized according to each item $i = 1 \ldots n$. For example, the rating data corresponding to the item $i_1$ in Fig. 6 is as follows:

$$< i_1; u_1\_A = 2 , u_3\_A = 3 , u_1\_B = 2 > \tag{11}$$

where the part after $i_1$ is the set of users ($u_1$ and $u_3$) who have rated $i_1$ along with their corresponding rating values (2 and 3). These users are divided into two types: original users, denoted by _A ($u_1\_A$ and $u_3\_A$), and significant users selected as initial base users, denoted by _B ($u_1\_B$).

The $n$ rating data samples will be processed in a distributed manner, each by a separate Hadoop mapper function. Specifically, for each pair of users consisting of one followed by _A and another followed by _B, the mapper will compute the product of their ratings. Additionally, for all users, the square of their ratings is also calculated. For example, considering the rating data sample in Eq. (11), the pair of users $u_3\_A$ and $u_1\_B$ will achieve the product of their ratings for the item $i_1$, denoted as $u_3\_A\_u_1\_B\_L$. Additionally, the square of $u_1$'s rating and $u_3$'s rating for the item $i_1$ will be calculated, denoted as $u_1\_D$ and $u_3\_D$. Therefore, the mapper function for the rating data sample in Eq. (11) will output the $<$ key; value $>$ pairs as follows:

$$\begin{array}{cc} < u_3\_A\_u_1\_B\_L = 2 * 3 = 6 > & < u_1\_A_{u_1}\_B_L = 2 * 2 = 4 > \\ < u_1\_D = 2 * 2 = 4 > & < u_3\_D = 3 * 3 = 9 > \end{array} \tag{12}$$

After all mappers have completed their computations, a reducer function calculates the sum of values with the same key. For example, keys such as $u_3\_A\_u_1\_B\_L$, $u_1\_D$, $u_3\_D$, etc., may be present. To compute the cosine between $u_3$ and $u_1$ according to Eq. (10), the sum of values with the key $u_3\_A\_u_1\_B\_L$ represents the value of the component $\sum_{i \in \mathbb{I}_{u_3} \cap \mathbb{I}_{u_1}} R_{i,u_3}.R_{i,u_1}$; the sum of values with the same key $u_1\_D$ represents the value of the component $\sum_{i \in \mathbb{I}_{u_1}} R_{i,u_1}^2$; and the sum of values with the key $u_3\_D$ represents the value of the component $\sum_{i \in \mathbb{I}_{u_3}} R_{i,u_3}^2$.

### 4.3   Data Ordering in the System Training

Because the rating scales are defined to be extensive, it's challenging for users to select a rating that accurately reflects their true preferences. Consequently, the accuracy of the observed ratings is often not high. This significantly impacts models trained on observed ratings. In Subsect. 4.2, users are selected to possess diverse experiences. Such experiences build trust in the accuracy of the ratings they provide.

Therefore, in this subsection, we propose prioritizing the observed ratings of users selected in Subsect. 4.2 to be fed into SGD first. Their prioritized learning helps the model quickly converge to crucial aspects of item characteristics and user preferences, avoiding overlooking them during the learning process. However, we still ensure the randomness of SGD in processing among the ratings of selected users and among the ratings of non-selected users to avoid overfitting during the learning process. Algorithm 2 outlines the modifications to the SGD presented in Algorithm 1.

Algorithm 2. The modifications to the SGD presented in Algorithm 1.

---

Get observed ratings of selected users: $\mathbb{Q}$
Get observed ratings of non-selected users: $\mathbb{E}$
While (Not satisfying the convergence criterion):
    Randomly shuffle the elements in $\mathbb{Q}$
    Randomly shuffle the elements in $\mathbb{E}$
    For each pair $(i, u)$ in $\mathbb{Q}$:
        Update the latent factor vector of $i$: $X_{row_i} \leftarrow X_{row_i} - \alpha.\nabla_{X_{row_i}} J(i, u)$
        Update the latent factor vector of $u$: $Y_{row_u} \leftarrow Y_{row_u} - \alpha.\nabla_{Y_{row_u}} J(i, u)$
    For each pair $(i, u)$ in $\mathbb{E}$:
        Update the latent factor vector of $i$: $X_{row_i} \leftarrow X_{row_i} - \alpha.\nabla_{X_{row_i}} J(i, u)$
        Update the latent factor vector of $u$: $Y_{row_u} \leftarrow Y_{row_u} - \alpha.\nabla_{Y_{row_u}} J(i, u)$

---

## 5 Experiment

### 5.1 Experiment Setup

The context of this study is solely using ratings and item categories. With these two types of data, as presented in Fig. 7, we implemented two initialization methods: UCInit, as proposed in Subsect. 4.2, and Random (RandInit). Regarding the objective function of the latent factor model, we employed two methods: MF [7] (refer to Eq. (3)), which relies on ratings, and SC2 [4] (refer to Eq. (5)), which incorporates both ratings and behaviors. Thus, if a user rates an item, he/she is assumed to make a behavior on it.

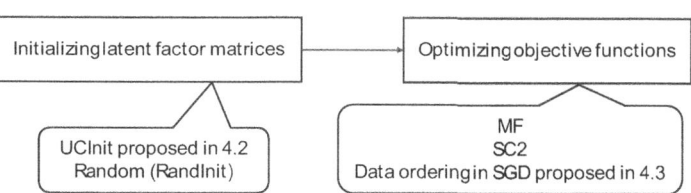

**Fig. 7.** Experimental methods.

To optimize the objective functions using Stochastic Gradient Descent (SGD), we set the learning rate to 0.001 and defined the convergence condition as reaching 500 iterations of variable updates. For UCInit, we fixed the number of users selected in the first filtering round, i.e., parameter $p$, at 300. Additionally, the number of users selected in the second filtering round was equal to the number of latent factors. UCInit utilizes a supervised feature selection method. In this study, we implemented common supervised feature selection methods: IG [13], CHI [14], CMFS [15], and OCFS [16].

## 5.2  Dataset

In the experiment, we used the following two datasets:

- Movielens 1M: 1,000,209 ratings of 6,040 users for 3,706 movies.
- Yahoo Webscope R4: 221,367 ratings of 7,642 users for 106,959 movies.

In these datasets, the ratings were divided into 80% for training and 20% for testing. The movie categories are retrieved from the website https://www.imdb.com/.

## 5.3  Measure

The latent factor matrices directly influence the precision of subsequent rating predictions. In this paper, we employ RMSE to assess the precision of these rating predictions. Its formula is as follows:

$$RMSE = \sqrt{\frac{\sum_{R_{ui} \in \mathbb{R}} \left( \widehat{R}_{ui} - R_{ui} \right)^2}{|\mathbb{R}|}} \tag{13}$$

where $R_{ui}$ and $\widehat{R}_{ui}$ are the rating in the testing set ($\mathbb{R}$) and its prediction, respectively.

## 5.4  Experiment Result and Discussion

Figure 8 illustrates the RMSE results of MF and SC2 initialized by UCInit and RandInit while varying the number of latent factors from 60 to 90. It can be observed that the methods initialized by UCInit outperform those initialized randomly (RandInit). Specifically, in the Movielens 1M dataset, with 60, 70, 80, and 90 latent factors, SC2 initialized by UCInit reduces RMSE by 0.013, 0.026, 0.030, and 0.031 respectively compared to SC2 initialized by RandInit. Recommendation systems often recommend a small number of items. Therefore, even a minor adjustment in prediction error can result in significant changes in the recommended items and their ordering within the recommendation set. This argument reinforces the effectiveness of our proposed initialization method UCInit.

Next, the top-performing methods mentioned above will be integrated with our method for determining the order of data fed into the training proposed in Subsect. 4.3. Figure 9 illustrates that our methods (UCInit+MF+Ordering and UCInit+SC2+Ordering) continue to decrease the RMSE of the rating predictions. The improvement in RMSE becomes more apparent as the number of latent factors increases. The reason for this is as follows: When the number of latent factors increases, the number of variables also significantly increases. With such a large number of variables, the risk of overfitting is high when training data is limited. At this point, rationality in initialization and data ordering during training will be most effective.

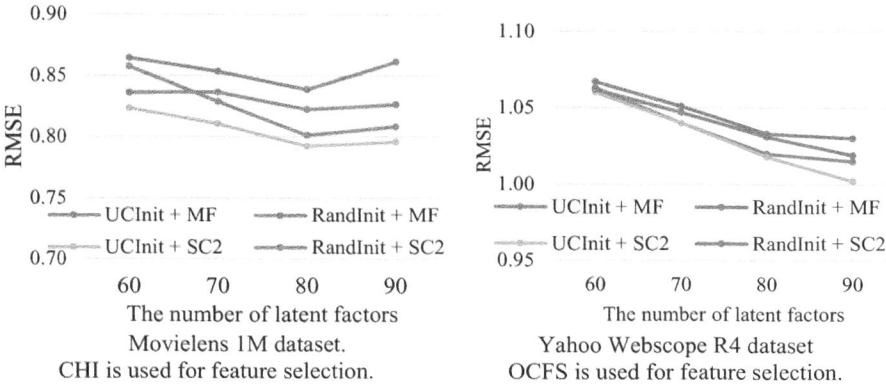

**Fig. 8.** RMSE results of MF and SC2 initialized by UCInit and RandInit

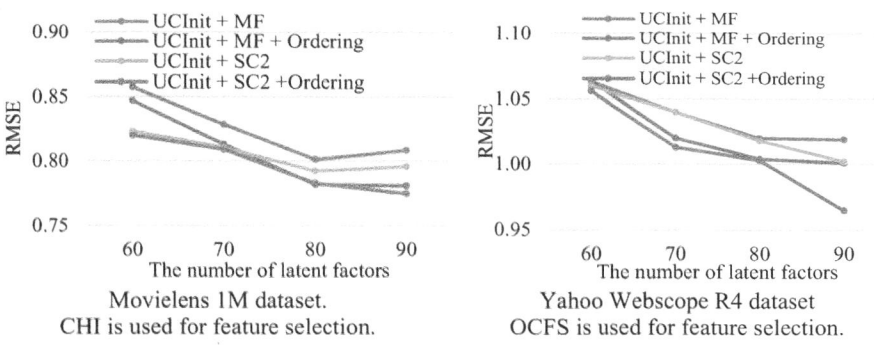

**Fig. 9.** RMSE results when data is ordered during system training.

Figure 10 depicts the average RMSE of MF and SC2 initialized by UCInit when employing feature selection methods such as IG [13], CHI [14], CMFS [15], and OCFS [16]. Among these four feature selection methods, CHI and OCFS yield the best results in Movielens 1M and Yahoo Webscope R4, respectively. Hence, we selected CHI and OCFS for the experiments above. In this paper, we employed filter-based feature selection methods. Their advantage lies in cost-effectiveness, as they consider each feature independently. However, a drawback is that the selected features may exhibit high correlations. In the future, our methods could be further enhanced by incorporating more sophisticated feature selection techniques, such as wrapper-based feature selections.

Finally, we evaluated the methods using t-test statistical analysis. The sample set for the t-test consists of the RMSE of all users using both MF and SC2. The results in Table 1 indicate that initializing with UCInit and ordering data during training improved the statistical effectiveness of rating predictions, as all p-values are less than 0.05.

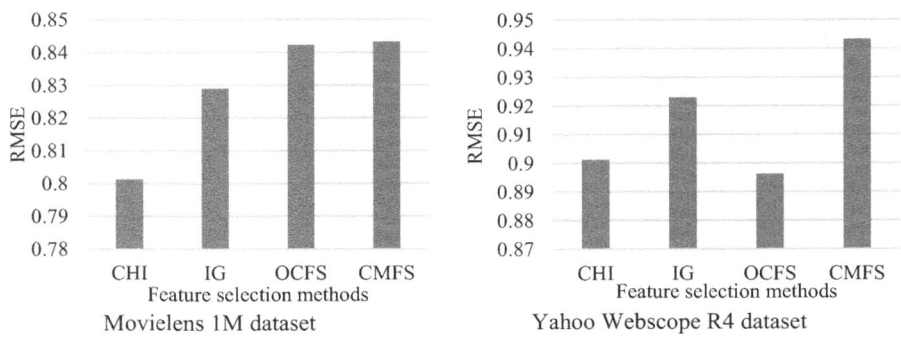

**Fig. 10.** Average RMSE results of UCInit when feature selection methods such as CHI, IF, OCFS, and CMFS are used.

**Table 1.** T-test result on RMSE results

|  | UCInit and RandInit | UCInit + ordering and UCInit |
|---|---|---|
| RMSE mean: p-value | 0.923 >> 0.942: 0.0061 | 0.855 >> 0.923: 0.0095 |
| Statistical Conclusion | UCInit >> RandInit | UCInit + Ordering >> UCInit |

## 6   Conclusion

In this paper, we have interpreted the meaning of the latent factor matrices in latent factor-based recommendation systems into base user matrix and user coordinate matrix. As a result, we propose a method to initialize them during the system training. Specifically, we select significant users from the original user set to initialize the base users in the base users matrix. Each entry in the initial user coordinate matrix is the cosine correlation between each original user and each selected significant user. Furthermore, we also prioritize the ratings collected from the selected significant users during system training due to their accuracy and diversity. This prioritization helps in achieving latent factor matrices that more accurately reflect the characteristics of items and users.

**Acknowledgments.** This research is funded by University of Science, VNU-HCM under grant number CNTT 2023–12.

## References

1. Ahmadian, M., Ahmadi, M., Ahmadian, S.: A reliable deep representation learning to improve trust-aware recommendation systems. Expert Syst. Appl. **197**, 116697 (2022)
2. Smith, B., Linden, G.: Two decades of recommender systems at Amazon. com. IEEE Int. Comput. **21**(3), 12–18 (2017)
3. Shen, R.P., Zhang, H.R., Yu, H., Min, F.: Sentiment based matrix factorization with reliability for recommendation. Expert Syst. Appl. **135**, 249–258 (2019)

4. Nam, L.N.H.: Latent factor recommendation models for integrating explicit and implicit preferences in a multi-step decision-making process. Expert Syst. Appl. **174** (2021)
5. Nam, L.N.H.: Incorporating textual reviews in the learning of latent factors for recommender systems. Electron. Comm. Res. Appl. **52** (2022)
6. Aggarwal, C.C.: An introduction to recommender systems. Recomm. Syst. Textbook 1–28 (2016)
7. Koren, Y., Bell, R., Volinsky, C.: Matrix factorization techniques for recommender systems. Computer **42**(8), 30–37 (2009)
8. Koren, Y.: Factorization meets the neighborhood: a multifaceted collaborative filtering model. In: Proceedings of the 14th ACM SIGKDD International Conference on Knowledge Discovery and Data Mining, pp. 426–434 (2008)
9. Wang, R., Cheng, H.K., Jiang, Y., Lou, J.: A novel matrix factorization model for recommendation with LOD-based semantic similarity measure. Expert Syst. Appl. **123**, 70–81 (2019)
10. Luo, X., Qin, W., Dong, A., Sedraoui, K., Zhou, M.: Efficient and high-quality recommendations via momentum-incorporated parallel stochastic gradient descent-based learning. IEEE/CAA J. Autom. Sinica **8**(2), 402–411 (2020)
11. Vy, H.T.H., Pham-Nguyen, C., Nam, L.N.H.: Integrating textual reviews into neighbor-based recommender systems. Expert Syst. Appl. 123648 (2024). https://doi.org/10.1016/j.eswa.2024.123648
12. Khan, Z., Iltaf, N., Afzal, H., Abbas, H.: Enriching non-negative matrix factorization with contextual embeddings for recommender systems. Neurocomputing **380**, 246–258 (2020)
13. Quinlan, J.R.: Induction of decision trees. Mach. Learn. **1**, 81–106 (1986)
14. Yang, Y., Pedersen, J.O.: A comparative study on feature selection in text categorization. In: ICML, vol. 97, no. 412–420, p. 35 (1997)
15. Yang, J., Liu, Y., Zhu, X., Liu, Z., Zhang, X.: A new feature selection based on comprehensive measurement both in inter-category and intra-category for text categorization. Inf. Process. Manage. **48**(4), 741–754 (2012)
16. Yan, J., et al.: OCFS: optimal orthogonal centroid feature selection for text categorization. In: Proceedings of the 28th Annual International ACM SIGIR Conference on Research and Development in Information Retrieval, pp. 122–129 (2005)

# Usability Assessment of the Use-Case Model Textual Specification Language

Bogumiła Hnatkowska[(✉)] and Joanna Pszon

Wrocław University of Science and Technology, 50-370 Wrocław, Poland
{bogumila.hnatkowska,joanna.pszon}@pwr.edu.pl

**Abstract.** The requirements specification is a key outcome of requirements engineering. It serves as input to all phases of software development, including design, implementation, and testing. On one hand, due to the diverse competencies of stakeholders, it should be written in a clear and understandable language, preferably natural. On the other hand, it should be structured and computationally processable. Controlled languages, with limited syntax, partially meet both demands, albeit restricting the freedom of requirement expression. This raises the question of whether this limitation affects the overall usability of the specification language from the perspective of its users. To address this question, a controlled experiment was conducted to assess the usability of the Use-Case Model Textual Specification Language (T-UFCL). Its usability was compared with alternative notations – informal (natural language) and formal (activity diagrams). The experiment results showed that the controlled language's usability is comparable to natural language and, in certain cases, higher than that of formal notations.

**Keywords:** Usability · Specification language · Use-case model · Requirement analysis

## 1 Introduction

Requirements analysis is an essential stage in requirements engineering. Its outcome – the Software Requirements Specification (SRS) – forms a contract between development process stakeholders, provides a basis for budget and schedule estimation, and a baseline for validation and verification [12]. Goot SRS should possess adequate quality, be well-structured, readable, internally consistent, and complete concerning user needs [15]. Maintaining structure and readability is aided by using templates, e.g., [11] or controlled language, which resembles natural language [12–14].

However, applying a template or controlled language does not determine the language's usability. Depending on the adopted syntax, the language may be easier or more difficult to learn, understand, and apply in typical scenarios.

Usability refers to the "quality of the interaction in terms of parameters such as time taken to perform tasks, number of errors made, and the time to become a competent user." It also refers to user satisfaction [7].

© The Author(s), under exclusive license to Springer Nature Switzerland AG 2024
N.-T. Nguyen et al. (Eds.): ICCCI 2024, CCIS 2166, pp. 378–390, 2024.
https://doi.org/10.1007/978-3-031-70259-4_29

This first research goal is to evaluate the usability of a controlled natural language (CNL) named Textual Use-case Flow Language (T-UCFL), as described in [15,16]. The language serves for use-case specification, where the specification is understood as a description of intended interactions between a system and its actors. The primary motivation for the T-UCFL was to enhance the reuse of specification fragments.

The language is in its early development stage. The paper [16] presented its genesis, including sources of inspiration for individual constructions, the meta-model of the language, and its textual syntax. A context-free grammar, as well as language application examples, are available at [18]. Before further language development, including the proposal of a graphical syntax and the development of validation tools for internal specification consistency, it was decided to conduct a usability study.

The second research goal is more general and aims to answer whether the application of NCL does not negatively influence language usability compared with other notations used for requirement specification.

Potential benefits of NCL include [12–14]: standardized format to achieve coherence and uniformity, simplified grammar, and predefined vocabulary with precise semantics to achieve better understandability and less ambiguity, automatic verifiability. However, NCL removes some degree of freedom in comparison to natural language. Its user needs to adjust to the predefined syntax; therefore, tool support is requested.

The concept of applying a controlled language for requirement specification is not new. Other examples are presented in [12–14]. The two first describe the ITLingo RSL Language, which covers use-case specification, business domain, and tests. The third is a combination of object-oriented concepts and requirement specification defined at a low level of abstraction. Unfortunately, the usability of none was checked and presented to a broader audience.

The usability study of the T-UCFL encompasses a controlled experiment focusing (similarly to [10]) on effectiveness and user satisfaction analysis. The experiment aimed not only to answer whether the language syntax has reached a maturity level enabling its practical application but also to compare it with other alternative notations used for use case specifications: natural language and activity diagrams.

The rest of the paper is structured as follows. Section 2 presents related works in the context of a specification language usability. Section 3 describes the experiment, while Sect. 4 discusses its results. Inferred conclusions are presented in the last section.

## 2   Related Works

Similarly to a system, requirement specification language can have inadequate usability because it is too difficult. Basic factors of usability include [4–8]:

- Learnability
- Effectiveness

- Efficiency relating to understanding
- Memorability
- Usage satisfaction

Those factors can be evaluated using several methods, e.g., usability inspections done by usability specialists, testing with users working on typical tasks or questionnaires, and surveys [8].

Except for the last, most factors can be assessed with a performance-based approach. Effectiveness can be assessed by the task completion rate, objectives achieved, errors in tasks, tasks with errors, error insensitivity, and help required; efficiency - by task time, time efficiency, and user productivity. Satisfaction is achieved through overall satisfaction with features, feature utilization, user pleasure, or physical comfort. Questionnaires can assess any aspect of the language, e.g., learnability, intuitiveness, or understanding [5,6,8].

The evaluation instruments can also be defined with a cognitive model. The cognitive activities with a specification language include syntax and semantic learning, syntax composition, syntax understanding and remembering, syntax debugging, and changing a function written by another developer [5,9].

A good practice when evaluating usability is the application of both quantitative and qualitative methods [6]. ISO/IEC 25062 Common Industry Format for usability test reports recommends, among others:

- User observation:
  - Qualitative: Observing user behavior to identify usability problems
  - Quantitative: Measures of user performance and responses to obtain data on effectiveness and efficiency
- Subjective information from users:
  - Qualitative: Problems, opinions, and impressions given during or after a usability evaluation
  - Quantitative: Measures of user satisfaction or perception

Quantitative methods may take the form of a controlled experiment in which individuals or teams conduct one or more tasks to compare different languages or tools [9]. Typical tasks for a language are: "sentence writing, sentence reading, comprehension, memorization and problem solving [9]." As evaluation of all of them could be very expensive, the process typically concerns only the most critical activities [9].

The System Usability Scale (SUS) is commonly used to measure perceived usability. There also exist other standardized questionnaires developed for the same purpose (e.g., CSUQ, UMUX), but their results are highly correlated, and SUS is pointed as the best default choice [2]. The SUS scores can serve as a tool for competitor comparison and be used together with usability testing methods to asses both actual user experience and perceived user satisfaction [1]. Moreover, some guidelines exist for the score interpretations (Sauro-Levis CGS). Practitioners with a valid reason to remove one item from the SUS can do so without significantly impacting the final SUS score, as long as they adjust the formula used to calculate the score accordingly [3].

# 3   Research Plan

## 3.1   Objectives

The objectives of T-UCFL usability evaluation were inspired by [10], where the evaluation of a formal specification language named Generalised Test Tables was presented.

The T-UCFL language, to be promoted first among students and next among requirement analysts, should be evaluated against its usability and seen in a broader perspective in the context of its competitors, such as natural language and activity diagrams. To accomplish such goals, three research questions were formulated:

- RQ1 - Is the T-UCFL mature enough for the use-case specification?
- RQ2 - Is the T-UCFL satisfactory for software engineers?
- RQ3 - How is the T-UCFL perceived in the context of alternative notations: (a) natural language, (b) activity diagrams?

The first research question aims to answer whether the T-UCFL serves the intended purpose. The second and third refer to the language's subjective usability, also in the context of alternative notations.

## 3.2   Participants

The bachelor students of Applied Computer Science at Wroclaw University of Technology, during their 5th semester, attend the Software Design course. The course continues another, Basics of Software Engineering, within which the students are taught different requirement specification techniques based on a use-case model, including scenario definitions in natural language and activity diagrams. The project in the 5th semester allows them to solidify acquired knowledge and skills. The project aims to specify, analyze, design, implement, and test a non-trivial software system for the university. Students work in 2–3 member teams. One of the tasks they perform is to specify the behavior of two use cases indicated by the teacher in selected notation.

In the winter semester of 2023, the students, except for natural language or activity diagrams, were offered the possibility to use T-UCFL as an alternative. Activity diagrams and T-UCFL specifications were awarded higher (maximum 7,5 points per use-case) than specifications in natural language (maximum 5 points per use-case) because of their complexity. Each student could decide on which notation to use on his/her own. It should be mentioned that it was the first student's contact with the T-UCFL. Students were offered limited guidelines on how to use the language. They got the link to the paper [16] and the link to language examples and grammar [18]. A total of 27 students took part in the project. The distribution of use case specification methods was as follows: 5 students employed T-UCFL, 8 utilized natural language, and 14 opted for activity diagrams.

### 3.3   Research Methods

We used a survey as a basic evaluation method. The survey was split into five sections devoted to different evaluation aspects:

a. Syntax learnability
b. Notation applicability in the business context
c. Notation efficiency
d. Tool support
e. Perceived satisfaction

All questions used in the survey are presented in Table 1, and its runnable version (in Polish) is available at [17].

The first section of the survey was intended to explore students' perception of T-UCFL syntax. This part targeted only students who actively employed T-UCFL during their use case specification tasks. The remaining part of the survey targeted all students.

A performance-based approach was used as a secondary evaluation method to assess notation effectiveness (section F). Normalized student grades were used as indicators of the error number in tasks.

**Syntax Learnability.** The syntax assessment section consisted of six statements, each rated on a 5-point Likert scale [20]. Additionally, four open-ended optional questions were included.

**Notation Applicability in the Business Context.** This section focuses on the capabilities of the chosen use case specification method. Questions had a particular emphasis on the applicability of a given method in a business context. This part of the survey evaluated the comprehensibility, accuracy, and flexibility of chosen use case specification methods.

**Notation Efficiency.** Two-choice questions were designed to evaluate student engagement with use case specification tasks. These questions aimed to gauge students' time spent learning about and applying the chosen use case specification method.

**Tool Support.** A range of tools can be employed based on the chosen method for use case specification to specify use cases. These tools vary from standard text editors to specialized applications. This section assessed the students' satisfaction with the tools used.

**Perceived Satisfaction.** A subset of the System Usability Scale (SUS) [19] questionnaire was employed to evaluate student satisfaction. While the SUS is primarily designed to assess system usability, certain statements were deemed

**Table 1.** T-UCFL Usability evaluation questionnaire

| No | Type | Statement |
|---|---|---|
| | | T-UCFL Syntax assessment |
| A1 | Likert | I found T-UCFL understandable |
| A2 | Likert | Using T-UCFL, I was able to precisely express everything I wanted |
| A3 | Open | What was missing in T-UCFL? |
| A4 | Likert | I found T-UCFL unnecessarily complicated |
| A5 | Likert | T-UCFL had too many inconsistencies |
| A6 | Open | Provide specific examples of inconsistencies |
| A7 | Open | Are there parts of the syntax that could be more intuitive? If so, which ones? |
| A8 | Open | Is it easy to update use case specification using T-UCFL? What difficulties may arise? |
| A9 | Likert | T-UCFL contributed to making the use case specifications more readable |
| A10 | Likert | The use of T-UCFL accelerates the process of identifying and solving problems in the use case specification process |
| | | Notation applicability in the business context |
| B1 | Likert | I believe that the chosen use case specification method is understandable to business representatives |
| B2 | Likert | I believe that the chosen use case specification method allows me to reflect the requirements of business representatives accurately |
| B3 | Likert | The chosen method of use case specification made me think more deeply about the functionality of the designed system |
| B4 | Likert | I believe that the chosen use case specification method simplifies teamwork on use case specification task |
| B5 | Likert | I find the chosen use case specification method to be flexible in adapting to different types of projects |
| | | Notation efficiency |
| C1 | Enum | Estimate time spent on performing use case specification task |
| C2 | Enum | Estimate time spent on learning how to use the chosen use case specification method |
| | | Tool support |
| D1 | Enum | What tool did you use to create the use case specification? |
| D2 | Likert | Was the tool helpful in the use case specification process? |
| D3 | Likert | Did you find the chosen tool-less cumbersome than other use case specification tools you know? |
| | | Perceived satisfaction |
| E1 | Likert | I think that I would like to use the chosen use case specification method frequently |
| E2 | Likert | I thought the chosen use case specification method was easy to use |
| E3 | Likert | I think that I would need the support of a technical person to be able to use the chosen use case specification method |
| E4 | Likert | I would imagine that most people would learn to use the chosen use case specification method very quickly |
| E5 | Likert | I found the chosen use case specification method very cumbersome to use |
| E6 | Likert | I felt very confident using the chosen use case specification method |
| E7 | Likert | I needed to learn a lot of things before I could get going with the chosen use case specification method |
| E8 | Likert | I found the chosen use case specification method easy to learn |

inappropriate for evaluating use case specification methods. Consequently, three statements from the standard SUS were omitted, and an additional statement was incorporated – see Table 1 for details.

Due to the standard SUS questionnaire deviation, the conventional method for calculating the SUS score was not applicable in this context. Consequently, direct comparisons with existing results within the same field were not feasible. However, by adhering to the standard SUS score calculation procedure, we could derive a numerical score that allows for comparative analysis of T-UCFL with other use case specification methods. The score was calculated as follows:

1. 1 was subtracted from the score for each positive statement
2. The score for each negative statement was subtracted from 5
3. Results for all questions were added
4. The sum of results was multiplied by 2.5

**Notation Effectiveness.** We used the students' grades to evaluate and compare the effectiveness of the notations. Each use-case specification got a numeric grade from 0 (the worst) to 5 or 7.5 (the best), depending on the notation used and the specification quality. The grades were normalized by the subtraction of 2.5 points for each use-case specification defined using the activity diagram or T-UCFL. After that, the points of 2 use-cases were summed up for each student.

### 3.4   Data Analysis

In the context of our research, varying methodologies for data analysis were adopted, contingent upon the data type under investigation. Syntax of the T-UCFL notation was assessed by analyzing open and closed questionnaire questions (section A).

The Kruskal-Wallis non-parametric test was employed to compare the usability of considered notations based on the factors evaluated using a Likert scale (sections B, C, and D).

SUS scores (section E) and academic grades (section F) exhibit numeric properties on an interval scale. Therefore, the Shapiro-Wilk normality tests were conducted to assess the distribution of this data, and Welch's t-tests were employed to evaluate the statistical significance of the mean differences between groups.

## 4   Research Results

The raw data from the usability questionnaire are publicity available at [18].

### 4.1   Syntax Learnability

Results of the initial section dedicated to students utilizing T-UCFL have been visualized in Fig. 1. The majority of students characterized T-UCFL as understandable (A1, A9), consistent (A5), and flexible (A2). The language is not very complex (A4) and generally not time-consuming (A10).

Noteworthy observations regarding T-UCFL included students expressing a desire for enhanced comment functionality within the language and highlighting the potential utility of incorporating a linter.

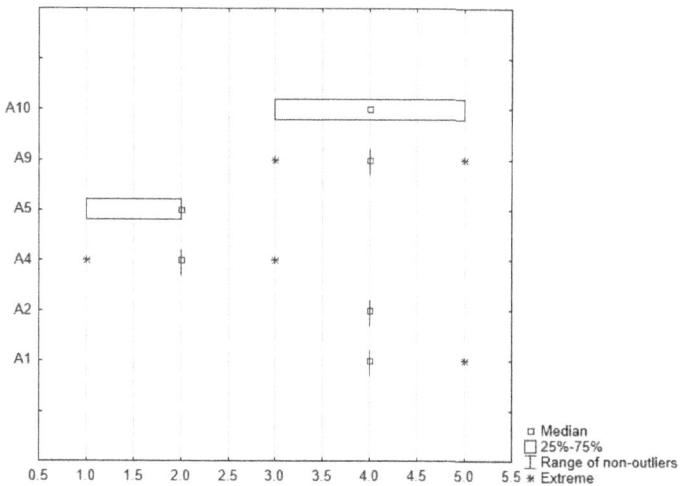

**Fig. 1.** Results of T-UCFL syntax assessment

### 4.2   Notation Applicability in the Business Context

The positive perception of T-UCFL was observed among students in the business context section - see Fig. 2. Its results are very similar to activity diagrams.

Although optimal answers were not consistently observed for every question, T-UCFL did not elicit predominantly negative responses (see Fig. 2). The responses to the final B5 question in this section merit attention. Specifically, the answers associated with T-UCFL exhibit remarkable consistency compared to other methods. Regarding natural language, students do not converge toward a consensus; their answers span the entire scale range. However, the Kruskal-Wallis test confirmed no statistically significant differences between the use-case specification methods.

### 4.3   Notation Efficiency

Notations' efficiency for all use case specification methods was tested for all groups using the Kruskal-Wallis test. Although time spent on performing tasks has not shown statistical significance, time spent on learning the use case specification notation had statistically significant differences for the pairs T-UCFL and natural language ($p = 0.038$), as well as natural language and activity diagram ($p = 0.013$), what confirmed the statement, that learning to write specifications

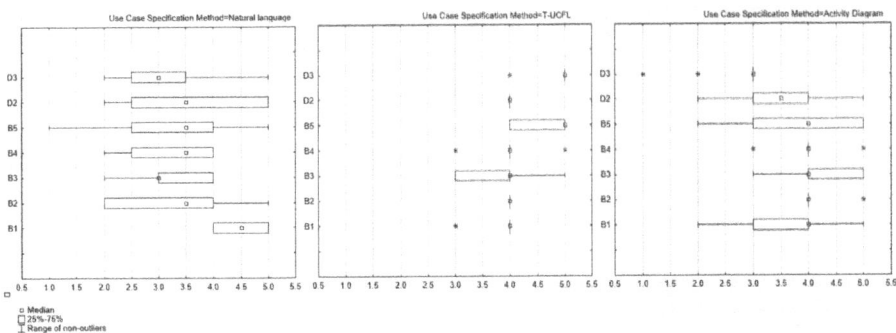

**Fig. 2.** Results for applicability in business context and tool support for different use case specification methods

in natural language is the least time-consuming. The median for each group looked as follows: natural language – 3.5, T-UCFL – 4, activity diagram – 3. Students need more time to learn T-UCFL, comparable to learning activity diagrams. However, it is a good result as they weren't offered training or help.

### 4.4   Tool Support

It is also worth noticing the importance of the quality of tooling in the use case specification process. A well-designed tool can yield several advantages for designers, including error detection, streamlined iteration, responsiveness to business requirements, and improved focus on substantive work. Conversely, suboptimal tools may hinder productivity and impede the learning process.

Students employing IntelliJ in conjunction with Antlr (a tool used to check the T-UCFL grammar) consistently reported positive feedback. However, when considering alternative methods such as word processors (e.g., Word, Overleaf, Google Docs) for natural language or specialized applications like Visual Paradigm for activity diagrams – responses were less uniform (see Fig. 2).

In the context of the Kruskal-Wallis test, the responses to the D3 question exhibited statistical significance for the pairs T-UCFL and activity diagrams ($p = 0.0038$) as well as T-UCFL and natural language ($p = 0.0333$). The Antlr tool and InteliJ were perceived as more convenient than competitive tools for specification purposes.

### 4.5   Perceived Satisfaction and Notation Effectiveness

The normality tests for SUS scores and students' grades have been performed to validate a hypothesis about the differences among considered notations. Based on Shapiro-Wilk's test results, the data had a normal distribution.

As the data had a normal distribution but unequal variances, Welch's t-test was conducted, which showed the disparity between the mean SUS score of the natural language and activity diagram to be statistically significant. Table 2 presents detailed results for each test.

The results obtained from the calculated SUS scores were scaled proportionally to align with the commonly used SUS Grade Scale. The scale was derived from the study titled "Determining What Individual SUS Scores Mean: Adding an Adjective Rating Scale" [21]. Thanks to that it was possible to assign the following grades to different use case specification methods. Specifically, the activity diagram method was assigned D grade, while the natural language and T-UCFL methods received C grade. Despite the generally low grades, it is evident that T-UCFL represents a significant improvement over the activity diagram approach.

Additionally, we checked the correlation between SUS scores and academic performance (grades). The null hypothesis, which posits no linear relationship between SUS and grades, cannot be rejected based on the available data.

**Table 2.** T-test results for SUS score and grade for different use case methods. AD – active diagrams, NL – natural language, M – mean, SD – standard deviation

|  | AD | | NL | | T-UCFL | | AD NL | AD T-UCFL | NL T-UCFL |
|---|---|---|---|---|---|---|---|---|---|
|  | M | SD | M | SD | M | SD | t-test | t-test | t-test |
| SUS | 45.54 | 12.14 | 57.50 | 9.82 | 56.00 | 13.30 | 2.37* | ns | ns |
| Grade | 7.50 | 1.41 | 8.44 | 1.24 | 7.80 | 1.15 | ns | ns | ns |

* $p < 0.05$

### 4.6  Conclusions

In the light of research results, the answer to RQ1 is positive. The T-UCFL can be effectively used for use-case specification. The language is understandable and flexible, and the support of the offered tools is good enough.

The perception of T-UCFL in the context of alternative specification methods is also positive (RQ2). Similarly to other notations, it can be applied to any software product (notation applicability). The language learnability is lower than the learnability of writing specifications in natural language, but that is unsurprising. Even without training, the productivity of novice users using T-UCFL was comparable to using other notations. The language effectiveness measured by grades was at the same level as for natural language (with the highest mean).

T-UCFL is satisfactory for software engineers (RQ3). SUS values it got were comparable to natural language (no statistical differences). There was no correlation between perceived usability and language effectiveness.

## 5  Threats to Validity

### 5.1  Internal Validity

Students themselves decided on the notation to be used. In the case of two notations, they were adequately trained. Regarding the controlled language, they received minimal support. The lack of support probably was the reason that few people chose this specification method. These individuals could have higher overall qualifications and enjoy challenges.

## 5.2   External Validity

The respondents of the study were third-year engineering students. This may affect the interpretation of the study results in terms of their generalizability. However, many students have already completed professional internships in companies and work in various positions, encountering different forms of requirements specifications.

A small number of students using the controlled language may impact the sample generalizability concerning notation comparison but it represents a lesser threat in evaluating the controlled language's usability.

## 6   Summary

The usability study of the Textual Use-case Flow Language (T-UCFL) has demonstrated its potential as a robust tool for software engineering. Our findings indicate that T-UCFL provides a clear and concise means of documenting software requirements, facilitating better understanding and stakeholder communication. Compared to traditional notations such as natural language descriptions and activity diagrams, T-UCFL offers a structured yet flexible approach that aligns with the cognitive processes of software engineers.

The participant feedback has indicated that, for most aspects, there are no statistically significant differences between various use case specification methods. However, it is worth emphasizing that the results obtained using the T-UCFL method exhibited remarkable consistency compared to other approaches in the business context. Furthermore, T-UCFL received exceptionally positive ratings in terms of tool support. Therefore, we believe that the application of NCL with a good structure does not negatively influence language usability.

The good results obtained by natural language in evaluating all usability factors, including performance, efficiency, and tool support, are not surprising, but natural language is difficult to be automatically processed. NCL forces a good specification structure and offers additional possibilities, e.g., checking internal specification consistency.

In the future, we are going to experiment with a graphical notation of the T-UCFL. Diagrams can be drawn by specification analysts or obtained automatically from textual documents. Another development direction is T-UCFL extension with data specification part.

# References

1. Drew, M.R., Falcone, B., Baccus, W.L.: What does the system usability scale (SUS) measure? In: Marcus, A., Wang, W. (eds.) DUXU 2018. LNCS, vol. 10918, pp. 356–366. Springer, Cham (2018). https://doi.org/10.1007/978-3-319-91797-9_25
2. Lewis, J.R.: Measuring perceived usability: the CSUQ, SUS, and UMUX. Int. J. Hum.-Comput. Interact. **34**(12), 1148–1156 (2018). https://doi.org/10.1080/10447318.2017.1418805
3. Lewis, J.R.: The system usability scale: past, present, and future. Int. J. Hum.-Comput. Interact. **34**(7), 577–580 (2018). https://doi.org/10.1080/10447318.2018.1455307
4. Lauesen S., Younessi, H.: Six styles for usability requirements. In: REFSQ 1998, Presses Universitaires de Namur (1998)
5. Poltronieri Rodrigues, I., de Borba Campos, M., Zorzo, A.F.: Usability evaluation of domain-specific languages: a systematic literature review. In: Kurosu, M. (ed.) HCI 2017. LNCS, vol. 10271, pp. 522–534. Springer, Cham (2017). https://doi.org/10.1007/978-3-319-58071-5_39
6. Bevan, N., Carter, J., Earthy, J., Geis, T., Harker, S.: New ISO standards for usability, usability reports and usability measures. In: Kurosu, M. (ed.) HCI 2016. LNCS, vol. 9731, pp. 268–278. Springer, Cham (2016). https://doi.org/10.1007/978-3-319-39510-4_25
7. Tomayess, I., Pedro, I.: Sustainable Design HCI, Usability and Environmental Concerns. Springer, London (2016). https://doi.org/10.1007/978-1-4471-6753-2
8. Gupta, S. Epiphaniou, G., Maple, C.: AI-augmented usability evaluation framework for software requirements specification in cyber-physical human systems. Internet Things 100841 (2023). https://doi.org/10.1016/j.iot.2023.100841
9. Barišic, A., Amaral, V., Goulão, M., Barroca, B.: Evaluating the usability of domain-specific languages. In: Management Association (eds.), Software Design and Development: Concepts, Methodologies, Tools, and Applications, pp. 2120–2141. Wiley (2021). https://doi.org/10.1049/cps2.12017
10. Cha S., Vogel-Heuser, B., Weigl, A., Ulbrich, M., Beckert, B.: Table-based formal specification approaches for control engineers-empirical studies of usability. In: IEC Cyber-Physical Systems: Theory/Applications, vol. 6, pp. 193–207. Wiley (2021). https://doi.org/10.1049/cps2.12017
11. da Silva, A.R., et al.: A pattern language for use cases specification. In: EuroPLoP 2015, pp. 1-18, Association for Computing Machinery, New York (2015). https://doi.org/10.1145/2855321.2855330
12. da Silva, A. R.: Linguistic patterns and linguistic styles for requirements specification (I): an application case with the rigorous RSL/business-level language. In: EuroPLoP 2017, pp. 1–27. Association for Computing Machinery, New York (2017). https://doi.org/10.1145/3147704.3147728
13. da Silva, A.R.: Rigorous specification of use cases with the RSL language. In: Siarheyeva, A., Barry, C., Lang, M., Linger, H., Schneider, C. (eds.) ISD2019 Proceedings. ISEN Yncréa Méditerranée, Toulon (2019)
14. Bugayenko, Y.: Combining object-oriented paradigm and controlled natural language for requirements specification. In: BCNC 2021, pp. 11–17, Association for Computing Machinery, New York (2021). https://doi.org/10.1145/3486949.3486963

15. Hnatkowska, B., Zabawa, P.: A reusability-oriented use-case model specification language. In: Ganzha, M., et al. (eds.) Proceedings of the 18th Conference on Computer Science and Intelligence Systems, pp. 567–576. IEEE, New York City (2023)

16. Hnatkowska, B., Zabawa, P.: A reusability-oriented use-case model: textual specification language. In: Jarzębowicz A., et al. (eds.) S3E 2023 Topical Area, KKIO 2023 and WAPL 2023, Held as Part of FedCSIS 2023 revised selected papers, Lecture Notes in Business Information Processing, pp. 35–62. Springer, Cham (2024). https://doi.org/10.1007/978-3-031-51075-5_2

17. https://forms.gle/66mnKh3oWSB5FYp96 . Accessed 15 Mar 2024

18. https://github.com/bhnatkowska/UCF/ . Accessed 15 Mar 2024

19. Brooke, J.: SUS-a quick and dirty usability scale. In: Usability Evaluation in Industry. CRC Press (1996)

20. Likert, R.: A technique for the measurement of attitudes. Arch. Psychol. (1932)

21. Bangor, A., Kortum, P., Miller, J.: Determining what individual SUS scores mean: adding an adjective rating scale (2009)

# Collective Intelligence in Healthcare

# A Multi-view Spatio-Temporal EEG Feature Learning for Cross-Subject Motor Imagery Classification

Adel Hameed[1,2], Rahma Fourati[1,3(✉)], Boudour Ammar[1],
Javier Sanchez-Medina[4], and Hela Ltifi[1,5]

[1] REsearch Groups in Intelligent Machines, National Engineering School of Sfax,
3038 Sfax, Tunisia
`adelhameedoleiwi.alhlaiki@enetcom.u-sfax.tn,`
`{boudour.ammar,rahma.fourati,hela.ltifi}@ieee.org`
[2] National School of Electronics and Telecommunications of Sfax, University of Sfax,
Sfax, Tunisia
[3] Université de Jendouba, Faculté des Sciences Juridiques, Economiques et de
Gestion de Jendouba, 8189 Jendouba, Tunisie
[4] Innovation Center for the Information Society, University of Las Palmas de Gran
Canaria, Las Palmas de Gran Canaria, Spain
`javier.sanchez.medina@ieee.org`
[5] Department of Computer Sciences, Faculty of Sciences and Techniques of Sidi
Bouzid, University of Kairouan, Kairouan, Tunisia

**Abstract.** This study introduces MV-FocalNet, a novel approach for classifying motor imagery from electroencephalography (EEG) signals. MV-FocalNet leverages multi-view representation learning and spatial-temporal modeling to extract diverse properties from multiple frequency bands of EEG data. By integrating information from multiple perspectives, MV-FocalNet captures both local and global features, significantly enhancing the accuracy of motor imagery task classification. Experimental results on two EEG datasets, 2a and 2b, show that MV-FocalNet accurately categorizes various motor movements, including left and right-hand activities, foot motions, and tongue actions. The proposed method outperforms existing state-of-the-art models, achieving substantial improvements in classification accuracy.

**Keywords:** Electroencephalography · Motor imagery · Multi-View representation · Focal Modulation Networks

## 1 Introduction

For real-world applications, data are usually manifested in multiple types of features or multiple modalities that are considered as multiple views [16]. In computer vision, multi-view images provide multiple views of the same object or scene, capturing it from various angles or viewpoints. Each view offers unique information about the object's appearance, structure, and spatial relationships, which can be valuable for tasks such as object recognition, 3D reconstruction, and scene understanding [13]. By integrating information from multiple views, multi-view images offer a more comprehensive representation of the underlying scene or object. This enhanced information can improve the accuracy and robustness of computer vision algorithms by providing additional context and reducing ambiguity.

In multi-view EEG analysis, EEG signals from different electrode views are filtered into different frequency bands to extract features that capture the underlying neural dynamics associated with Motor Imagery (MI) tasks. Feature extraction methods such as Power Spectral Density (PSD) estimation or Common Spatial Pattern (CSP) are applied to each frequency band view separately to obtain frequency-specific information. Integration of information from multiple frequency bands allows for a more comprehensive characterization of brain activity during MI, leading to improved decoding performance and understanding of neural processes involved in motor control and imagery [22] [3].

PSD and CSP are examples of handcrafted features extracted from EEG trials [6]. Despite handcrafted feature extraction offering interpretability and computational efficiency but may be limited in capturing complex spatio-temporal patterns present in EEG data [5]. Deep learning feature learning, on the other hand, automatically learns hierarchical representations from raw EEG signals, potentially achieving higher classification performance but requiring large amounts of labeled data and computational resources. In recent years, deep feature extraction or feature/ representation learning learns to extract the most discriminative features automatically during the training phase which boosted the performance of classification tasks [8].

In this paper, we propose a novel method for MI classification of EEG signals with the following innovative aspects.

– A multi-view of raw EEG data is performed by filtering the frequency bands to capture specific features within each band.
– A spatio-temporal module based on VideoFocalNet [20] is adapted after segmenting EEG signals into 4D structure (number of segments, timesteps, number of EEG channels, number of EEG bands) shape to capture correlation in the time domain and extract features in the spatial domain.
– Raising the subject-independent challenge which ensures the generalization of the proposed model.

## 2   Related Work

This paper focuses on the feature extraction methods based on multi-view information in EEG signals. Therefore, we only highlight the EEG-based MI methods yielding multi-view concepts.

Ma *et al.* [14] extracted deeper features associated with MI EEG signals using a multi-branch hybrid neural network model that contains structures with different frequency EEG signal inputs. Temporal, depthwise, and separable convolutions are handled sequentially to capture temporal and spatial information. Validation on 2a dataset is performed in a subject-dependent scheme where the authors proved that swapping EEG segments between subjects is beneficial. Jiao *et al.* [9] introduced a novel algorithm named multi-scale optimization of spatial patterns, which aims to enhance the classification performance of MI tasks by jointly learning from multi-block EEG data. The algorithm predefines several subsets of EEG channels considered as multiple views to obtain multiple spatial patterns from raw EEG data across overlapping filter bands. Subsequently, a multi-view learning model incorporating $L_{2,1}$-norm regularization is employed to partition the CSP features into groups, facilitating the automatic determination of optimal sub-feature groups. The validation with subject-dependent scheme on the 3a dataset achieved $89.6 \pm 13.9$.

Mane *et al.* [15] introduced FBCNet, which used a multi-view data representation followed by spatial filtering to extract spectro-spatially discriminative features. This multistage approach facilitates efficient network training, particularly in scenarios with limited training data availability. Notably, they introduced a novel Variance layer that adeptly aggregates EEG time-domain information allowed to reach 79.03 in a 10-fold CV scheme on the 2a dataset. Ghimire *et al.* [4] investigated the conversion of 1D temporal EEG signals into 2D spatiotemporal EEG image sequences, as well as their integration into a proposed multi-view hierarchical deep learning framework for recognition. The model comprises Conv2D layers arranged in a hierarchical structure, where decisions are made independently at each level, leveraging the decisions from the preceding level.

Fan *et al.* [3] introduced a novel algorithm by incorporating two modules into CNN. The first module is the Filter Band Combination (FBC) Module, designed to retain as many frequency domain features as possible while preserving the time domain characteristics of EEG signals. The second module is a Multi-View structure capable of extracting features from the output of the FBC module. The average accuracy on 2a dataset is $70.52 \pm 4.19\%$. Hu *et al.* [7] introduced a cross-space fusion algorithm tailored to address four classification tasks of MI-EEG. Their approach involved multi-view feature extraction in the time domain, frequency domain, and spatial domain (both detailed and global) to extract more comprehensive and diverse MI-EEG information. Validation on the 2a dataset in a subject-dependent scheme showed $90.37\%$ as accuracy.

Zhang and Li [22] proposed a novel method that involves creating multiple views or sub-band signals from EEG data using bandpass filters. These sub-bands represent different frequency ranges capturing diverse information related to MI tasks. The study fed these multi-view inputs into temporal convolution, spatial convolution, and dense convolution blocks. Thus, the model can effectively propagate information and extract features from the sub-band signals. Furthermore, extracted features are integrated using concatenation fusion before final classification, enhancing the model's performance in decoding MI tasks. Experimental results on a 2a dataset demonstrate significant improvements in accuracy, with an average accuracy enhancement of 72.45±14.10. Wang *et al.* [19] introduced a novel weighted multi-branch (WMB) structure to address the challenge of handling multisubject data. In this approach, each branch is tasked with fitting a pair of source-target subject data, and adaptive weights are employed to integrate all branches or select branches with the highest weights to make the final decision. The average accuracies on 2a and 2 b datasets are $84.14 \pm 10.94$ and $90.23 \pm 8.09$, respectively.

According to Table 1, several attempts at muti-view processing of EEG signals are presented. Most of the work done relies upon band filtering followed by temporal convolutional and spatial convolution sequentially with a subject-dependent classification scheme. In our work, a novel method taking EEG signals into 4D structure processed by temporal and spatial branches in parallel is proposed in a subject-independent scheme which is more challenging than the subject-dependent scheme.

**Table 1.** EEG feature extraction in existing works on MI

| Year | Ref. | Structure | Views | Evaluation | Performance (%) |
|------|------|-----------|-------|------------|-----------------|
| 2020 | [9] | $(2M\, N_b, S)$ | S channels subset | 3a dataset | $89.60_{\pm}13.90$ |
| 2022 | [14] | $(1, C, T)$ | $\mu[8\text{--}12\,\text{Hz}]$, $\beta_{low}[16\text{--}20\,\text{Hz}]$, $\beta_{med}[20\text{--}24\,\text{Hz}]$, $\beta_{high}[24\text{--}28\,\text{Hz}]$ | 2a dataset | $83.91 \pm 9.09$ |
| 2022 | [4] | $(T, C, L)$ | 9 levels | PhysioNet | 69.08 |
| 2023 | [15] | $(N_b, C, T)$ | 9 bands in [4–40 Hz] | 2a dataset | 79.03 |
| 2023 | [22] | $(C, T, N_b)$ | $\delta[1\text{--}5\,\text{Hz}]$, $\theta[4\text{--}8\,\text{Hz}]$, $\alpha[7\text{--}13\,\text{Hz}]$, $\beta[12\text{--}32\,\text{Hz}]$ | 2a dataset | $72.45 \pm 14.10$ |
| 2023 | [3] | $(N_b, C, T)$ | $\delta[1\text{--}3\,\text{Hz}]$, $\theta[4\text{--}7\,\text{Hz}]$, $\alpha[8\text{--}12\,\text{Hz}]$, $\beta[13\text{--}30\,\text{Hz}]$ | 2a dataset | $70.52 \pm 4.19$ |
| 2023 | [19] | $(C, T, (X, Y))$ | Number of trials and Classes | 2a dataset | $84.14 \pm 10.94$ |
| 2024 | Ours | $(Ns, T, C, N_b)$ | $\theta[4\text{--}8\,\text{Hz}]$, $\alpha[8\text{--}16\,\text{Hz}]$, $\beta[16\text{--}32\,\text{Hz}]$, $\gamma[32\text{--}45\,\text{Hz}]$ | 2a dataset | 93.32 |

$M$: CSP spatial filters $(C, T)$: Channels, Timesteps
$N_s$, $N_b$: Number of segments, Number of bands $(X, Y)$: Trials, Labels

## 3   Materials and Methods

In this section, the different steps of our methodology, as depicted in Fig. 3 are explained. We begin by describing the considered datasets. After that, the transformation of EEG trials from 2D into a 4D structure is presented. Finally, the spatial and temporal branches of our FocalNet are presented in detail.

### 3.1   Considered Datasets

**2a Dataset** [2]: This dataset includes recordings from nine subjects with 22 electrodes and a sample rate of 250 Hz. It is divided into four distinct groups based on MI tasks: Class 1 for left-handed activities, Class 2 for right-handed activities, Class 3 for activities involving both feet, and Class 4 for activities including the tongue. Each participant's data is separated into two sessions,

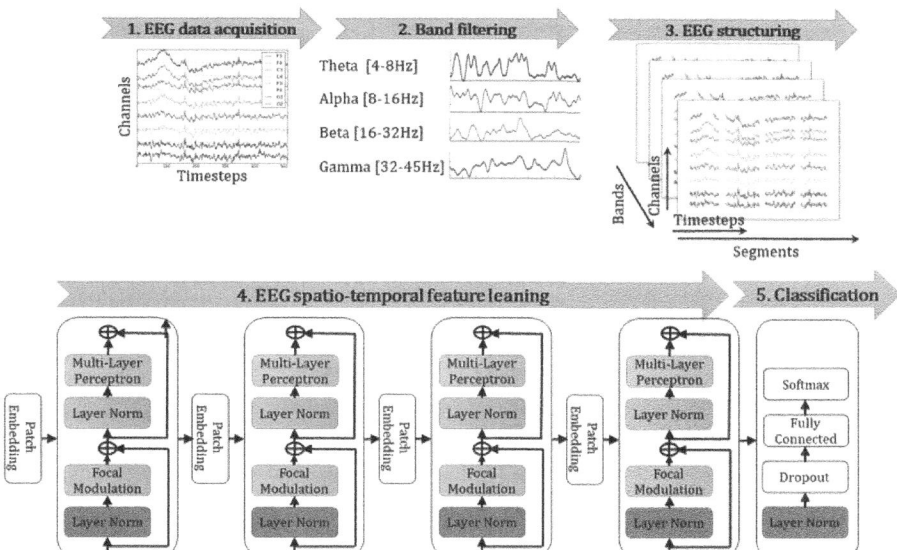

**Fig. 1.** The flowchart of our proposed methodology

each of which has 288 trials. The signal is divided into periods ranging from one to six seconds. The dataset is divided into eighteen files, each labeled 'T' for the training set and 'E' for the evaluation set.

**2b Dataset** [12]: This dataset contains EEG recordings from nine people and was captured at a rate of 250 Hz using three bipolar electrodes. It is divided into two basic categories: left-hand activities and right-hand activities. The dataset includes five sessions, each with 320 trials. The data in this collection is segmented every 1 to 7 s. Class labels are available for the first three sessions. The dataset is divided into eighteen files, labeled 'T' for training and 'E' for evaluation purposes.

## 3.2  Preprocessing

Our preprocessing strategy focuses on extracting relevant insights from raw EEG data by using a 4th-order Butterworth filter to identify four critical frequency bands: $\theta[4\text{–}8\,\text{Hz}]$, $\alpha[8\text{–}16\,\text{Hz}]$, $\beta[16\text{–}32\,\text{Hz}]$, and $\gamma[32\text{–}45\,\text{Hz}]$. By focusing on these frequency ranges, the model will capture the subtle brain activity that underpins motor imaging tasks. Previous research detailed in Sect. 2 shows that this preprocessing step improves signal-to-noise ratio, reduces dimensionality, and leads to more accurate analysis.

### 3.3   Spatio-Temporal FocalNet Module

In this work, we propose the MV-FocalNet, drawing inspiration from the notion of focal modulation in image recognition, introduced by Yang et al. [21], and then adapted for video recognition by [20]. Our primary goal is to improve efficiency and performance by successfully collecting both local and global settings during the decoding of MI EEG signals.

The representation of the EEG signals is denoted by $X_{st} \in \mathbb{R}^{T \times T_e \times C \times B}$, where $T$ represents the number of segments, $T_e$ signifies the timesteps, $C$ denotes the number of EEG channels, and $B$ reflects the number of bands. The EEG signal data is processed using a patch embedding layer and a convolutional layer with filter size and stride set to 4. It proceeds through four stages of focal modulation blocks, indexed as $i \in \{1, 2, 3, 4\}$, with each stage including $N_i$ focal modulation layers. Following each level, we insert another patch embedding layer. This layer reduces the feature map's spatial size while gradually increasing the last dimension. As the model progresses through its levels, the last dimension increases from 8 to 16, 32, and eventually 64. Our approach purposely augments the last dimension over layers while concurrently diminishing the time dimension: $T \to \frac{T}{4} \to \frac{T}{8} \to \frac{T}{16} \to \frac{T}{32}$. Finally, the average output from the last stage is computed and routed to a classification layer via a fully connected layer. The MV-FocalNet is comprised of two main components: hierarchical contextualization and gated aggregation as depicted in Fig. 1.

**Hierarchical Contextualization.** The process begins by projecting the spatial-temporal feature map $X_{st} \in \mathbb{R}^{T \times T_e \times C \times B}$ using two linear layers, resulting in $Z_0^s$ and $Z_0^t$.

$$Z_0^s = f_{z,s}(X_{st}) \in \mathbb{R}^{T \times T_e \times C \times B},$$
$$Z_0^t = f_{z,t}(X_{st}) \in \mathbb{R}^{T \times T_e \times C \times B}. \tag{1}$$

where the spatial and temporal linear projection layers are denoted, respectively, by $f_{z,s}$ and $f_{z,t}$. The spatially and temporally projected inputs, $Z_0^s$ and $Z_0^t$, are then subjected to a succession of $L$ depth-wise convolutions (DWConv) and point-wise convolutions (PWConv) along their corresponding dimensions. This produces the following outputs for each focal level $\ell \in \{1, ..., L\}$: $Z_\ell^s$ and $Z_\ell^t$.

$$Z_\ell^s = f_{\ell a,s}(Z_{\ell-1}^s) \triangleq \text{GeLU}(\text{DWConv}(Z_{\ell-1}^s)) \in \mathbb{R}^{T \times T_e \times C \times B},$$
$$Z_\ell^t = f_{\ell a,t}(Z_{\ell-1}^t) \triangleq \text{GeLU}(\text{PWConv}(Z_{\ell-1}^t)) \in \mathbb{R}^{T \times T_e \times C \times B}. \tag{2}$$

The GeLU activation is employed by the temporal and spatial contextualization functions, denoted by $f_{\ell a,s}(\cdot)$ and $f_{\ell a,t}(\cdot)$ correspondingly. Global average pooling along the spatial and temporal dimensions is applied to $Z_L^s$ and $Z_L^t$ respectively to get the overall representation.

$$Z_{L+1}^s = \text{Avg-Pool}(Z_L^s),$$
$$Z_{L+1}^t = \text{Avg-Pool}(Z_L^t). \tag{3}$$

**Gated Aggregation.** Next, we compress the respective spatial and temporal feature maps, $Z_\ell^s$ and $Z_\ell^t$, into their corresponding spatial and temporal modulators through a gating mechanism. The spatial and temporal gating weights, denoted as $G_s = f_{g,s}(X_{st}) \in \mathbb{R}^{T_e \times C \times (L+1)}$ and $G_t = f_{g,t}(X_{st}) \in \mathbb{R}^{T \times (L+1)}$, are obtained using the linear projection layers $f_{g,s}$ and $f_{g,t}$. This is followed by a dot product operation between the feature maps and their respective gates, yielding the aggregated spatial and temporal feature maps $Z_{\text{out}}^s$ and $Z_{\text{out}}^t$, as shown in Eq. (4).

$$
\begin{aligned}
Z_{\text{out}}^s &= \sum_{\ell=1}^{L+1} G_\ell^s \odot Z_\ell^s \in \mathbb{R}^{T_e \times C \times B}, \\
Z_{\text{out}}^t &= \sum_{\ell=1}^{L+1} G_\ell^t \odot Z_\ell^t \in \mathbb{R}^{T \times B}.
\end{aligned}
\tag{4}
$$

where $Z_{\text{out}}^s$ and $Z_{\text{out}}^t$ represent the single aggregated spatial and temporal feature maps respectively. Here, $G_\ell^s \in \mathbb{R}^{T_e \times C \times 1}$ and $G_\ell^t \in \mathbb{R}^{T_e \times 1}$ denote slices of $G_s$ and $G_t$ for the level $\ell$.

To facilitate communication across different channels, another set of linear layers, $h_s(\cdot)$ and $h_t(\cdot)$, are employed to derive the spatial modulator $(M_s = h_s(Z_{\text{out}}^s) \in \mathbb{R}^{T \times T_e \times C \times B})$ and temporal modulator $(M_t = h_t(Z_{\text{out}}^t) \in \mathbb{R}^{T \times T_e \times C \times B})$ respectively. Consequently, the spatial-temporal focal modulation process can be expressed as:

$$
y_i = q(x_i) \odot h_s \left( \sum_{\ell=1}^{L+1} g_{\ell i,s} \cdot z_{\ell i,s} \right) \odot h_t \left( \sum_{\ell=1}^{L+1} g_{\ell i,t} \cdot z_{\ell i,t} \right)
\tag{5}
$$

where $z_{\ell i,s}/z_{\ell i,t}$ and $g_{\ell i,s}/g_{\ell i,t}$ represent the spatial/temporal visual feature and spatial/temporal gating value at location $i$ of $Z_\ell^s/Z_\ell^t$ and $G_\ell^s/G_\ell^t$ respectively.

## 4    Experimental Results and Discussion

### 4.1    Experimental Setup

The model was developed using the PyTorch framework with an Nvidia RTX 3060 GPU, an AMD Ryzen 5 5500 CPU, and 16 GB of RAM. The model was trained across 200 epochs with a batch size of 32 using a subject-independent methodology and optimized using Adam with a learning rate of 1e-3. 5-fold cross validation (CV) was investigated which splits the dataset into five folds, with the model trained on four and validated on the fifth. To boost training efficiency and minimize overfitting, we implemented an early stopping strategy that ends training if validation accuracy does not improve for 10 consecutive epochs.

## 4.2   Scoring Performance

Performance metrics including F1-score, specificity, recall, accuracy, and precision are calculated for each category to provide a thorough evaluation. These indicators are presented to provide a comprehensive analysis of the model's performance, acknowledging the difficulty in achieving a perfect balance among them in real-world scenarios. Table 2 presents MV-FocalNet's performance on datasets 2a and 2b with a 5-fold CV strategy.

**Table 2.** Performance analysis using a 5-fold CV scheme on 2a and 2b datasets

| Class | F1-score | Precision | Recall | Specificity |
|---|---|---|---|---|
| 2a dataset | | | | |
| 0 | 93.33 | 94.30 | 92.37 | 97.48 |
| 1 | 91.98 | 91.52 | 92.45 | 97.48 |
| 2 | 92.50 | 93.35 | 91.67 | 97.24 |
| 3 | 92.79 | 91.51 | 94.10 | 98.02 |
| 2b dataset | | | | |
| 0 | 87.21 | 87.09 | 87.33 | 87.30 |
| 1 | 87.18 | 87.30 | 87.06 | 87.09 |

In the 2a dataset, Class 0 (representing left-hand activities) achieves the best results compared to other classes, with an F1-score of 93.33% and Precision of 94.30%. Class 1 (right-hand activities) exhibits an F1-score of 91.98%, Precision of 93.35%, Recall of 92.45%, and the same specificity of 97.48% as in Class 0. Class 2 (both foot motions) achieves a better F1-score of 92.50% compared to Class 1, as well as a slightly similar specificity of 97.24% as in Classes 0 and 1. The precision and recall metrics are 93.35% and 91.67%, respectively. Class 3 (representing tongue actions) has the lowest precision at 91.51% and the highest recall at 94.10%. The model achieves high specificity for all classes. The average metric values of F1-score, Precision, Recall, and Specificity are 92.65%, 92.67%, 92.64%, and 97.55%, respectively.

For the 2b dataset, MV-FocalNet retains its performance in classification tasks. In class 0, the model has a relatively high F1-score of 87.21%, balanced accuracy and recall values of 87.098% and 87.33%, respectively, and a specificity of 87.30. Similarly, for class 1, MV-FocalNet has a comparable F1-score of 87.18%, along with balanced precision and recall metrics and a moderate specificity of 87.09%. These results reveal that MV-FocalNet performs well on the 2b dataset, demonstrating its capacity to categorize instances across many classes.

Figure 2 shows the confusion matrices for EEGFocal-Nets' performance in categorizing movements using 5-fold cross-validation on the 2a and 2b datasets. In the 2a dataset, the model achieves 92% accuracy for left-hand, right-hand,

and foot movements, as well as 94% for tongue movements. Similarly, in the 2b dataset, the model achieves 87% accuracy for both left and right-hand movements.

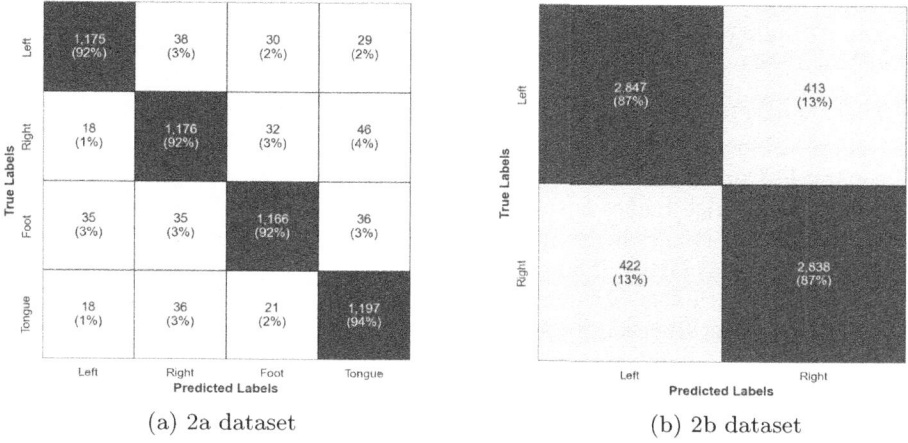

(a) 2a dataset                                    (b) 2b dataset

**Fig. 2.** Confusion matrices for MV-FocalNet using 5-fold CV

Moreover, Fig. 3 illustrates the progression of accuracy and loss across 200 epochs for the 2a dataset. Both figures indicate that our model does not display overfitting, as the validation data closely resembles the patterns observed in the training data for accuracy and loss (CrossEntrpy).

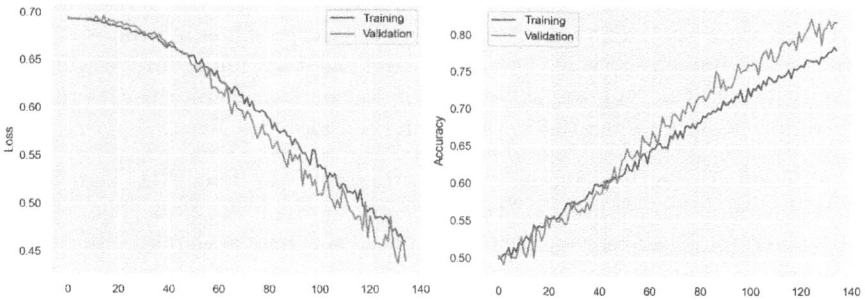

**Fig. 3.** Loss and accuracy curves over 200 epochs for 2a dataset

## 4.3   Comparative Study Using Different Classification Schemes

In our research, we conducted a detailed comparison analysis, using well-established baseline models as references. Our approach is designed primarily for subject-independent classification problems. To assess its performance, we ran two experiments on 2a and 2b datasets using a 5-fold CV.

The selection of methods involved in the comparison is based on the classification type and the handling of the multi-view aspect. In this paper, a novel method called "MV-FocalNet" processes EEG signals in time and frequency domains of different views for subject-independent classification tasks. Selected models with the same configuration are detailed as follows.

- **EEGNet** [11]: It is a compact convolutional neural network composed of a temporal block with Conv1D layers and a spatial block with depthwise and separable Conv2D layers. EEGNet model reduced the number of trainable parameters which makes it suitable for training on small datasets.
- **CNN** [18]: It is enriched with three Mega Blocks composed of a sequence of convolutional layers. These blocks can repeated several times. The number of convolution layers, learning rate, momentum, and regularization are optimized using Bayesian hyperparameter optimization.
- **MIN2Net** [1]: It is an end-to-end multi-task autoencoder performing feature learning and classification simultaneously. The model is optimized using mean squared error between the original inputted EEG signals and the reconstructed output of the decoder. A triplet loss is used to measure the distance between anchor, positive, and negative latent codes generated by the encoder block. A fully connected layer is stacked over the encoder to learn to minimize the cross-entropy loss function.
- **SCBNN-MOPED** [17]: It is a Bayesian neural network with Conv2DFlipout layers trained by placing MOPED's prior distribution [10] over the weights. The model is optimized by minimizing the uncertainty quantification of the CNN prediction.
- **TST-ICA** [5]: A TST model is proposed to capture spatio-temporal information from EEG signals after an Independent Component Analysis (ICA) preprocessing step. The self-attention module aggregates information across channels to improve spatial correlations and classification accuracy. Moreover, the inclusion of a temporal self-attention mechanism facilitates the encoding of comprehensive sequential information on a global scale.

Compared to earlier models, the proposed MV-FocalNet significantly improved accuracy on datasets 2a and 2b, as shown in Table 3. The MSCNet [23] obtained 69.77% accuracy on the 2a dataset, whereas the TST-ICA model [5] reached an improved accuracy of 88.75% with a 6 s trial period. However, the MV-FocalNet model outperformed both prior models with an accuracy of 92.06%. This is a significant improvement in classification accuracy for the 2a dataset. Similarly, with the 2b dataset, the CNN model [18] obtained 67.78% accuracy with a 2 s trial length. The SCBNN-MOPED model [17] achieved a better accuracy of 76.06% than the CNN model [18] whereas the TST-ICA model [5] reached an improved accuracy of 88.75% with a 6 s trial period. Our MV-FocalNet with a 6 s trial period boosted the performance to get an accuracy of 87.19%.

**Table 3.** Subject-independent comparative analysis on 2a and 2b datasets

| Year | Trial duration | Method | Multi-view | Accuracy (%) |
|------|---------------|--------|-----------|-------------|
| 2a dataset | | | | |
| 2018 | — | EEGNet [11] | | 64.26 |
| 2021 | — | MIN2Net [1] | | 65.23 |
| 2023 | 4 s | FBCNet [15] | ✓ | 79.03 |
| 2024 | 6 s | TST-ICA [5] | | 88.75 |
| 2024 | 6 s | **MV-FocalNet** | ✓ | **92.06** |
| 2b dataset | | | | |
| 2020 | 2 s | CNN [18] | | 67.78 |
| 2023 | 4 s | SCBNN-MOPED [17] | | 76.06 |
| 2024 | 6 s | TST-ICA [5] | | 84.20 |
| 2024 | 6 s | **MV-FocalNet** | ✓ | **87.19** |

## 5  Discussion

The suggested MV-FocalNet model has shown good results in the categorization of MI tasks, particularly on datasets 2a and 2b. By using a multi-view strategy to capture diverse characteristics from multiple frequency bands, MV-FocalNet effectively integrates spatial and temporal information, thus boosting the model's ability to recognize complicated patterns in EEG data related to MI tasks. In comparison to existing state-of-the-art models, our MV-FocalNet outperforms them by achieving 93.32% accuracy on dataset 2a and 76.49% on dataset 2b. This significant improvement demonstrates the suggested methodology's usefulness in tackling the challenges of subject-independent MI classification. The robustness of MV-FocalNet is further demonstrated by its constant performance across multiple classes within each dataset, as evidenced by good precision, recall, specificity, and F1-scores. The model's ability to accurately classify a wide range of MI tasks, including left- and right-hand activities, foot motions, and tongue actions, demonstrates its adaptability and application in real-world circumstances. The absence of overfitting, as evidenced by the near alignment of training and validation accuracy and loss curves, improves the reliability and generalization ability of MV-FocalNet. This ensures that the model can effectively generalize to previously unseen data while maintaining consistent performance across various experimental circumstances.

## 6  Conclusion

In this paper, we introduce MV-FocalNet, a novel deep neuronal model for classifying MI of EEG signals. This model combines multi-view representation learning and spatial-temporal modeling. By leveraging information from various frequency bands, MV-FocalNet captures both local and global characteristics in

EEG data, enabling accurate classification of MI tasks. Experimental results on datasets 2a and 2b demonstrate that MV-FocalNet is successful and robust in categorizing EEG signals associated with various motor movements. The achieved performance indicators, such as precision, recall, specificity, and F1-score, confirm the model's capability to reliably identify EEG data across multiple motion categories. Furthermore, the subject-independent training scheme ensures that the model applies to a wide range of individuals, thereby enhancing its practical utility in real-world applications.

This study is limited by the choice of the considered EEG bands as views. An investigation of the sufficient number of bands as well as the most relevant band affecting the global performance should be handled.

**Acknowledgment.** The research leading to these results has received funding from the Ministry of Higher Education and Scientific Research of Tunisia under grant agreement number LR11ES48.

# References

1. Autthasan, P., et al.: Min2net: end-to-end multi-task learning for subject-independent motor imagery EEG classification. IEEE Trans. Biomed. Eng. **69**(6), 2105–2118 (2021)
2. Brunner, C., Leeb, R., Müller-Putz, G., Schlögl, A., Pfurtscheller, G., Competition, B.: Graz data set a, provided by the institute for knowledge discovery (laboratory of brain–computer interfaces). Graz University of Technology (2008)
3. Fan, Z., et al.: Joint filter-band-combination and multi-view CNN for electroencephalogram decoding. IEEE Trans. Neural Syst. Rehabil. Eng. **31**, 2101–2110 (2023)
4. Ghimire, A., Sekeroglu, K.: Classification of EEG motor imagery tasks utilizing 2D temporal patterns with deep learning. In: IMPROVE, pp. 182–188 (2022)
5. Hameed, A., et al.: Temporal-spatial transformer based motor imagery classification for BCI using independent component analysis. Biomed. Signal Process. Control **87**, 105359 (2024)
6. Hu, W., Geng, X., Yue, M., Wang, L., Zhang, X.: Feature extraction of motor imagery EEG signals based on PSD CSP fusion. In: Intelligent Computing Technology and Automation, pp. 66–72. IOS Press (2024)
7. Hu, Y., et al.: A cross-space CNN with customized characteristics for motor imagery EEG classification. IEEE Trans. Neural Syst. Rehabil. Eng. **31**, 1554–1565 (2023)
8. Jawed, S., Faye, I., Malik, A.S.: Deep learning-based assessment model for real-time identification of visual learners using raw EEG. IEEE Trans. Neural Syst. Rehabil. Eng. **32**, 378–390 (2024)

9. Jiao, Y., Zhou, T., Yao, L., Zhou, G., Wang, X., Zhang, Y.: Multi-view multi-scale optimization of feature representation for EEG classification improvement. IEEE Trans. Neural Syst. Rehabil. Eng. **28**(12), 2589–2597 (2020)
10. Krishnan, R., Subedar, M., Tickoo, O.: Specifying weight priors in bayesian deep neural networks with empirical bayes. In: The Thirty-Fourth AAAI Conference on Artificial Intelligence, AAAI 2020, The Thirty-Second Innovative Applications of Artificial Intelligence Conference, IAAI 2020, The Tenth AAAI Symposium on Educational Advances in Artificial Intelligence, EAAI 2020, New York, NY, USA, 7–12 February 2020, pp. 4477–4484. AAAI Press (2020). https://doi.org/10.1609/AAAI.V34I04.5875
11. Lawhern, V.J., Solon, A.J., Waytowich, N.R., Gordon, S.M., Hung, C.P., Lance, B.J.: Eegnet: a compact convolutional neural network for EEG-based brain-computer interfaces. J. Neural Eng. **15**(5), 056013 (2018)
12. Leeb, R., Brunner, C., Müller-Putz, G., Schlögl, A., Pfurtscheller, G.: Bci competition 2008-Graz data set b. Graz University of Technology, Austria, pp. 1–6 (2008)
13. Luo, H., Zhang, J., Liu, X., Zhang, L., Liu, J.: Large-scale 3D reconstruction from multi-view imagery: a comprehensive review. Remote Sens. **16**(5), 773 (2024)
14. Ma, W., et al.: A novel multi-branch hybrid neural network for motor imagery EEG signal classification. Biomed. Signal Process. Control **77**, 103718 (2022)
15. Mane, R., et al.: Fbcnet: a multi-view convolutional neural network for brain-computer interface. arXiv preprint arXiv:2104.01233 (2021)
16. Mena, F., Arenas, D., Nuske, M., Dengel, A.: Common practices and taxonomy in deep multi-view fusion for remote sensing applications. IEEE J. Sel. Topics Appl. Earth Obs. Remote Sens. **17**, 4797–4818 (2024)
17. Milanés-Hermosilla, D., et al.: Robust motor imagery tasks classification approach using bayesian neural network. Sensors **23**(2), 703 (2023)
18. Roy, S., Chowdhury, A., McCreadie, K., Prasad, G.: Deep learning based inter-subject continuous decoding of motor imagery for practical brain-computer interfaces. Front. Neurosci. **14**, 918 (2020)
19. Wang, H., Jiang, J., Gan, J.Q., Wang, H.: Motor imagery EEG classification based on a weighted multi-branch structure suitable for multisubject data. IEEE Trans. Biomed. Eng. **70**, 3040–3051 (2023)
20. Wasim, S.T., Khattak, M.U., Naseer, M., Khan, S., Shah, M., Khan, F.S.: Video-focalnets: spatio-temporal focal modulation for video action recognition. In: Proceedings of the IEEE/CVF International Conference on Computer Vision, pp. 13778–13789 (2023)
21. Yang, J., Li, C., Dai, X., Gao, J.: Focal modulation networks. Adv. Neural. Inf. Process. Syst. **35**, 4203–4217 (2022)
22. Zhang, J., Li, K.: A multi-view CNN encoding for motor imagery EEG signals. Biomed. Signal Process. Control **85**, 105063 (2023)
23. Zhao, R., et al.: A mutli-scale spatial-temporal convolutional neural network with contrastive learning for motor imagery EEG classification. Med. Novel Technol. Dev. **17**, 100215 (2023)

# A New MLEM Reconstruction Algorithm for Ultra-low Dose PET

Robert Cierniak[✉]

Department of Intelligent Computer Systems, Czestochowa University of Technology,
Czestochowa, Poland
robert.cierniak@pcz.pl
https://www.kisi.pcz.pl

**Abstract.** This study introduces a novel ML-EM estimation method for reconstructing images in positron emission tomography. The concept proposed here utilizes a continuous-to-continuous data model, with the reconstruction problem expressed as a shift-invariant system. The primary objective of this research is to illustrate the methodology founded on probabilistic principles, emphasizing the consideration of statistical characteristics of PET signal data. The central focus of this paper is to establish that our method is grounded in statistical theory, offering alternative strategies to improve image resolution in low-dose PET scans.

**Keywords:** ultra-low dose PET · model-based iterative reconstruction · image reconstruction from projections

## 1 Introduction

The reconstruction method presented here relates to one of the most popular medical imaging techniques, which falls under the category of emission tomography, specifically positron emission tomography (PET). The crucial purpose of this technique is to determine the distribution of the radiotracer in the tissues of the investigated region of the body based on a set of radiation measurements from various so-called lines of responses (LOR) obtained by a PET scanner. This burden of ionizing radiation is a major concern in PET imaging, hindering the application of this technique in many situations. Radiation doses related to PET procedures may result in stochastic effects, which is related to the level of exposure. As a consequence, the probability of these detrimental effects caused by PET examinations performed over many years accumulate [2]. Moreover, individuals with allergies, asthma, heart disease, dehydration, blood cell disorders such as sickle cell anemia, polycythemia vera, multiple myeloma, kidney disease, or those on a drug regimen including beta-blockers, nonsteroidal anti-inflammatory drugs (NSAIDs), or interleukin-2 (IL-2) should consult with their doctor regarding potential risks of allergic reactions to the iodine tracer. These efforts aim to achieve ultra-low-dose PET imaging while maintaining image quality. Due to the limited number of annihilations observed in a single LOR, the statistical nature of these measurements strongly influences the process and should be taken into account. The standard reconstruction method used in PET is the maximum likelihood-expectation maximization (ML-EM) algorithm [1,9,12,13]. It is important to note that the image processing methodology applied in this algorithm aligns with a discrete-to-discrete (D-D) data model, where the reconstructed image is a priori divided into homogeneous blocks representing pixels. In this conception, individual elements of the system matrix are determined separately for every pixel, and for every annihilation event detected along the given LOR. In this case, the reconstruction problem is formulated using huge matrices. The forward model formulated for this reconstruction approach is commonly employed in deep learning-based PET image reconstructions, for example in a deep image prior (DIP)-based fully 3D image reconstruction method (see e.g. [7,10] or [8]) or in a denoising CNN-based method integrated within the iterative PET reconstruction framework (see e.g. [11]).

In this paper, we propose a new statistical approach to image reconstruction, consistent with a continuous-to-continuous (C-C) data model. An algorithm of this form was initially proposed in [4]. The forward model for this problem can be defined as an approximate discrete 2D reconstruction problem, in the form of a shift-invariant system. The preliminary conception of this kind of image reconstruction from projections strategy for computed tomography (CT) [3,5,6].

These origins help in avoiding many of the drawbacks associated with the standard D-D method. Although the proposed reconstruction method requires establishing appropriate coefficients, these can be pre-calculated and stored in memory. The use of a shift-invariant system allows for the implementation of an FFT algorithm during the most demanding calculations. This significantly accelerates the calculations required for the image reconstruction process. The primary aim of this paper is to demonstrate that our approach is based on statistical principles and enables the provision of alternative solutions to enhance image quality for low-dose PET imaging.

## 2    Forward Model Formulation

If both a measurement system and a reconstructed image in continuous spaces during the reconstruction problem formulation are defined, then this problem is considered as a continuous-to-continuous type. Conceptually, let the function $f(x, y)$ denote the unknown continuous image representing the distribution of a radiopharmaceutical molecule in a given cross-section of the human body, which is utilized for cancer diagnostics. During reconstruction, the image $f(x, y)$, defined as a function $f : \mathbb{R}^2 \to \mathbb{R}$, will be obtained based on measurements referred to as projections $g(s, \alpha)$. These projections are performed in the hypothetical measurement system illustrated in Fig. 1.

Hypothetically, it is possible to conduct a scanning process along an axis $s$ continuously at every possible angle $\alpha$. The function $p(s, \alpha)$ defines a measurement carried out at a distance $s$ from the origin when this scanning is made at

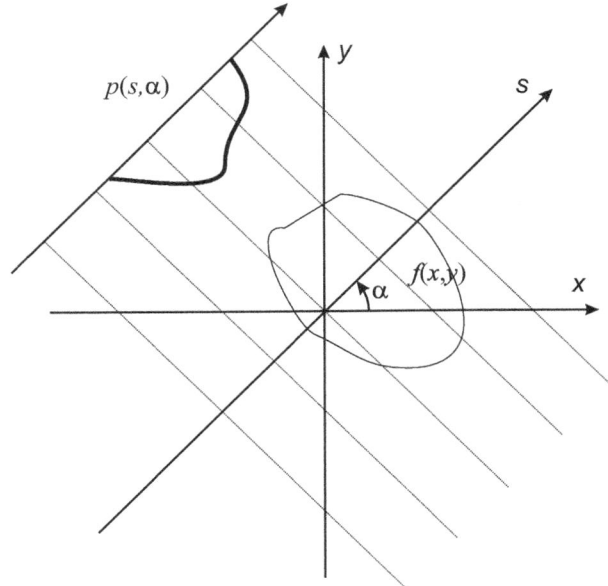

**Fig. 1.** The continuous measurement system

a given angle $\alpha$. The entire process is called the Radon transform and can be written formally as:

$$p(s,\alpha) = \int\limits_{-\infty}^{+\infty} \int\limits_{-\infty}^{+\infty} f(x,y) \cdot \delta(x\cos\alpha + y\sin\alpha - s)\, dx dy. \tag{1}$$

Because the proposed reconstruction method induces the problem of image reconstruction from projections which is formulated here, we propose firstly to present a scheme for this method in Fig. 2.

If we consider the scheme of the reconstruction algorithm shown in Fig. 2 we perform the first step of the reconstruction procedure, i.e. the back-projection operation. This operation is described using the following relation:

$$\tilde{f}(x,y) = \int\limits_{-\pi}^{\pi} \int\limits_{-\infty}^{+\infty} p(\bar{s},\alpha)\, int(s - \bar{s})\, d\bar{s} d\alpha. \tag{2}$$

The continuous function $int(\Delta s)$ is related to an interpolation whose utilization is necessary during the back-projection operation performed in implementations of this reconstruction method.

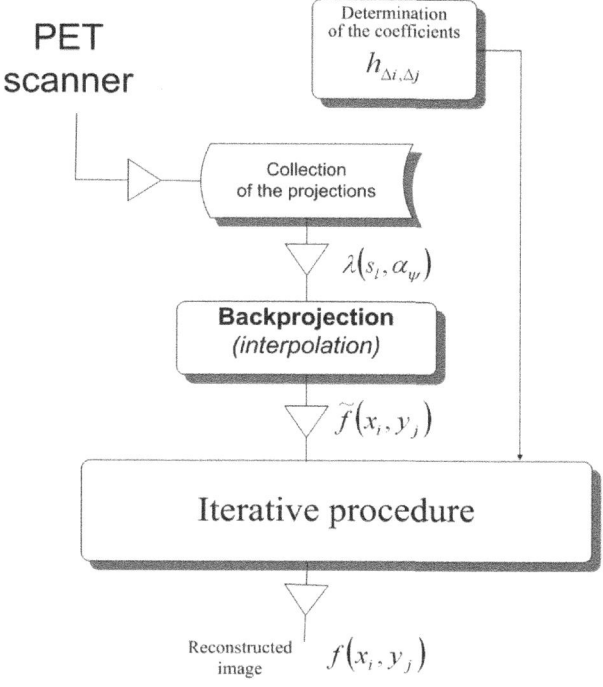

**Fig. 2.** Scheme of the proposed reconstruction method.

We can transform relation (2), taking into account definition (1), into the following form:

$$\tilde{f}(x,y) = \int_{-\pi}^{\pi} \int_{-\infty}^{+\infty} \int_{-\infty}^{+\infty} \int_{-\infty}^{+\infty} f(\bar{x}, \bar{y}) \cdot \delta(\bar{x}\cos\alpha + \bar{y}\sin\alpha - \bar{s}) \, d\bar{x}d\bar{y} \cdot int(s - \bar{s}) \, d\bar{s}d\alpha.$$

$$(3)$$

After some modifications of the Eq. (3) we get a more compact form:

$$\tilde{f}(x,y) = f(x,y) * h(x,y) = \int_{-\infty}^{+\infty} \int_{-\infty}^{+\infty} f(\bar{x}, \bar{y}) h(\bar{x} - x, \bar{y} - y) \, d\bar{x}d\bar{y}, \quad (4)$$

where:

$$h_{\Delta x, \Delta y} = \int_{-\pi}^{\pi} int((\bar{x} - x)\cos\alpha + (\bar{y} - y)\sin\alpha) \, d\alpha. \quad (5)$$

It should be noted the image $\tilde{f}(x,y)$ obtained after back-projection operation is in fact an original image $f(x,y)$ convolved with the kernel $h(\Delta x, \Delta y)$.

Therefore, the reconstruction of the image $f(x, y)$ relies on deconvolution process based on the transformed measurements fixed in $\tilde{f}(x, y)$ and the known kernel $h(\Delta x, \Delta y)$.

## 3    Statistical Considerations

It is commonly accepted by physicists that the number of decays of the nucleons in the reconstructed cross-section during PET examination, and subsequent annihilation events follow the inhomogeneous Poisson point process. We assume that the probability that in a reconstructed plane $\lambda$ annihilation events are observed is:

$$P\{\Lambda = \lambda\} = e^{\lambda^*} \frac{(\lambda^*)^\lambda}{\lambda!}, \tag{6}$$

where expectation value of the random variable $\Lambda = \lambda^*$.

In this case, it is fully justified to use the ML method for estimation of the expected value $\lambda^*$ having measurement $\lambda^*$, according to the following expression:

$$l_1(\lambda) = \ln(P(\Lambda = \lambda)) \approx \lambda \ln \frac{\lambda^*}{\lambda} - \lambda^* + const, \tag{7}$$

or alternatively:

$$l_2(\lambda) = H\left(\lambda \ln \frac{\lambda^*}{\lambda} - \lambda^* + const\right), \tag{8}$$

where constant $H = \int_x \int_y h(x, y)\, dx dy$.

In the consequence of symmetry of the kernel $h$ it is possible to rearrange the first and the second terms in $l_2$ from the Eq. (8) to the following form, respectively:

$$\int_{\bar{x}} \int_{\bar{y}} f^*(\bar{x}, \bar{y}) \int_x \int_y h(x - \bar{x}, y - \bar{y})\, dx dy d\bar{x} d\bar{y} =$$
$$\int_x \int_y \left(\int_{\bar{x}} \int_{\bar{y}} h(x - \bar{x}, y - \bar{y}) f^*(\bar{x}, \bar{y})\, d\bar{x} d\bar{y}\right) dx dy \tag{9}$$

(it obeys Campbell's theorem), and

$$H\lambda =$$
$$\int_w \int_v h(w, v)\, dw dv \sum_{k=1}^{\lambda} 1(s, \alpha)\, \delta(s - s_k)\, \delta(\alpha - \alpha_k), \tag{10}$$

It is worth noting that the first term in $l_2$ from formula (8) is linearized using the approximation $ln(u) \approx u - 1$.

As a consequence, we obtain the following expression:

$$l_3\left(f\right) = \int_x \int_y \tilde{f}\left(x,y\right) \ln \frac{\tilde{f}^*\left(x,y\right)}{\tilde{f}\left(x,y\right)} - \tilde{f}^*\left(x,y\right) + const\, dx dy, \tag{11}$$

wherein:

$$\tilde{f}^*\left(x,y\right) = \int_{\bar{x}} \int_{\bar{y}} h\left(x - \bar{x}, y - \bar{y}\right) f^*\left(\bar{x}, \bar{y}\right) d\bar{x} d\bar{y} \tag{12}$$

means a function describing an expected value for all points in an image obtained after the back-projection operation, and the equation

$$\tilde{f}\left(x,y\right) = \sum_{k=1}^{\lambda} int\left(\left(x - x_k\right)\cos \alpha_k + \left(y - y_k\right)\sin \alpha_k\right) \tag{13}$$

relates to points in an image obtained after a back-projection operation. This latter function can be understood as equivalent to a set of direct measurements in the standard ML-EM method.

In our approach, the reconstruction from projections operation is carried out by directly recovering the image from the blurred image $\tilde{f}\left(x,y\right)$ using the maximum likelihood - expectation maximization (ML-EM) estimation method, according to the following formula:

$$f_{min}\left(x,y\right) = \arg \min_{f^*\left(x,y\right)} E\left(l_3\left(\tilde{f}^*\left(x,y\right), \tilde{f}\left(x,y\right)\right)\right), \tag{14}$$

This method is formulated as an optimization problem based on an objective $l_3\left(\tilde{f}^*\left(x,y\right), \tilde{f}\left(x,y\right)\right)$ tailored according to the statistical conditions of measurements obtained in PET scanners.

After using a gradient method to find the optimum for $l_3$, i.e., $\frac{\partial l_3}{\partial f^*\left(x,y\right)} = 0$, we obtain the main relation as follows:

$$f^{t+1}\left(x,y\right) = f^t\left(x,y\right) \frac{1}{H} \int_{\bar{x}} \int_{\bar{y}} \frac{\tilde{f}\left(\bar{x}, \bar{y}\right)}{\int_{\bar{\bar{x}}} \int_{\bar{\bar{y}}} f^t\left(\bar{\bar{x}}, \bar{\bar{y}}\right) h_{\Delta x, \Delta y} d\bar{\bar{x}} d\bar{\bar{y}}} h_{\Delta x, \Delta y} d\bar{x} d\bar{y}, \tag{15}$$

wherein kernel $h_{\Delta x, \Delta y} d\bar{x} d\bar{y}$ obeys relation (5).

Formula (5) is consistent with the continuous-to-continuous data model. Of course, to perform calculations using a computer, it is necessary to discretize this formula, and it is easy to obtain the following form:

$$f^{t+1}\left(x_i, y_j\right) = f^t\left(x_i, y_j\right) \frac{1}{g_{ij}} \sum_{\bar{i}}^{I} \sum_{\bar{j}=1}^{I} \frac{\tilde{f}\left(x_{\bar{i}}, y_{\bar{j}}\right)}{\sum_{\bar{\bar{i}}} \sum_{\bar{\bar{j}}} f^t\left(x_{\bar{\bar{i}}}, y_{\bar{\bar{y}}}\right) h_{\Delta i, \Delta j}} h_{\Delta i, \Delta j} \tag{16}$$

wherein $g_{ij}$ is a sum of all coefficients $h_{\Delta i, \Delta j}$ taken into account at the calculation of a given expression $\frac{\tilde{f}\left(x_{\bar{i}}, y_{\bar{j}}\right)}{\sum_{\bar{\bar{i}}} \sum_{\bar{\bar{j}}} f^t\left(x_{\bar{\bar{i}}}, y_{\bar{\bar{y}}}\right) h_{\Delta i, \Delta j}} h_{\Delta i, \Delta j}$, and the kernel $h_{\Delta i, \Delta j}$ is established according to the formula

$$h_{\Delta i, \Delta j} = \Delta_\alpha \sum_{\psi=0}^{\Psi-1} int \left( \Delta i \cos \psi \Delta_\alpha + \Delta j \sin \psi \Delta_\alpha \right), \tag{17}$$

where $\Psi$ is a determined number of virtually performed projections for a half-revolution of the projection system and $\Delta_\alpha = \frac{\pi}{\Psi}$; $int \left( \Delta s \right)$ is an interpolation function with the same form as has been used during back-projection operation according to the relation (2).

The iterative reconstruction procedure is illustrated schematically in the diagram in Fig. 3. It is worth noting that in this diagram, the part responsible for determining the convolution kernel matrix $h_{\Delta i, \Delta j}$ has been excluded from the iterative area because it is invariant through all iterations. This means that it can be established before starting the iterative reconstruction procedure. Moreover, if we assume a constant image resolution, the kernel $h_{\Delta i, \Delta j}$ can be kept in the file to speed up the calculations. Because a forward model in the reconstruction problem (4) is formulated as a shift invariant system, it is possible to move the most demanding calculations into the frequency domain. For this purpose, a frequency representation $F$ of the reconstructed image $f$ is used in the iterative procedure, and similarly, a frequency representation $H$ of the kernel $h$. In this way, the computational complexity of every convolution performed is reduced from $I^4$ to $8 log_2 4 I^2$, where $I$ is the dimension of the reconstructed image.

## 4    Experimental Results

In our experimentation, we modified the popular Shepp-Logan mathematical head phantom by scaling all values by $10^{-3}$. We utilized parallel projections with 512 virtual detectors on a virtual screen. The number of parallel views per half-rotation was set to 728, and the image size was maintained at $I \times I = 512 \times 512$ pixels. With these parameters established, we could simulate virtual measurements with a relatively high noise level and generate all necessary parallel projections for the lines of response (LORs). By applying appropriate rebinning procedures, we carried out the back-projection operation to acquire an image $\tilde{f}ij$, serving as the reference for the reconstruction process. The reconstructed images after 1,000 iterations are displayed in Figs. 4, 5, and 6. Before commencing the iterative reconstruction, all elements $h\Delta i, \Delta j$ were pre-determined. The iterative reconstruction process involved performing convolution operations in the frequency domain. Additionally, we provide a comparison of the reconstructed images using a reference reconstruction algorithm based on the D-D data model, showcasing results for low, medium, and high noise levels in Figs. 4, 5 and 6, respectively.

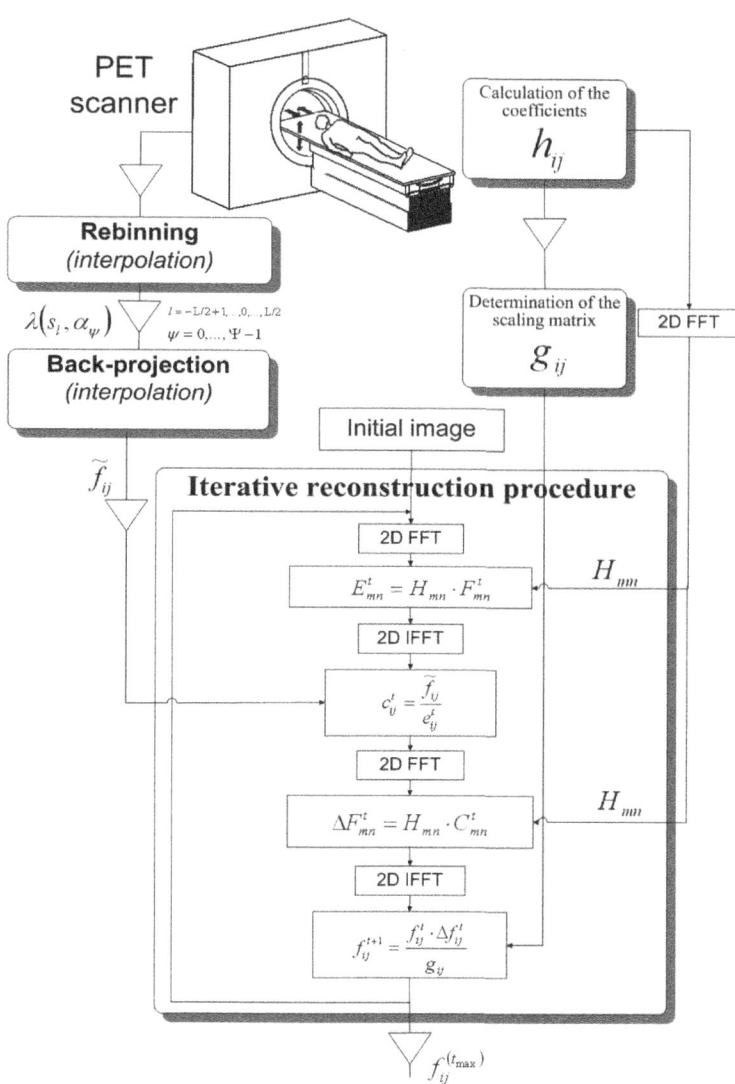

**Fig. 3.** Diagram of the entire reconstruction method with the iterative reconstruction procedure.

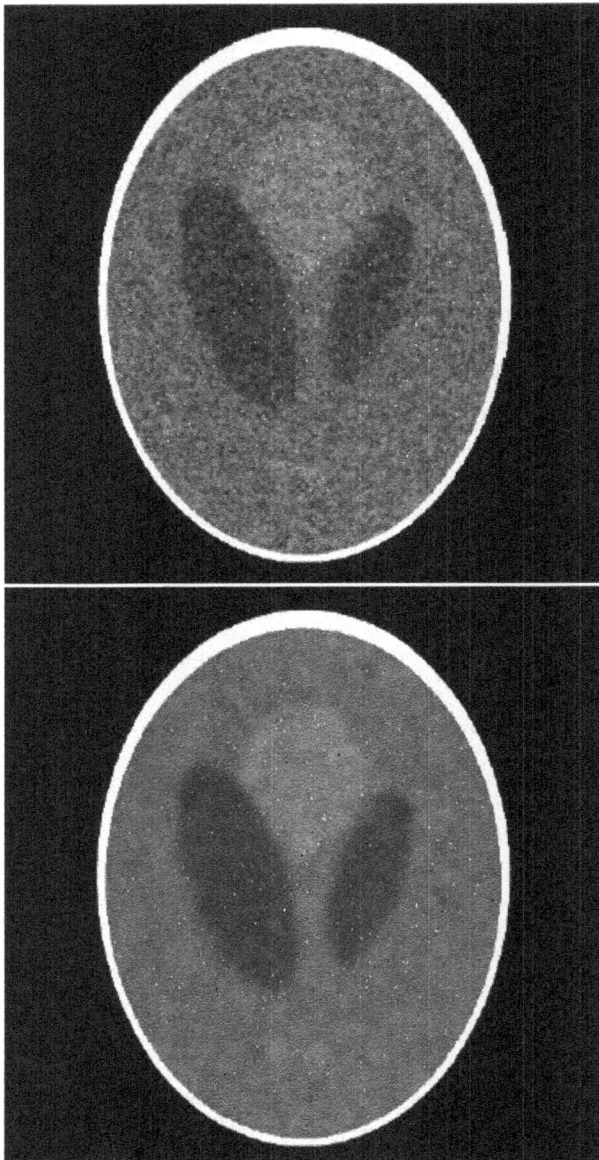

**Fig. 4.** Views of the images (window center $C = 1.05 \cdot 10^{-3}$, window width $W = 0.1 \cdot 10^{-3}$): reconstructed image using the referential statistical approach described by (1) after 50 iterations ($MSE = 5.13 \cdot 10^{-7}$) (at the top); reconstructed image using the statistical approach presented in this paper obtained after 1 000 iterations ($MSE = 5.21 \cdot 10^{-7}$) (at the bottom).

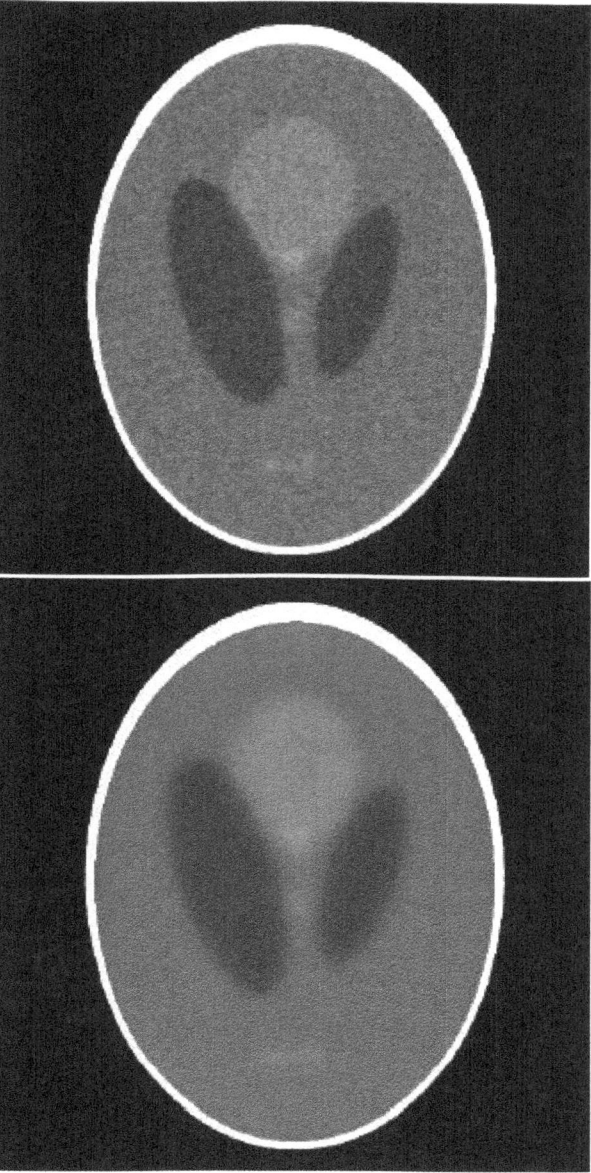

**Fig. 5.** Views of the images (window center $C = 1.05 \cdot 10^{-3}$, window width $W = 0.1 \cdot 10^{-3}$): reconstructed image using the referential statistical approach described by (1) after 50 iterations ($MSE = 6.98 \cdot 10^{-7}$) (at the top); reconstructed image using the statistical approach presented in this paper obtained after 1 000 iterations ($MSE = 6.56 \cdot 10^{-7}$) (at the bottom).

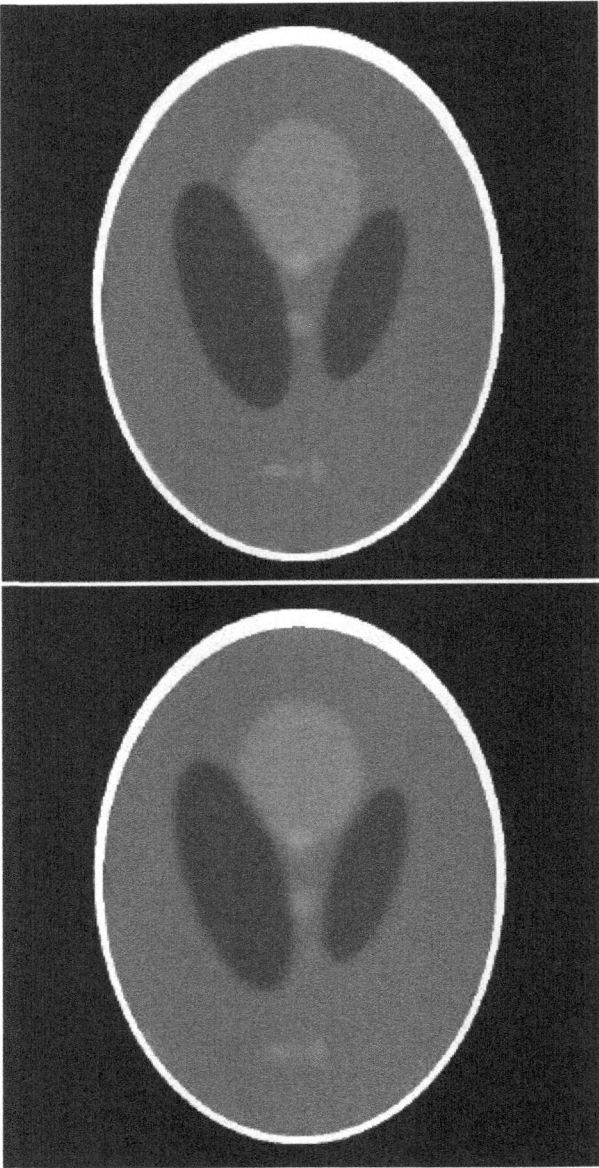

**Fig. 6.** Views of the images (window center $C = 1.05 \cdot 10^{-3}$, window width $W = 0.1 \cdot 10^{-3}$): reconstructed image using the referential statistical approach described by (1) after 50 iterations ($MSE = 7.89 \cdot 10^{-7}$) (at the top); reconstructed image using the statistical approach presented in this paper obtained after 1 000 iterations ($MSE = 7.03 \cdot 10^{-7}$) (at the bottom).

# 5   Conclusion

This research paper introduces a statistical reconstruction method for PET based on the C-C data model. We have developed a practical statistical reconstruction ML-EM algorithm. Our experimental results demonstrate that our reconstruction technique is notably fast, thanks to FFT utilization, and produces satisfactory results with minimized noise, particularly at high noise levels. While the computational complexity for 2D reconstruction geometries (such as parallel rays) typically scales with $I^4$ for each D-D reconstruction iteration, our innovative approach requires approximately $8I^2 \log_2 (2I)$ operations. Consequently, it becomes feasible to reduce the radiation dose of radiopharmaceuticals administered to a patient during scanning without compromising the quality of the reconstructed image.

# References

1. Boudjelal, A., Elmoataz, A., Attallah, B., Messali, Z.: A novel iterative MLEM image reconstruction algorithm based on beltrami filter: application to ECT images. Tomography **7**, 286–300 (2021). https://doi.org/10.3390/tomography7030026
2. Brix, G., Nekolla, E., Nosske, D.: Risks and safety aspects related to PET/MR examinations. Eur. J. Nucl. Med. Mol. Imaging **36**, 131–138 (2009). https://doi.org/10.1007/s00259-008-0937-4
3. Cierniak, R.: Analytical statistical reconstruction algorithm with the direct use of projections performed in spiral cone-beam scanners. In: Proceedings of the 5th International Meeting on Image Formation in X-Ray Computed Tomography, Salt Lake City, pp. 293–296 (2018)
4. Cierniak, R., Dobosz, P., Grzybowski, A.: EM-ML algorithm based on continuous-to-continuous model for pet. In: Proceedings of the 15th International Meeting on Fully Three-Dimensional Image Reconstruction in Radiology and Nuclear Medicine, Philadelphia, Proceedings of SPIE, vol. 11072 (2019)
5. Cierniak, R., Pluta, P., Kaźmierczak, A.: A practical statistical approach to the reconstruction problem using a single slice rebinning method. J. Artif. Intell. Soft Comput. Res. **10**, 137–149 (2020)
6. Cierniak, R., Pluta, P., Waligóra, M., Szymański, Z., Grzanek, K., Piuri, F.: A new statistical reconstruction method for the computed tomography using an x-ray tube with flying focal spot. J. Artif. Intell. Soft Comput. Res. **11**, 271–286 (2021)
7. Gong, K., Catana, C., Qi, J., Li, Q.: Pet image reconstruction using deep image prior. IEEE Trans. Med. Imaging **38**, 1655–1665 (2019). https://doi.org/10.1109/TMI.2018.2888491
8. Gong, K., Kim, K., Cui, J., Wu, D., Li, Q.: The evolution of image reconstruction in pet: from filtered back-projection to artificial intelligence. PET Clin. **16**, 533–542 (2021). https://doi.org/10.1016/j.cpet.2021.06.004
9. Green, P.: Bayesian reconstructions from emission tomography data using a modified EM algorithm. IEEE Tran. Med. Imag. **9**, 84–93 (1990)
10. Hashimoto, F., Onishi, Y., Ote, K., Tashima, H., Yamaya, T.: Fully 3D implementation of the end-to-end deep image prior-based pet image reconstruction using block iterative algorithm. ArXiv arxiv:2212.11844 (2022)

11. Kim, K., Wu, D., Gong, K., Dutta, J., Kim, J., Son, Y.: Penalized pet reconstruction using deep learning prior and local linear fitting. IEEE Trans. Med. Imaging **37**, 1478–1487 (2018)
12. Shepp, L.A., Vardi, Y.: Maximum likelihood reconstruction for emission tomography. IEEE Tran. Med. Imag. **MI-1**, 113–122 (1982)
13. Zheng, X., Qu, G., Zhou, J.: Accelerated strategy for the mlem algorithm. J. Xray Sci. Technol. **29**, 135–149 (2021). https://doi.org/10.3233/XST-200749

# Cerebral Cortex Extraction Methods Based on a Priori Knowledge for T1-Weighted MRI Images

Hajer Ouerghi$^{(\boxtimes)}$, Olfa Mourali, and Ezzeddine Zagrouba

LR16ES06 Laboratoire de Recherche en Informatique, Modélisation et Traitement de l'Information et de La Connaissance (LIMTIC), Université de Tunis el Manar, Institut Supérieur d'Informatique el Manar, Research Team SIIVA, 2 Rue Abou Rayhane Bayrouni, 2080 Ariana, Tunisia
hajer.ouerghi@fst.utm.tn, olfa.mourali@isi.utm.tn,
e.zagrouba@gmail.com

**Abstract.** Cerebral Cortex Extraction (CCE) plays a significant role in clinical applications such as pre-surgical planning and tumor segmentation. However, designing an efficient CCE technique is still a challenging task. In this work, we propose two efficient methods for CCE from T1-weighted MRI images. The first method (named CCE-AK) is divided in two phases: Pretreatment phase and CCE phase. Indeed, the input image is firstly filtered by a Gaussian filter to smooth the image and reduce noise. Thereafter, we apply the anisotropic diffusion to improve the texture quality on the filtered image. Thus, a binary image is obtained after the integration of a priori knowledge and the thresholding steep using the Otsu's method to simplify treatment and eliminate non-brain portions. After that, we start the second phase by eroding the image via a structuring element to eliminate the outer brain parts. In order to extract the Cerebral Cortex (CC), we look for the Largest Connected Component (LCC) in the eroded image. Finally, we use the dilation operation to preserve the totality of the CC region. However, the LCC concept failed in few slices to identify the CC correctly. To address this issue, we introduce a second method (CCE2), which makes use of information in the adjacent slices. To assess the performance, experiments are conducted on different MRI datasets collected from the Surgical Planning Laboratory (SPL). The proposed methods achieve better results in both visual effects and objective criteria than three popular methods (SPM, BET and BSE).

**Keywords:** Cerebral cortex · A priori knowledge · T1-weighted MRI images · Anisotropic diffusion · Morphological operations · Thresholding · Overlap test

## 1 Introduction

Magnetic resonance imaging (MRI) is a non-invasive medical tool that brings an important revolution in the medical field, especially in oncology domain [1]. The major advantage of MRI is its ability to provide 3D image sequences that reflect the structural information of an organ with high spatial resolution without using ionizing radiation. In the

© The Author(s), under exclusive license to Springer Nature Switzerland AG 2024
N.-T. Nguyen et al. (Eds.): ICCCI 2024, CCIS 2166, pp. 419–431, 2024.
https://doi.org/10.1007/978-3-031-70259-4_32

literature, MRI images are used in several medical neuroimaging applications [2]. Cerebral Cortex Extraction (CCE) is the process of removing non-brain tissues from input data without removing any part of the brain [3]. The CCE process is a preliminary step to perform multiple treatments such as pre-surgical planning, cortical reconstruction, automatic tumor detection and analysis [4]. Therefore, the precise definition of the cerebral cortex (CC) can greatly minimize errors in the several neuro-oncology studies and decisions. However, designing an efficient CCE technique is still a challenging task. Machine learning (ML) have been widely adopted for skull stripping analysis [5]. However, the major limitation of ML models is the time-consuming due to several parameters and complicated function mechanisms. In a similar vein, deep learning approaches are used for brain detection and extraction [6], but they also suffer from the high time-consuming limitation and the significant demand for computational power. Despite a plethora of skull stripping approaches in the literature, few are robust, efficient and accurate for MRI processing and clinical treatment analysis. In this paper, we propose two improved CCE methods in T1-weighted MRI. The core contribution of the first proposed method is to highlight the effect of a priori Knowledge based on the intensity of pixels in 2D-MRI data for brain extraction approaches. A sequence of morphological operations and the Largest Connected Component (LCC) analysis are done. The second method makes use of information in the adjacent slices to overcome the failure of LCC in some slices to identify the CC correctly. The rest of this paper is organized as follows. Firstly, recent literature works associated with skull stripping techniques are presented in Sect. 2. Then, we describe the proposed methods in Sect. 3. Thereafter, experimental results and discussed are presented in Sect. 4. Finally, conclusions are given.

## 2  Related Works

Over the years, several brain extraction approaches (BEA) from MRI images have been proposed in neuroimaging research, especially for image processing applications such as image segmentation [7], image classification [8], and image fusion [9]. Edge-based BEA such as Brain Surface Extractor (BSE) [10], use predetermined sets of parameters to separate brain and non-brain tissue via the use of region-growing or morphological operations. Intensity-based methods such as Statistical Parametric Mapping (SPM) [11] and MRI watershed [12], that rely on intensity variations to find the edge of the brain. In Ref. [13], the authors have proposed an MRI-brain extraction method with (SE = 96.4%, and SP = 98.5%) using Dense-Vne and SPM. v12, Similarly, the authors in Ref. [14] have used morphological operations and histogram equalization based enhancement to propose an automatic BEA. In a similar vein, an abnormality brain identification method is proposed in Ref. [15] using convolutional network (CNN) and SPM from FDG-PET imaging. In Ref. [16], a BEA technique from T1-Weighted MRI data is defined based on three steps: data acquisition, pre-processing and LCC extraction. Recently, a simple protocol for fMRI image pre-processing is adopted in Ref. [17], using on SPM. v12 and MATLAB. On the hand, deformable surface-based methods are described in the literature. They use image gradient to fit an active contour/curve to the brain. As examples, we found Brain Extraction Tool (BET) [18], BET2 [19] and SMHASS [20]. Besides, Atlas-based methods like MAPS [21] and ANTs [22] define the boundaries of the brain by registering images to one or many atlases to improve the accuracy of the BEA. Patch-based methods such as BEaST [23] and SPECTRE [24] are an extension of atlas-based BEA in which image-to-atlas registration is performed on non-local image patches. Likewise, hybrid BEA integrate numerous of the above methods to obtain enhanced results. As examples, we found MONSTR [25] and ROBEX [26].

Several studies compared the performance of the most commonly used algorithms [7, 27]. The accuracy and robustness of BEA are key in their adoption, these two measures often being counter-balanced. Hybrid methods to be superior in accuracy at the cost of time-efficiency. Intensity-based and edge-based methods tend to be fast because of their simplicity, but their accuracies tend to fluctuate across heterogeneous datasets with varying levels of image resolutions, noise, and artefacts [28]. On the other hand, atlas-based methods are designed for healthy subjects, but they fail in the presence of large pathological tissue on the image like high grade glioma tumors. Overall, BEA efficiently extract the brain but, they have some weaknesses. Indeed, they are limited to specific orientation, type or datasets and forced the user to compromise either on the processing speed or the accuracy. In addition, their results often include other non-brain tissues. Thus, most of BEA algorithms have produced accurate results, but they required a parameters initialization strategy (like BET or BSE).

## 3 Proposed Methods

In this work, we propose two cortex cerebral extraction (CCE) methods from T1-weighted MRI images. Our method (CCE-AK) contains two phases. The first one named pretreatment phase, is based on a priori knowledge (AK). The second phase is called CCE phase and it is based on morphological operations. The result of the first phase is a binary image, that will be used as input image in the second phase. The block diagram of CCE-AK method is given in Fig. 1.

### 3.1 Phase 1: Pretreatment

Like any other acquisition techniques, MRI has some flaws, such as artifacts related to motion, partial volume effect and noise [1]. These artifacts can distort the anatomical image and degrade the quality of the images. For this reason, we firstly use a low pass filter to smooth the input image and reduce noise. To improve the texture quality of the filtered image, we apply the diffusion anisotropic. Subsequently, we incorporate a priori knowledge to remove some non-brain parts in the image. In addition, we use the thresholding technique based on Otsu method to generate the binary image of the brain.

**Low Pass Filter.** MRI images can be noisy with unwanted pixels that may affect the useful information. Sometimes, we notice that the gray levels of the same tissue in the same area are not homogeneous. For this reason, we first treated the MRI image with a low-pass filter, it is applied to smooth the image, remove small details that appear in the background and enhance important features. We choose the Gaussian filter with a size convolution mask [5 × 5] to improve the quality of our image. It is a special linear isotropic with very specific mathematical properties. It corrects the noise in homogeneous parts of the image and provides less image degradation than other filters such as averaging and median filters [29].

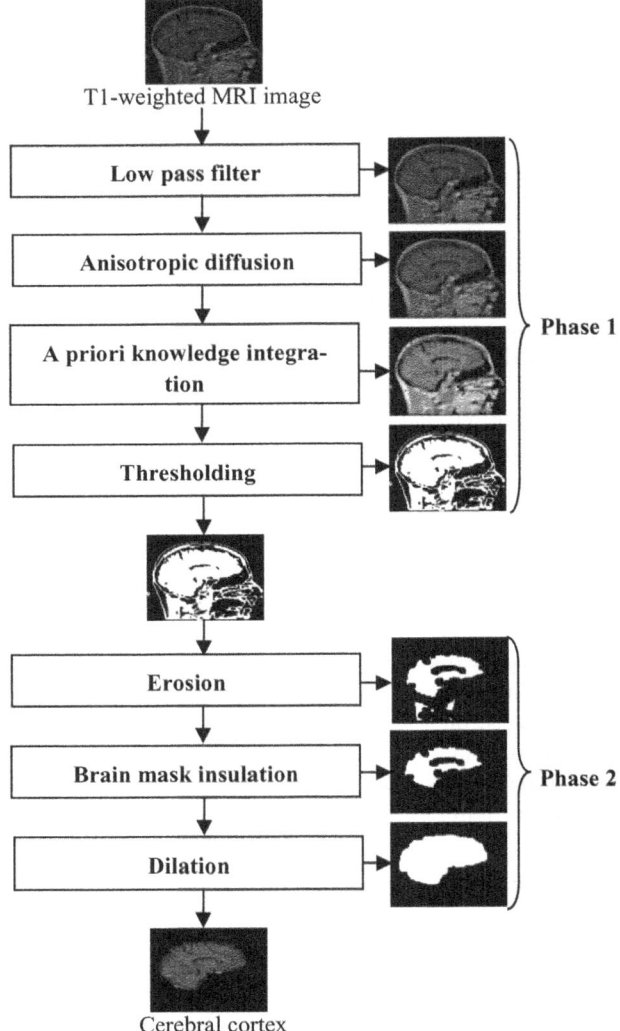

**Fig. 1.** Block diagram of the proposed CCE-AK method.

**Anisotropic Diffusion.** For the filtered image, we apply the diffusion anisotropy process. It is a derivative of the convolution with a Gaussian function. It allows to improve image textures quality and preserve the edges and details in extreme views. The anisotropic diffusion equation is given by Perona and Malik [30]:

$$\frac{\partial I}{\partial t} = \nabla(C\nabla I) \tag{1}$$

where $\nabla I$ is a local image gradient, $I$ is the image and $C\nabla I$ is the diffusion function. We have chosen the diffusion function given by [30] as:

$$C(\nabla I) = \exp\left(-\left(\frac{|\nabla I|}{K}\right)^2\right) \tag{2}$$

$K$ is a diffusion constant. Equation 1 can be defined using the four nearest neighbours as:

$$I_{i,j}^{n+1} = I_{i,j}^n + \nabla t(C_N\nabla_N I + C_S\nabla_S I + C_E\nabla_E I + C_W\nabla_W I)_{i,j}^n \tag{3}$$

where $N$, $S$, $E$ and $W$ represent north, south, east and west direction, respectively. $\nabla t$ is an iteration constant and $n$ is the number of iterations, $\nabla I$ is the local gradient. It is calculated using nearest neighbor differences. In this work, we have set $K$ to 60 that is the nominal edge gradient at cerebral CSF junctions in T1-weighted images.

**A Priori Knowledge Integration.** In the medical field, all existing information in the images provides several meanings especially during diagnosis [1]. Indeed, several studies in medical imaging have shown that in MRI head images, scalp tissue and brain are brighter than the cerebrospinal fluid CSF, skull and the bottom [2, 6]. Our goal is to extract the CC region based on a priori knowledge (AK) learned from the expertise that only depends on the anatomy of the brain and the intensity of T1-weighted MRI images. AK information can target structures to extract, improve outcomes and facilitate the analysis and interpretation of MRI data. Using the expertise above, we will modify the image using two values. The first value (*Valmin*) denotes the minimum intensity below which the pixel belongs to dark areas such as skull, CSF, where these areas are generally characterized by low values of intensity. The second value (*Valmax*) denotes the maximum intensity above which the pixel belongs to the bright areas such as the scalp. We will change to zero, the intensity of pixels which are above to *Valmax* and below to *Valmin*. The AK is incorporated to the image I as follows:

$$I(i,j) = \begin{cases} 0 \text{ if } I(i,j) > Valmax \text{ or } I(i,j) < Valmin \\ \quad I(i,j) \quad otherwise \end{cases} \tag{4}$$

**Thresholding.** To facilitate the next steps, we generate a binary image from the already diffused image and improved by a priori knowledge [31, 32]. For this, an optimal intensity threshold value ($T_{opt}$) for $I(i,j)$ is calculated using Otsu method [33]. A binary image $T(i,j)$ is obtained using Eq. 5. This binary image is taken as a coarse brain mask and it is passed as input for the phase-2.

$$\forall_{i,j} \in N \times M, T(i,j) = \begin{cases} 1 & if \ I(i,j) > T_{opt} \\ 0 & otherwise \end{cases} \tag{5}$$

### 3.2   Phase 2: Cerebral Cortex Extraction

In this phase, we will use two popular morphological operations to extract the CC region. The function of these operations is strongly depended on a structuring element. we chose a structuring element with a disk shape. For this, three pixels in each corner of the structing element are disabled by setting the value to zero.

**Erosion.** The image T obtained in phase 1 is eroded by a structuring element (*SE*) with 7 × 7 size, to get the eroded image E. The erosion is used to separate the weakly connected regions and to remove small non-brain objects such as the skull, cerebrospinal fluid and leather hair. Let ⊖ be the erosion operation. The image E is obtained as:

$$E = T \ominus SE \tag{6}$$

**Brain Mask Insulation.** After the erosion, we obtained several connected regions in the image E, Let $n$ denotes the number of the regions ($R(i)$, $i = 1...,n$). In this work, the area $R_c(i)$ of the $i^{th}$ region $R(i)$ is calculated. $R_c(i)$ is the total number of pixels in the region $R(i)$. For each slice, we are looking for the Largest Connected Component (LCC) in the image E. The obtained LCC region is treated as the brain region $R_{LCC}$. LCC is performed by initially detecting the largest element in the image and then by extracting it through mathematical morphology operators [16]. We get the $R_{LCC}$ as:

$$R_{LCC} = R(max_{1<i<n}(R_c(i))) \tag{7}$$

The brain selection is done as follows:

$$LCC(x, y) = \begin{cases} 1 & if \ (x, y) \in R_{LCC} \\ 0 & otherwise \end{cases} \tag{8}$$

**Dilation.** We apply the dilation with the same structuring element used in erosion's step. In our approach, the dilation operation is used to recapture the CC tissues that were lost in the erosion or thresholding steps. The dilated image D is obtained as:

$$D = LCC \oplus SE \tag{9}$$

where ⊕ represents dilation operation. The image D is the final binary mask of the cerebral cortex and it's used to extract the CC from the original MRI scan. The final cerebral cortex is obtained as:

$$Cortex(x, y) = \begin{cases} I(x, y) & if \ D(x, y) = 1 \\ 0 & otherwise \end{cases} \tag{10}$$

### 3.3 CCE2

The assumption that the cerebral cortex is the LCC in the MRI image failed in certain slices. In some brain volumes, few slices might contain more than one connected component and yet correspond to the CC region. The brain can be divided in 3 volume types according to the slice position. First, we find upper slices (near the top of the head) where the cerebral hemispheres appear as two regions. So, the LCC results contain one region. Then, we find the middle slices, due to partial volume effect that splits the brain into two or more regions especially in the skull or the head. Finally, there are the very lower slices, where the temporal and frontal lobes are separated from the cerebellum. They may appear in more than two regions. In MRI head scans, there is a continuity of the CC portion between two adjacent slices. So, we can explode the neighboring slices to select regions corresponding to the cortex. We introduce an additional process in our method CCE-AK and we call the extended method as CCE2. The block diagram of the CCE2 is shown in Fig. 2.

**Dice Coefficient.** The similarity between two successive slices is checked by computing the dice coefficient (DC) [34]. The DC is a measurement of similarity between the CC masks of the current slice M1 and the previous slice M2. The DC is given by:

$$DC(M_1, M_2) = 2\frac{|M_1 \cap M_2|}{|M_1| + |M_2|} \tag{11}$$

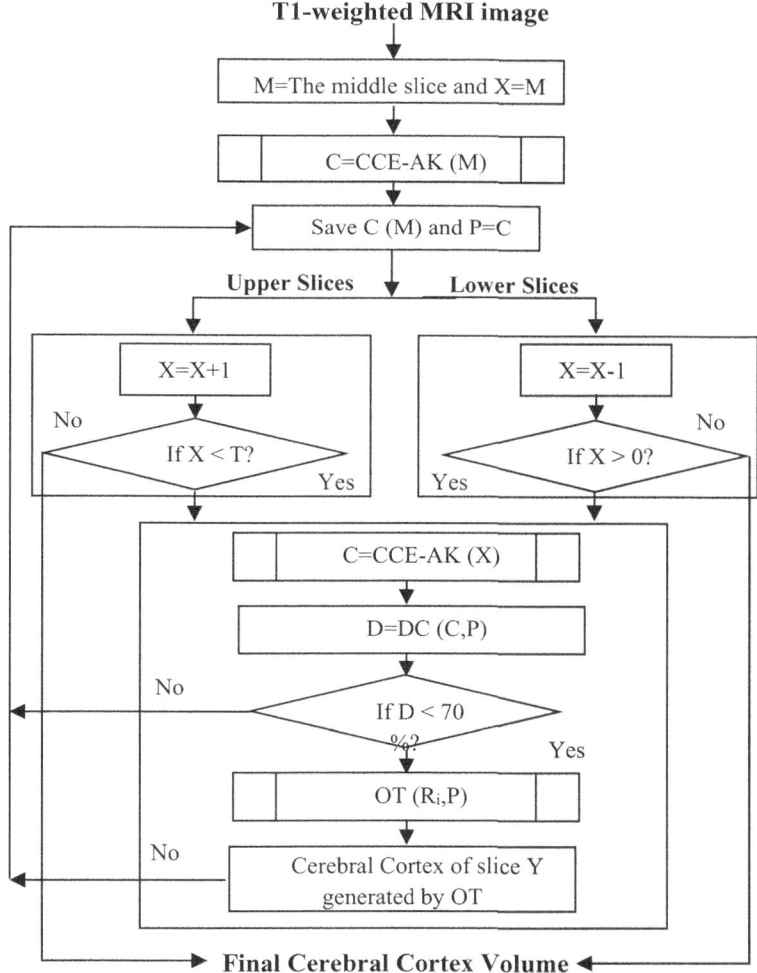

**Fig. 2.** Block diagram of the proposed CCE2 method.

If DC value is greater than 70%, the CC is assumed to have been correctly extracted from the slice, otherwise an overlap test (OT) is performed.

**Overlap Test.** The Overlap Test (OT) is used to choose the missed or disconnected CC regions. This procedure starts with the eroded image E. For each region $R_i$ of E, we calculate the OT with the cerebral cortex $M$ of the previous slice by:

$$OT(R_i, M) = \frac{|R \cap M|}{|R_i|} \tag{12}$$

If OT > 95% then the region $R_i$ is treated as a cortex portion, and added to the CC region, Otherwise, it will be discarded. Hence for the image E, with more than one connected component, Eq. 8 is modified as:

$$Cortex(x, y) = \begin{cases} 1 & if \ (x, y) \in R_i and OT_i > 95\% \\ 0 & otherwise \end{cases} \tag{13}$$

In our work, we choose the same value (95%) of OT used in [35]. While, the value 70% for DC is selected after running the method on several datasets with different DC values. In MRI head scans, the middle slices contain only one connected region, where the LCC works without ambiguity. The CCE2 starts with the middle slice M for each brain volume, at approximately T/2 position, where T is the total number of slices. Our processing will propagate from the center slice and move to lower slices and then from center slice to upper slices, one direction at a time, and produce the brain mask of each slice. These CC masks are used to produce the final CC portion of each T1-weighted MRI image. Finally, we used Eq. 9 and Eq. 10 to select the CC region (cortex).

## 4 Experimental Results and Discussion

To determine the overall performance of the proposed methods, extensive experiments are conducted on T1-weighted MRI images. All used images were taken from the Website of the Surgical Planning Laboratory (SPL) Dataset (https://spl.harvard.edu/). SPL contains ten 3D-MRI cases acquired from 10 patients. Each volume (case) consisted of 124 slices, 256 x 256 pixels per slice. The total number of used images is 700 (70 images per case). These images are accompanied by the ground truth of the manual brain mask. Further, we adopted three popular comparative techniques: BET, BET and SPM version 11. To execute BET and BSE algorithms, we used Medical Image Processing, Analysis, and Visualization (MIPAV) software (https://mipav.cit.nih.gov/).

### 4.1 Evaluation Metrics

In this study, four objective metrics are adopted: (i) Dice coefficient (DC) that evaluates the similarity between the final image and the ground truth mask. (ii) Sensitivity (SE) which is the percentage of CC pixels recognized by the algorithm. (iii) Specificity (SP) that is the percentage of background pixels recognized by the method. (iv) Error rate (ER) is calculated from the SE and SP using a priori probability of pixels [36].

| Image35 | Image45 | Image55 | Image65 | Image75 | Image85 | Image95 |

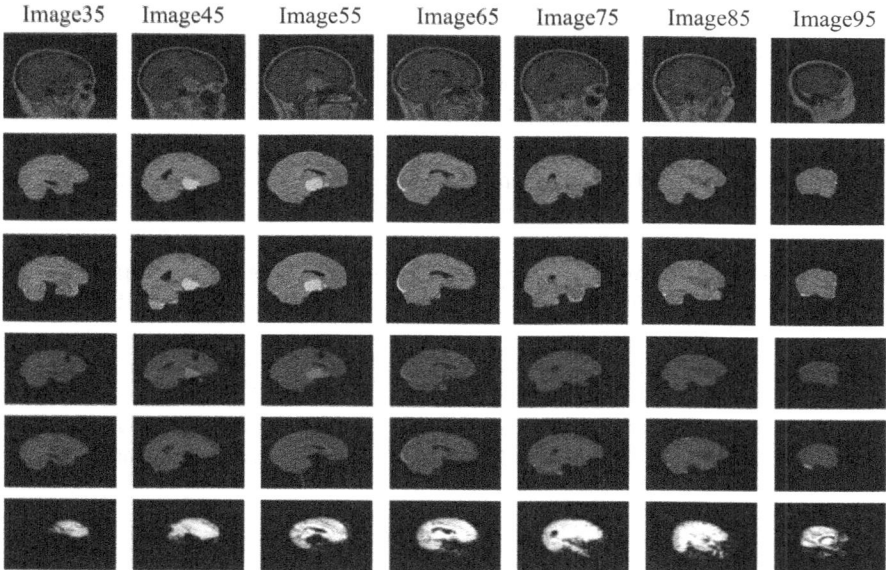

**Fig. 3.** Comparative visual results obtained from different methods applied to case 2, Row 1 shows original T1-scans and rows 2–6 show CCE results by CCE2, CCE-AK, BET, BSE and SPM, respectively.

## 4.2   Results and Discussions

Some examples of CCE results for case 2 are given in Fig. 3. BSE and BET are two interactive tools, where the parameters should be adjusted to obtain a good segmentation result. Table. 1 provide average quantitative performance in terms of DC, SE, SP and ER for the 10 datasets. It should be noted that a larger measure implies better result (except ER). The bold values in the table show the highest value of the metric. With SPM, the average value of DC is very low for the datasets 1, 2, 7 and 9. The reason is due to the inability to remove excess portions like neck, eyes and other non-brain tissues, especially in case 1. The DC value obtained by all algorithms except CCE2 was lower for case 7, due to the intensity variation among the slices, image contrast and the problem of partial volume effect. Adjustments of parameters were done, so as to get the best result. BSE produced higher average values of SP than the proposed methods, it keeps all brain and sometimes even neck portions, but it unable to identify the background that is represented by the number of true negatives.

Looking at the results generated by CCE2, CCE-AK and BET, we observe that they are often similar (with average ER equal to 4%). The ER results are due to the type of tissue, the removal of some brain portions or the keep of non-brain portions. For BET, we have lower average values of ER due to the high number of iterations that it is fixed at 500 iterations. BSE and SPM provide higher average values of ER. The average ER for BSE is equal to 8%, which could be caused by the partial volume effect that prevents BSE to detect edges and find the correct edges of different brain tissues. For SPM, the average value of ER is 14.95%, because sometimes SPM adds same non-brain tissues or removed some of the brain tissue.

In pursuit of the clear comparison, average time costs of the compared algorithms for the 10 cases are illustrated in Table 2. It can be observed that BET and SPM consume lot more time (35s and 4mn) compare to others. SPM estimate a high time cost because the role of this tool is not only the extraction of the brain, but also the separation between the components of the brain (grey matter, white matter and CSF), which round his task more difficult. Further, the CCE-AK method is taking less average time about 10 s (approximately 0.15s / slice) to produce result on the platform implemented in Matlab2018b on a PC with Intel core2 Duo CPU at 2.2 GHz and 4 GB of RAM. This is because its easy implementation and the advantage of the use of the a priori knowledge, which has a very low computational cost.

To conclude, experimental results using 10 datasets indicated that our methods obtained a competitive objective and subjective results, which have been extensively applied in neuroimaging applications like image fusion or tumor classification.

**Table 1.** Performance results for the comparative methods.

| Method | Images | | | | | | | | | | Emerge value |
|---|---|---|---|---|---|---|---|---|---|---|---|
| | Case 1 | Case 2 | Case 3 | Case 4 | Case 5 | Case 6 | Case 7 | Case 8 | Case 9 | Case10 | |
| Dice coefficient (DC) | | | | | | | | | | | |
| CCE2 | 0.8932 | 0.9467 | 0.8675 | 0.8952 | 0.9075 | 0.8534 | 0.8132 | 0.9415 | 0.8514 | 0.9642 | **0.8934** |
| CCE-AK | 0.8727 | 0.9039 | 0.8168 | 0.8230 | 0.9094 | 0.8355 | 0.7360 | 0.9291 | 0.7850 | 0.9440 | 0.8555 |
| BET | 0.8485 | 0.8761 | 0.8173 | 0.8362 | 0.8796 | 0.7769 | 0.7517 | 0.8878 | 0.8166 | 0.9039 | 0.8515 |
| BSE | 0.8439 | 0.8635 | 0.8345 | 0.8765 | 0.8742 | 0.8494 | 0.7440 | 0.9023 | 0.8199 | 0.8906 | 0.8499 |
| SPM | 0.2042 | 0.6818 | 0.7359 | 0.7306 | 0.7458 | 0.7230 | 0.6313 | 0.8045 | 0.6847 | 0.7877 | 0.6730 |
| Sensitivity (SE) | | | | | | | | | | | |
| CCE2 | 0.9865 | 0.9748 | 0.9812 | 0.9978 | 0.9821 | 0.9912 | 0.9990 | 0.9943 | 0.9890 | 0.9818 | 0.9878 |
| CCE-AK | 0.9730 | 0.9750 | 0.9780 | 0.9915 | 0.9753 | 0.9861 | 0.9980 | 0.9825 | 0.9607 | 0.9795 | 0.9800 |
| BET | 0.9965 | 0.9861 | 0.9955 | 0.9995 | 0.9975 | 0.8368 | 0.9946 | 0.9993 | 0.9891 | 0.9895 | 0.9785 |
| BSE | 0.9606 | 0.9808 | 0.9955 | 0.9998 | 0.9994 | 0.9997 | 0.9983 | 0.9934 | 0.9862 | 0.9998 | **0.9913** |
| SPM | 0.2409 | 0.6457 | 0.8882 | 0.9533 | 0.8030 | 0.9654 | 0.7378 | 0.9918 | 0.9208 | 0.9551 | 0.8102 |
| Specificity (SP) | | | | | | | | | | | |
| CEE2 | 0.9675 | 0.9684 | 0.9478 | 0.9574 | 0.9765 | 0.9614 | 0.8990 | 0.9818 | 0.9485 | 0.9801 | **0.9588** |
| CCE-AK | 0.9539 | 0.9551 | 0.9120 | 0.9055 | 0.9560 | 0.9219 | 0.8722 | 0.9653 | 0.8963 | 0.9747 | 0.9313 |
| BET | 0.9449 | 0.9482 | 0.9166 | 0.9230 | 0.9398 | 0.9403 | 0.8971 | 0.9410 | 0.9063 | 0.9501 | 0.9307 |
| BSE | 0.9471 | 0.9297 | 0.9158 | 0.9371 | 0.9295 | 0.9248 | 0.8785 | 0.9474 | 0.9107 | 0.9399 | 0.9261 |
| SPM | 0.9164 | 0.9584 | 0.8933 | 0.8518 | 0.9189 | 0.8564 | 0.8919 | 0.8836 | 0.8414 | 0.8935 | 0.8906 |
| Metric: Error rate (ER) | | | | | | | | | | | |
| CEE2 | 0.0290 | 0.0304 | 0.0495 | 0.0487 | 0.0255 | 0.0401 | 0.0574 | 0.0518 | 0.0813 | 0.0174 | **0.0431** |
| CCE-AK | 0.0366 | 0.0350 | 0.0550 | 0.0515 | 0.0344 | 0.0460 | 0.0649 | 0.0715 | 0.1044 | 0.0229 | 0.0444 |
| BET | 0.0293 | 0.0328 | 0.0439 | 0.0387 | 0.0313 | 0.1114 | 0.0541 | 0.0299 | 0.0523 | 0.0302 | 0.0454 |
| BSE | 0.1902 | 0.0448 | 0.0425 | 0.2092 | 0.0356 | 0.0378 | 0.0616 | 0.1765 | 0.0516 | 0.0302 | 0.0880 |
| SPM | 0.4214 | 0.1979 | 0.1093 | 0.0974 | 0.1390 | 0.0891 | 0.1851 | 0.0623 | 0.1189 | 0.0757 | 0.1496 |

**Table 2.** Average time cost comparison for the 10 cases.

| Method | CCE2 | CCE-AK | BET | BSE | SPM |
|---|---|---|---|---|---|
| Average time cost | 12.01s | 10.6s | 12,27s | 35,85s | 4.2mn |

## 5  Conclusion

In this paper, we propose an efficient 2 phases-method (CCE-AK) for cerebral cortex extraction from T1-weighted MRI scans. The first phase is based on a priori knowledge, while, the second phase used morphological operations and LLC technique. Based on CCE-AK, the method CCE2 is proposed for 3D-MRI volumes using dice coefficients and a proposed overlap test. This method makes use of information in the adjacent slices. Overall, we remained competitive with literature methods. As prospects, we opt to improve the reliability of our approaches. Firstly, our method CCE-AK is based on a priori knowledge, if that knowledge value changes, the result will also change. We will seek to improve the a priori knowledge specification. Secondly, a generalized algorithm for different MRI modality (T2-weighted or proton density) will be adopted.

## References

1. Li, C., et al.: Artificial intelligence in multiparametric magnetic resonance imaging: a review. Med. Phys. **49**(10), e1024–e1054 (2022)
2. Hu, J., et al.: Diagnostic performance of magnetic resonance imaging–based machine learning in Alzheimer's disease detection: a meta-analysis. Neuroradiology **65**(3), 513–527 (2023)
3. Qin, C., Li, B., Han, B.: Fast brain tumor detection using adaptive stochastic gradient descent on shared-memory parallel environment. Eng. Appl. Artif. Intell. **120**, 105816 (2023)
4. Ranjbarzadeh, R., et al.: Brain tumor segmentation of MRI images: a comprehensive review on the application of artificial intelligence tools. Comput. Biol. Med. **152**, 106405 (2023)
5. Amin, J., et al.: Brain tumor detection and classification using machine learning: a comprehensive survey. Complex Intell. Syst. **8**(4), 3161–3183 (2022)
6. Balasubramanian, S., et al.: RF-ShCNN: a combination of two deep models for tumor detection in brain using MRI. Biomed. Signal Process. Control **88**, 105656 (2024)
7. Kumar, A.: Study and analysis of different segmentation methods for brain tumor MRI application. Multimedia Tools Appl. **82**(5), 7117–7139 (2023)
8. Ouerghi, H., Mourali, O., Zagrouba, E.: Glioma classification via MR images radiomics analysis. Vis. Comput., 1–15 (2022)
9. Premalatha, R., Dhanalakshmi, P.: Robust neutrosophic fusion design for magnetic resonance (MR) brain images. Biomedical Sig. Process. Control **84**, 104824 (2023)
10. Shattuck, D.W., et al.: Magnetic resonance image tissue classification using a partial volume model. Neuroimage **13**(5), 856–876 (2001)
11. Penny, W.D., et al., Statistical parametric mapping: the analysis of functional brain images. Elsevier (2011)
12. Tarhini, G.M. and R. Shbib, Detection of brain tumor in MRI images using watershed and threshold-based segmentation. nternational Journal of Signal Processing Systems, 2020. 8(1): p. 19–25

13. Ranjbar, S., et al.: Robust automatic whole brain extraction on magnetic resonance imaging of brain tumor patients using dense-Vnet. arXiv preprint arXiv:02627 (2020)

14. Ullah, Z., Lee, S.-H., An, D.: Histogram equalization based enhancement and mr brain image skull stripping using mathematical morphology. Int. J. Adv. Comput. Sci. Appl. **11**(3), 569–577 (2020)

15. Whi, W., et al.: Fully automated identification of brain abnormality from whole-body FDG-PET imaging using deep learning-based brain extraction and statistical parametric mapping. EJNMMI Phys. **8**(1), 1–10 (2021)

16. Duarte, K.T.N., et al.: Brain extraction in multiple T1-weighted magnetic resonance imaging slices using digital image processing techniques. IEEE Lat. Am. Trans. **20**(5), 831–838 (2022)

17. Di, X., Biswal, B.B.: A functional MRI pre-processing and quality control protocol based on statistical parametric mapping (SPM) and MATLAB. Front. Neuroimaging **1**, 1070151 (2023)

18. Smith, S.M.: Fast robust automated brain extraction. Hum. Brain Mapp. **17**(3), 143–155 (2002)

19. Jenkinson, M., Pechaud, M., Smith, S.: BET2: MR-based estimation of brain, skull and scalp surfaces. In: Eleventh Annual Meeting of the Organization for Human Brain Mapping. Toronto (2005)

20. Galdames, F.J., Jaillet, F., Perez, C.A.: An accurate skull stripping method based on simplex meshes and histogram analysis for magnetic resonance images. J. Neurosci. Methods **206**(2), 103–119 (2012)

21. Leung, K.K., et al.: Brain MAPS: an automated, accurate and robust brain extraction technique using a template library. Neuroimage **55**(3), 1091–1108 (2011)

22. Avants, B.B., et al.: A reproducible evaluation of ANTs similarity metric performance in brain image registration. Neuroimage **54**(3), 2033–2044 (2011)

23. Eskildsen, S.F., et al.: BEaST: brain extraction based on nonlocal segmentation technique. Neuroimage **59**(3), 2362–2373 (2012)

24. Carass, A., et al.: Simple paradigm for extra-cerebral tissue removal: algorithm and analysis. Neuroimage **56**(4), 1982–1992 (2011)

25. Roy, S., Butman, J.A., Pham, D.L.: Robust skull stripping using multiple MR image contrasts insensitive to pathology. Neuroimage **146**, 132–147 (2017)

26. Iglesias, J.E., et al.: Robust brain extraction across datasets and comparison with publicly available methods. IEEE Trans. Med. Imaging **30**(9), 1617–1634 (2011)

27. Mahalaxmi, G., et al.: A comparison and survey on brain tumour detection techniques using MRI images. Curr. Signal Transduct. Ther. **18**(1), 14–23 (2023)

28. Gull, S., Akbar, S.: Artificial intelligence in brain tumor detection through MRI scans: advancements and challenges. Artif. Intell. Internet of Things, 241–276 (2021)

29. Sagheer, S.V.M., George, S.N.: A review on medical image denoising algorithms. Biomedical Signal Process. Control **61**, 102036 (2020)

30. Perona, P., Malik, J.: Scale-space and edge detection using anisotropic diffusion. IEEE Trans. Pattern Anal. Mach. Intell. **12**(7), 629–639 (1990)

31. Sharma, S.R., et al.: Hybrid multilevel thresholding image segmentation approach for brain MRI. Diagnostics **13**(5), 925 (2023)

32. Jardim, S., António, J., Mora, C.: Image thresholding approaches for medical image segmentation-short literature review. Procedia Comput. Sci. **219**, 1485–1492 (2023)

33. Kumar, V.V., Prince, P.G.K.: Magnitude normalized and OTSU intensity based brain tumor detection using magnetic resonance images. IETE J. Res. **69**(8), 5079–5089 (2023)

34. Shamir, R.R., et al.: Continuous dice coefficient: a method for evaluating probabilistic segmentations. arXiv preprint arXiv:11031 (2019)
35. Somasundaram, K., Kalaiselvi, T.: Fully automatic brain extraction algorithm for axial T2-weighted magnetic resonance images. Comput. Boil. Med. **40**(10), 811–822 (2010)
36. Wang, Z., Wang, E., Zhu, Y.: Image segmentation evaluation: a survey of methods. Artif. Intell. Rev.Intell. Rev. **53**, 5637–5674 (2020)

# Modelling of Drug-Induced Liver Injury with Multiple Machine Learning Algorithms

Wojciech Lesiński[1]([✉])[ID], Agnieszka Kitlas Golińska[1][ID],
and Witold R. Rudnicki[1,2,3][ID]

[1] Faculty of Computer Science, University of Białystok, Białystok, Poland
w.lesinski@uwb.edu.pl
[2] Computational Center, University of Białystok, Białystok, Poland
[3] Interdisciplinary Centre for Mathematical and Computational Modelling,
University of Warsaw, Warsaw, Poland

**Abstract. Motivation:** Drug-induced liver injury (DILI) is one of the primary problems in drug development. The current study aims to build predictive models using the physical and chemical properties of compounds causing DILI. Early prediction of DILI may result in redirecting drug development towards compounds with a lower risk of DILI and thus significantly reduce the risk of market failure.

**Methods:** Research was performed on the FDA DILI Rank data set. All compounds were labelled in three DILI classification schemes: FDA DILI concern classification, DILI severity score, and commercial status of the drug. We used five classifiers to build cross-validated predictive models for different binary splits of drugs into high- and low-risk classes of DILI.

**Results:** The best models obtained (AUC = 0.81, MCC = 0.46) discern between harmless and DILI-causing compounds, regardless of DILI severity. The models for other splits of compounds along DILI scales are worse, showing that prediction of the severity of DILI is more difficult.

The filter based on the classifier can be used to assess lead compounds in drug development to decrease DILI risk in new drugs.

**Keywords:** Drug Induced Liver Injury · feature selection · machine learning · classification · drugs development

## 1 Background

Bringing a new therapeutic molecule to the market is a complex and expensive process – it can take 12 years and cost 2.6 billion US dollars [3,10]. Once a drug has been introduced to the market, the risk of Drug-induced liver injury (DILI) is one of the main problems since nearly all classes of medications can cause liver disease [7]. An estimated 1000 drugs have been implicated in causing liver disease on more than one occasion [12]. Some drugs can injure the liver;

N.-T. Nguyen et al. (Eds.): ICCCI 2024, CCIS 2166, pp. 432–444, 2024.
https://doi.org/10.1007/978-3-031-70259-4_33

in extreme cases, therapy can be more dangerous than the disease itself. DILI accounts for approximately half of the cases of acute liver failure and is the leading cause of drug withdrawals from the market [20].

DILI has diverse symptoms – it mimics all forms of acute and chronic liver disease. Although, except for rare cases, DILI subsides after cessation of treatment with the drug, it represents a significant diagnostic and therapeutic challenge for physicians [12].

Many researchers approached the problem of DILI prediction with the help of bioinformatical and cheminformatical tools using published biological and chemical data [16]. For example, Hong et al. [11] used a decision forest [25] based on FDA's Liver Toxicity Knowledge Base for DILI prediction.

Muller and coworkers [19] used standard Machine learning to predict DILI, relying on *in vivo* models of DILI of organic molecules. Alternatively, DILI can be predicted in relation to oral doses and blood concentrations, such as in a study by Albrecht and coworkers [1].

**DILI in CAMDA Challenges.** The DILI prediction problem was also investigated in three challenges organised by the CAMDA (Critical Assessment of Massive Data Analysis) community in 2018, 2019 and 2020. Prediction of DILI was based on three types of data: human cell lines gene expression, chemical descriptors, and cell images. The DILI definition was based on FDA DILI classification [4]. Results of these works were described in [13, 15, 23].

The current study was an extension of our research in the CAMDA challenges [13, 14]. In those works, we obtained AUC=0.75 to distinguish between 'DILI' and 'no DILI' compounds. Our previous work has focused on integrating different types of data. Here we try to predict DILI based on the chemical structure of drugs on a data set more extensive than those available in the CAMDA challenges.

## 2   Methods

### Data

The current study is based on the FDA DILI Rank data set [4]. It provides a list of 1036 compounds categorised based on their DILI potential, severity and market status. The SMILES (Simplified molecular-input line-entry system) descriptors [26] for 967 compounds were identified and used as input for the molecular descriptor calculator Mordred [18]. This procedure resulted in a data set comprising 967 compounds described with 1613 descriptors.

All compounds were labelled using three different DILI classification schemes:

- FDA DILI concern classification with four classes:
  1. 'vNo-DILI-Concern' (276 compounds);
  2. 'Ambiguous DILI-concern' (238 compounds);
  3. 'vLess-DILI-Concern' (271 compounds);

    4. 'vMost-DILI-Concern' (182 compounds);
- Decision based on the commercial status of the drug with the following classes:
    1. 'no match' (301);
    2. 'adverse event' (328);
    3. 'warning and precaution' (245);
    4. 'box warning' (32);
    5. 'withdrawn' (53);
- DILI severity score (from 0 to 8) (see Table 1).

**Table 1.** Number of objects in DILI severity score classification

| DILI severity | DILI category | Number |
|---|---|---|
| 0 | no DILI | 300 |
| 1 | Steatosis | 3 |
| 2 | Cholestatsis; Steatohepatitis | 46 |
| 3 | Liver aminotransferases increase | 294 |
| 4 | Hyperbilirubinemia | 50 |
| 5 | Jaundice | 61 |
| 6 | Liver necrosis | 5 |
| 7 | Acute liver failure | 54 |
| 8 | Fatal hepatotoxicity | 154 |

**Feature Selection**

Classifiers achieve the best accuracy when all used variables are informative. This is especially important when the number of variables is huge since both model quality and computational performance are degraded when a large number of variables are used. Our data consist of more than 1500 variables with non-zero variance, so we used feature selection algorithms. The identification of informative variables was performed using two methods, **Welch t-test** for differences in sample means, and a multidimensional filter based on information theory developed in our laboratory [17, 21] and implemented in the R package **MDFS**. MDFS allows the identification of variables that take part in multivariate synergistic interactions. In this study, MDFS 1D (i.e. univariate mutual information filter) and 2D (i.e. filter for pairwise synergies) were tried.

A non-redundant subset of relevant variables was used for model building. To construct such a set, variables were ordered by their relevance, and then variables that were highly correlated with more relevant ones were removed from the description. The threshold for removal was set at a correlation coefficient equal to 0.7.

**Classification**

We used popular classifiers for modelling: **Random Forest (RF)** algorithm [2],

**k-Nearest Neighbours (k-NN)** [24], **Support Vector Machine (SVM)** [6], **Deep Neural Network** [22] and **Naive Bayes.** Random Forest is based on decision trees and works well *out of the box* on most data sets [9]. SVM [6] is a machine learning algorithm based on statistical learning theory. DNN is a neural network with some level of complexity, with at least two hidden layers. K-NN is a simple and widely known method based on distances between objects. Finally, Naive Bayes is a simple algorithm that may work well for simple additive problems. Two methods for data normalisation for the k-NN classifier were used: z-score- and rank-normalisation. Random Forest was used with the default set of hyperparameters (ntree = 500 and $mtry = \sqrt{d}$, where d is the number of variables). The quality of SVM models depends strongly on the parameters used. Therefore, we tested the performance of various kernel functions and performed grid optimisation of gamma, degree and coef0 hyperparameters. The radial kernel function with gamma equal 0.02 have the best results and was used in subsequent models. Deep Neural Network was based on Keras library [5]. Our final network consisted of 4 hidden layers with 20 neurons in each (dropout ratio equal 0.2, ReLU function was used as activation). Models were trained for 50 epochs, and Adam optimisation was used.

Seven measures were utilised for the assessment of classifiers' performance: error, precision, recall, F-measure, MCC (Matthews Correlation Coefficient), AUC (area under the ROC curve) and AUPR (area under the precision-recall curve). The two latter measures provide the overall performance measure for the varying balance between precision and recall, whereas the former five are point estimates at the selected cutoff value. The following discussion of results is therefore based mainly on the AUC measure, which is well correlated with AUPR in most cases.

**Cross-Validation**

Machine learning methods frequently generate models that are overfitted to the training set. In particular, the selection of variables used for modelling can result in strong overfitting. This was avoided by performing a cross-validation procedure. In each repeat, the data set was split randomly into five parts in a stratified manner. Five models were then developed for each split, with four parts used as a training set and one as a validation set. The entire procedure was repeated ten times to consider the influence of variance due to different data splits.

## 3   Results and Discussion

Previous works have shown that DILI prediction is difficult even if defined as a binary classification problem. Our experience has shown that models developed in the multi-class setting were significantly worse than those obtained for a binary problem based on the same data. Therefore, we have transformed the multi-class problem into a series of simpler binary problems.

In the case of the FDA DILI Rank classification data was aggregated in 2 ways:

- Class 1 as 'no DILI concern', classes 2, 3, and 4 as 'DILI concern';
- Classes 1 as 'no DILI concern', classes 3 and 4 as 'DILI concern' (compounds belonging to the 'Ambiguous DILI-concern' class were removed);

In the DILI severity score and DILI label section, we have examined the performance of classifiers for all possible binary divisions into the low- and high-risk classes. All tests were performed in the repeated, fully cross-validation scheme.

**Feature Selection**
The informative variables were identified using the Welch t-test and MDFS in 1 and 2 dimensions. The best results were achieved by MDFS in 1-dimensional mode. For binary DILI concern classification, our filtering methods found about 200 relevant variables (about 100 uncorrelated) in each cross-validation fold. Scanning of severity level and market status gave varied results, depending on the cut-off point. In two other DILI scales, when the divisions were similar to the 'no DILI' – 'DILI' classification above, the number of informative descriptors was also close to 200. The number of relevant variables decreased for unbalanced classes. For example, when the binary division was defined so that compounds causing high severity of DILI (severity level 8) were compared with all others (severity level 0–7), only about 20 informative variables were identified in each fold. Similar results were obtained in binary decisions based on the market status of drugs.

The variable selection process was stable, with similar feature sets obtained from different cross-validations folds. The most important physical and chemical properties include the following:

- averaged and centred Moreau-Broto auto correlations;
- Geary coefficient;
- eigenvalue of Burden matrix;
- MOE descriptors.

**Binary 'DILI Concern' Classification**
Our first tests were based on FDA 'DILI concern' classification, with all compounds assigned either to the 'DILI concern' or 'no DILI concern' class. As described earlier, classes 'Ambiguous DILI-concern', 'Less-DILI-Concern', and 'Most-DILI-Concern' were aggregated to the 'DILI concern' class. Classifiers were built to discern this class from the 'no DILI concern' class. Our main evaluation measures were AUC and MCC, and we treated the others as auxiliaries.

The Random Forest classifier obtained the best results in all but one measure. Comparable, if slightly worse, results were obtained with SVM. The SVM had a marginally higher recall, tied with RF on F-Measure, but was worse at other measures. The results obtained with DNN and both variants of k-NN (k = 7) classifiers were generally similar to each other, significantly worse than those of RF and SVM, but better than Naive Bayes. Finally, Naive Bayes produced the worst results. The statistical significance of the results difference was confirmed by the DeLong test for two AUC ROCs [8]. P-values for hypothesis "true difference in AUC is not equal to 0" were not higher than 0.00015.

**Table 2.** Binary 'DILI concern' classification

|  | AUC | MCC | Recall | F1 | Precision | Error | AUPR |
|---|---|---|---|---|---|---|---|
| Data with ambiguous class | | | | | | | |
| RF | 0,77 | 0.41 | 0.92 | 0.85 | 0.79 | 0.23 | 0.88 |
| SVM | 0.75 | 0.37 | 0.95 | 0.85 | 0.76 | 0.25 | 0.86 |
| DNN | 0.72 | 0.32 | 0.86 | 0.82 | 0.79 | 0.27 | 0.65 |
| Bayes | 0.69 | 0.23 | 0.9 | 0.82 | 0.76 | 0.28 | 0.83 |
| k-NN zscore | 0.71 | 0.33 | 0.88 | 0.83 | 0.78 | 0.26 | 0.78 |
| k-NN rank | 0.72 | 0.32 | 0.93 | 0.83 | 0.79 | 0.26 | 0.85 |
| Data without ambiguous class | | | | | | | |
| RF | 0.81 | 0.46 | 0.85 | 0.81 | 0.77 | 0.25 | 0.86 |
| SVM | 0.78 | 0.43 | 0.84 | 0.80 | 0.76 | 0.26 | 0.82 |
| DNN | 0.74 | 0.36 | 0.80 | 0.77 | 0.75 | 0.29 | 0.58 |
| Bayes | 0.72 | 0.30 | 0.89 | 0.78 | 0.70 | 0.31 | 0.79 |
| k-NN zscore | 0.74 | 0.36 | 0.82 | 0.78 | 0.75 | 0.28 | 0.80 |
| k-NN rank | 0.75 | 0.4 | 0.84 | 0.79 | 0.75 | 0.27 | 0.83 |

In the second analysis, the compounds belonging to the 'Ambiguous DILI-concern' class were removed from the data set. This class contained compounds with unclear DILI status. Removal of these compounds leads to a smaller set with better-defined decision classes. As expected, the quality of predictive models was slightly better in this case. Both AUC and MCC were better for all classifiers. Interestingly, the AUPR measure was worse for all classifiers than the former case. As before, the Random Forest turned out to be the best classifier. The ROC curves for all classifiers are displayed in Fig. 1.

Full results for all classifiers are displayed in Table 2. Figures 2 and 3 show the distribution of AUC and MCC in cross-validation folds.

It is visible that the Random Forest classifier generates models outperforming all other algorithms. Also, in further analyses, Random Forest consistently outperformed all other algorithms. Therefore, the results are discussed for Random Forest only.

**DILI Severity Level Scanning**
All possible binary divisions for DILI severity levels were explored in the next stage. Each step of the scan represented a shift towards discerning more severe DILI cases. As in the previous case, the best performance was obtained with the help of the Random Forest classifier, slightly worse with SVM, whereas DNN, k-NN and naive Bayes classifiers generated significantly worse models. The results obtained with the help of the RF classifier for all stages are collected in Table 3. The best results for all classifiers were obtained when binary classes were set as 'level 0' versus 'rest' ('no DILI' versus all DILI severity levels). This division is nearly identical to the 'DILI concern' and 'no DILI concern' split from

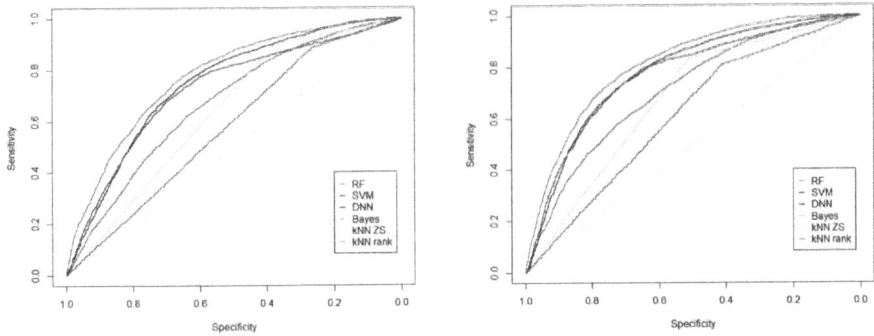

**Fig. 1.** Typical ROC curves for 'DILI concern' versus 'no DILI concern' classification for all compounds (left panel) and data set with ambiguous compounds removed (right panel). Differences between ROC curves were confirmed by the DeLong test.

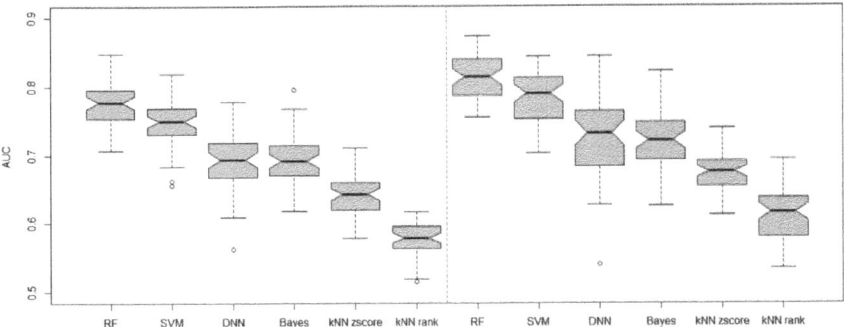

**Fig. 2.** Distribution of AUC in cross-validation obtained for 'DILI concern' versus 'no DILI concern' classification. Results are from 10 repetitions of 5-fold cross-validation. All compounds are on the left panel, and ambiguous compounds are removed on the right panel.

the previous analysis. The performance of the models degraded steadily with moving the separation threshold towards more severe cases of DILI. One should notice that while the AUC drops from the 0.75–0.80 range for low thresholds towards the 0.65–0.70 range for high thresholds, the other quality measures degrade monotonously, with recall dropping even below 0.10. Similar, if slightly better, results were observed for the data set limited to the compounds with non-ambiguous DILI status. The inconsistency of the classification error should be noted; it first increases, only to decrease with the rising separation threshold. This results from a strong class imbalance at later stages. It perfectly showcases why the overall classification error is an inferior quality measure, particularly for imbalanced data.

**Scanning of the Market Status of Drugs**
The final analysis was performed for possible binary divisions for DILI market

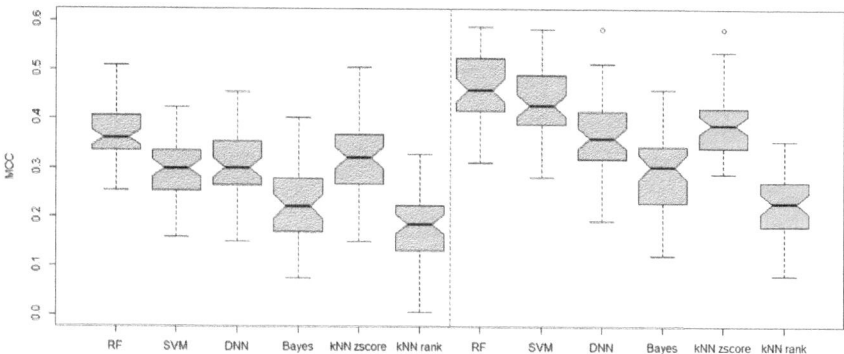

**Fig. 3.** Distribution of MCC in cross-validation obtained for 'DILI concern' versus 'no DILI concern' classification. Results are from 10 replicates of 5-fold cross-validation. All compounds on the left panel and ambiguous ones are removed on the right panel.

status. The data was divided as follows: 'no match' versus rest, 'no match' and 'adverse event' versus rest, 'no match', 'adverse event' and 'warning and precaution' versus 'box warning', and all milder cases versus 'withdrawn'. The results are collected in Table 4.

The first division is again nearly identical to the 'DILI concern' and 'no DILI concern' split from the first analysis. The best results, in this case, were also obtained for the first split. Interestingly, similar results were obtained for the second split, where mild cases of DILI were merged with DILI and contrasted with more severe cases. These results were obtained for roughly 2:1 split between 'no & mild DILI' and 'severe DILI' classes. This split corresponds roughly to the separation between classes 0–4 vs 5–7 in the previous analysis, yet the results for the current study are significantly better. This happens even though the former analysis is based purely on medical evidence. In contrast, the market status also considers other factors, including the balance between the risk of DILI and the beneficial action of the drug. ML models based on chemical descriptors prove superior in reflecting these complex effects.

### Enrichment of 'Non-DILI' Class

To test the practicality of our study, we performed the following analysis. We first divided compounds in the 'low-risk' and 'high-risk' classes using a fraction of votes of the Random Forest classifier for the 'non-DILI' class. The 'low-risk' class was assigned to the compounds with a fraction of votes for the 'non-DILI' class above the cut-off level and 'high-risk' for the remaining ones; see Table 5. Then we analysed the enrichment of the 'non-DILI' class in the 'low-risk' class compared to the 'high-risk one. We checked three cut-off levels: 60%, 50% and 35% votes for the 'non-DILI' class. Promising results were obtained for the classifier, developed on the data set excluding compounds of ambiguous DILI status. At the cut-off level set at 60%, only 20% of all compounds were assigned to the 'low-risk' class. Among them, 80% belong to 'non-DILI' and 20% to the 'DILI' class, respectively.

**Table 3.** Random Forest severity prediction. 'Rec' means 'Recall', 'Prec' - 'Precision', and 'cp' - 'class proportion'. In 'class proportion', the left number means '0' class (no or lighter DILI).

| Level | AUC | MCC | Rec. | F1 | Prec. | Error | AUPR | 'cp' |
|---|---|---|---|---|---|---|---|---|
| Data with ambiguous class | | | | | | | | |
| >0 | 0.76 | 0.40 | 0.89 | 0.83 | 0.78 | 0.25 | 0.86 | 300–667 |
| >1 | 0.76 | 0.41 | 0.89 | 0.83 | 0.78 | 0.25 | 0.86 | 303–664 |
| >2 | 0.74 | 0.36 | 0.84 | 0.79 | 0.74 | 0.29 | 0.82 | 349–618 |
| >3 | 0.69 | 0.28 | 0.32 | 0.41 | 0.57 | 0.31 | 0.50 | 643–324 |
| >4 | 0.67 | 0.23 | 0.20 | 0.28 | 0.51 | 0.28 | 0.43 | 693–274 |
| >5 | 0.66 | 0.20 | 0.10 | 0.16 | 0.46 | 0.23 | 0.36 | 754–213 |
| >6 | 0.68 | 0.22 | 0.11 | 0.18 | 0.48 | 0.22 | 0.37 | 759–208 |
| >7 | 0.69 | 0.21 | 0.07 | NA | 0.41 | 0.16 | 0.31 | 813–154 |
| Data without ambiguous class | | | | | | | | |
| >0 | 0.83 | 0.46 | 0.82 | 0.79 | 0.76 | 0.26 | 0.83 | 300–429 |
| >1 | 0.82 | 0.48 | 0.83 | 0.79 | 0.76 | 0.25 | 0.84 | 301–428 |
| >2 | 0.80 | 0.46 | 0.79 | 0.77 | 0.75 | 0.27 | 0.81 | 321–408 |
| >3 | 0.71 | 0.32 | 0.40 | 0.47 | 0.57 | 0.32 | 0.54 | 470–259 |
| >4 | 0.71 | 0.29 | 0.31 | 0.40 | 0.57 | 0.29 | 0.51 | 503–226 |
| >5 | 0.67 | 0.27 | 0.19 | 0.27 | 0.49 | 0.26 | 0.41 | 543–186 |
| >6 | 0.69 | 0.23 | 0.20 | 0.28 | 0.48 | 0.26 | 0.42 | 545–184 |
| >7 | 0.70 | 0.23 | 0.13 | 0.21 | 0.53 | 0.20 | 0.38 | 584–145 |

**Table 4.** Random Forest label section prediction. Labels: 1: 'no match', 2: 'adverse event', 3: 'warning and precaution', 4: 'box warning', 5: 'withdrawn'. 'Rec' means 'Recall', 'Prec' - 'Precision', and 'cp' - 'class proportion'. In 'class proportion' left number means '0'.

| Level | AUC | MCC | Rec. | F1 | Prec. | Error | AUPR | 'cp' |
|---|---|---|---|---|---|---|---|---|
| Data with ambiguous class | | | | | | | | |
| 1:2345 | 0.77 | 0.40 | 0.89 | 0.83 | 0.78 | 0.25 | 0.86 | 301–658 |
| 12:345 | 0.77 | 0.40 | 0.42 | 0.52 | 0.68 | 0.27 | 0.63 | 629–330 |
| 123:45 | 0.72 | 0.16 | 0.02 | 0.11 | 0.46 | 0.09 | 0.21 | 874–85 |
| 1234:5 | 0.74 | 0.14 | 0.03 | 0.17 | 0.39 | 0.06 | 0.17 | 906–53 |
| Data without ambiguous class | | | | | | | | |
| 1:2345 | 0.81 | 0.47 | 0.82 | 0.79 | 0.76 | 0.26 | 0.83 | 301–420 |
| 12:345 | 0.77 | 0.40 | 0.46 | 0.55 | 0.69 | 0.26 | 0.65 | 471–250 |
| 123:45 | 0.71 | 0.16 | 0.02 | 0.11 | 0.38 | 0.12 | 0.25 | 636–85 |
| 1234:5 | 0.72 | 0.15 | 0.04 | 0.18 | 0.41 | 0.07 | 0.21 | 668–53 |

**Table 5.** Enrichment of 'non-DILI' class. The low-risk and high-risk groups were defined according to the results of Random Forest classification for three diverse cutoff levels. In columns are, respectively: 1 - cutoff level, 2 - a fraction of all compounds selected into the low-risk group, 3 - a fraction of non-DILI compounds selected into the low-risk group, 4 - a fraction of DILI compounds selected into the low-risk group, 5 - a fraction of non-DILI compounds in the selected group.

| cutoff level | 2 | 3 | 4 | 5 | non-DILI/DILI odds in low-risk group | non-DILI/DILI odds in high-risk group | odds ratio |
|---|---|---|---|---|---|---|---|
| Without *ambiguous* class | | | | | | | |
| 0.6 | 20% | 42% | 7% | 80% | 3.93:1 | 1:2.66 | 10.4 |
| 0.5 | 31% | 59% | 14% | 71% | 2.49:1 | 1:3.43 | 8.5 |
| 0.35 | 51% | 80% | 33% | 58% | 1.46:1 | 1:5.54 | 8.1 |
| With *ambiguous* class | | | | | | | |
| 0.6 | 10% | 24% | 4% | 70% | 2.40:1 | 1:3.10 | 7.5 |
| 0.5 | 17% | 39% | 8% | 66% | 1.90:1 | 1:3.78 | 7.3 |
| 0.35 | 34% | 64% | 22% | 54% | 1.16:1 | 1:5.50 | 6.4 |

At this cut-off level, 93% of compounds from the 'DILI' class were filtered out to the 'high-risk' class. However, only 42% of compounds from the 'non-DILI' class were retained in the 'low-risk' class. One should note that the fraction of the 'non-DILI' class in the database is 38%; therefore, at this cut-off level, one obtains more than two-fold enrichment of 'non-DILI' in the 'low-risk' class. On the other hand, the odds of 'non-DILI' to 'DILI' compounds in the 'low-risk' class is 3.93:1, vs 12.66 in the 'high-risk' class, resulting in the odds ratio of 10.4.

At the other end of the spectrum, setting the cut-off level at 35% of votes for 'non-DILI' resulted in the assignment of 50% of compounds to the 'low-risk' class. Among them, 58% belongs to 'non-DILI' and 42% to the 'DILI' class, respectively. At this cut-off level, 80% of 'non-DILI' compounds are retained in the 'low-risk' class, whereas 67% of 'DILI' compounds are filtered out to the 'high-risk category. The enrichment of 'non-DILI' in the 'low-risk' class is only roughly 1.5-fold. However, the odds of 'non-DILI' to DILI in the 'low-risk' class are still favourable (1.46:1). At the same time, in the 'high-risk' category, they are significantly lower (1:5.54), resulting in the odds ratio of 8.1.

The results for the intermediate cut-off level of 0.5 are intermediate between two former choices; see Table 5. The severity of DILI measured on any scale does not correlate with the probability of assignment to the low- and high-risk classes at all three examined cut-off levels. That result agrees with the results from scanning along the severity scales. While we can, to some extent, predict the risk of DILI for a compound. However, the molecular descriptors alone are not sufficient for building models for assessing the severity of DILI.

The results obtained for the classifier developed on the complete data set are worse. This is expected since the presence of the ambiguous class has a negative effect both on model building and then also for predictions. In the model development, compounds from the ambiguous class were assigned to the 'DILI' category, although some may be harmless. The detrimental effects are twofold - wrongly labelled compounds misguide the algorithm. Moreover, adding all these compounds to the 'DILI' class significantly influences the proportion of 'non-DILI' to 'DILI' compounds. This results in a lower proportion of compounds predicted as 'non-DILI' at each cutoff level. Moreover, the presence of additional compounds labelled as 'DILI' in the data set leads to a higher proportion of 'DILI' compounds in the 'low-risk' class at all cutoff levels.

## 4   Conclusion

Models for three definitions of DILI were constructed using only the molecular properties of drugs. For all three definitions, relatively good models (AUC=0.81, AUPR=0.86) were obtained for discerning harmless compounds from compounds with some level of DILI concern. These results were better than those obtained by integrating various data during the CAMDA challenges. The improvement was most likely achieved due to using a more extensive data set. Best classification models were achieved by the Random Forests classification algorithm, slightly worse by SVM. Other methods were significantly worse. Interestingly, the model discerning severe and mild cases of DILI defined by the market status of a drug was significantly better than the analogous models developed on the DILI scale determined by pure DILI severity.

The quality of the models developed in the current study cannot answer whether a particular substance will be harmful or not. The classification of DILI depends on multiple factors, including the severity of harmful effects, the severity of the medical condition for which the drug is applied, and the length of treatment. A drug may have minor effects in the short term but, when used for years, may have much stronger cumulative DILI effects than a drug that is taken only sporadically. Nevertheless, the models are informative and can be used as a guideline for development as one of the factors contributing to funnelling the design process towards the most promising targets.

Assuming that the classifier developed in the current study is applied to the filtering of lead compounds in the drug development process, setting the filter at 60% of votes for the 'non-DILI' class results in the aggressive filtering of the 'non-DILI' class, at the price of significant reduction of a number of all compounds and also removing from considerations also most of the harmless compounds. On the other hand, setting the filter at 35% of votes for the 'non-DILI' class results in moderate filtering that allows the retention of most of the harmless compounds at the price of letting through a substantial number of DILI-causing ones. Selection of the filtering level should therefore reflect the composition of the initial database and the severity of the targeted condition. Aggressive filtering should be applied for the mild condition when many potential lead compounds are predicted as 'non-DILI' and mild when the disease is severe. Most lead compounds are expected to belong to the 'DILI' class.

**Acknowledgments.** Computations were performed at the Computational Centre of the University of Białystok.

# References

1. Albrecht, W., et al.: Prediction of human drug-induced liver injury (DILI) in relation to oral doses and blood concentrations. Arch. Toxicol. **93**(6), 1609–1637 (2019). https://doi.org/10.1007/s00204-019-02492-9
2. Breiman, L.: Random forests. Mach. Learn. **45**, 5–32 (2001)
3. Chan, H.S., Shan, H., Dahoun, T., Vogel, H., Yuan, S.: Advancing drug discovery via artificial intelligence. Trends Pharmacol. Sci. **40**(8), 592–604 (2019)
4. Chen, M., Suzuki, A., Thakkar, S., Yu, K., Hu, C., Tong, W.: Dilirank: the largest reference drug list ranked by the risk for developing drug-induced liver injury in humans. Drug Disc. Today **21**(4), 648–653 (2016). https://doi.org/10.1016/j.drudis.2016.02.015
5. Chollet, F., et al.: Keras (2015). https://github.com/fchollet/keras
6. Cortes, C., Vapnik, V.: Support-vector networks. Mach. Learn. **20**(3), 273–297 (1995)
7. David, S., Hamilton, J.: Drug-induced liver injury. US Gastroenterol. Hepatol. Rev. **1**(6), 73–80 (2010)
8. DeLong, E.R., DeLong, D.M., Clarke-Pearson, D.L.: Comparing the areas under two or more correlated receiver operating characteristic curves: a nonparametric approach. Biometrics, 837–845 (1988)
9. Fernández-Delgado, M., et al.: Do we need hundreds of classifiers to solve real world classification problems. J. Mach. Learn. Res. **15**(1), 3133–3181 (2014)
10. Godoy, P., et al.: Recent advances in 2D and 3D in vitro systems using primary hepatocytes, alternative hepatocyte sources and non-parenchymal liver cells and their use in investigating mechanisms of hepatotoxicity, cell signaling and adme. Arch. Toxicol. **87**(8), 1315–530 (2013). https://doi.org/10.1007/s00204-013-1078-5
11. Hong, H., Thakkar, S., Chen, M., Tong, W.: Development of decision forest models for prediction of drug-induced liver injury in humans using a large set of FDA-approved drugs. Sci. Rep. **7**(17311) (2017). https://doi.org/10.1038/s41598-017-17701-7
12. Kaplowitz, N.: Drug-induced liver injury. Clin. Infect. Dis. **38**(Supplement2), S44–S48 (2004)
13. Lesiński, W., Mnich, K., Kitlas Golińska, A., Rudnicki, W.: Integration of human cell lines gene expression and chemical properties of drugs for drug induced liver injury prediction. Biol. Direct **16**(1), 2 (2021). https://doi.org/10.1186/s13062-020-00286-z
14. Lesiński, W., Mnich, K., Rudnicki, W.R.: Prediction of alternative drug-induced liver injury classifications using molecular descriptors, gene expression perturbation, and toxicology reports. Front. Genetics **12** (2021). https://doi.org/10.3389/fgene.2021.661075

15. Liu, A., et al.: Prediction and mechanistic analysis of drug-induced liver injury (DILI) based on chemical structure. Biol. Direct **16**(6) (2021). https://doi.org/10.1186/s13062-020-00285-0
16. Minerali, E., Foil, D.H., Zorn, K.M., Lane, T.R., Ekins, S.: Comparing machine learning algorithms for predicting drug-induced liver injury (DILI). Mol. Pharm. **17**(7), 2628–2637 (2020). https://doi.org/10.1021/acs.molpharmaceut.0c00326
17. Mnich, K., Rudnicki, W.R.: All-relevant feature selection using multidimensional filters with exhaustive search. Inf. Sci. **524**, 277–297 (2020). https://doi.org/10.1016/j.ins.2020.03.024
18. Moriwaki, H., et al.: Mordred: a molecular descriptor calculator. J. Cheminf. **10**(1), 4 (2018). https://doi.org/10.1186/s13321-018-0258-y
19. Muller, C., et al.: Prediction of drug induced liver injury using molecular and biological descriptors. Comb. Chem. High Throughput Screen. **18**, 315–322 (2015). https://doi.org/10.2174/1386207318666150305144650
20. Olson, H., et al.: Concordance of the toxicity of pharmaceuticals in humans and in animals. Regul. Toxicol. Pharmacol. **32**(1), 56–67 (2000). https://doi.org/10.1006/rtph.2000.1399
21. Piliszek, R.: MDFS: MultiDimensional feature selection in R. R J. **11**(1), 198–210 (2019)
22. Schmidhuber, J.: Deep learning in neural networks: an overview. Neural Netw. **61**, 85–117 (2015). https://doi.org/10.1016/j.neunet.2014.09.003
23. Sumsion, G., et al.: Diverse approaches to predicting drug-induced liver injury using gene-expression profiles. Biol. Direct **15**(1) (2020). https://doi.org/10.1186/s13062-019-0257-6
24. Cover, T., Hart, P.: Nearest neighbor pattern classification. IEEE Trans. Inf. Theory **13**(1), 21–27 (1967). https://doi.org/10.1109/TIT.1967.1053964
25. Tong, W., Hong, H., Fang, H., Xie, Q., Perkins, R.: Decision forest: combining the predictions of multiple independent decision tree models. J. Chem. Inf. Comput. Sci. **43**, 525–31 (2003). https://doi.org/10.1021/ci020058s
26. Weininger, D.: Smiles, a chemical language and information system. 1. Introduction to methodology and encoding rules. J. Chem. Inf. Comput. Sci. **28**(1), 31–36 (1988). https://doi.org/10.1021/ci00057a005

# Temporal Focal Modulation Networks for EEG-Based Cross-Subject Motor Imagery Classification

Adel Hameed[1,2], Rahma Fourati[1,3(✉)], Boudour Ammar[1],
Javier Sanchez-Medina[4], and Hela Ltifi[1,5]

[1] REsearch Groups in Intelligent Machines, National Engineering School of Sfax,
3038 Sfax, Tunisia
{boudour.ammar,hela.ltifi,rahma.fourati}@ieee.org

[2] National School of Electronics and Telecommunications of Sfax, University of Sfax,
Sfax, Tunisia

[3] Faculté des Sciences Juridiques, Economiques et de Gestion de Jendouba,
Université de Jendouba, 8189 Jendouba, Tunisia

[4] Innovation Center for the Information Society, University of Las Palmas de Gran
Canaria, Las Palmas de Gran Canaria, Spain
javier.sanchez.medina@ieee.org

[5] Department of Computer Sciences, Faculty of Sciences and Techniques of Sidi
Bouzid, University of Kairouan, Kairouan, Tunisia

**Abstract.** Motor Imagery (MI) EEG decoding is crucial in Brain-Computer Interface (BCI) technology, facilitating direct communication between the brain and external devices. However, accurately capturing temporal dependencies in MI EEG signals, especially in subject-independent MI-BCIs, remains a persistent challenge.

In this paper, we present Temporal-FocalNets, a novel framework designed to address this challenge by leveraging focal modulation techniques. Temporal-FocalNets efficiently prioritize temporal dynamics, thereby enhancing the accuracy and robustness of MI EEG decoding models. Through comprehensive experiments on benchmark datasets (2a and 2b), Temporal-FocalNets demonstrates superior performance compared to established baseline models. This innovation marks a significant advancement in subject-independent MI-BCIs, offering new possibilities for individuals with motor disabilities to interact with their environment using brain signals.

**Keywords:** Electroencephalography · Motor imagery · Transformer · Focal Modulation Networks

## 1 Introduction

Brain-computer interfaces (BCIs) establish links between the human brain and external devices, finding applications in medical, neurobiological, and psychological fields. Motor imagery electroencephalography (MI-EEG) is a popular BCI

N.-T. Nguyen et al. (Eds.): ICCCI 2024, CCIS 2166, pp. 445–457, 2024.
https://doi.org/10.1007/978-3-031-70259-4_34

type known for its simplicity and adaptability. In a MI-BCI system, EEG signals collected during specific motor imagery tasks are utilized for content identification and control of peripheral devices [10, 25].

Subject-dependent MI-BCIs play a crucial role in the rehabilitation of motor disorders, necessitating personalized data collection and training. However, this approach results in weak generalization across different subjects [2, 15]. On the contrary, subject-independent MI-BCIs offer a more convenient and advanced solution, catering to multiple subjects without the need for additional calibration [27]. Consequently, this research centers on subject-independent MI-BCI.

Motor Imagery (MI) involves mentally simulating movement without physical execution. Decoding MI-EEG signals is vital for understanding evolving brain activity during imagined motor tasks, with temporal dynamics capturing dynamic EEG changes over time.

In recent times, the utilization of Deep Learning (DL) methods for MI-EEG task classification has surged. Convolutional Neural Networks (CNNs), known for their adept feature extraction, have been extensively applied for decoding motor imagery from EEG signals [6] [4]. However, CNNs face challenges in handling long-range dependencies and larger datasets [4].

To address CNNs' limitations, researchers have turned to Recurrent Neural Networks (RNNs), especially using Long Short-Term Memory (LSTM) networks to capture temporal features of EEG signals [11]. Despite their successes, RNNs face challenges with extensive data and efficiency due to limited parallelization in RNN steps [34].

Transformers have gained attention for capturing spatial and temporal dependencies across domains [5, 12, 14, 19, 21, 22, 28, 29, 33]. Despite their effectiveness, concerns exist regarding their efficiency, especially with a large number of tokens, leading to escalated computation time and cost. This heightened complexity stems from the dual-step self-attention operation, involving query-key interaction followed by aggregation over context values. The computationally intensive step entails calculating token-to-token attention scores using dot-product, as queries and keys lack information about the surrounding context.

To overcome self-attention challenges, we present Temporal-FocalNets, an inventive framework inspired by focal modulation [31] in image recognition. This approach prioritizes temporal dynamics to improve Motor Imagery EEG signal decoding while ignoring the spatial aspects. Spatial modeling tends to involve higher dimensionality, leading to increased computational complexity. Prioritizing temporal aspects may offer a more computationally efficient solution. Our key contributions can be summarized as follows:

1. Introducing Temporal-FocalNets, a new framework that enhances the decoding of Motor Imagery EEG signals using a temporal focal modulation system inspired by image recognition techniques.
2. Addressing computational and efficiency challenges in self-attention mechanisms of transformer models, effectively modeling both local and global contexts in EEG signal decoding.

3. Describing the architecture of Temporal-FocalNets, which combines depthwise and pointwise convolutions within a hierarchical context aggregation framework tailored for capturing temporal patterns in motor imagery classification.

4. Comparative studies highlight Temporal-FocalNets' superior performance over other models, particularly in terms of faster training and inference times, making it highly suitable for real-time brain-computer interface (BCI) applications.

## 2    Related Work

This section provides a review of deep learning approaches for Motor Imagery (MI) recognition, with a specific focus on transformers in the context of deep learning. It explores the advancements in classifying Motor Imagery from EEG signals.

### 2.1    Deep Learning Approaches for Motor Imagery (MI) Recognition

Deep learning models for Motor Imagery (MI) classification, such as Kwon et al. [16], employ advanced Convolutional Neural Networks (CNNs) with intelligent frequency band selection. Zhang et al. [32] introduce a convolutional recurrent attention model for multisubject MI recognition, outperforming existing methods. Hermosilla et al. [13] enhance subject-independent MI recognition with an improved Shallow Convolutional Network (SCN).

Autthasan et al. [1] propose MIN2Net, an end-to-end multitask learning model, achieving a compact and discriminative feature representation. In neurorehabilitation, Niazi et al. [24] introduce a template-based method for subject-independent movement-related cortical potentials detection.

Regarding model architectures, Lawhern et al. [17] present EEGNet, a compact CNN, while Zhao et al. [35] propose MSCNet, a multi-scale spatio-temporal CNN. Dong et al. [8] introduce MSAENet, a dual-branch multiscale autoencoder network. Dolzhikova et al. [7], Roy et al. [26], and Milanes et al. [23] contribute diverse CNN architectures for subject-independent MI classification, showcasing advancements in EEG-based BCI systems.

### 2.2    Transformer-Based Models

Transformer-based Deep Learning (DL) models, introduced by Vaswani et al. [30], have demonstrated effectiveness in both Computer Vision (CV) and Natural Language Processing (NLP). D'Ascoli et al. [9] emphasize transformers' impact on performance limits, showcasing their ability to extract global information through Self-Attention (SA) mechanisms.

In Brain-Computer Interface applications like MI-EEG decoding, Ma et al. [22] and Song et al. [28] explore hybrid CNN-Transformer models, while Tao et al. [29] enhance performance with a gating mechanism. However, challenges in computational costs persist.

Studies like Zhang et al. [33], Luo et al. [21], and Hu et al. [14] present approaches applicable across subjects, integrating local and global transformers with CNNs for comprehensive feature extraction. In [12], Independent Component Analysis is leveraged alongside a transformer-based approach for enhanced Motor Imagery (MI) EEG feature extraction.

In transformer-based studies, initial research introduced self-attention [30] for effective long-range dependency modeling in visual inputs. However, self-attention's computational demands led to recent innovations like focal modulation [31]. Focal modulation, utilizing hierarchical context aggregation, integrates information from different-sized receptive fields efficiently. Our proposed Temporal-FocalNets framework, inspired by focal modulation in image recognition, aims to enhance MI-EEG decoding by balancing global context modeling with computational efficiency.

## 3   The Proposed Methodology

In our research, we introduce a new framework called "Temporal-FocalNet," drawing inspiration from focal modulation in image recognition [31]. We aim to improve efficiency and performance while accurately capturing temporal dependencies in decoding Motor Imagery (MI) EEG Signals. To address temporal dimensions, we introduce a temporal focal modulation block. The overall architecture is depicted in Fig. 1, with the design of the temporal focal modulation detailed in Fig. 2.

The model is specifically tailored for the analysis of Motor Imagery EEG signals, denoted as $X_{st} \in \mathbb{R}^{T \times T_e \times C \times 1}$. Here, $T$ represents the number of segments, $T_e$ signifies the time steps, $C$ denotes the number of EEG channels, and the last dimension reflects the absence of a color space, in contrast to Image data. The EEG signal data undergoes processing through a patch embedding layer employing a convolutional layer with a filter size and stride set to 4.

The model progresses through Three stages of temporal focal modulation blocks, each designated as $i \in \{1, 2, 3, 4\}$, with each stage encompassing $N_i$ temporal focal modulation layers. Following each stage, we introduce another patch embedding layer. This layer effectively reduces the temporal size of the feature map while maintaining the channel dimension. Specifically, the temporal dimension undergoes a reduction: $T \rightarrow \frac{T}{4} \rightarrow \frac{T}{8} \rightarrow \frac{T}{16} \rightarrow \frac{T}{32}$. Importantly, our model is intentionally designed to maintain the channel dimension as it progresses through various layers. Simultaneously, the average output from the final stage is calculated and passed on to a classification layer using a fully connected layer.

The temporal focal modulation comprises two crucial stages: Hierarchical Contextualization and Gated Aggregation. Figure 3 provides an in-depth visual representation, offering a detailed illustration of the processes involved in Hierarchical Contextualization and Gated Aggregation.

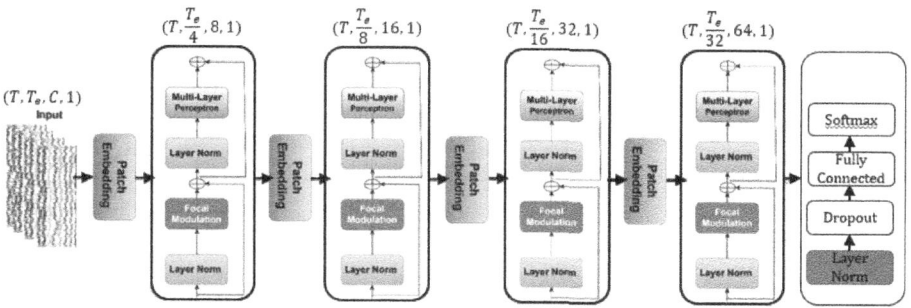

**Fig. 1.** The architecture of Temporal-FocalNets, as described in [20], consists of four stages that include Temporal-FocalNet blocks and patch embedding. These blocks use a technique called Focal Modulation, which is similar to transformer blocks [30], which deviates from ordinary self-attention and improves the model's capacity for EEG signal interpretation.

***Hierarchical Contextualization:*** We initiate the process by projecting the temporal feature map $X_t$ using two linear layers, generating $Z_0^t$:

$$Z_0^t = f_{z,t}(X_{st}) \in \mathbb{R}^{T \times T_e \times C \times 1}. \tag{1}$$

We then employ a series of $L$ depth-wise and point-wise convolution operations to process the temporal projected input $Z_0^t$ across its dimension:

$$Z_\ell^t = f_{\ell a,t}(Z_{\ell-1}^t) \triangleq \text{GeLU}(\text{PWConv}(Z_{\ell-1}^t)) \in \mathbb{R}^{T \times T_e \times C \times 1}. \tag{2}$$

Finally, we apply global average pooling to $Z_L^t$ along temporal dimensions:

$$Z_{L+1}^t = \text{Avg-Pool}(Z_L^t). \tag{3}$$

***Gated Aggregation:*** In the next step, we consolidate the temporal feature maps, denoted as $Z_t^\ell$, into their corresponding temporal modulators using a gating mechanism. We derive temporal gating weights, denoted as $G_t = f_{g,t}(X_t)$, in the temporal domain (with dimensions $T \times (L+1)$), employing linear projection layers $f_{g,t}$. Following this, we perform a dot product operation between the feature maps and their respective gates:

$$Z_{\text{out}_t} = \sum_{\ell=1}^{L+1} G_t^\ell \cdot Z_t^\ell. \tag{4}$$

To facilitate communication across different channels, we employ another set of linear layers denoted as $h_t(\cdot)$ to obtain the temporal modulator $M_t = h_t(Z_{\text{out}_t})$ with dimensions $T \times C \times 1$.

The temporal focal modulation process is defined as follows:

$$y_i = q(x_i) \odot h_t(Z_{\text{out}_t}) \tag{5}$$

Here, $q(\cdot)$ serves as a query projection function, while $\odot$ represents element-wise multiplication. $h_t(\cdot)$ operates as a context aggregation function, with its output denoted as the temporal modulator.

The temporal focal modulation efficiently captures hierarchical temporal contextualization and gated aggregation, addressing the unique temporal characteristics of EEG signals while ignoring spatial information.

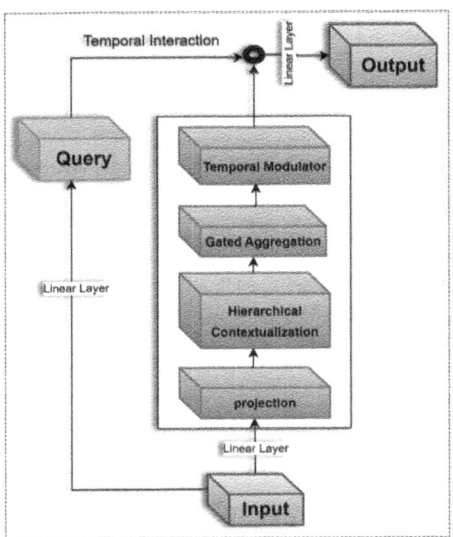

**Fig. 2.** The temporal focal modulation layer.

## 4    Experimental Results and Discussion

### 4.1    Considered Datasets

The EEG datasets utilized in our study provide valuable information for deep learning models in the Brain-Computer Interface domain. We employed the well-known 2a dataset [3] and 2b dataset [18].

**2a Dataset:** Recorded from nine individuals using 22 electrodes at a 250 Hz sampling rate, this dataset involves four distinct classes for motor imagery tasks: left-hand actions (class 1), right-hand actions (class 2), both feet actions (class 3), and tongue actions (class 4) visualization. Each subject's data is split into two sessions, each containing 288 trials. The data is segmented into time windows of 1 to 6 s and saved in the General Data Format (GDF), organized into 18 files labeled 'T' for the training set and 'E' for the evaluation set.

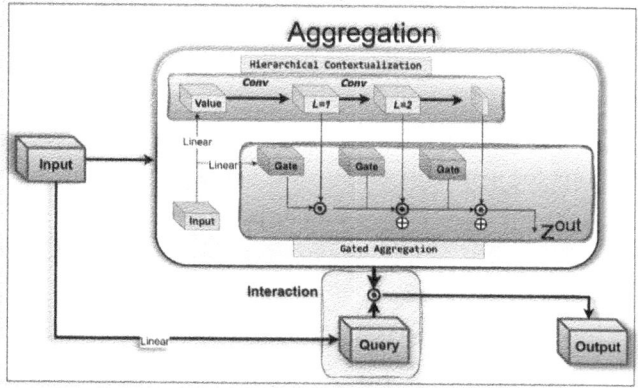

**Fig. 3.** Detailed illustration of hierarchical contextualization and gated aggregation.

**2b Dataset:** This dataset comprises EEG data from nine patients recorded using three bipolar electrodes at a 250 Hz sampling rate. It features two classes: left-hand movement and right-hand movement. The data is collected across five sessions, each containing 320 trials. The segmentation window for this dataset ranges from 1 to 7 s. Similar to Dataset 2a, it includes 18 MAT files, designated as 'T' for the training set and 'E' for the evaluation set.

## 4.2   Implementation Details

The model, developed using the PyTorch framework, was executed on a system with an AMD Ryzen 5 5500 CPU, 16 GB of RAM, and an Nvidia RTX 3060 GPU. During preprocessing, electrooculogram channels were directly eliminated, and no additional artifact removal techniques were employed. The model training utilized a subject-independent scheme, spanning 200 epochs with a fixed batch size of 32. The Adam optimizer minimized the Cross-Entropy loss function with a learning rate set to 1e-3.

In our evaluation, 5-fold cross-validation (CV) was employed. This process divides the dataset into five folds, training the model on four and validating on the fifth fold, repeated five times. To enhance training efficiency and prevent overfitting, we integrated an early stopping technique which stops training if validation accuracy fails to improve for 10 consecutive epochs.

## 4.3   Scoring Performance

In a comprehensive evaluation, various performance indicators, including accuracy, precision, recall, specificity, and F1-score, are computed for each category:

**Precision:** Proportion of true positive predictions relative to true positive and false positive predictions for a specific class.

$$Precision = \frac{TP}{(TP + FP)} \tag{6}$$

**Recall:** Proportion of true positive predictions relative to true positive and false negative predictions for a specific class.

$$Recall = \frac{TP}{(TP + FN)} \tag{7}$$

**Specificity:** Proportion of true negative predictions relative to true negative and false positive predictions for a specific class.

$$Specificity = \frac{TN}{(TN + FP)} \tag{8}$$

**F1-Score:** Harmonic mean of precision and recall, providing a balanced measure of the model's performance.

$$F1 - Score = \frac{2 * Precision * Recall}{(Precision + Recall)} \tag{9}$$

These indicators offer a comprehensive analysis of the model's performance, considering the challenges in achieving a perfect balance in real-world scenarios.

Table 1 provides a concise evaluation of Temporal-FocalNets performance on datasets 2a and 2b using a 5-fold cross-validation scheme. The precision, recall, specificity, and F1-score for each class are presented, showcasing the model's proficiency in differentiating between various movements.

**Table 1.** Temporal-FocalNets Performance on 2a and 2b Datasets (5-fold CV)

| Class | Precision | Recall | Specificity | F1-score |
|---|---|---|---|---|
| 2a dataset | | | | |
| 0 | 87.22 | 85.30 | 95.14 | 86.25 |
| 1 | 86.80 | 81.13 | 93.84 | 83.87 |
| 2 | 82.47 | 84.67 | 94.84 | 83.55 |
| 3 | 82.21 | 87.19 | 95.64 | 84.62 |
| 2b dataset | | | | |
| 0 | 81.73 | 83.01 | 82.74 | 82.36 |
| 1 | 82.74 | 81.44 | 81.73 | 82.08 |

Figure 4 depict the confusion matrices, offering visual insights into **Temporal-FocalNets** classification performance for 5-fold validation on datasets 2a and 2b. The models demonstrate robust accuracy in identifying different hand and limb activities, showcasing their effectiveness in diverse scenarios.

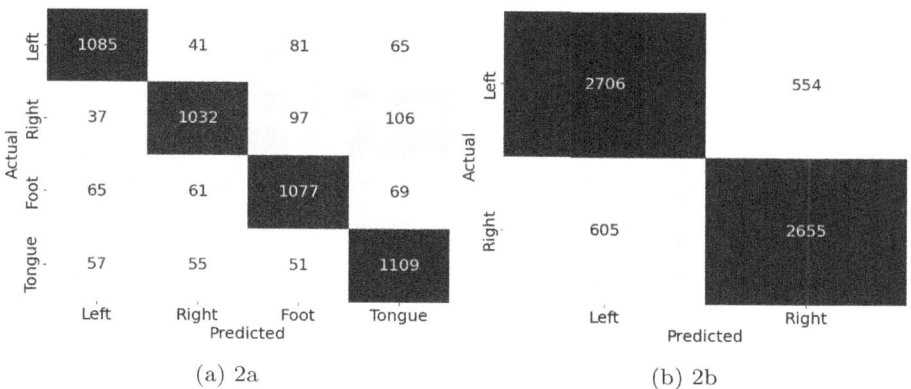

**Fig. 4.** Confusion matrices for Temporal-FocalNets on test datasets (a) 2a and (b) 2b, using 5-fold cross-validation

## 4.4 Comparative Study with Established Models

In our study, we conducted a comprehensive comparative analysis, incorporating established baseline models as benchmarks. Given our focus on subject-independent classification, experiments were conducted on both datasets 2a and 2b using 5-fold cross-validation schemes.

We selected recent models for comparison, outlined as follows:

- **EEGNet by Lawhern et al.** [17]: Utilizes a lightweight CNN framework for extracting temporal and spatial features from EEG data. It employs convolutional kernels to capture temporal and spatial information, followed by a separable convolution process to optimize feature map outputs and reduce parameters.
- **MIN2Net by Autthasan et al.** [1]: Incorporates an autoencoder network with a momentum learning mechanism, excelling in subject-independent scenarios. It consists of an encoder and decoder using linear convolution kernels for spatial-temporal feature extraction. Back-propagation occurs through the momentum learning mechanism and error reconstruction functions.
- **MSCNet by Zhao et al.** [35]: A multi-scale spatial-temporal convolutional neural network that integrates contrastive learning to enhance the robustness and discriminative power of embedding vectors, improving feature extraction and classification.

– **CNN by Roy et al.** [26]: Enriched with three Mega Blocks, each consisting of a sequence of convolutional layers. The architecture allows for the repetition of specific blocks, and hyperparameters like the number of convolution layers, learning rate, momentum, and regularization are optimized using Bayesian hyperparameter optimization.

Detailed results of our comparative analysis, including accuracy metrics, are presented in Table 2.

**Table 2.** Performance comparison on 2a and 2b datasets with 5-fold CV scheme

| Year | Method | Accuracy (%) |
|------|--------|--------------|
| 2a dataset | | |
| 2018 | EEGNet [17] | 64.26 |
| 2021 | MIN2Net [1] | 65.23 |
| 2023 | MSCNet [35] | 69.77 |
| 2024 | **Temporal-FocalNets** | **84.57** |
| 2b dataset | | |
| 2023 | CNN [26] | 67.78 |
| 2024 | **Temporal-FocalNets** | **82.22** |

## 4.5 Discussion

Our study introduces an innovative approach, Temporal-FocalNets, designed to enhance the decoding of Motor Imagery EEG signals by prioritizing temporal aspects using a focal modulation system. The methodology and model architecture are thoroughly explained, providing a clear understanding of the proposed framework. The experimental results demonstrate the effectiveness of Temporal-FocalNets in classifying motor imagery tasks on datasets 2a and 2b, showcasing its proficiency in handling diverse scenarios.

The results of our experiments on datasets 2a and 2b demonstrate the efficacy of Temporal-FocalNets in decoding MI EEG signals. In the 5-fold cross-validation scheme, Temporal-FocalNets achieved an accuracy of 84.57% on the 2a dataset and 82.22% on the 2b dataset, outperforming established baseline models such as EEGNet, MIN2Net, MSCNet, and CNN. These results highlight the superiority of Temporal-FocalNets in capturing the intricate temporal dynamics present in MI EEG signals, thereby enabling more accurate classification.

Moreover, the performance metrics detailed in Table 1, including precision, recall, specificity, and F1-score for each class, provide a comprehensive assessment of Temporal-FocalNets' classification capabilities. The high precision and recall values across different classes, particularly for the 2a dataset where Temporal-FocalNets achieved precision values ranging from 82.21% to 87.22%

and recall values from 81.13% to 87.19%, underscore the model's proficiency in differentiating between various hand and limb activities.

The visualization of confusion matrices, as depicted in Figs. 4a and 4b, provides additional insights into Temporal-FocalNets' classification performance. These visualizations illustrate not only the overall accuracy of Temporal-FocalNets but also how well the model performs in distinguishing between different classes. The high values along the diagonal of the confusion matrices indicate strong performance in correctly classifying instances across various motor imagery tasks, further corroborating the superior performance of Temporal-FocalNets compared to baseline models.

## 4.6   Conclusion

In this study, Temporal-FocalNets, an innovative framework for decoding Motor Imagery EEG signals, Temporal-FocalNets effectively capture temporal dependencies, resulting in accurate classification. With impressive accuracies of 84.57% on the 2a dataset and 82.22% on the 2b dataset, Temporal-FocalNets show promise in advancing Brain-Computer Interface technology. Further research in this direction holds the potential for enhancing the accessibility and effectiveness of Brain-Computer Interface systems in real-world applications.

# References

1. Autthasan, P., et al.: Min2net: end-to-end multi-task learning for subject-independent motor imagery EEG classification. IEEE Trans. Biomed. Eng. **69**(6), 2105–2118 (2021)
2. Autthasan, P., et al.: Min2net: end-to-end multi-task learning for subject-independent motor imagery EEG classification. IEEE Trans. Biomed. Eng. **69**(6), 2105–2118 (2022). https://doi.org/10.1109/TBME.2021.3137184
3. Brunner, C., Leeb, R., Müller-Putz, G., Schlögl, A., Pfurtscheller, G.: Bci competition 2008–Graz data set A. Inst. Knowl. Disc. (Lab. Brain-Comput. Interfaces), Graz Univ. Technol. **16**, 1–6 (2008)
4. Dai, G., Zhou, J., Huang, J., Wang, N.: HS-CNN: a cnn with hybrid convolution scale for EEG motor imagery classification. J. Neural Eng. **17**(1), 016025 (2020)
5. Deny, P., Cheon, S., Son, H., Choi, K.W.: Hierarchical transformer for motor imagery-based brain computer interface. IEEE J. Biomed. Health Inf. (2023)
6. Ding, Y., Robinson, N., Zhang, S., Zeng, Q., Guan, C.: Tsception: capturing temporal dynamics and spatial asymmetry from EEG for emotion recognition. IEEE Trans. Affect. Comput. **14**, 2238–2250 (2022)
7. Dolzhikova, I., Abibullaev, B., Sameni, R., Zollanvari, A.: An ensemble CNN for subject-independent classification of motor imagery-based EEG. In: 2021 43rd Annual International Conference of the IEEE Engineering in Medicine & Biology Society (EMBC), pp. 319–324. IEEE (2021)
8. Dong, Y., et al.: Subject-independent EEG classification of motor imagery based on dual-branch feature fusion. Brain Sci. **13**(7), 1109 (2023)
9. d'Ascoli, S., Touvron, H., Leavitt, M.L., Morcos, A.S., Biroli, G., Sagun, L.: Convit: improving vision transformers with soft convolutional inductive biases. In: International Conference on Machine Learning, pp. 2286–2296. PMLR (2021)

10. Edelman, B.J., et al.: Noninvasive neuroimaging enhances continuous neural tracking for robotic device control. Sci. Rob. **4**(31), eaaw6844 (2019)
11. Fourati, R., Ammar, B., Sanchez-Medina, J., Alimi, A.M.: Unsupervised learning in reservoir computing for EEG-based emotion recognition. IEEE Trans. Affect. Comput. **13**(2), 972–984 (2022). https://doi.org/10.1109/TAFFC.2020.2982143
12. Hameed, A., et al.: Temporal-spatial transformer based motor imagery classification for BCI using independent component analysis. Biomed. Signal Process. Control **87**, 105359 (2024)
13. Hermosilla, D.M., et al.: Shallow convolutional network excel for classifying motor imagery EEG in BCI applications. IEEE Access **9**, 98275–98286 (2021)
14. Hu, L., Hong, W., Liu, L.: Msatnet: multi-scale adaptive transformer network for motor imagery classification. Front. Neurosci. **17**, 1173778 (2023)
15. Jiao, Y., et al.: Sparse group representation model for motor imagery EEG classification. IEEE J. Biomed. Health Inform. **23**(2), 631–641 (2019). https://doi.org/10.1109/JBHI.2018.2832538
16. Kwon, O.Y., Lee, M.H., Guan, C., Lee, S.W.: Subject-independent brain-computer interfaces based on deep convolutional neural networks. IEEE Trans. Neural Netw. Learn. Syst. **31**(10), 3839–3852 (2019)
17. Lawhern, V.J., Solon, A.J., Waytowich, N.R., Gordon, S.M., Hung, C.P., Lance, B.J.: Eegnet: a compact convolutional neural network for EEG-based brain-computer interfaces. J. Neural Eng. **15**(5), 056013 (2018)
18. Leeb, R., Brunner, C., Müller-Putz, G., Schlögl, A., Pfurtscheller, G.: Bci competition 2008-graz data set b. Graz University of Technology, Austria, pp. 1–6 (2008)
19. Liang, G., Cao, D., Wang, J., Zhang, Z., Wu, Y.: Eisatc-fusion: inception self-attention temporal convolutional network fusion for motor imagery eeg decoding. TechRxiv (2023). https://doi.org/10.36227/techrxiv.24003582.v1
20. Liu, Z., et al.: Swin transformer: hierarchical vision transformer using shifted windows. In: Proceedings of the IEEE/CVF International Conference on Computer Vision, pp. 10012–10022 (2021)
21. Luo, J., et al.: A shallow mirror transformer for subject-independent motor imagery BCI. Comput. Biol. Med. **164**, 107254 (2023)
22. Ma, Y., Song, Y., Gao, F.: A novel hybrid cnn-transformer model for EEG motor imagery classification. In: 2022 International Joint Conference on Neural Networks (IJCNN), pp. 1–8. IEEE (2022)
23. Milanés-Hermosilla, D., et al.: Robust motor imagery tasks classification approach using bayesian neural network. Sensors **23**(2), 703 (2023)
24. Niazi, I.K., Jiang, N., Jochumsen, M., Nielsen, J.F., Dremstrup, K., Farina, D.: Detection of movement-related cortical potentials based on subject-independent training. Med. Biol. Eng. Comput. **51**, 507–512 (2013)
25. Penaloza, C., Nishio, S.: Bmi control of a third arm for multitasking. Sci. Rob. **3**(20), pmid: 33141729 (2018)
26. Roy, S., Chowdhury, A., McCreadie, K., Prasad, G.: Deep learning based inter-subject continuous decoding of motor imagery for practical brain-computer interfaces. Front. Neurosci. **14**, 918 (2020)
27. Shanechi, M.M.: Brain-machine interfaces from motor to mood. Nat. Neurosci. **22**(10), 1554–1564 (2019)
28. Song, Y., Zheng, Q., Liu, B., Gao, X.: EEG conformer: convolutional transformer for EEG decoding and visualization. IEEE Trans. Neural Syst. Rehabil. Eng. **31**, 710–719 (2022)

29. Tao, Y., et al.: Gated transformer for decoding human brain EEG signals. In: 2021 43rd Annual International Conference of the IEEE Engineering in Medicine & Biology Society (EMBC), pp. 125–130. IEEE (2021)
30. Vaswani, A., et al.: Attention is all you need. Adv. Neural Inf. Process. Syst. **30** (2017)
31. Yang, J., Li, C., Dai, X., Gao, J.: Focal modulation networks. Adv. Neural. Inf. Process. Syst. **35**, 4203–4217 (2022)
32. Zhang, D., Yao, L., Chen, K., Monaghan, J.: A convolutional recurrent attention model for subject-independent EEG signal analysis. IEEE Signal Process. Lett. **26**(5), 715–719 (2019)
33. Zhang, J., Li, K., Yang, B., Han, X.: Local and global convolutional transformer-based motor imagery EEG classification. Front. Neurosci. **17** (2023)
34. Zhang, N.: Learning adversarial transformer for symbolic music generation. IEEE Trans. Neural Netw. Learn. Syst. (2020)
35. Zhao, R., et al.: A mutli-scale spatial-temporal convolutional neural network with contrastive learning for motor imagery EEG classification. Med. Novel Technol. Dev. **17**, 100215 (2023)

# Blood Glucose Prediction in Type 1 Diabetes Based on Long Short-Term Memory

Bogdan-Petru Butunoi[1]([✉]) [iD], Cristina Stolojescu-Crisan[3] [iD],
and Viorel Negru[1,2] [iD]

[1] Computer Science Department, West University of Timisoara, 4 Blv. Vasile Parvan,
300223 Timisoara, Romania
bogdan.butunoi94@e-uvt.ro
[2] E-Austria Institute, Timisoara, Romania
[3] Communication Department, Politehnica University of Timisoara,
2 Blv. V. Parvan, 300223 Timisoara, Romania

**Abstract.** The management of Type 1 Diabetes (T1D) has always been a complex task, requiring patients to continuously monitor their blood glucose levels and adjust insulin doses accordingly. This paper explores the potential of Artificial Intelligence (AI), specifically Long Short-Term Memory (LSTM) networks, to revolutionize the way Type 1 Diabetes is managed. This research takes a novel approach by focusing solely on blood glucose levels, aiming to determine if Artificial Intelligence can effectively predict and recommend insulin doses based on this singular data point. In this paper, we compare three variants of Long short-term memory (LSTM) models, with the purpose of blood glucose level prediction: unidirectional LTSM, stacked LTSM and Bi-directional LTSM. The data used for this study is unique in that it is sourced from a single individual, ensuring consistency and eliminating inter-individual variability. The findings presented in this paper could serve as a foundation for future research and the development of Artificial Intelligence driven diabetes management systems.

**Keywords:** Continuous glucose monitoring · Type 1 diabetes · Time series forecasting · LSTM

## 1 Introduction

Diabetes is a chronic metabolic disorder that affects millions of people worldwide. It is characterized by high blood glucose levels (BGLs) due to insufficient or ineffective insulin production or action. The insulin is a hormone that controls the uptake of glucose from the blood into the cells for energy. When the insulin is not working properly or there is an insulin deficiency, the glucose accumulates in the blood, leading to various complications, such as cardiovascular diseases, kidney failure, eye damage, or foot ulcers, among others [1].

N.-T. Nguyen et al. (Eds.): ICCCI 2024, CCIS 2166, pp. 458–469, 2024.
https://doi.org/10.1007/978-3-031-70259-4_35

There are two main types of diabetes: type 1 and type 2. T1D is an autoimmune condition that destroys the insulin-producing beta cells in the pancreas. People with T1D need to take insulin injections or use an insulin pump to maintain normal BGL. T1D accounts for about 5–10% of all cases of diabetes and can occur at any age [2].

Managing diabetes is challenging and requires constant monitoring and adjustment of BGLs, diet, exercise, and medication. People with T1D need to measure their BGL several times a day using a finger-prick test or a continuous glucose monitor (CGM). They also need to follow a balanced diet that limits carbohydrates and sugars, engage in regular physical activity that lowers BGLs and improves insulin sensitivity, and take medications that lower BGLs or increase insulin production or action. However, these tasks are often burdensome, inconvenient, and costly. Also, an optimal glycemic control may not always be achieved due to various factors such as human error, variability in blood glucose response, stress, illness, or lack of adherence [3].

People with T1D need to use a CGM and an insulin pump to measure their BGLs and deliver insulin accordingly. However, the current CGM and insulin pump systems are not fully integrated and automated, and they require manual intervention and calibration from the user. This can lead to suboptimal glycemic control, increased risk of hypoglycemia or hyperglycemia, and reduced quality of life for people with T1D.

Both hyperglycemia and hypoglycemia can have short-term and long-term consequences for people with diabetes. Some of the short-term effects of hyperglycemia include increased thirst, frequent urination, fatigue, blurred vision, headache, and nausea. If hyperglycemia is not treated promptly, it can lead to diabetic ketoacidosis (DKA), a life-threatening condition that occurs when the body breaks down fat for energy and produces ketones, which are acidic substances that can harm the organs. DKA can cause symptoms such as fruity breath, abdominal pain, vomiting, confusion, and coma.

Some of the short-term effects of hypoglycemia include shakiness, sweating, hunger, dizziness, irritability, anxiety, and difficulty concentrating. If hypoglycemia is not treated quickly, it can lead to severe hypoglycemia, a dangerous condition that occurs when the brain does not get enough glucose to function properly. Severe hypoglycemia can cause symptoms such as loss of coordination, slurred speech, blurred vision, seizures, loss of consciousness, and even death.

Both hyperglycemia and hypoglycemia can also have long-term effects on the health of people with diabetes. Chronic hyperglycemia can damage various organs and tissues in the body, leading to complications such as cardiovascular disease, kidney disease, nerve damage, eye damage, foot ulcers, gum disease, sexual problems, and increased risk of certain cancers. Chronic hypoglycemia can impair the brain's ability to sense low blood sugar levels and trigger appropriate responses, a condition known as hypoglycemia unawareness. This can increase the risk of severe hypoglycemia and its consequences.

Therefore, it is important for people with diabetes to monitor their blood sugar levels regularly and keep them within a target range that is recommended by their health care provider. This can help prevent or delay many of the complications of diabetes and improve their quality of life.

Various methods have been investigated in the literature with the purpose of predicting BGL. The idea of predicting BGL based on the past values was first presented in [4] and then it was taken by other researchers and improved by using increasingly sophisticated prediction models. One of the most widely used techniques for predicting time series, including blood glucose prediction (BGP) is based on the Autoregressive Integrated Moving Average (ARIMA) model [5,6].

Predicting BGL for T1D using machine learning (ML) has gained a lot of attention in the recent years. ML algorithms are diverse and capable of achieving complex tasks, such as to predict future BGL using, for example, a CGM device. Support vector regression (SVR) is a classic ML algorithm used for regression analysis and has been used for blood glucose level prediction in [7,8]. Another ML method that has been extensively used by previous studies for BGP is based on classical Artificial Neural Networks (ANNs) [9,10].

Deep learning (DL) models [11–13], like RNNs or Convolutional Neural Networks (CNNs), have a higher complexity than classical ANNs (contain multiple hidden layers) and are able to detecting complicated systems' dynamics. However, these methods are usually very demanding, in terms of training data and computational resources [11]. There are other studies that uses simulated data to train and test the methods.

Long short-term memory (LSTM) networks [14] represent an evolution of the classic Recurrent Neural Network (RNN) architecture widely used in DL, especially for time series prediction, thanks to their versatility and flexibility [11]. They have gained a lot popularity for BGP after 2019, since it has been observed that LSTM are successful in both short-term and long-term BGP. Therefore, many authors used variants of LSTM networks in their research for solving BGP tasks [11,12,15–17], showing promising results as compared to other approaches.

The main objective of this paper is to design and implement three LSTM models that can predict the blood glucose trend for the next half an hour or more. This paper will contribute to the field of diabetes care by providing a new solution that can improve glycemic control, reduce human error, and enhance user convenience and satisfaction. This work is part of a more complex project that, in the end, will provide people with T1D who use CGM and insulin pump systems with a more reliable, accurate, and user-friendly system that can automate their diabetes management and reduce their burden.

## 2    Prediction Models

In this paper, we propose an approach to predicting BGL with three variants of LSTM: classical LTSM, stacked LTSM and Bi-directional LTSM. In previous papers, we motived our approach on the filtering and smoothing techniques and the comparison between a mathematical model (ARIMA) and a basic RNN.

## 2.1   Long Short-Term Memory (LSTM)

LSTM networks were introduced by Hochreiter and Schmidhuber [18] in 1997 and represent a powerful type of RNN, capable of learning long-term dependencies in sequential data, such as time series, speech, and text [19]. Unlike classical ANNs, such as Feed-Forward, where the information flows in one direction and each layer is characterized by a different set of parameters, RNNs are represented by multiple layers of recurrent units that share the same parameters [11]. The structure of the LSTM network is very similar to RNN, but a different function is used to compute the hidden layer [20]. In addition, LSTM networks are capable of handling the vanishing gradient problem faced by classical RNNs and thus, to avoid long-term dependency problems. At each computational step, both current input and previous input are taken into consideration, which make LSTM suitable for time series forecasting.

Figure 1 illustrates the structure of an LSTM cell [20].

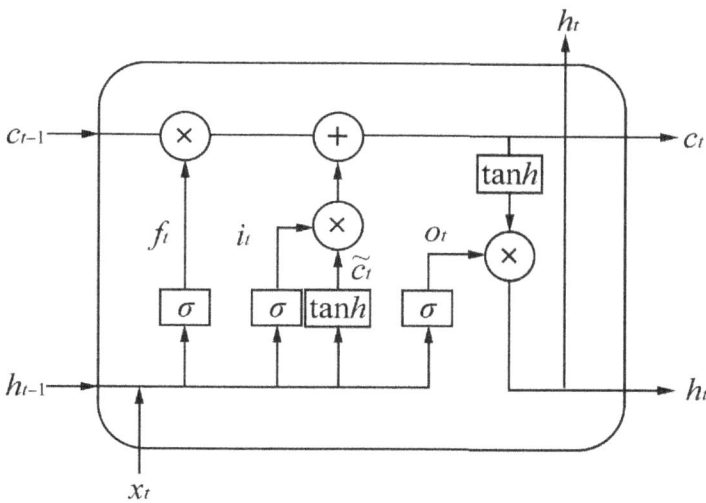

**Fig. 1.** Structure of an LSTM cell

More theoretical aspects regarding the functioning of LSTMs are given in [19,20].

LSTM networks have been widely used for various applications, such as machine translation, speech recognition, natural language processing, and image captioning. One of the challenging problems that LSTM networks can address is the prediction of BGL for diabetes patients. BGP is important for diabetes management but it is also difficult, as it depends on many factors, such as food intake, insulin injection, physical activity, stress, and individual variability. Therefore,

LSTM networks can be a suitable method for BGP, as they can capture the temporal dynamics and nonlinear relationships of blood glucose and its influencing factors.

## 2.2   Bidirectional Long Short-Term Memory

Bidirectional LSTMs (BiLSTMs) [21] are composed by two unidirectional LSTMs and process the input data in two directions, forward and backward, thus capturing dependencies in both directions. BiLSTMs have two layers, one for processing the input information in the forward direction and the second for processing in the backward direction. In this way, the network is able to access simultaneously the past and the future information sequences and to better understand the relationship between them. This is especially salient in glucose prediction, where both past patterns and imminent events (like meals or exercise) can influence glucose dynamics. BiLSTMs improve the network's ability to capture long-term dependencies and to preform more accurate predictions for complex time series.

Figure 2 illustrates the general structure of BiLSTMs [22].

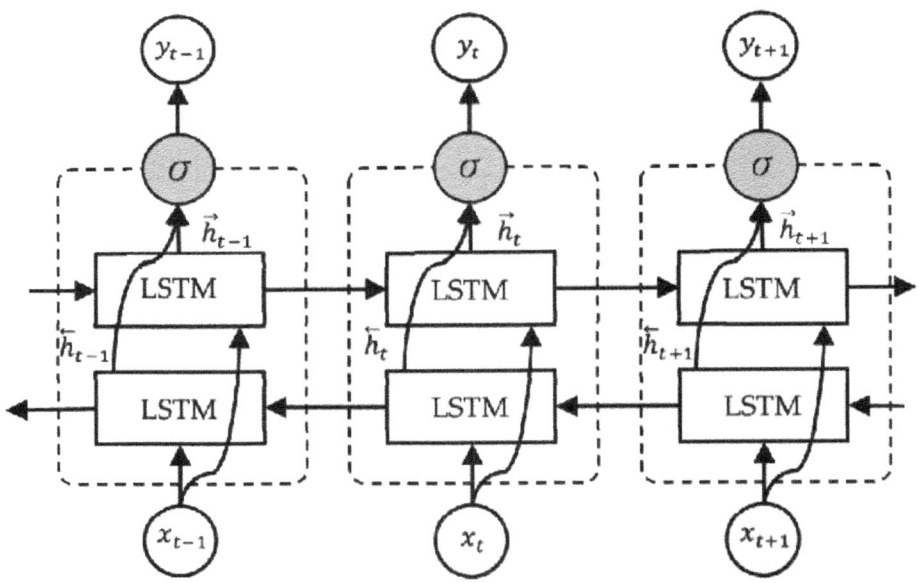

**Fig. 2.** The general structure of BiLSTMs

## 2.3   Stacked Long Short-Term Memory

The Stacked LSTM (SLSTM) network, also called Deep LSTM, introduced by Graves, et al. in [23], is a model that consists of multiple LSTM layers stacked on top of each other, as shown in Fig. 3 and each layer contains multiple memory cells. This layered approach improve the model's capacity to learn and represent intricate patterns.

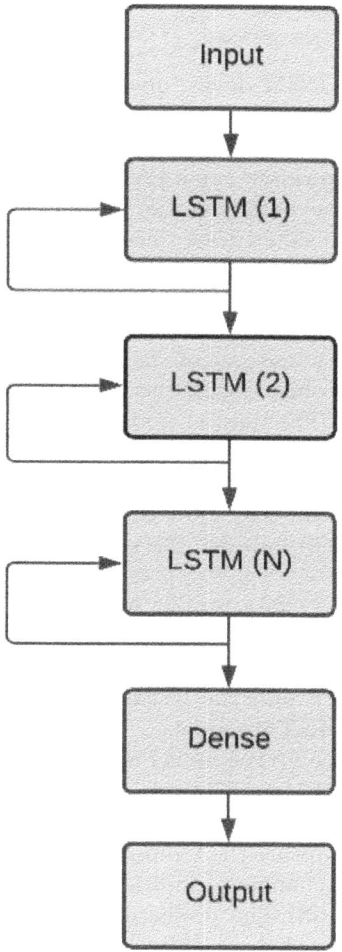

**Fig. 3.** The architecture of a SLSTM

SLSTM networks undeniably herald a promising frontier in the realm of blood glucose prediction, offering a depth and granularity that was previously challenging to achieve. Their innate ability to retain long-term dependencies and capture intricate temporal hierarchies positions them as potent tools for this critical healthcare application.

## 3    System Architecture

The goal of this work is to collect BGLs from a CGM device, store them in a time series, and use them to predict the optimal insulin or sugar intake for a diabetic patient. The proposed system architecture consists of the following steps:

1. **Data collection**: The CGM device measures the blood glucose level every few minutes and sends the data to a cloud server. The server stores the data in a time series database, which can be queried by the workflow.
2. **Data ingestion**: The workflow can ingest historical data from the time series database, or use real-time data from the CGM device. The data is formatted as a sequence of glucose values with timestamps.
3. **Data preparation**: The data is pre-processed to be used in the prediction algorithm. The pre-processing steps include:
   - **Data augmentation**: The data is augmented with synthetic samples to increase the diversity and size of the dataset. This can help improve the generalization and robustness of the LSTM model.
   - **Data smoothing**: The data is smoothed using a moving average filter to reduce the noise and outliers in the data. This can help improve the accuracy and stability of the LSTM model.
   - **Data normalization**: The data is normalized using min-max normalization. This can help improve the convergence and performance of the LSTM model.
   - **Data splitting**: The data is split into three subsets: 60% for training, 20% for validation, and 20% for testing. The training set is used to train the LSTM model, the validation set is used to tune the hyperparameters of the LSTM model, and the testing set is used to evaluate the final performance of the LSTM model.
4. **LSTM prediction**: The LSTM model is a type of RNN that can learn from sequential data. The LSTM model takes as input a sequence of glucose values and outputs a sequence of predicted glucose values for the next time steps. The LSTM model can capture the temporal dependencies and patterns in the data, and can adapt to changing conditions over time.
5. **Performance valuation**: The prediction is evaluated using the following metrics: the Mean Squared Error (MSE), the Mean Absolute Error (MAE), the Mean Absolute Percentage Error (MAPE), and the Symmetric Mean Absolute Percentage Error (sMAPE) error. The interpretation is very simple: the smaller the error, the better the model's performance.

### 3.1    Data Set

The data used in this paper comes from a CGM device, namely the Dexcom G6. The Dexcom G6 is a small wearable device that measures the glucose levels in the interstitial fluid under the skin every five minutes. The device transmits the glucose readings wirelessly to a compatible smart device or a receiver. The

Dexcom G6 also provides alerts for low and high glucose levels, as well as an urgent low soon alert that warns the user when their glucose level is projected to be at or below 55 mg/dL within 20 min.

The Dexcom G6 system consists of a sensor (inserted under the skin of the abdomen or the upper arm) which measures the glucose levels in the interstitial fluid every five minutes and sends the data to the transmitter, a transmitter that attaches to the sensor and wirelessly communicates with the display device, and a display device. that can be a compatible smart device (for example a smartphone or a smartwatch) or a dedicated Dexcom receiver. The display device shows the current glucose level, the trend arrow, and the historical data in easy-to-read graphs. It can also alert the user when the glucose level is too high or too low, or when it is changing rapidly.

The data collection process involved only one patient with type 1 diabetes who was using the Dexcom G6 device for a period of five years, from the 5th of July 2019 to the 2nd of March 2023. Over this extensive period, a total of 355,890 data points were meticulously compiled, providing a substantial dataset that encompasses various blood glucose patterns and fluctuations. This exhaustive data pool serves as an invaluable resource, providing us with a wide-ranging overview of blood glucose dynamics.

The data was retrieved from the xDrip app, which is an open-source software that allows users to view and analyze their glucose trends and statistics. The app also allows users to enter additional information, such as insulin doses, carbohydrate intake, exercise, and other events that may affect their glucose levels.

The data obtained from the xDrip app consisted of timestamped glucose values in mg/dL, along with event types and values for insulin and carbohydrate entries. The data was pre-processed to remove any missing or erroneous values, such as "low" or "high" readings that indicate glucose levels below 40 mg/dL or above 400 mg/dL.

## 3.2   Clustering External Data

Blood glucose levels are influenced by a plethora of factors including diet, physical activity, medication, hormonal fluctuations, and even psycho-social stressors. Consequently, two patients might display markedly different glucose dynamics despite having similar profiles. When diverse patient data is pooled into a single training set without differentiation, LSTMs struggle to discern generalized patterns, leading to sub-optimal performance.

Using unsupervised machine learning, patient data can be grouped based on similarity in glucose dynamics, insulin sensitivity, lifestyle habits, or other relevant metrics. Clustering algorithms like K-means or DBSCAN can segment patients into homogenous groups, ensuring that each cluster captures a specific, distinct glucose behavior pattern.

Post clustering, separate LSTM models are trained for each patient cluster. By focusing on the unique characteristics and patterns inherent to each cluster, these models can capture the nuanced temporal relationships more effectively than a generalized model.

Once clusters are established and individual models trained, new patient data can be assigned to the most fitting cluster. This ensures that the LSTM model used for prediction is already tailored to the specific characteristics of that patient subgroup.

## 4   Results

The first experiment is centered around the LSTM model. The network was created from scratch, no pre-training done. The result are shown in Table 1.

**Table 1.** Performance Metrics for LSTM Configurations

| Configuration | MSE | MAE | MAPE (%) | sMAPE (%) |
|---|---|---|---|---|
| Initial (ReLU, 1 layer, 50 neurons) | 452.3543 | 15.5132 | 11.3514 | 10.6158 |
| Activation switched to tanh | 292.5748 | 11.2585 | 7.3335 | 7.4949 |
| 100 neurons per layer | 239.0717 | 10.1737 | 7.7975 | 7.5406 |
| 100 neurons, trained for 100 epochs | 134.0968 | 7.2285 | 4.7631 | 4.8343 |

The initial configuration involved one layers with 50 neurons, and ReLU as activation function. Then we switched to tanh activation function and 100 neurons per layer and different number of training epochs. Analyzing the results in Table 1, we can conclude that the best results (smaller errors) were obtained using 100 neurons per layer, trained for 100 epochs.

Encouraging outcomes obtained from the initial LSTM configurations set a promising premise for delving into more complex designs, such as bidirectional and stacked LSTMs, which will be the focus of subsequent experiments.

The stacked version of LSTM was the center piece of the most extensive experiment, the results are shown in Table 2.

The initial configuration involved three layers with 50 neurons each, Adam optimizer, MSE loss, and 20 epochs with a batch size of 32. Subsequent configurations included varying the number of neurons, activation functions, layers, and inclusion of dropout. The best model turned out to be the one with 4 layers, tanh activation function, and 50 neurons.

The Bi-LSTM model allows information to flow in both forward and backward directions, capturing past and future contexts respectively. Starting with 50 neurons activated by ReLU, the model trained for 20 epochs with a batch size of 32. Leveraging insights from our previous LSTM experiment, the Bi-LSTM was optimized with 100 neurons, tanh activation, and was trained for 100 epochs. The results are shown in Table 3

We can observe that for the second experiment, with Bi-LSTM optimized with 100 neurons and tanh activation the values of the errors are lower than in the first experiment.

**Table 2.** Performance Metrics for Stacked LSTM Configurations

| Configuration | MSE | MAE | MAPE (%) | sMAPE (%) |
|---|---|---|---|---|
| 3 layers, ReLU, 50 neurons | 170.085 | 9.151 | 6.218 | 6.017 |
| 3 layers, tanh, 50 neurons | 212.380 | 9.736 | 7.870 | 7.401 |
| 3 layers, tanh, 100 neurons | 200.423 | 9.353 | 6.731 | 6.498 |
| 4 layers, tanh, 50 neurons | 197.309 | 9.114 | 6.346 | 6.293 |
| 4 layers, tanh, 50 neurons + 0.2 dropout | 319.214 | 11.537 | 7.957 | 7.840 |
| 5 layers, tanh, 50 neurons | 209.041 | 9.973 | 6.870 | 6.984 |
| 5 layers, tanh, 100 neurons, 100 epochs | 959.679 | 21.699 | 17.360 | 15.778 |

**Table 3.** Performance Metrics for Bi-LSTM Configurations

| Configuration | MSE | MAE | MAPE (%) | sMAPE (%) |
|---|---|---|---|---|
| ReLU 50 neurons | 285.66 | 11.74 | 9.18 | 8.71 |
| tanh, 100 neurons | 233.80 | 9.44 | 6.28 | 6.25 |

## 5   Conclusions

LSTM networks are a suitable method for blood glucose prediction, as they can capture the temporal dynamics and nonlinear relationships of blood glucose and its influencing factors, such as food intake, insulin injection, physical activity, stress, and individual variability.

LSTM networks can be combined with other methods, such as bidirectional LSTM and stacked LSTM, to improve the performance and robustness of blood glucose prediction. However, the complexity and interpretability of the models may also increase as more layers and components are added.

Although it was difficult from a legal and biological point of view, currently we managed to obtain similar data from several patients and we are going to verify hypotheses that represent some weak points of the present work. Such experiments are trying to predict on new data without retraining or a possible clustering of patients and using transfer learning to reuse the existing model[1].

**Acknowledgments.** This research was partially supported by MOISE grant number 240/2020, ID 911 POC/398/1/1, and by AI4EUROPE+ grant number 24PHE, PN-IV-P8-8.1-PRE-HE-ORG-2023-0072, financed by the Romanian government.

**Disclosure of Interests.** The authors have no competing interests to declare that are relevant to the content of this article.

---

[1] If EquinOCS, our proceedings submission system, is used, then the disclaimer can be provided directly in the system.

# References

1. Magliano, D.J., Boyko, EJ.: IDF Diabetes Atlas, 10th edn. International Diabetes Federation, Brussels (2021). https://www.ncbi.nlm.nih.gov/books/NBK581934/
2. Lucier, J., Weinstock, R.: Type 1 Diabetes. StatPearls Publishing, Treasure Island (2023). https://www.ncbi.nlm.nih.gov/books/NBK507713/
3. Mathew, T.K., Zubair, M., Tadi, P.: Blood Glucose Monitoring. StatPearls Publishing, Treasure Island (2023). https://www.ncbi.nlm.nih.gov/books/NBK555976/
4. Bremer, T., Gough, D.A.: Is blood glucose predictable from previous values? a solicitation for data. Diabetes **48**(3), 445–451 (1999)
5. Yang, J., Li, L., Shi, Y., Xie, X.: An ARIMA model with adaptive orders for predicting blood glucose concentrations and hypoglycemia. IEEE Biomed. Health Inf. **23**(3), 1251–1260 (2018)
6. Saravanan, R., Mahmud, F.: Blood glucose prediction based on ARIMA time-series machine learning model. Evol. Electr. Electron. Eng. **4**(2), 457–463 (2023)
7. Bunescu, R., Struble, N., Marling, C., Shubrook, J., Schwartz, F.: Blood glucose level prediction using physiological models and support vector regression. In: Proceedings of the 12th International Conference on Machine Learning and Applications (ICMLA), vol. 1, pp. 135–140 (2013)
8. Hamdi, T., Ali, J.B., Di Costanzo, V., Fnaiech, M.E.F., Ginoux, J.-M.: Accurate prediction of continuous blood glucose based on support vector regression and differential evolution algorithm. Biocybern. Biomed. Eng. **38**(2), 362–372 (2018)
9. Ali, J.B., Hamdi, T., Fnaiech, N., Di Costanzo, V., Fnaiech, F., Ginoux, J.-M.: Continuous blood glucose level prediction of type 1 diabetes based on artificial neural network. Biocybern. Biomed. Eng. **38**(4), 828–840 (2018)
10. Pappada, S.M., et al.: Neural network-based real-time prediction of glucose in patients with insulin-dependent diabetes. Diab. Technol. Therapeut. **13**(2), 135–141 (2011)
11. Aliberti, A., et al.: A multi-patient data-driven approach to blood glucose prediction. IEEE Access **7**, 69311–69325 (2019)
12. Martinsson, J., Schliep, A., Eliasson, B., Mogren, O.: Blood glucose prediction with variance estimation using recurrent neural networks. Int. J. Healthc. Inf. Syst. Inform. **4**, 1–18 (2020)
13. Mhaskar, H.N., Pereverzyev, S.V., van der Walt, M.D.: A deep learning approach to diabetic blood glucose prediction. Front. Appl. Math. Statist. **3**, 14 (2017)
14. Greff, K., Srivastava, R.K., Koutník, J., Steunebrink, B.R., Schmidhuber, J.: LSTM: a search space odyssey. IEEE Trans. Neural Netw. Learn. Syst. **28**(10), 2222–2232 (2017)
15. Song, W., Cai, W., Li, J., Jiang, F., He, S.: Predicting blood glucose levels with EMD and LSTM based CGM data. In: Proceedings of the 6th International Conference on Systems and Informatics (ICSAI), Shanghai, China, pp. 1443–1448 (2019)
16. Q. Sun, Q., Jankovic, M. V., Bally, L., Mougiakakou, S. G.: Predicting blood glucose with an LSTM and Bi-LSTM based deep neural network. In: Proceedings of 14th Symposium Neural Network Application (NEUREL), Belgrade, Serbia, pp. 1–5 (2018)
17. Idriss, T. El., Idri, A., Abnane, I., Bakkoury, Z.: Predicting blood glucose using an LSTM neural network. In: Federated Conference on Computer Science and Information Systems (FedCSIS), Leipzig, Germany, pp. 35–41 (2019)
18. Hochreiter, S., Schmidhuber, J.: Long short-term memory. Neural Comput. **9**(8), 1735–1780 (1997)

19. Xiao, C., Sun, J.: Introduction to Deep Learning for Healthcare. Springer, Cham (2021). https://doi.org/10.1007/978-3-030-82184-5
20. Olah, C.: Understanding LSTM networks. Technical Report. https://colah.github.io/posts/2015-08-Understanding-LSTMs/. Accessed 29 Jan 2024
21. Schuster, M., Paliwal, K.K.: Bidirectional recurrent neural networks. IEEE Trans. Signal Process. **45**(11), 2673–2681 (1997)
22. Li, Y., Harfiya, L.N., Purwandari, K., Lin, Y.-D.: Real-time cuffless continuous blood pressure estimation using deep learning model. Sensors **20**(19), 5606 (2020)
23. Graves, A., Mohamed, A.-R., Hinton, G.: Speech recognition with deep recurrent neural networks. In: Proceedings of the International Conference on Acoustics, Speech, and Signal Processing (ICASSP), Vancouver, BC, Canada (2013)

# Multi-method Analysis for Early Diagnosis of Alzheimer's Disease on Magnetic Resonance Imaging (MRI) Using Deep Learning and Hybrid Methods

Dhouha Guesmi[1]([✉]), Hasna Njah[1,2], and Yassine Ben Ayed[1]

[1] Multimedia, Information Systems and Advanced Computing Laboratory: MIRACL, University of Sfax, Sfax, Tunisia
dhouha.guesmi@enetcom.u-sfax.tn
[2] Higher Institute of Computer Sciences and Multimedia, University of Gabes, Zrig Eddakhlania, Tunisia

**Abstract.** Alzheimer's disease (AD) is a progressive and irreversible brain disorder that leads to cognitive impairment and an inability to perform daily tasks. It is the primary cause of dementia, accounting for the majority of cases. Abnormal protein deposits in the brain are responsible for AD, resulting in the death of brain cells. While there is currently no cure, Computer Assisted Diagnostics (CAD) with Magnetic Resonance Imaging (MRI) has facilitated disease detection. In this paper, we aimed to improve Alzheimer's disease identification by employing ensemble-modified transfer learning techniques and hybrid approaches that combine deep learning with machine learning. We utilized four pre-trained models (AlexNet, ResNet-50, VGG16, and InceptionV3) and applied batch normalization and regularization techniques. Additionally, various machine learning algorithms (Support Vector Machine, Logistic Regression, Random Forest, and K Nearest Neighbors) were used, achieving high performance in diagnosing dementia. Our results indicate that the hybrid methods combining deep learning and machine learning outperformed the modified deep learning models. Particularly, the AlexNet-M + SVM hybrid model demonstrated an accuracy, precision, sensitivity, specificity and F-measure of 91.41%, 91.41%, 91.41%, 100.00% and 91.39%, respectively. In summary, our study highlights the importance of hybrid approaches for AD diagnosis, showcasing the superior performance of the hybrid model AlexNet-M + SVM. The findings contribute to the advancement of Computer Assisted Diagnostics for the early detection and management of this debilitating neurological condition.

**Keywords:** Alzheimer's disease · MRI · hybrid techniques · machine learning · OASIS · transfer learning

N.-T. Nguyen et al. (Eds.): ICCCI 2024, CCIS 2166, pp. 470–487, 2024.
https://doi.org/10.1007/978-3-031-70259-4_36

# 1 Introduction

A kind of dementia called Alzheimer's disease (AD) can permanently harm memory cells and is a significant public health concern. AD, a type of dementia, poses a significant public health concern due to its detrimental impact on memory cells. This condition affects a substantial number of individuals, particularly the elderly. Globally, approximately 55 million people suffer from dementia, with AD accounting for 60–70% of these cases, as reported by the World Health Organization (WHO) [1]. The prevalence of AD is projected to surge in the coming years, underscoring the crucial importance of early detection for improved treatment outcomes. Traditional approaches to AD diagnosis have relied on clinical assessment and cognitive testing. However, recent research suggests that neuroimaging scan image processing may offer a more reliable and sensitive method. Unfortunately, incorrect prescriptions and delayed diagnoses hinder the achievement of early and accurate AD detection [2].

Convolutional Neural Networks (CNNs) present a more effective alternative for image classification tasks. These networks leverage convolutional layers to extract features from raw input images and subsequently reduce dimensionality through pooling layers [3]. Transfer learning, a concept employed in CNNs, utilizes pre-trained models, such as VGG16 and VGG19, which have been trained on large datasets like ImageNet, containing over 1.2 million real photos categorized into more than 1000 different classes [4].

Recent research efforts have focused on deep learning models and approaches to directly extract features from medical images. Deep learning models, predominantly applied in multi-classification tasks, have demonstrated significant advancements in the analysis of various medical imaging modalities, including mammography, X-rays, CT scans, and MRIs [5]. Among these modalities, structural Magnetic Resonance Imaging (MRI) presents an intriguing diagnostic tool for AD due to its non-invasive nature and its correlation with alterations in brain morphology. The Open Access Series of Imaging Studies (OASIS) standardized dataset provides a wealth of relevant data for analysis. Deep Convolutional Neural Networks (CNNs) and machine learning-based methods offer potential solutions to address challenges in brain imaging data processing, with CNNs demonstrating promising performance on large-scale datasets. Their capacity to achieve high accuracy in medical image classification tasks has made CNNs widely recognized in the field of deep learning [6].

In this study, we focus on MRI as a non-invasive imaging modality that enables the acquisition of two or three-dimensional images with various contrasts. Neuroimaging has become integral to AD research, serving both as a tool for basic science investigations and as a diagnostic aid. We refer to the work by Yagis et al. (2020) [7], who constructed a 3D VGG variant convolutional network to analyze the classification accuracy based on publicly available datasets, ADNI and OASIS. Their model achieved accuracies of 73.4% ± 0.04 and 69.9% ± 0.06 for categorizing AD participants from healthy controls in the ADNI and OASIS datasets, respectively, without employing feature extraction stages. Frenzel et al. (2020) [8] employed supervised machine learning to develop a structural MRI-based biomarker for in vivo AD diagnosis, achieving accuracies of 89% (AUC = 95%) and 87% (AUC = 93%) in distinguishing patients from healthy controls in the ADNI-1 and OASIS-1 datasets using the proposed AD score. JONSSON et al.

(2020) [9] conducted a study utilizing the VGGNet-16 convolutional neural network to accurately classify AD using MRI data from the Alzheimer's Disease Neuroimaging Initiative (ADNI) database, achieving AD identification accuracies ranging from 66.6% to 74.8% depending on the training method.

In this paper, we aim to benchmark the impact of pre-training on the diagnostic classification of AD by evaluating various deep learning network architectures and pre-training scenarios. The key contributions of this research are as follows:

- Exploring suitable feature representations for AD classification by evaluating pre-training models on the OASIS and MRI Alzheimer's datasets.
- Introducing modifications such as batch normalization and regularization to enhance the performance of these models.
- Evaluating multiple deep learning architectures and comparing hybrid methods combining deep learning and machine learning techniques.

Overall, our study seeks to advance the field of AD classification by leveraging deep learning methodologies, thereby enhancing the accuracy and reliability of diagnostic models for this debilitating disease.

## 2  Methodology

In this study, we aimed to predict Alzheimer's disease on the basis of brain MRI images, which indicated the absence or presence of Alzheimer's disease. The proposed system model's application-level representation is shown in Fig. 1 and Fig. 2, that shows our workflow of the proposed methods from beginning to end.

### 2.1  Datasets Selection

In this study, two datasets were used, which called OASIS and Alzheimer's MRI for the early detection of AD.

#### A. OASIS Dataset

We used the OASIS open-access dataset for our research. Daniel S. Marcus from the Neuroimaging Informatics Analysis Center (NIAC) at Washington University School of Medicine prepared these datasets for investigation [10]. The dataset is generated by data obtained from brain MRI images, collected from the publicly available database of OASIS-2. This set consists of a longitudinal collection of 150 subjects aged 60 to 96 for a total of 373 imaging sessions [10]. The exploratory data analysis (EDA) stage was based on investigating the relationship between the various aspects of the MR assessments and the patient's dementia rating. The architecture may be less accurate due to the various image sizes. The time required to train neural network models is shortened by image scaling. Figure 2 describes a sample of the OASIS MR image dataset for AD.

#### B. MRI Alzheimer's Dataset

The MRI dataset for AD was obtained from the open source Kaggle website. The MRI dataset contains 6400 images separated into four classes, namely, mild dementia disease (896 images), moderate dementia disease (64 images), non-dementia (3200 images)

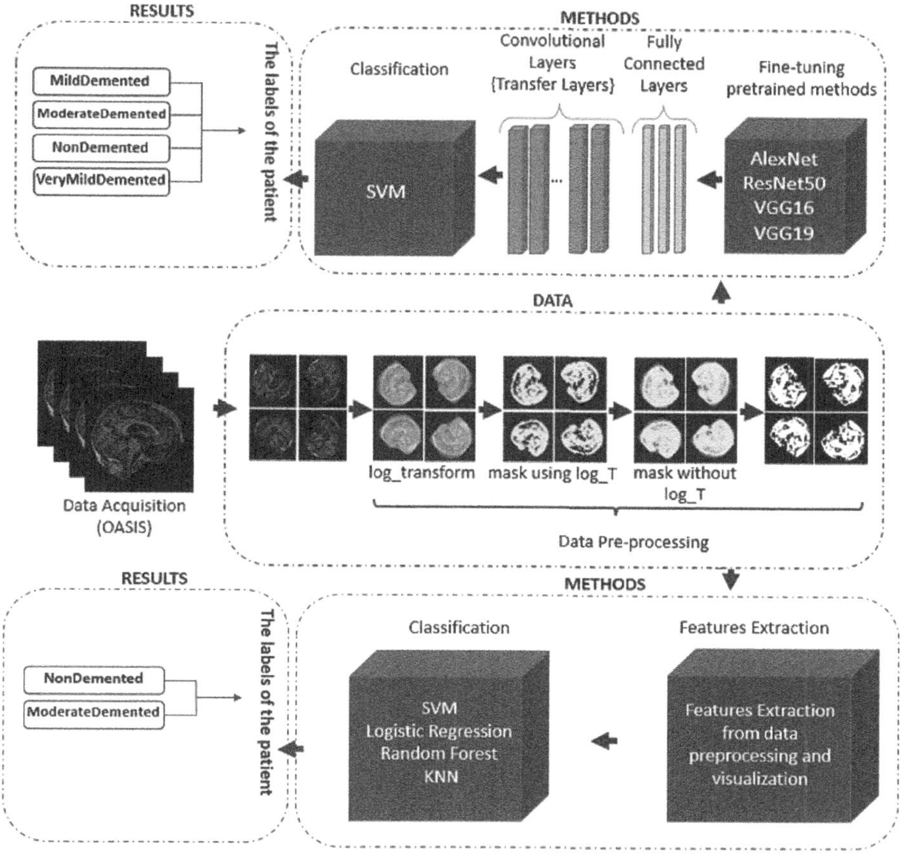

**Fig. 1.** Methodology for classifying the OASIS and MRI Alzheimer's datasets.

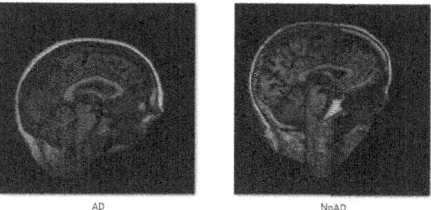

AD                                    NoAD

**Fig. 2.** Examples of OASIS MRI image data. (1) Patient with moderate AD; (2) patient with no dementia.

and very mild dementia disease (2240 images). Figure 3 describes a sample of the MRI dataset for AD. https://www.kaggle.com/tourist55/alzheimers-dataset-4-class-of-images (accessed on 25 May 2021) [11].

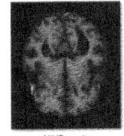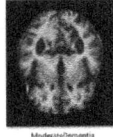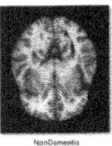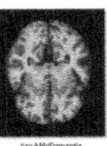

**Fig. 3.** Examples of Alzheimer's MRI data. (1) Patient with mild AD; (2) patient with moderate AD; (3) patient with no dementia; (4) patient with very mild AD.

## 2.2 Image Preprocessing

During the photo capturing process, some undesired information was added to the image as a result of nonlinear light intensity, or noise. Non-linear light intensity in particular affects the effectiveness and accuracy of image processing [12]. Since the finished photographs from this procedure had greater contrast and accurate light distribution, contrast stretching was employed to widen the dynamic range of light intensity. Images in the OASIS repository were enhanced using linear contrast stretching for better performance on following stages. During the photo capturing process, some undesired information was added to the image as a result of nonlinear light intensity, or noise. Non-linear light intensity in particular affects the effectiveness and accuracy of image processing [13]. Contrast stretching was used to expand the dynamic range of light intensity since the final photos from this method had better contrast and proper light distribution. Utilizing linear contrast stretching, images in the OASIS repository were improved for improved performance on later stages.

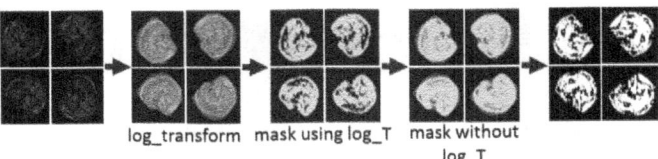

log_transform   mask using log_T   mask without log_T

**Fig. 4.** Image enhancement.

# 3  Data Augmentation

In neuroimaging, a large number of scans related to AD patient's availability are a major issue because few hundreds of image samples are available. It is a common thing for a deep learning model to provide more effective results on more data. In medical research, the classification of cancer and AD are problematic due to lack of availability of data.

Obviously, overfitting is a main risk. To overcome this issue, we need more data to enhance the effective accuracy in our proposed model. We used the augmentation technique to create 10 more images on each available MRI image. It can be increased by rotating, translating and reflecting the existing images, using the socalled label preserving transformations [14]. In Table 1, data augmentation is described for the parameters used

**Table 1.** Data augmentation.

| Rotation Range | 10 selected angles $\Theta$; in the interval $0 \leq \Theta \leq 360$ |
| --- | --- |
| Zoom range | 0.5, 1.5 |
| Shear range | 0.15 Degree |
| Height shift range | 0.1 Degree |
| Width shift range | 0.1 Degree |
| Channel shift range | 150.0 |

for augmentation, where in our case, each image is randomly rotated by 10 selected angles $\Theta$; in the interval $0 \leq \Theta \leq 360$, width and height shift range 0.1 degrees, and shear range 0.15 degree.

## 4 Data Characterization

After pre-processing, we labeled the OASIS MRI data for binary classification and the Alzheimer's MRI data for multi classification then, the dataset comprises features relevant to Alzheimer's disease, including demographic attributes such as age, education, and socioeconomic status, alongside clinical assessments like Mini-Mental State Examination (MMSE) scores. Additional features encompass brain volume metrics (eTIV, nWBV, ASF) and categorical variables such as gender. This comprehensive set of features provides a multifaceted view of Alzheimer's disease progression, facilitating insightful analysis and predictive modeling.). Next, we divided the dataset into a 8:1:1 ratio based on random selection. That means that 80% of the data used for training, 10% of the data used for verification and 10% of the data used for testing.

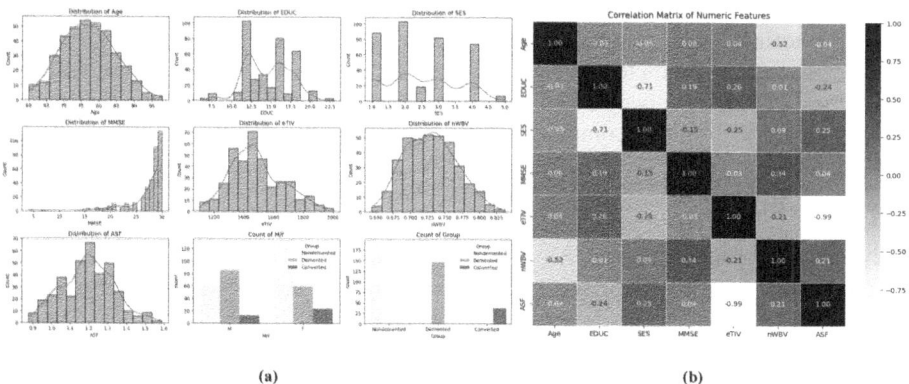

(a)    (b)

**Fig. 5.** (a) Plot of Univariate Analysis for Numeric and Categorical Features, (b) Correlation Matrix of Numeric Features.

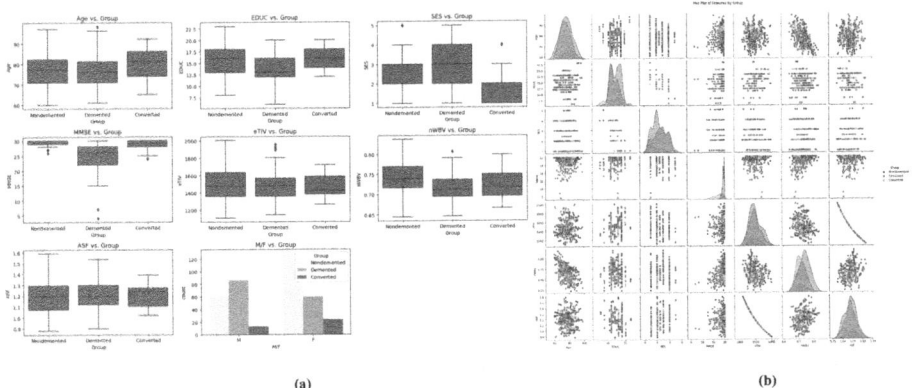

(a)　　　　　　　　　　　　　　　　　(b)

**Fig. 6.** (a) Plot of Bivariate Analysis for Numeric and Categorical Features, (b) Plot of selected features, colored by 'Group'.

## 5　Classification Model Training

Transfer learning is a technique used widely in training Deep learning models for new tasks nowadays. In this transfer learning, we use the knowledge and training of previously pre-trained models as the base model of the new task. Where we fine-tune the parameters according to our current model. The major reason we use transfer learning is it reduces our training time and improves our accuracy and model performance.

### 1) Deep learning models

#### A. *AlexNet-S*

AlexNet is a convolutional neural network architecture. It was one of the first deep neural networks to achieve state-of-the-art performance on the ImageNet dataset. AlexNet consists of eight layers, including five convolutional layers, two fully connected layers, and a final softmax layer for classification. It also includes local response normalization and dropout regularization to prevent overfitting. One of the key innovations of AlexNet was the use of ReLU activation functions, which allowed for faster training and better generalization performance compared to traditional activation functions like the sigmoid function [15].

#### B. *AResNet50-S*

ResNet50-S is a Deep Learning Model we implement in using Keras. The model uses the pre-trained weights of ImageNet. ResNet50 has 50 layers. ResNet has resolved the problem of vanishing gradient which was not possible before that in dense neural networks. As the network grows large by stacking up layers the gradient starts to vanish. ResNet was able to resolve the problem of vanishing gradient by adding the concept of skip connection in ResNet we not only stacked the layers together like we did in other networks but we also added the original input to every layer making ResNet produce very good results in even bare minimum epochs [16].

## C. *VGG16-S*

VGG16-S is a convolutional neural network model proposed by K. Simonyan and A. Zisserman from the University of Oxford in the paper "Very Deep Convolutional Networks for Large-Scale Image Recognition". The model achieves 92.7% top-5 test accuracy in ImageNet, which is a dataset of over 14 million images belonging to 1000 classes [17]. It was one of the famous model submitted to ILSVRC-2014. It makes the improvement over AlexNet by replacing large kernel-sized filters (11 and 5 in the first and second convolutional layer, respectively) with multiple $3 \times 3$ kernel-sized filters one after another. VGG16 was trained for weeks and was using NVIDIA Titan Black GPU's.

## D. *InceptionV3-S*

InceptionV3 is a deep convolutional neural network architecture that improves computational efficiency while maintaining high accuracy. It is part of the Inception family of networks, which use carefully crafted modules to reduce the number of parameters [18]. InceptionV3 has 48 layers, including convolutional layers, pooling layers, and fully connected layers. It employs techniques like factorized convolutions, aggressive regularization, and auxiliary classifiers. These design choices enable the model to achieve state-of-the-art performance on various image classification tasks while being less computationally demanding. InceptionV3 is often used in transfer learning to leverage pre-trained weights for new tasks.

### 2) *Proposed modified models*

This study employed four DNN models, namely AlexNet-S, ResNet50-S, VGG16-S, and InceptionV3-S, for Alzheimer's disease identification, implementing transfer learning. Each pre-trained model underwent modifications by adding fully connected layers and dropout layers with a 50% dropout rate to enhance classification accuracy. The classifiers, comprising fully connected layers fed by features from the convolutional base, concluded with a softmax layer for determining image classification probabilities. The selected architectures, winners in prior challenges, demonstrated superior precision and efficiency in computer-aided diagnostic tasks. Specifically, the VGG16 network, featuring 16 convolutional layers, 5 max-pooling layers, and three fully-connected layers with a softmax layer, underwent adjustments for optimal problem-solving. Unlike typical transfer learning approaches, this model incorporated modifications to both fully connected and convolutional layers, utilizing distinct hyperparameters outlined in Table 2.

While training the models, we want to get the best possible result according to the chosen metric. At the same time, we want to keep a similar result on the new data. But the cruel truth is that we can't get 100% accuracy. Even if we did, the result is still not without errors. There are simply too few test situations to find them.

To address overfitting, we can at the first time apply weight regularization to all the models. This will add a cost to the loss function of the network for large weights (or parameter values). As a result, we get a simpler model that will be forced to learn only the relevant patterns in the train data.

Batch normalization allows us to not only work as a regularizer but also reduce training time by increasing a learning rate. The problem is that during a training process the distribution on each layer is changed. So we need to reduce the learning rate that

**Table 2.** Hyper-parameters for the proposed method, used during training and testing, ReLU (Rectified Linear Unit).

| HYPERPARAMETERS | |
|---|---|
| Activation Function | ReLU |
| Base Learning Rate | 1e5 |
| Epochs | 20 |
| Batch Size | 32 |
| Optimizer | Adam |
| Loss Function | Categorical Cross Entropy |

slows our gradient descent optimization. But, if we will apply a normalization for each training mini-batch, then we can increase the learning rate and find a minimum faster.

Batch normalization is a technique used in deep learning to improve the training of artificial neural networks. It involves normalizing the inputs of each layer to have zero mean and unit variance, which can help mitigate the problem of internal covariate shift and improve the generalization performance of the model [4].

Transfer learning is used as a feature extractor for the MRI images. The output of the CNN is a set of high-level feature maps, which are then fed into a fully connected (FC) layer for classification. The mean and variance of each feature map are computed over the entire training set, and these statistics are used to normalize the feature maps. The CNN is frozen, meaning that its weights are not updated during training.

The equation for batch normalization is as follows:

Given a mini-batch of size m, with inputs $x_1$, $x_2$,..., $x_m$, and a layer with k-dimensional output, the batch normalization operation computes the mean and standard deviation of the input along each dimension, denoted by mu and sigma, respectively:

$$mu = (1/m) * sum(x_i) \#mean \qquad (1)$$

$$sigma = sqrt((1/m) * sum((x_i - mu)^2)) \#standard\ deviation \qquad (2)$$

These values are then used to normalize the input $x_i$ for each dimension:

$$x_i\_normalized = (x_i - mu)/sigma \qquad (3)$$

Finally, the normalized values are scaled and shifted using learned parameters gamma and beta, respectively, to allow the model to learn the optimal scale and location for the output:

$$y_i = gamma * x_i\_normalized + beta \qquad (4)$$

These parameters are learned during training using backpropagation, and are updated in the same way as other model parameters. By normalizing the inputs, batch normalization can reduce the dependence of the model on the initialization of the weights, improve the gradient flow during training, and help prevent overfitting.

The equation for batch normalization in this context is the same as the general equation for batch normalization I provided earlier, but with x_i being a 3D tensor representing a feature map from the CNN, and the mean and variance being computed over the entire training set instead of the mini-batch. The gamma and beta parameters are learned during fine-tuning of the FC layer and the batch normalization layer.

### 3) Classification Methods

This section introduces the classification methods that were used in this study.

### A. Classification Algorithm of the OASIS Dataset

The classification algorithms that were used with the OASIS dataset are:

**Support Vector Machines (SVM)**

SVM is a machine learning algorithm used for regression analysis and to solve classification issues for linearly and nonlinearly separable classes. SVM finds a hyperplane that best separates the data into different classes. The hyperplane is chosen to maximize the margin between the classes, making the algorithm highly effective in handling complex and non-linear datasets. SVM is widely used in image classification, text classification, and bioinformatics applications [19]. One of the key advantages of SVM is its ability to handle high-dimensional datasets and perform well even with limited data.

**Decision Trees**

Decision Trees are a popular machine learning algorithm used for classification and regression analysis. They work by recursively partitioning the data into subsets based on the most informative features, creating a tree-like structure of decision rules. Each node in the tree represents a decision based on a feature, and each branch represents the possible outcome of that decision. Decision Trees are easy to interpret, and the resulting tree can be visualized and easily understood by non-experts. Trees are quick to learn and quick to predict. They are also often accurate in diagnosing and categorising the dataset to be solved [20].

**Random Forest**

Random Forest is an ensemble learning algorithm combining multiple decision trees to enhance model accuracy and robustness. Utilizing random subsets of data and features for each tree, it excels in handling noisy, high-dimensional datasets. Widely applied in finance, healthcare, and natural language processing, it provides feature importance scores and robustness against overfitting, missing data, and outliers. Overall, Random Forest stands out as a versatile and powerful machine learning tool applicable across diverse domains [21].

**K Nearest Neighbours (KNN)**

K Nearest Neighbors (KNN) is a popular machine learning algorithm used for classification and regression analysis. The algorithm works by finding the k-nearest points in the training set to a given test point, based on a similarity measure such as Euclidean distance. The class or value of the test point is then determined by a majority vote or weighted average of the k-nearest neighbors. KNN is a simple and effective algorithm that is easy to implement, interpret, and understand. It can handle nonlinear relationships and is robust to noisy data. KNN has been used in various applications such as image recognition, natural language processing, and recommendation systems. One of the drawbacks of KNN is its computational complexity, especially with large datasets. Furthermore, KNN is sensitive to the choice of distance metric and the value of k, which can affect the performance of the algorithm [22].

### B. Classification Algorithm of the MRI Dataset

A hybrid technique that combines deep learning and machine learning has been developed to diagnose Alzheimer's disease (AD) [23]. This approach utilizes deep learning models to extract deep feature maps from input images, which are then used to train the models. The hybrid technique consists of two blocks: the first block involves the use of CNN models to extract feature maps, which serve as the input for the second block and the second block utilizes the SVM algorithm of machine learning to classify the deep feature maps [24]. The next steps implement the hybrid techniques. The dataset is then split into 70% for training, 20% for validatiom and 10% for testing. CNN models are applied to the training dataset to extract deep feature maps through convolutional layers, while the SVM algorithm is used to classify the test dataset for AD diagnosis. This approach offers benefits such as fast implementation, the ability to solve complex computational problems, and effective diagnostic accuracy.

## 6 Experimental Result and Discussion

### 6.1 Splitting Dataset

The classification was performed on training, validation, and test datasets. We divided the dataset for training 80%, validation 10% and testing 10. We used full test data and validated the model. We used early stopping and applied the SGD training with a minibatch size of 64, a learning rate of 0.01, a weight decay of 0.06 and a momentum factor of 0.9 with Nesterov optimization.

### 6.2 Model Evaluation

The performance of each model on the OASIS (medical records) dataset was evaluated using five statistical measures. The MRI dataset was also evaluated by four CNN models and hybrid techniques CNN with machine learning (SVM classifier).

In order to analyze the performance of our proposed models, we calculated five different statistical metrics. Their equations are given as follows:

- Accuracy

This is defined as the ratio of the number of correctly classified instances to the total number of instances.

$$\text{Accuracy} = \frac{TP + TN}{TP + TN + FP + FN} * 100\% \tag{5}$$

- Precision

Precision is the evaluation measure that is used to determine, of all the expected positive observations, how many are actually positive. Mathematically, it looks like this:

$$\text{Precision} = \frac{TP}{TP + FP} * 100\% \tag{6}$$

- Sensitivity–Recall

Recall is the evaluation metric that is used to determine among all the true positive observations how well were predicted successfully. Mathematically, it is given by:

$$\text{Sensitivity–Recall} = \frac{TP}{TP + FN} * 100\% \tag{7}$$

- Specificity

This is defined as the ratio of expected true negative cases to all negative cases in the baseline information. Gives an evaluation of the performance of instances with negative metrics. Mathematically, it is given by:

$$\text{Specificity} = \frac{TN}{TN + FP} * 100\% \tag{8}$$

- F-measure

F-measure is a measure of evaluation of the accuracy of the test. Score F achieves the best value, which means perfect accuracy and recall, at a value of 1.

$$\text{F} - \text{measure} = 2.\frac{Precision.Recall}{Precision + Recall} * 100\% \tag{9}$$

The OASIS dataset consists of 373 records for two classes namely, AD and NoAD. The second dataset, Alzheimer's MRI Dataset, consists of 6400 images divided into mild dementia disease (896 images), moderate dementia disease (64 images), non-dementia (3200 images) and very mild dementia disease (2240 images). The both datasets were balanced by the data augmentation technique. Table 3 shows the division of the OASIS and Alzheimer's MRI datasets after balancing during the training and selection phases for dementia and non-dementia patients for the first dataset and for Alzheimer's patients with varying severity from mild to moderate to severe for the second dataset.

**Table 3.** Splitting the two datasets.

| Class | Non-Dementia | Dementia | Non_Demented | Very_Mild_ Demented | Mild_ Demented | Moderate_ Demented |
|---|---|---|---|---|---|---|
| Training | 114 | 26 | 2562 | 1775 | 734 | 49 |
| Testing | 32 | 11 | 329 | 83 | 224 | 4 |

### 6.3 Analyzing the Outcomes of the OASIS Dataset

Analyzing the outcomes of the OASIS dataset involves evaluating the performance of different machine learning classifiers such as Support Vector Machines (SVM), K-Nearest Neighbors (KNN), Logistic Regression, and Random Forest. Each classifier's performance is assessed based on metrics like precision, sensitivity-recall, specificity, F-measure, and accuracy. Among these metrics, SVM consistently demonstrates high precision and sensitivity, indicating its effectiveness in correctly identifying positive cases. KNN also performs well across several metrics. However, Logistic Regression shows lower specificity compared to other classifiers, and Random Forest exhibits slightly lower precision. Overall, SVM appears to be the most promising classifier for predicting outcomes in the OASIS dataset, but considerations regarding dataset characteristics and specific requirements are crucial in making a final decision (Figs. 4, 5 and 6).

**Table 4.** Results of dementia diagnosis by using machine learning for each class.

| Classifiers | SVM Testing (10%) | KNN Testing (10%) | Logistic Regression Testing (10%) | Random Forest Testing (10%) |
|---|---|---|---|---|
| Precision % | **89.00** | 84.00 | 84.00 | 85.00 |
| Sensitivity–Recall % | **89.00** | 84.00 | 85.00 | 85.00 |
| Specificity % | **55.00** | 45.00 | 27.00 | 45.00 |
| F-measure % | **89.00** | 84.00 | 83.00 | 84.00 |
| Accuracy % | **89.00** | 84.00 | 85.00 | 85.00 |

### 6.4 Analyzing Alzheimer's MRI Dataset Using CNN Models

The problem of the dataset for unbalanced MRI and overfitting was overcome by the technique of data augmentation. Table 2 describes the tuning of CNN models in terms of the optimiser, learning rate, Mini Batch Size, maximum epoch, validation frequency and training time. The deep feature maps extracted by AlexNet-M, ResNet50-M, VGG16-M and InceptionV3-M models were evaluated for the diagnosis of AD. Figures 7, 8, 9, 10 shows the confusion matrix for models that contains all correctly classified (TP and TN) and incorrectly classified (FP and FN) cases of AD. Table 4 shows the results achieved by the models. Analyzing Alzheimer's MRI dataset using Convolutional Neural Network (CNN) models involves evaluating the performance of different architectures like AlexNet-M, ResNet50-M, VGG16-M, and InceptionV3-M. Each model's effectiveness is measured through metrics such as accuracy, precision, sensitivity-recall, specificity, and F-measure. VGG16-M stands out with the highest accuracy at 90.46%, suggesting it is the most effective at correctly classifying images overall. While all models demonstrate high precision and specificity, indicating they accurately identify positive cases and correctly recognize negative cases, their sensitivity-recall scores, which reflect their ability to capture true positive cases, are slightly lower. InceptionV3-M and AlexNet-M show similar performance across most metrics. The high specificity across all models highlights their strength in correctly identifying non-Alzheimer's cases. In conclusion, VGG16-M emerges as the most promising model for analyzing Alzheimer's MRI data, but the choice of model might also depend on other factors such as computational efficiency and the specific clinical application.

## 6.5 Evaluating the Findings of the MRI Dataset Using Hybrid CNN Models with SVM

Evaluating the findings of the MRI dataset using hybrid CNN models with SVM reveals significant insights into model performance. Transfer learning enhances model robustness, as observed when comparing poor results in disease identification without it. The regularization model, particularly VGG19-M, displayed superior performance, achieving 99.1% accuracy and the lowest testing loss.

**Table 5.** Comparison on test OASIS/ MRI Alzheimer's data between the four standard and modified models.

| Models | | Accuracy on testing data | | | | |
|---|---|---|---|---|---|---|
| | | Accuracy | Precision | Sensitivity– Recall | Specificity | F- measure |
| Modified Deep Learning | AlexNet-M | 80.15% | 85.45% | 75.99% | 91.17% | 80.44% |
| | ResNet50-M | 81.25% | 85.79% | 74.80% | 91.37% | 79.92% |
| | VGG16-M | 90.46% | 85.72% | 76.61% | 91.36% | 80.22% |
| | InceptionV3-M | 81.25% | 85.59% | 75.81% | 91.41% | 80.40% |
| A Hybrid between Machine Learning and Modified Deep Learning | AlexNet-M | **91.41%** | 91.41% | **91.41%** | **100.00%** | 91.39% |
| | ResNet50-M | 78.75% | 79.59% | 78.75% | 100.00% | 78.84% |
| | VGG16-M | **91.41%** | **91.43%** | **91.41%** | 96.34% | **91.42%** |
| | InceptionV3-M | 80.31% | 80.62% | 80.31% | 100.00% | 79.74% |

Overall results (Table 5) highlighted VGG19-M's effectiveness with 98.91% accuracy, 99.37% precision, 99.15% sensitivity-recall, 99.80% specificity, and 99.40% F-measure, making it ideal for diverse classifications. In this context, VGG16-M also performed well with 97.80% accuracy. The current dataset results show that AlexNet-M, ResNet50-M, VGG16-M, and InceptionV3-M vary in their performance. AlexNet-M and VGG16-M achieved a high accuracy of 91.41%, while ResNet50-M and InceptionV3-M showed lower accuracy at 78.75% and 80.31%, respectively. Hybrid models, combining SVM with deep learning, slightly outperformed in accuracy, with VGG16-M + SVM reaching 99.00%. Modified deep learning models excelled in precision, while hybrids demonstrated superior sensitivity-recall. Both approaches showed similar specificity, favoring modified deep learning in identifying negative instances. The results indicate the effectiveness of both approaches, with the choice depending on task-specific requirements and the precision-sensitivity-recall trade-off. Further analysis is crucial for optimal selection in varied classification scenarios.

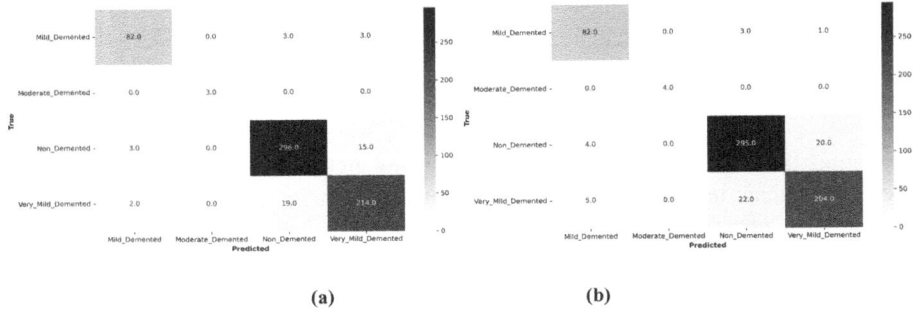

(a)                              (b)

**Fig. 7.** Confusion matrices. (a) Confusion matrix for the AlexNet-M to evaluate Alzheimer's disease. (b) Confusion matrix for the AlexNet-M + SVM to evaluate Alzheimer's disease.

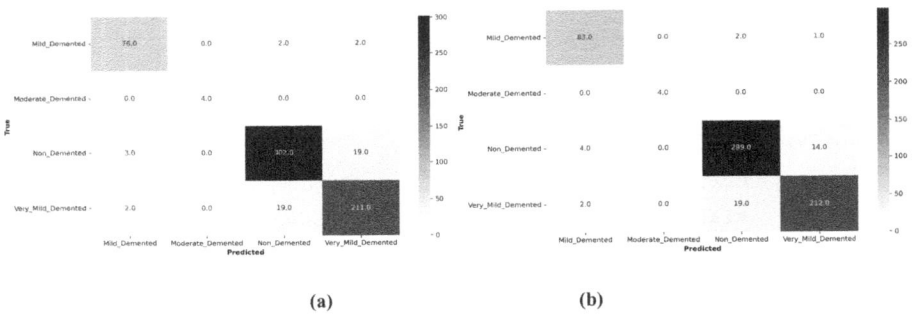

(a)                              (b)

**Fig. 8.** Confusion matrices. (a) Confusion matrix for the ResNet50-M to evaluate Alzheimer's disease. (b) Confusion matrix for the ResNet50-M + SVM to evaluate Alzheimer's disease.

### 6.6   Comparing Performance: Deep Learning vs. Hybrid Deep and Machine Learning Approaches

Comparing the performance of deep learning models with hybrid approaches that integrate Support Vector Machines (SVM) reveals significant enhancements in classification metrics for Alzheimer's MRI data analysis. The inclusion of SVM with CNN architectures such as AlexNet-M, ResNet50-M, VGG16-M, and InceptionV3-M results in notable improvements across accuracy, precision, sensitivity-recall, and F-measure. Specifically, AlexNet-M and VGG16-M achieve an accuracy of 91.41% when paired with SVM, surpassing their standalone performance.

The specificity reaches 100% for AlexNet-M, ResNet50-M, and InceptionV3-M, indicating a perfect identification of true negative cases, which is crucial in minimizing

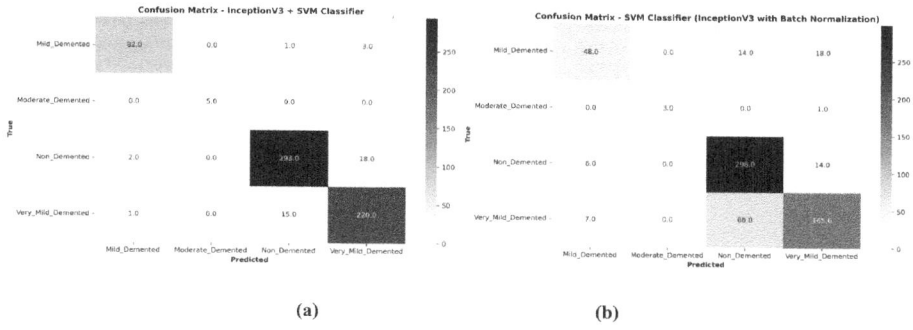

**Fig. 9.** Confusion matrices. **(a)** Confusion matrix for the VGG16-M to evaluate Alzheimer's disease. **(b)** Confusion matrix for the VGG16-M + SVM to evaluate Alzheimer's disease.

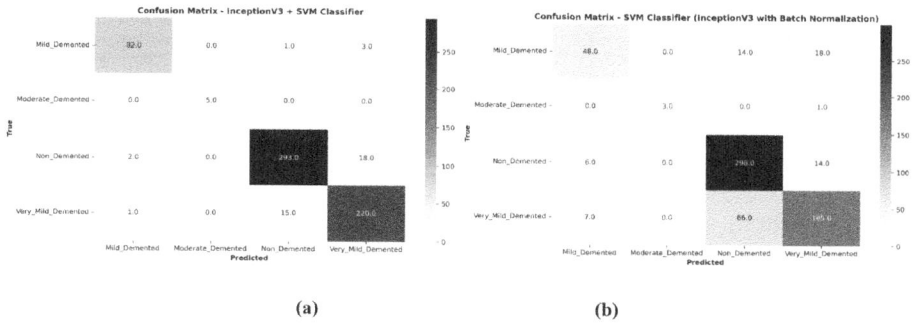

**Fig. 10.** Confusion matrices. **(a)** Confusion matrix for the InceptionV3-M to evaluate Alzheimer's disease. **(b)** Confusion matrix for the InceptionV3-M + SVM to evaluate Alzheimer's disease.

false positives. The hybrid models demonstrate superior precision and recall balance, as evidenced by higher F-measure scores. In contrast, models without SVM integration show reduced performance, highlighting the synergistic effect of combining deep learning's feature extraction capabilities with SVM's robust classification. These findings underscore the potential of hybrid approaches to enhance diagnostic accuracy and reliability in medical imaging applications.

## 7 Conclusion

With the rapid advancement of artificial intelligence, computer vision has emerged as an increasingly invaluable tool in the identification of Alzheimer's disease. In recent years, deep learning technology has prominently dominated the field of medical imaging, exhibiting remarkable efficacy in automating the detection of AD through the analysis of medical images. Specifically, a deep network model based on transfer learning has been developed to facilitate the classification of patients with Alzheimer's disease, enabling early diagnosis.

In the current study, we have presented findings that compare different pre-training methods for the detection of Alzheimer's disease using brain MRI scans. Among all the models evaluated, AlexNet-M + SVM achieved an outstanding accuracy rate of 91.41%. Furthermore, our proposed methodology exhibits substantial potential for applying convolutional neural networks (CNNs) to other domains with limited datasets, thereby offering significant benefits for the early-stage diagnosis of AD.

# References

1. WHO.: World Health Statistics (2008). http://books.google.tn
2. Alberdi, A., et al.: On the early diagnosis of Alzheimer's disease from multimodal signals. Artif. Intell. Med. **71**, 1–29 (2016). https://www.sciencedirect.com/science/article/abs/pii/S0933365716300732
3. Khan, A., Sohail, A., Zahoora, U., et al.: A survey of the recent architectures of deep convolutional neural networks. Artif. Intell. Rev. **53**, 5455–5516 (2020). https://doi.org/10.1007/s10462-020-09825-6
4. Guesmi, D., Salah, F., Ayed, Y.B.: Recognition of Alzheimer's disease based on transfer learning approach using brain MR images with regularization. In: Nguyen, N.T., et al. (eds.) Advances in Computational Collective Intelligence. ICCCI 2023. Communications in Computer and Information Science, vol. 1864, pp. 1–12. Springer, Cham (2023). https://doi.org/10.1007/978-3-031-41774-0_12
5. Wang, L.: Mammography with deep learning for breast cancer detection. Front. Oncol. **14**, 1281922 (2024). https://doi.org/10.3389/fonc.2024.1281922
6. Bernal, J., et al.: Deep convolutional neural networks for brain image analysis on magnetic resonance imaging: a review. Artif. Intell. Med. **95**, 64–81 (2019). ISSN 0933–3657. https://doi.org/10.1016/j.artmed.2018.08.008
7. 3D convolutional neural networks for diagnosis of Alzheimer's disease via structural MRI. In: Proceedings of the 2020 IEEE 33rd International Symposium on Computer-Based Medical Systems (CBMS) (2020). https://doi.org/10.1109/CBMS49503.2020.00020
8. Frenzel, S., et al.: A biomarker for Alzheimer's disease based on patterns of regional brain atrophy. Front. Psychiatry (2020). https://doi.org/10.3389/fpsyt.2019.00953
9. Lim, B.Y., et al.: Deep learning model for prediction of progressive mild cognitive impairment to Alzheimer's disease using structural MRI. Front. Aging Neurosci. **14**, 876202 (2022). https://doi.org/10.3389/fnagi.2022.876202. PMID:35721012;PMCID:PMC9201448
10. Open Access Series of Imaging Studies (OASIS). https://sites.wustl.edu/oasisbrains/
11. Alzheimer's Dataset (4 class of Images): Images of MRI Segmentation. Kaggle. https://www.kaggle.com/datasets/tourist55/alzheimers-dataset-4-class-of-images
12. Prijatna, D., Muhaemin, M., Wulandari, P., Herwanto, T., Muhammad, S., Sugandi, W.: A study of light level effect on the accuracy of image processing-based tomato grading. IOP Conf. Ser.: Earth Environ. Sci. **147**, 012005 (2018). https://doi.org/10.1088/1755-1315/147/1/012005
13. Park, G.-H., Cho, H.-H., Choi, M.-R.: A contrast enhancement method using dynamic range separate histogram equalization. IEEE Trans. Consumer Electron. **54**, 1981–1987 (2008). https://doi.org/10.1109/TCE.2008.4711262
14. Oza, P., Sharma, P., Patel, S., Adedoyin, F., Bruno, A.: Image augmentation techniques for mammogram analysis. J. Imaging **8**(5), 141 (2022). https://doi.org/10.3390/jimaging8050141. PMID:35621905;PMCID:PMC9147240
15. Singh, I., Goyal, G., Chandel, A.: AlexNet architecture based convolutional neural network for toxic comments classification. J. King Saud Univ. - Comput. Inf. Sci. **34**(9), 7547–7558 (2022). ISSN 1319–1578. https://doi.org/10.1016/j.jksuci.2022.06.007

16. Rezende, E., Ruppert, G., Theophilo, A., Carvalho, T.: Exposing computer generated images by using deep convolutional neural networks. Signal Process. Image Commun. 66 (2017). https://doi.org/10.1016/j.image.2018.04.006

17. Simonyan, K., Zisserman, A.: Very deep convolutional networks for large-scale image recognition. (4 Sep 2014, revised 10 Apr 2015). https://arxiv.org/abs/1409.1556

18. Szegedy, C., Vanhoucke, V., Ioffe, S., Shlens, J., Wojna, Z.B.: Rethinking the inception architecture for computer vision. In: 2016 IEEE Conference on Computer Vision and Pattern Recognition (CVPR), pp. 2818–2826 (2016). https://doi.org/10.1109/CVPR.2016.308

19. Byvatov, E., Schneider, G.: Support vector machine applications in bioinformatics. Appl. Bioinform. **2**, 67–77 (2003)

20. Song, Y.Y., Lu, Y.: Decision tree methods: applications for classification and prediction. Shanghai Arch. Psychiatry **27**(2), 130–135 (2015). https://doi.org/10.11919/j.issn.1002-0829. 215044. PMID:26120265;PMCID:PMC4466856

21. Liu, Y., Wang, Y., Zhang, J.: New machine learning algorithm: random forest. In: Proceedings of the 11th International Conference on Intelligent Data Engineering and Automated Learning (IDEAL 2012), Lecture Notes in Computer Science, vol. 7473, pp. 246–252. Springer, Berlin (2012). https://doi.org/10.1007/978-3-642-34062-8_32

22. Zhang, S.: Challenges in KNN classification. IEEE Trans. Knowl. Data Eng., 1 (2021). https:// doi.org/10.1109/TKDE.2021.3049250

23. Balaji, P., Chaurasia, M.A., Bilfaqih, S.M., Muniasamy, A., Alsid, L.E.G.: Hybridized deep learning approach for detecting Alzheimer's disease. Biomedicines **11**(1), 149 (2023). https:// doi.org/10.3390/biomedicines11010149. PMID:36672656;PMCID:PMC9855764

24. Rashmi, P., Singh, M.: Convolution neural networks with hybrid feature extraction methods for classification of voice sound signals. World J. Adv. Eng. Technol. Sci. **8**, 110–125 (2023). https://doi.org/10.30574/wjaets.2023.8.2.0083

25. Che, D., et al.: Decision tree and ensemble learning algorithms with their applications in bioinformatics. In: Arabnia, H., Tran, QN. (eds) Software Tools and Algorithms for Biological Systems. Advances in Experimental Medicine and Biology 696 (2011). https://doi.org/10. 1007/978-1-4419-7046-6_19

# Author Index

© The Editor(s) (if applicable) and The Author(s), under exclusive license
to Springer Nature Switzerland AG 2024
N.-T. Nguyen et al. (Eds.): ICCCI 2024, CCIS 2166, pp. 489–491, 2024.
https://doi.org/10.1007/978-3-031-70259-4

# GPSR Compliance

*The European Union's (EU) General Product Safety Regulation (GPSR) is a set of rules that requires consumer products to be safe and our obligations to ensure this.*

*If you have any concerns about our products, you can contact us on ProductSafety@springernature.com*

In case Publisher is established outside the EU, the EU authorized representative is:

Springer Nature Customer Service Center GmbH
Europaplatz 3
69115 Heidelberg, Germany

The manufacturer's authorised representative in the EU is Springer
Nature Customer Service Centre GmbH, Europaplatz 3, 69115 Heidelberg,
Germany. If you have any concerns regarding our products, please
contact ProductSafety@springernature.com

Printed and bound by CPI Group (UK) Ltd, Croydon, CR0 4YY
24/04/2026
02096358-0015